W9-CLD-379

ECOSPIRIT

TRANSDISCIPLINARY THEOLOGICAL COLLOQUIA

Theology has hovered for two millennia between scriptural metaphor and philosophical thinking; it takes flesh in its symbolic, communal, and ethical practices. With the gift of this history and in the spirit of its unrealized potential, the Transdisciplinary Theological Colloquia intensify movement between and beyond the fields of religion. A multivocal discourse of theology takes place in the interstices, at once self-deconstructive in its pluralism and constructive in its affirmations.

Hosted annually by Drew University's Theological School, the colloquia provide a matrix for such conversations, while Fordham University Press serves as the midwife for their publication. Committed to the slow transformation of religio-cultural symbolism, the colloquia continue Drew's long history of engaging historical, biblical, and philosphical hermeneutics, practices of social justice, and experiments in theopoetics.

ECOSPIRIT

Religions and Philosophies
for the Earth

EDITED BY LAUREL KEARNS

AND CATHERINE KELLER

FORDHAM UNIVERSITY PRESS ❖ NEW YORK ❖ 2007

Library of Congress Cataloging-in-Publication Data

Ecospirit : religions and philosophies for the
earth / edited by Laurel Kearns and Catherine
Keller.—1st ed.

   p.   cm.

  Includes index.

  ISBN 978-0-8232-2745-7 (cloth : alk.
paper)—ISBN 978-0-8232-2746-4 (pbk. : alk.
paper)

    1. Nature—Religious aspects.   2. Human
ecology—Religious aspects.   I. Kearns,
Laurel.   II. Keller, Catherine, 1953–

   BL65.N35E26   2007

   201′.77—dc22

                  2007014464

Printed in the United States of America
09 08 07 5 4 3 2 1
First edition

CONTENTS

PREFACE

We earthlings may be approaching our ecological "tipping point." That phrase indicates the transitional moment when small changes make huge differences, when predictable processes give way to nonlinear and irreversible amplification.[1] This rhetoric of climate change and its "boiling point"[2] belongs to the attempt to conceptualize and dramatize the impending planetary shifts heralded by mounting CO_2 levels and melting polar caps.

Yet hope for a sustainable earthling future—for a *green shift*—may depend upon a dramatic nonlinear transition as well. All effort on behalf of ecosocial justice thrives on the hope that the achingly small beginnings of movement may precipitate a "butterfly effect" of change, a tipping *counter*point, an avalanche of responsible action. The green shift seems to require a root change of human outlook, a mutation of collective philosophy, a spiritual phase transition. Certainly the ecoapocalypytic rhetoric of threat alone will not do the trick.

This anthology springs from the hope, the difficulty, and the transdisciplinarity of this possible shift. Its authors attempt to ground thinking—thinking in the most *spirited* sense—in the earth. Thought roots in the elemental, it *matters*, it bears fruit—or it dies. But this rooting is in the Deleuzian sense "rhizomatic"; it moves, branches, folds, and bifurcates multidimensionally, like the earth itself. Indeed such thinking roots and wings its way to change—it butterflies.[3] Its potentiality, which always exceeds our knowing, may be called *ecospirit*. In this volume, ecospirit moves with a predictions, rhythm, oscillating among and between the

disciplines of philosophy, theology, sociology, religious studies, and ecological ethics. These disciplines fold into a broad, not readily nameable,
transdisciplinarity that may be called "ecospirituality," or in a self-critically widened sense of the theological, "ecotheology." Thus this collection of essays shares not only theoretical and activist wisdoms, but also
literary and liturgical cadences. It is an invitation, in the words of the
comic Swami Beyondananda, to "a Great Upwising."[4]

We do not ask for all thinking people, secular and religious, to stop
whatever they are doing and become environmentalists. We might, however, all be called to ground whatever we are doing in the life of the
earth—with intention, with care, with grit in our thinking. For despite
the now-deafening scientific evidence of climate change, much of the
intellectual leadership of United States culture still ignores such a call.
Not just our flagrantly culpable heads of state and corporation, but academic and religious leaders, too, are still willing to tuck the crisis (like
the excess CO_2 that *should* have remained tucked deep in the earth) into
the lower layers of consciousness. Isn't ecology someone else's concern,"
after all? Besides, all this earth "stuff," isn't it a bit boring, a bit beneath
us? Indeed from this volume's peculiar perspective, the proudly secular
cultural elite of the Western world is not far removed from that mass of
churchgoers who expect come Sunday or doomsday to transcend
"earthly matters." Both want to escape any sense of ecological crisis.

In the face of the immensity and the uncertainty of the collective threat
to the carrying capacity of the earth, we get caught in a cycle of sleepy
denial that alternates with apocalyptic insomnia. How, then, might we
each deconstruct the indifference and the denial in which we cannot
help but participate? The soporific effects of economic convenience and
normalized greed combine, ironically, with ancient codes that trigger
cravings not for the things of this world, but for a supernatural world, a
world beyond decay and death. The present volume seeks to disentangle
our thinking from this dangerous and pervasive contradiction. Economic
and spiritual ideologies collude to reduce the nonhuman to a subordinate, exploitable background, seen as material means to rational human
ends. Our essays presume that such religious symbolism, conscious and
unconscious, comprises a significant part of the problem—especially
within the context of a brittle secular democracy menaced by the convergence of neoliberal economics with theocratic antiliberalism. A solution

will not stem from mere deconstruction, but rather will root, in part, in a transformation of all our theologies and practices.

One sign of hope is the multigenerational maturation of varieties of eco-Christianity, eco-Judaism, eco-Buddhism, indigenous activism, and so forth.[5] Contributing to the developing movement of ecotheology, including and open to many who never think of themselves as theologians, *Ecospirit* works to transmute simple *emergency* into complex *emergence*. Complexity always emerges in the phase transition scientists call "the edge of chaos." As in the format of the conference that generated the volume, such emergence as a collective practice joins vigorous activism with rigorous thinking. Its complexity requires a poetics of hope. When language grounds and moves, when it pulses and inspires, it not only refreshes activism, it makes a nonliteralist, open-ended theology possible—a *theopoetics*. Its theological self-understanding is radically pluralist and autodeconstructive. That is, religious language as deconstructive theopoetics takes responsibility for its own constructions. It recognizes its shifting and overlapping historical contexts as the basis for an eco-hermeneutics, a textuality that cannot be abstracted from the spirited materiality of the earth. In its root intuition of the divine relation to the world, a relation in which all earthlings in our ecosocial lives are called to participate, it confesses the holiness of diversity, the goodness of the nonhuman, the multiplicity of truth. Therefore even in a discussion such as this, not centered on interfaith dialogue or religious survey, partners of different religious, spiritual, and social ways remain at every level indispensable.

Many important anthologies of religion and ecology have become available over the past twenty years. The situation changes rapidly, and needs timely responses. But what is unique in this collection is how the desire for a green theology generates an intensive dialogue with philosophy. Of course philosophy and theology have always been joined at the hip, for good and for ill. They are co-implicated in the disdain for the nonhuman. This volume invites that dispiriting complicity into an ecospirited complexity. Of course, much so-called postmodern theory has remained stubbornly modernist in its liberty from specific ecosocial contexts. It has tended to delegitimize any language in which we might witness to the reality or truth of climate crisis and to the ground—not to be confused with foundation—that all terrestrials share. The indubitable interest of poststructuralist thought in justice and multiplicity has so far remained largely anthropomorphic and anthropocentric. But isn't this true of most Eurocentric thinking, including most of our own thoughts? Most of the time?

HOW TO READ *ECOSPIRIT*

We are hopeful that this interchange generates a renewable energy for discourse across and beyond disciplines. Because our purposes are ecological, and not merely *inter*disciplinary but *trans*disciplinary, they aim beyond academic fields and into earthly actions. So this collection begins and ends with essays that serve handily to introduce the practices of ecotheology. Essays by Jay McDaniel, Anna Peterson, Rosemary Radford Ruether, and then Mary Evelyn Tucker and Fletcher Harper, are well suited by their clarity and concreteness for the greening of new audiences. We include ecumenical liturgies developed for the "Ground for Hope" conference at Drew University as resources for a style of ecological practice that can inspire and not merely warn. Other pieces investigate vivid cross-sections of current concern: biodiversity, the felling of forests, the treatment of animals, the Alaska National Wildlife Refuge, global warming.

Many of the essays engage some form of postmodern thought, and so will require scholarly interest or guidance. Thus in its hybridity of thought-forms as well as of fields, the volume itself practices a diversified intellectual ecology. Interdependent conversations nest together for the nurture of distinct if overlapping populations. We mean to offer a multi-layered resource for scholars, students, and activists within the secular or the theological academy. Scholars concerned with the ecological crisis hope to be surprised and stimulated in their own thinking; but they also seek ways to introduce the topic afresh to as broad a public as possible. With this in mind, we have brought together many of the conversations into one complex textual locus, opening to future readers the golden autumnal space of the originating conversation under the oaks of Drew University.

ECOSPIRITUAL THANKS

Scholars barefoot in the grass, planting trees, reciting poetry? Telling stories of their significant animal others, singing songs of the creation? Such images linger from the event at which we tried to combine practice and theory—even literally grounding theory as we planted a time-capsule, part of an ecological art project.[6] Foremost thanks go to all who gathered in the willingness to think hard and yet exceed their scholarly selves, to ritualize and to struggle toward further commitments, to become their greening selves. That September gathering took the form of an activist event entitled "Ground for Hope: Faith, Justice, and the Earth," coupled with the scholarly conversation dubbed "Ecosophia." Together these composed the Fifth Annual Transdisciplinary Theological Colloquium (TTC) at Drew University, from which this volume results. This colloquium series depends upon the gracious support of the Theological School of Drew University. Deans Maxine Beach and Anne Yardley said yes from the beginning, and kept saying yes as the vision expanded—getting greener, and so costlier—and as the volume was born. Our new president, Robert Weisbuch, opened the conference with erudite insights from children's literature and visions for a transdisciplinary university. The commitment, resources, and involvement of GreenFaith, particularly the Reverend Fletcher Harper and Rabbi Larry Troster, as well as the gift given by Leigh and Charles Merinoff, were essential in the expansion of TTC5 to include the public conference, with its keynote addresses, interfaith worship, and wide spectrum of workshops (from "Solar Panels for your Sanctuary" to "Process Kabbalistic Theology"). All the while a host of devoted graduate students could be found working behind the scenes. Special thanks are due to Antonia Gorman for her conference management, and to Luke Higgins, Rick Bohannon, and Jane Ellen Nickell for their assistance in planning, design and maintenance of the Web site, and overall execution. Nicholas Stepp and Matthew Westbrook, with later help from Lydia York, have, with immense skill and stoic persistence, collaborated in the deep editing of this book. Laurel Kearns would like to thank the thirty students enrolled in her conference-related Theological School class or her related PhD seminar for their energizing presences and subsequent reflections. We recognize that no such project happens without the underthanked support of administrative and editorial staff and the unnamed, all-too-often unnoticed people who make universities tick. And from the start we have been graced by the guidance of a visionary editor, Helen Tartar, of Fordham University Press. Thanks also to production editor Nick Frankovich and others who assisted in the preparation of this volume. And for pretending (most of the time) not to notice how often we weren't "there," we are so grateful for those grounding spirits in our lives: Bob McCoy, Claire and Christopher Kearns-McCoy, and Jason Starr.

Finally, we'd better not go any further without saluting all our animal soulmates, butterflies included, and particularly Derrida's cat, Lutece, who makes a spectral appearance in this volume. And we extend special thanks to the particular manifestation of felinity that tried all too often to walk across the keyboard.

ECOSPIRIT

✧ Introduction: Grounding Theory—Earth in Religion and Philosophy

CATHERINE KELLER AND LAUREL KEARNS

A *New Yorker* magazine cartoon displays a sporty little flying saucer flitting away from the earth. One extraterrestrial is commenting to the other: *"The food's OK, but the atmosphere is terrible."* Of course a lot of us terrestrials (not only New Yorkers) zip about tasting the aesthetic variety of our gifted planet. We relish our global interconnectedness amid the sheer abundance of available options. But occasionally we recall the shadow side of that interconnectedness: the atmospheric changes our cosmopolitan species is inadvertently cooking up, the mounting concentration of CO_2 in the atmosphere that is turning the planet into a pressure cooker, with nowhere for us to flit away to.

With a mere three-, at most nine-, degree rise in our atmospheric temperature anticipated by midcentury—a range smaller than most days' temperature swings—the proverbial flap of the butterfly's wing might quite soon tip the earth into conditions unprecedented in the past two million years. How can we *bear* such information? How can we even *think* such eerie possibilities? Compared to what interests us day by day, climate change seems at once too flat in its realism and too dramatic in its rhetoric, too factual and too speculative, too complex and too immense to bear in mind. The model seems to wax season by season with a negative grandiosity, quietly unifying us all in a metanarrative of mounting futility. Such dystopic probability emits its own apocalyptic heat. It wilts the edges of our imagination, desiccating subtler distinctions. It gathers everyone, every creature, into a rough collectivity that mocks all our differences, even difference itself. And yet at the same time it demands a care for difference wildly exceeding recent discourses of difference, of the Other, of the others.

Ecological difference pushes the encounter with the Other over the edge, into the infinity of nonhumans, into the engulfing differences of biodiversity. We may be accustomed in our philosophy or theology to an ancient stretch beyond the human, indeed beyond the earth's atmosphere. Nonetheless the present stretch, this self-extension into the full terrestrial spectrum of pressing, vulnerable life: this feels inhuman. We defer. We despair. We deny. We return to thought as usual.

At any rate, this climate change and everything it touches—imprecisely called "nature"—is not only aggressively ignored across the policy spectrum, but oddly undertheorized across the academic disciplines. "The environment? . . . Not my issue." Ecological crisis gets relegated to the vulnerable subdisciplines of "environmental studies," at risk in every budget cut or change in leadership.[1] An odd coincidence: just as the science about climate change and other anthropogenic threats to the biosphere becomes indisputable—and the role of organized religion, even evangelical Christians, in calling for a response splashes occasionally into the news—support for grassroots-activist environmental networks, denominational and academic programs, and projects related to religion and ecology seems in many contexts to be sinking in arguments about priorities.[2] But there are many exceptions, including the fecund microatmosphere in which this volume was able to take root. These various exceptions are gaining strength, the roots have become rhizomes, the ground activates a future that may yet have a chance to elude the dour trends.

We hope, if we hope, for a common terrestrial future. But does the very notion of a common future, of "the common" or of what is "to come," not have a hard time *mattering* in late-capitalist culture? Among those economically privileged earthlings primarily responsible for climate change, the thought of the future appears increasingly reducible to an individual life-span, projected ahead, if at all, through a child or two, and measured by an accumulation of wealth and private heritage. Ecologically sensitive theologians will rightly diagnose an inadequate eschatology. Yet most Christians care little more about the planetary future of the earth—at least of *this* earth—than do the most voracious secular corporations. If standard Christian eschatology does extend the individual life-span into an endless future, its trajectory at death breaks sharply *out* of the earth's atmosphere—sometimes with body, sometimes without (sans flying saucers in either case). So the systemic indifference to the shared planetary future is compounded of both spiritual and economic habits.

Precisely because of this planetarity, in which everything is at stake, spiritual language and religious actions have become indispensable parts of the repertory in the struggle to awaken commitment to a just and sustainable planetary future.[3] As the *Earth Charter* avows, "The protection of earth's vitality, diversity, and beauty is a sacred trust." Indeed "environmental, economic, political, social and spiritual challenges are interconnected, and together we can forge inclusive solutions."[4] In that hope and that insistence, this volume meditates on the sacrality of the terrestrial trust. Its broadly ecumenical spirituality, sometimes called ecotheology, relates a global movement to the theological symbol of the spirit. Perhaps that subtle energy, that holy spirit—*ruach* or *pneuma*, breath and wind of life—is also suffering from our all too literal climate change. It names the in-spiration and con-spiracy of a shift both metaphorical and maternal in collective atmosphere.

The spirit—as the theologians in this volume reveal—did not in the Bible signify a way out of the creation, but rather a way to live gratefully *within* it. The mindful and moving spatiality of spirit, which names our collective endeavors, suggests the fluid medium for the sort of transdisciplinary thinking represented in this book. Indeed, transdiciplinarity itself, aiming not just at dialogue *between* disciplines but action *beyond* them, supplies a crucial part of the grounding strategy.

The structure of this volume enacts such a strategy: we start from the premise that a living and shared *ground*—the matrix of present relations from which a shared future emerges—has become difficult to think, to theorize. The spirited work of ecotheology opens, stimulates and depends upon such a ground. Yet the suspicion will reasonably persist that any notion of common ground hides one more totalizing ploy, one more unification that levels diversity. For as Bruno Latour warns us, the shared ground of the universe has been overwritten by the Western history of "Nature."[5] As the nonhuman recedes into a totalized background, *nature* itself may morph into a rigid foundation. To understand—indeed just to think—this planetary condition requires a certain philosophical deconstruction, but perhaps only one translatable, with David Wood's help, into an *econstruction*.[6] The *doctrines* of the theological tradition are being reconstructed in the place opened by deconstruction, especially by the autocritique of Christianity. As the spiritual traditions learn anew to attend reverently to their material context, they help to reveal *place* itself as an ecology of vibrant, vulnerable interrelation. Finally, to ground our thinking in the shifting and shared finitudes of present places enacts the *hope* of an intentionally common future through collective action and celebration. *Grounds, natures, constructions, doctrines, spaces, hopes*: it is under these section headings that we have organized the multilevel transdisciplinarity of this volume.

NEITHER NATURE NOR SUPERNATURE

There is no single explanation for the difficulty of sustaining ecological movements proportionate to the severity of the situation. Multiple forces are at work in the intersections of spiritual and economic habit: the ideological drift toward the Right; the privileged cosmopolitanism of progressives; a fervent, advertising-driven consumerism; the "below the belt" obsession of Christian conservatives; and the unavoidable distraction by more humanly pressing issues like war, genocide, systemic poverty, and threats to the constitutional freedoms by which strategies of justice, peace, and sustainability may be pursued in the first place. Good environmental intentions are damaged by discouragement after too many setbacks. A kind of apocalyptic exhaustion, a collective depression at how much irreversible loss has already taken place, coupled with failures of strategy and solidarity on the part of the environmental movement itself, makes it possible for some to pronounce (conveniently) the "death of environmentalism."[7] Within this toxic atmosphere operate the dual longterm motives of an *economic ideology* that exploits and discards the nonhumans along with the majority of the humans, and the *spiritual ideologies* that legitimate it by collusion or default.

Predominant among these spiritual ideologies is a Christendom that has tended to trade its own body-affirmative potentials—encoded in the doctrines of creation, Incarnation, and Resurrection—for body-denigrating priorities. It has intensified human "dominion" over the other creatures by way of a naturalized dualism of spirit over flesh, of a super-natural heaven over a material earth. Yet the biblical phrase "heaven and earth" can be well translated from the Hebrew as "atmosphere and earth." And if "God so loved the world/*kosmos*" (John 3:16), as is so frequently noted by evangelicals, in the context of that gospel's *logos*, theology is a gloss on God's love of the entire creation: a creation that in Genesis (1:31) is pronounced loveable—"very good"—only as the collective of all the species together.

The present volume is the fruit of the sort of theological education that fosters the ecosocial metamorphosis of the biblical heritage. In such a context, Christianity contextualizes itself. It maintains an ongoing critique of its own ideologies and an analysis of its own shifting social habits. As one religion amidst many, it knows itself incapable of evolution apart from attention to its own interdependence with multiple religions, philosophies, and practices. A range of perspectives within religious studies and constructive theology thus collude within this volume to disclose religious pathologies as well as redemptive potentialities. If disproportionately many of the essays embody or analyze Christian ecological habits—this is not just because of the originating context of this particular conversation. We are also aware that the world's sole superpower, operating with a disturbing aura of Christian legitimation, carries the lion's share of responsibility for climate change. Yet the essays all presume a pluralist milieu and engagement. The theological approaches represented here seek the difference of other religions, other spiritualities, secular philosophies and environmental ethics as resources for Christian self-deconstruction and metamorphosis.

Given the perennial force of philosophy in the formation of theology, the renegotiation of the theological tradition requires critical engagement of *live* philosophical options. The Hellenistic philosophies that formed Western theology imported our body-denigrating hermeneutics. Indeed, Western thought is so thoroughly habituated to the matter-transcending theory of Hellenized Christianity that a formative *idealism* underlies the most rapacious *materialism*. The tenacity of this paradox is illumined by Whitehead's "fallacy of misplaced concreteness." As demonstrated in John B. Cobb Jr.'s ecological critique of "economism," the quantifying abstractions of capital replace the spirited bodies of planetary life, rendering them mere "externalities" to the working of economics. The fallacy has been itself "naturalized."[8] Thus the peculiar transdisciplinarity of this volume involves the contributions of philosophers working largely from within the European tradition to deconstruct the modern effects of the dualistic metaphysics of the "ontotheologies" and supernaturalisms of the "God of the philosophers." They thereby craft theory supportive of ecological awareness across the disciplines.

We are therefore interested in the language, the discourse, the operative theory by which the values of sustainability with justice will or will not be taught and practiced, and new leaders for social change will or will not be inspired and trained. This volume thus explores the practices that both require and produce greener theory. Hence our conversations on deconstruction and social constructivism oscillate with discussions of constructive social, environmental, and spiritual *practices* suitable for academic and religious institutions. In other words, this collective of authors was called together partly in response to a problem with theory itself.

Over the crucial decades of environmental degradation, critical discourses have developed that deconstruct presumptions about God, Man and Nature, all the while perforating the boundaries between relevant disciplines. That loose cluster of philosophical and literary postmodernism, feminism and postcolonialism, poststructuralism and social constructivism (all contested yet persistent methods) has heated up and reshaped the terms of thought among and between academic disciplines. But here's the rub: the key vocabularies in social, philosophical, and cultural theory for exposing the constructed character of the status quo—indeed, of all knowledge, and so of all legitimations of epistemic or social domination—invariably marginalize or eliminate ecological questions. We are not the first to notice this. Some environmentalists even charge that in their deconstruction of "nature" postmodernists have lent support to those on the Right hostile to environmental concerns.[9]

These problems may be "academic," but not unimportant. An intellectual hornet's nest obstructs the needed development of cultural leadership for a planetary green shift. Both the emancipatory and the deconstructive discourses that have exposed the academy to its own multiple social contexts, and thus to its responsibilities for justice, remain on the whole relentlessly anthropocentric. Poststructuralism tends toward a hyperbolic *antinaturalism* in its antiessentialism—targeting not nonhuman nature as the problem but rather social control in the name of nature. But any rhetoric, let alone cosmology or ontology, supportive of ecological consciousness and activism is likely to miscarry in an atmosphere of antinaturalism. As it is, discourses of social difference remain barely able to negotiate *différance* between the embattled groupings of our own species—let alone between those (also) natural others, the antecedent nonhumans, and those proliferating "hybrids and imbroglios" in which the human increasingly participates.[10] Indeed, in the effort to expose the social constructedness of the category nature we do not yet have an adequate vocabulary for naming that reality that *is* us and is *more* than us, that *something* in which we are embedded and which remains, however we (re)construct it, irreducible to us. We lapse into the anthropomorphic language of "nonhuman," "other than human," and "more than human," once again (replicating the A/not A erasure that "female" or "woman" or "nonwhite" performs) losing commonality in order to stress difference.

GROUNDLESS DECONSTRUCTION?

Most postmodern theory—constructivist or deconstructive—runs against the grain of ecological sensibility, however inadvertently. For example, the early bon mot of Jacques Derrida, *"Il n'ya pas de hors texte,"* may have been distorted in its U.S. dissemination.[11] But the misunderstanding was effective. It contributed to the inhibition of a transgenerational and transdisciplinary cultivation of ecological responsibility. Derrida was himself later at pains to insist that of course a *real world* exists outside the text, to which the text does *refer.*[12] Indeed, he had already in *Of Grammatology* alluded to how contemporary biologists speak "of writing and pro-gram in relation to the most elementary processes of information within the living cell."[13] This sense of "the text" surely should deter any cryptoidealism. And as David Wood will demonstrate, the later work of Derrida yields rare but precious potentials for the greening of deconstruction. The problem lies less with originative thinkers than with North American and literary disciples busy resisting the naturalized identities of fellow progressives. In the movement of postmodernism, the hip new privilege of textuality could all too smoothly mirror and reinforce an ancient Western anthropocentrism, its disembodied spirit ever transcending and inscribing the superficies of all flesh. If intellectual interest centered exclusively on discourse, the "body" did often make its appearance, and with emancipatory *éclat*. But this has been the merely human body, read as an inscribed surface, abstracted aggressively from its nonhuman ground. Indeed, it is often precisely from nature that the body is being emancipated, and for good reason, as this nature is compounded of the reductive scientific physicalism of modernity and the patriarchal natural law traditions of the premodern. The trope of "denaturalization" so well stimulated by Judith Butler, for example, is no more driven by antienvironmentalism than by environmentalism.[14] Yet the cultural effect of the trope may be a condescending boredom with the nonhuman materialities in which the human materializes.

When ecological postulates irrupt out of an activist sense of urgency, sometimes in the raw hyperboles of emergency, they are not likely to be refined and strengthened within an atmosphere of high theory. Rather they may be simply ignored, or if they intrude within the citadels of high theory, criticized for their naturalism, essentialism, objectivism, or naïve scientism; they may even be chastised for presuming that nature is something clearly enough demarcated to be "saved" as such. Such criticisms may be correct, theoretically, as far as they go. But combined with the pressure of discouraging national trends, they have the toxic effect of isolating, marginalizing, innervating and intellectually malnourishing the ecological paradigm. With the institutionalization over three decades of overt antinaturalisms, or antirealisms, a celebratory groundlessness has managed within the academy to overwhelm a self-grounding tradition of metaphysical foundationalism. The effect has been largely liberating. Unfortunately, however, the language needed for ecology would seem to lie on the conservative side of these polarities. Yet the effort to *conserve* the integrity of the nonhuman basis of our collectively constructed life requires a radical disruption of the global status quo—and certainly a thorough deconstruction of its conceptual foundations. Such a restorative disruption demands a concerted cultural effort, involving great intellectual creativity as well as persistent activism. Such an effort will not get off the ground as long as the cultural leadership remains trapped in its binaries of realism versus antirealism, ecology versus deconstruction, practice versus theory: dualisms reflecting the binary matter/mind foundations of the regnant religioeconomic ideology.

Poststructuralism, however, had developed in tandem and in tension with the movement of more explicitly progressive-activist perspectives, themselves necessarily transdisciplinary. These movement-perspectives forms, have been productive of a sturdy cultural resistance to sexism, racism, ethnocentrism, heterosexism, classism, ableism, economism, militarism, imperialism, and indeed also to anthropocentrism. Many productive cross-fertilizations have arisen, such as ecofeminist theology, or the focus on environmental justice and racism. Ecological activism, especially in the religious world, derived much of its theoretical structure from these liberationist projects, expanding concern for oppression and exploitation of vulnerable human groups to all beings, and seeking to build coalitions (at least in theory) across the social movements. Nonetheless, the attempt to tack "the environment" onto lists of the oppressed has always strained the rhetoric of resistance. It is not just that, for instance, ecology could be read as a pretext for privileged white hikers to enjoy escapes from the cities of suffering humans. Even as the awareness dawned across the spectrum of religious and secular progressivism that the earth is not just one more issue represented by one more social context but rather, as Jay McDaniel insists, the "context of contexts" for all issues, collaboration remains begrudging. This is in part because the very terms of resistance are derived from a spectrum of prophetic humanisms—Enlightenment, Marxist, and various post-1960s liberation movements.[15] Poststructuralism has from the start both amplified and deconstructed that humanism. In the complex development of the politics of difference, of postcolonial theory and its planetary interstitiality, the difference of nonhuman creatures and gaseous atmospheres remains, well, a bit too different so far.

To think the *human in common* is itself a tremendous challenge in the face(s) of any politics of difference. But to situate the commonly human within a terrestrial *commons*, one that is not a mere surround, an environment, but is ground, context, and future—is this stretch simply implausible? It seems too late to reorganize the human population into holistic tribes living in totemic synergy with the nonhuman populations. We organize for a possibility yet to come. Might we collect our unwieldy human commonality into Latour's "collective of humans and nonhumans," for which a radicalization of *democracy* is required?[16] In such a political process, the scientists, activists, and politicians—perhaps even the shamans, ministers, clerics, and philosophers he omits—might learn to speak as expert witnesses on behalf of the nonhuman and nonlinguistic agents.

COMMON CREATURES, CREATURELY COMMONALITY?

Does it remain a utopian fancy or a naturalist dream to imagine our-selves, all of us creatures, as a planetary collective? Or is such a re-collec-tion of the shareable earth—we might call it the "genesis collective" —the only plausible path to a future humane even for humans? Perhaps the sustainable *common* future is a matter not of *compromising* human well being, as those with only economics on their minds believe, but of *complexifying* the democratic and differential tendencies of the humane. Doesn't the hope of progressive politics lie in more persuasive, ipso facto noncoercive, integration of diversity into a common life, a life that in its asymmetrical reciprocities—and not from "trickling down" or being redistributed from above—resists uniformity?

Perhaps every attempt to cross-fertilize a discourse of social liberation with ecology has reinforced this hope, this possibility of a sustainable genesis collective. Take for example the discipline of theology. It has the advantage of carrying with it an ancient transdisciplinarity, always dependent on philosophy yet aimed at practice. The theological interlink-age of ecological analysis with prophetic discourse on behalf of the vul-nerably human has produced prodigies of planetarity: for instance, the Brazilian liberation ecology of Leonardo Boff and ecofeminist Catholi-cism of Ivone Gebara, or the ecowomanism of Delores Williams and Karen Baker-Fletcher. Each has crossed the especially poignant divide between a passionately anthropocentric liberation theology and an envi-ronmentalism always in danger of overriding the human, especially the great blocks of the populous and the poor.[17] And from the beginning of the period in question, the multidimensional analysis of Rosemary Rad-ford Ruether has displayed the bondage of Christianity to sexism, anti-Semitism, racism, militarism, classism, and the overriding anthropocen-trism that hinders constructing a radically democratic response. Others have worked on similar projects—Larry Rasmussen in his circular jour-neys in *Earth Ethics, Earth Community*, while his colleague James Cone asks "Whose Earth Is It Anyway?"—to bring home the justice of environ-mental interconnectedness. Sallie McFague's development of the model of the earth as God's body leads to a "super, natural Christianity."[18]

Process theology, as crafted by John B. Cobb Jr., has all along been disseminating its own prophetic notion of the "postmodern" based on an ecologically vivid cosmology.[19] It has made available Whitehead's philosophical framework with which to embed the classical Christian doctrines, as well as contemporary social struggles, within a pluralistic universe in an open-ended process of interlinkage.[20] With the guidance of such influential figures as these, the churches "got it" (at least at the denominational staff level) comparatively early, with the World Council of Churches in 1983 encapsulating such concerns in the theme "Justice, Peace, and the Integrity of Creation" and the U.S. National Council of Churches forming an Eco-justice Working Group in the early 1980s, with 1970s antecedents.

Is it possible that by a great ecological stretch of difference—whereby we read human diversity always as an instance and intensification of the dizzying multiplicity of species—we may find the key to *human* coexistence? Indeed, now that humanity has covered the planet, a species thinking that lets us, perhaps for the first time, ground ourselves in our common heritage and future as one among many planetary species may represent the only path of interhuman peace. The often fearful reaction against any such common material ground—as though any discourse of the common must reduce and level, must inflict one group's agenda on the rest, as though there is not a difference between homogeneity and commonality—begins to appear as itself a symptom of colonization by postmodern capitalism, by what we may call *laissez-faire difference*. And the economically caused scarcities of jobs, education, food, and water all too vividly expose the *indifference* of such difference.

Perhaps biodiversity and transnational pluralism begin to converge in a new *ecodifference*. In the meantime, in the interest of democratic politics, Bruno Latour brings a poststructuralist antinaturalism full circle—to its ecological potential. He argues that "Nature" is the *problem* for political ecology, not its *reason*. For our operative language of "nature" historically, and in its persistent effects, stems from a Platonic abstraction, before which only the expert—the Philosopher first, then the Priest, now the Scientist—can mediate the relationship of reason to *physis*. "Political ecology has nothing whatsoever to do with nature, this jumble of Greek philosophy, French Cartesianism, and American parks."[21] If *nature* became the conceptual tool of a naturalized hierarchy of being that rigidly controls the spontaneities of embodied life, then it is not the central concern, but the opponent of ecological politics. Latour suggests a new bicameral "parliament of the humans and nonhumans." One hears echoing in reply the grim warning of Thomas Berry two decades ago: "If there were a parliament of creatures, its first decision might well be to vote the humans out of the community, too deadly a presence to tolerate any further. We are the affliction of the world, its demonic presence. We are the violation of the earth's most sacred aspects."[22] Latour, however, is not suggesting that humans give away our votes, even if we could. Meanwhile Berry is still holding out hope for the ecozoic age. His close colleague Mary Evelyn Tucker wisely advocates, as does the Earth Charter, an "earth-community." As do we.

MAPPING ECOSPIRIT

Amidst the mingling atmospheres of an ecological apocalypse, an upbeat activism and a greener cosmopolitanism, this volume ekes out its ecosophical vision. "EcoSophia," the nickname for the conference held at Drew University in which the present volume originated, betrays how steeped the editors are in the specifically ecofeminist dimension of theology and religious studies. The width, hybridity, and challenge to single-issue sex or gender politics that ecofeminism has cultivated restimulates *Hochma-Sophia*, the biblical symbol of the immanent divine wisdom, the only depiction of God as female that ever gained significant traction within the scriptural patriarchy.

The papers in this volume display the wide range of the conference discussions; we hope a feeling of joy comes through along with the tension. We are keenly aware that there are many missing voices who on principle should have been among us; and, happily, there were voices present—Continental philosophers and an Australian Germanist, for example—who would not predictably take part in the religion/ecology interdiscipline. We think the diversity of disciplines and intentions represented here creates its own ecology of thought. It gives theory back its ground. The grounding of theory is an aporetic notion: a strategic paradox. By this ground, we mean the moving incarnation of thought in the shared flesh of its spirited thinkers; a refusal of the identification of living ground with abstract foundation; and an ecological test for every ethical and spiritual value. Unless theory wants to continue its flight from the earth—its idealism, its supernaturalism, or its ontotheology—this grounding is neither a punishment nor a stasis.[23] It will rest neither in a reduction to "mere matter" nor in the literalism or objectivism of an unreconstructed nature. Any theory that recognizes its ground in the earth traces its own "lines of flight," to borrow a formulation of Deleuze. If they tend to follow avian rather than extraterrestrial trajectories, it is because they move—as does Mark Wallace's earth-bird Holy Spirit— within the figurative and physical atmospheres in which we continue to breathe. Thus we can call the book *Ecospirit* in good faith: the ancient Hebrew *ruach* was never an immaterial force but an earth-breath, at once grounding of and grounded in the creation.[24] Even in supernaturalist Christianity, the Holy Spirit represents divine immanence, if unfortunately subordinated to the transcendent Father. The trope of spirit signifies a field of activity ecumenical, mobile, and polyvalent enough to traverse the multiple disciplines and methods of *EcoSpirit*.

The first cluster of essays, *Ecogrounds*, articulates a shared "atmosphere and earth" for this multilayered volume, setting in motion ways of thinking, talking, and walking the spirit-ground. The capacity of ecological concern to awaken a new planetarity pushes toward a fresh model of interreligious encounter. So Jay McDaniel teaches in his contribution, which reaches out to a public much wider than the academy or church. The possible common ground for an ecoecumenism is within this volume always understood as involving theory for common practice. Anna Peterson thus calls upon us to extricate ourselves from the intellectual puzzles of ecology and religion. The needed values are at hand, waiting to be put into practice. Ground for hope is found in "talking the walk" of a practice-based environmental ethic. Rosemary Radford Ruether's essay (the conference's opening address) establishes the cultural pluralism of a developing ecofeminism as a ground for the greening of faith and thought. She offers a close reading of three instances of ecofeminist theology, displaying its multicultural, and therefore multidimensional, potentiality. For the plight of the earth and of the poor of the earth remain inextricably correlated within the colonized bodies of women. How do we walk, or not walk, the talk? Eschewing foundationalism while insisting upon common ground, Catherine Keller's "Talking Dirty: Ground Is Not Foundation" refuses the binary of ecology and deconstruction. Might a "dirty ground" for thought—even for the airy Christian trinity—

emerge from the deterritorializing forces of the earth itself? With its immense and ever-specific planetarity beginning to take place in thought, the volume means to get down and dirty.

In *Econatures,* the essays undertake a multifaceted investigation of current constructions of nature in the interstices of theology, science, and philosophy. Laurel Kearns documents the twists and turns of claims of faith, economics, and scientific "knowledge" in the religiously charged atmosphere of climate-change activism, including the dangerous collusions undermining the discussion. Hiding behind a seemingly classic science/religion standoff may lurk, for instance, those for whom "cooking the truth" in their allegiance to economics is part of the attempt to avoid responsibility for global warming. Glen Mazis considers strategies for healing the bifurcation of technology and nature, rather than merely idealizing a pretechnological organicism. The blurred boundaries of humans, animals, and machines structure a new sense of nature, of the ground of human and nonhuman interrelation. Barbara Muraca, engaging Whitehead, Latour, Heidegger, and Vattimo, sketches a path through and beyond a bifurcating nature toward "possible postmodern ontologies of nature(s)." Kevin O'Brien exhorts ethicists to engage the science of biodiversity and the practices of ecological restoration for insights into the postmodernity of a not-so-static nature. John Grim pushes ecoecumenism into an often neglected encounter with "indigenous knowing," with its acute sense of local place and relations to "nature" (a word that may not even make sense in these different contexts). Here the wisdom of interspecies interconnection signifies not "simply the ashes of an extinct, failed way of knowing, but the embers that indigenous elders are rekindling to confront their peoples with awareness of deeper purpose."

Next, under the heading of *Econstructions,* the volume steers right into the turbulent confluence of poststructuralism and ecological sensibility. As David Wood puts it, "environmentalism finds itself in an often problematic and aporetic space of posthumanistic displacement with which deconstruction is particularly well equipped to find guidance." Ecology might embolden deconstruction to embrace at least "a strategic materialism." Similarly, Kate Rigby articulates a postmodern ecopoetics. In a lively rereading of Goethe, she analyzes the background of romantic modernity, with its dual quest for scientific mastery over, and poetic renewal through, nonhuman nature. This group of essays considers postmodern tropes such as "the gift," "the democracy to come," and "the molecular versus the molar" as possible sites for the greening of theory. Can, for instance, "the gift" stretch to signify, as Anne Primavesi proposes, the "preoriginal gift" of life itself—not as a supernatural donation but as a grace of Gaia? Will Derrida's "democracy to come" grow with Wood's help into the "parliament of the living"? Or, as Luke Higgins proposes, might Deleuze and Guattari's "spirit-dust" become the deterritorialized matter of an ecopneumatology?

Already these poststructuralist reflections find themselves turning to the focus of the next section, theology, though hardly to dogmatic business as usual. *Ecodoctrines* offers a veritable garden of green microtheologies. Each contribution opens a doctrinal site both for the deconstruction of a rigid bit of orthodoxy and for an ecological reconstruction. First come two highly concentrated ecopneumatologies. Mark Wallace lets a green spirit reinhabit a theology of sacred land. For Sharon Betcher, a grounded spirit animates our humus/humanity/humility. Next, drawing on kabbalah and the work of Hans Jonas, as well as recent theories of cosmology and evolutionary biology, Larry Troster outlines a Jewish creation theology for an age of ecological trauma. Also clearing the way for an ecotheological revision of the creation narrative, Whitney Bauman analyzes the use of the Christian doctrine of *creatio ex nihilo* in the early modern colonization of *terra nullius*. Antonia Gorman performs an archaeology of nineteenth-century vivisection, excavating some startling uses of the dogma of the atoning sacrifice of Christ—all the more surprising for their contemporary resonance. Finally, for "the hope of the earth," Seung Gap Lee uses a dramatic protest narrative from contemporary South Korea to illustrate his process theological ecoeschatology.

The doctrinal tradition that grew under Hellenistic influence from biblical metaphors stimulated a radical deracination of the sacred in the interest of a universal mission. A spiritual deterritorialization led to a global reterritorialization. Transcendence tended to dissociate itself from the specificities of a place, with its density of creatures, relations, traditions. *Ecospaces* therefore offers several studies of specific places as sites of spiritual-ecological desecration. But in each ravaged case they are also potential places of renewed sacrality. It begins with Daniel Spencer's examination of how ecological restoration is "a particularly promising component of an ethic of sustainability rooted in place." His notion of ecosocial location is well exemplified in the particular locations that follow. In her journeys with Episcopal Athabascans, located in endangered territory in Alaska, Marion Grau finds "tragically embodied what often seems oracular and abstract postcolonial jargon: hybridity, ambivalence, and mimicry manifest in double-edged, often crazy-making incarnations that all facilitate survival." She paints the possibility of a theology of Arctic place. Anne Daniell's poignantly timed study of carnival and ecology in New Orleans releases the rhythms of a new estuarine metaphor of the incarnation. Rick Bohannon decodes the ecological architecture of a chapel in the Arkansas woods using French sociologists Bruno Latour and Pierre Bourdieu to expand our understanding of relations with the nonhuman. With other woods in mind, Nicole Roskos investigates the ancient Christian tradition of felling sacred groves, turned now into a weapon in the armory of the antienvironmentalist Right. She points toward a hopeful rereading of our relations—even Christian relations—to trees.

Trees for us became a sign of "Ecohope," the heading comprising the final grouping of essays. The conference was punctuated by a ritual tree planting outside the brand new wing of our seminary—itself a metaphor of seed. Drew University sits in an urban oak forest, and cannot, in truth, avoid a certain druidism. As it happens, "truth" roots etymologically in the Old English *dreugh*, from which stem "druid," "troth," and "trust," qualities of uprightness presumably modeled on the trunks of trees. The growth of truth in a living ground signifies earthen sturdiness in space and time, something worlds apart from the stability of an abstract foundation in changeless truth. Trees stretch forth to effect their own atmospheric change. We therefore include a narrative of the tree-planting ritual, performed by Heather Elkins and David Wood, as well as a more formal, ecumenical liturgy, created by Jane Ellen Nickell and Larry Troster, from the conference, hoping they may be useful models for ecotheological enactments. These ritual practices support the wider practices of environmental activism and help to activate or perform hope in the face of despair. Toward this same end, in the opening essay of the section, Mary Evelyn Tucker articulates her vision of the common ground on which "to build a multiform planetary civilization inclusive of both cultural and biological diversity," as articulated in her work with John Grim and the Forum on Religion and Ecology,[25] and of course, in the Earth

Charter. The rituals are preceded by the wisdom gleaned from the "on-the-ground" activism of Fletcher Harper, the executive director of GreenFaith, as it spreads its interfaith branches throughout New Jersey and beyond. We end, just as we did our formal conversations at the conference, with Karen Baker-Fletcher's generous gift of ecological poetry, itself liturgically potent, as it circulates through the catastrophic hurricane experiences of Katrina and Rita that closely preceded our conference.

But just as we gathered poems as an ending to share at the lightly hallowed site of the tree planting, the poetic end can become a beginning. We did not deify our local dirt. But we do hope that these essays might plant themselves fruitfully in the ecoground of your own work, your econatures, your econstructions, your ecodoctrines, your ecoplaces, and indeed your ecohopes.

❧ Ecogrounds: Language, Matrix, Practice

✦ Ecotheology and World Religions

JAY MCDANIEL

INTRODUCTION

My aim in this essay is to discuss a social and spiritual movement called "ecotheology." I want to provide an example of how it can be practiced among Christians and discuss its relevance to the many world religions. In addition, I will briefly introduce aspects of a philosophical foundation for ecotheology, showing how, in some instances, philosophy and spirituality can be companions to a process of social transformation.

I write for the religiously interested general reader. This general reader ultimately motivates all the essays in the present volume. Directly or indirectly, the authors of the essays collected in this anthology are all involved in encouraging people to participate in a movement to respect the community of life on earth. This includes a respect for human beings and a concern for their well-being, as well as a respectful concern for other living beings and the earth itself. The ecotheology movement seeks to promote this respect.

The ecotheology movement is found in many different communities around the world, even as it lacks central organizing authority or formal structure. This absence of a central organizing authority is part of its creativity. We sense the presence of this movement in conversations among people who are advocates for women, children, the elderly, the poor, animals, and the earth, and among people who sense that these forms of advocacy are all connected in some deep way. We sense it, too, among those who feel alienated from consumerism and fundamentalism, and who, as a result, take vows of voluntary simplicity and practice post-materialistic ways of living in the world. And, again, we sense it from people who have a hunger to be connected, not only with the sufferings and joys of other people's lives, but also with the beauty of the earth. Ecotheology is like a river coursing across the world, amid which there are many currents, fed by many tributaries. This river is but one of many rivers and some of them—consumerism and fundamentalism, for example—are much larger. But it is, I believe, the most promising river we have.

Of course the word "ecotheology" is not especially euphonious, and most people do not use the word. Some of them might refer to themselves as environmentalists, but it seems to me that ecotheology is not quite the same as environmentalism, because it is concerned as much with people as with the earth. Others might speak of themselves as human rights advocates, but the ecotheology movement includes more than a concern for human rights, important as they are. Thus, we might call ecotheology *the web-of-life movement* insofar as it takes the well-being of life as a whole—rather than ever-increasing economic growth—as the central organizing principle of its social vision. In any case I use the term "ecotheology" with some reluctance, and I will explain below why I think that while it is rightly called a "theology," I hope readers will feel free to plug in other terms if they so desire.

How old, then, is this movement? In some ways it is very old. Indeed, some of its attitudes are embodied by preindustrial agricultural peoples and rural residents still today. Agricultural peoples typically had a sense of the web of life, because their lives depended on being integrated into its rhythms. When they awoke in the morning, it was because the sun had come up; when they went to bed, it was because the sun had gone down. If they ate meat, they knew that an animal had been slaughtered who desired to live. If they tilled the soil, they knew that their lives depended on its health. What is new about contemporary or postindustrial ecotheology is that it is attractive to many people in urban settings, heirs to the Enlightenment and industrial period in human history, who are not in touch with more natural rhythms and who feel a spiritual need to feel connected not only with other people, but also with animals and the earth. Contemporary ecotheology speaks to a hunger of the heart, a hunger to be connected to something more than machines. This does not mean that the ecotheological movement is against technology. Its participants make full use of computers and the Internet. But it *is* against mechanistic ways of understanding the world. It sees the world on the analogy of a living whole.

It is for this reason that some ecotheologians in the West are drawn to the scientifically influenced yet organic philosophy of Alfred North Whitehead.[1] He offers a way of understanding the web of life as just that—a web of life—and not simply a vast, complicated machine. Much of what I have to say in the following is indebted to Whitehead, and more specifically to an intellectual movement that is indebted to him: *process thought.*[2] A brief word about the Whiteheadian approach is in order to give readers a sense of what one philosophical approach in this vein might look like. I must emphasize at the outset that there are, and can be, many different approaches, of which process thought is but one. The ecotheology movement can be grounded in Buddhist, Marxist, or evangelical Christian ways of thinking, and none of them need mention process thought. Nevertheless, the process approach is unique in that it can underlie and support these various views. In fact there are Buddhist and Marxist and Evangelical process thinkers, all of whom are indebted to Whitehead, even as they interpret process thought in slightly different ways. My own understanding of process thought is shaped by Buddhist and Christian ways of thinking, and also by the emerging tradition of Chinese Process Thought.[3] There are now twelve centers for process studies in China, and I have been active in helping organize some of them. I include references to China within this essay because this nation is so obviously important in the world today and that to come.

PROCESS THOUGHT: A PHILOSOPHICAL FOUNDATION FOR ECOTHEOLOGY

In China and in the United States, some scholars understand Whitehead-ian thought as a form of *constructive* postmodernism.[4] They use the phrase "constructive postmodernism" deliberately as an alternative to deconstructive postmodernism, but not as its enemy. They appreciate the impulse within deconstructivism to critique hidden assumptions in inherited ways of thinking, particularly as those assumptions validate or support arrangements of power that oppress human beings and denigrate the earth. Some of the other essays in this volume show the power of this approach to help free humans and the earth from harm. By constructive postmodernism, though, process thinkers have in mind a movement that can build upon, but also move beyond, the merely critical approach. Constructive postmodernism seeks to build upon the best of modern ways of thinking and also move beyond its worst aspects.

Of course, what is "best" and what is "worst" will be a matter of debate, and any evaluations will be profoundly shaped by the social location of the evaluators. For many who are influenced by Whiteheadian thought, though, a general consensus has emerged. The best of modernity includes (1) its reliance on science and empirical inquiry as modes of reasoning, (2) its emphasis on settling disputes by dialogue rather than violence or appeal to special revelation, (3) its emphasis on lifting people from poverty, and (4) its emphasis on the importance of people participating in the decisions that affect their lives. Many people in both China and the United States appreciate these aspects of modern industrial societies, and process thinkers join them.

The consensus holds the worst of modernity to include (1) its neglect of the environment amid its emphasis on material progress, (2) its affirmation of self-interest over the interests of family and community, (3) its rejection of premodern agricultural traditions and rural values in the pursuit of material progress, (4) its privileging of scientific ways of knowing over aesthetic ways, (5) its reaffirmation of a single, "rational" way of being in the world at the expense of cultural diversity, (6) its tendency to reduce all categories of value to economic value, (7) its assumption that almost all social problems can be solved by means of economic growth, and (8) its tendency to assume that all models of human development must follow a Western paradigm. Along with many Chinese and also many Americans, process thought rejects these attitudes and tendencies.

A constructively postmodern culture will affirm the best and move beyond the worst. It will be a culture in which people appreciate reason, science, progress, and participation. And it will be a culture in which people respect the earth, remember the role family and community play in human life, recognize that other people deserve respect quite apart from the money they make or their productivity in a growing economy, remember that alternative options for living in the world—scientific and spiritual, for example—can be complementary rather than contradictory, and understand there are multiple paths to development, not simply that of the West. Process thinkers believe that a constructively postmodern culture is relevant to many people in the world, especially, perhaps, those of China and the United States. It can help the former remember the richness of its own heritage as it brings its pursuit of a prosperous, socialist society into the twenty-first century. And it can help the latter move forward in creative partnership with other nations, China included, who play their important roles on the world stage.

Of course a constructively postmodern culture is an ideal to be approximated, not a utopia to be fully realized. It will be embodied in different ways by the various nations, each adding its unique cultural heritage to the mix. There is not one kind of constructively postmodern culture, but many. Process thought offers a worldview and points to a way of living that can contribute to the emergence of diverse, constructively postmodern cultures.

Whitehead's philosophy holds the universe to be a dynamic and evolving whole in which every event is related to every other event. A momentary energy-event within the depths of an atom would be an example of such an event, and so would a moment of experience in the life of a human being. Both are activities in which, as Whitehead contends, the many of the universe become one in a particular activity. Of course, most of us do not experience something as grand as "the universe" in the comings and goings of our daily lives. We may look up at the stars and get a sense of a greater whole, but most often we are looking across the street, or into the faces of people we encounter in the workplace, or at the food on our table. We meet the universe most concretely in the immediacy of local settings: in our homes, at work, in our villages and cities, our natural landscapes, and in our friends, family members, even the strangers we pass on the street.

Process thought says that our task in life is to live in creative harmony with people and other living beings in these local settings, adding our own distinctive kinds of beauty to the larger whole. Here "beauty" does not refer to how things appear on the outside, but to who we are on the inside. Beauty consists of the harmony and intensity of our own subjective lives as we interact with others. In process thought there are many kinds of beauty: love and courage, wisdom and compassion, creativity and laughter, faith and hope, struggle and peace. All are forms of harmony and intensity.

Process thought, then, adds that even as we add our own beauty to the larger whole, we can be enriched by the beauty of others: the hills and rivers, the plants and animals, the trees and stars. Human experience is not only active, but also receptive. We receive others into our lives through experience in the mode of causal efficacy, and the enjoyment of their presence is one of the gifts of life. Thus, the value of our lives is not simply what we *make of* them. It is also what we *receive from* them. In process thought, value is not simply a human projection onto a blank slate. It is a certain kind of beauty that is found in other people and also in the natural world.

Thus, Whitehead seeks to overcome the separation of fact and value that is so common in modern thinking. Often in the modern world we assume that "facts" are part of objective reality and that "values" are mere projections onto that reality. Certainly there is truth in this. Sometimes humans do indeed impute values to things that do not deserve them. But Whitehead believes that there is also a certain kind of value in the world even apart from human projections: namely, that which other living beings have in and for themselves as they struggle to survive with satisfaction. For example, when an animal struggles to find food, it experiences "survival" as a value and the acquisition of food is itself a realization of value, at least for the animal. The Whiteheadian view holds that this kind of value was part of the earth long before humans evolved, and thus that there are values in nature even if humans do not recognize them. Part of living in creative harmony with the wider world is to recognize this value in nature and, of course, the value of other people as well. It is to treat other persons—plants and animals, too—as ends in themselves and not simply as means to other ends. Process thought emphasizes the need for humans to build communities that respect and care for the larger community of life; it follows that such communities would be socially just, ecologically sustainable, and spiritually satisfying, with none left behind. Thus process thought provides an illustration of one kind of philosophy that can help ground the ecotheology movement.

Ecotheology, though, is *not* a philosophy, even of the Whiteheadian sort. Nor is it a religion. It is an orientation toward life and a way of living that can be embodied from many different religious points of view by people who seek a creative alternative to consumerism and fundamentalism. Most ecotheologians do not refer to themselves as such. Instead, they self-identify as Christians or Buddhists or Muslims, Africans or Asians or Americans, poets or painters or plumbers. Some say they are "spiritual but not religious" because they feel no identity with any particular religious point of view. If placed within a single room, they would not know each other, and they might have many important differences. If there were evangelicals among them, they might even argue with one another about who is or is not "saved." Still, they would share a common hope for the world and a common spiritual orientation. Their hope can be captured in the phrase, "Let there be communities of respect and care." Their common spiritual intuition can be captured in the phrase, "Let us find the sacred in the world."

A COMMON HOPE: LET THERE BE COMMUNITIES OF RESPECT AND CARE

The hope of ecotheology is that people in different nations and cultures can live with respect and care for the community of life, and in so doing develop communities that are just, sustainable, participatory, nonviolent, and compassionate. I borrow this language from the Earth Charter.[5] The Earth Charter is a short document, created in the last decade of the twentieth century, which is now being used by educators around the world in schools, institutions of higher education, and in community and professional development. Successive drafts of the Charter were circulated around the world for comment and debate by nongovernmental organizations, professional societies, international experts in many fields, and by representatives of the many world religions: Confucianism, Taoism, Sikhism, Hinduism, Jainism, Buddhism, Judaism, Islam, Christianity, and many indigenous traditions. As people gathered together, they listened to one another and jointly asked, "What are the deepest needs of the world today and how might we constructively respond?" Thus the Earth Charter may embody the most inclusive process of multireligious discernment the world has ever seen. It consists of a preamble and sixteen ethical principles, the foundation of which is the idea that people of all religions—and people without religion as well—should live with "respect and care for the community of life."

Some of the words used above—"just" and "sustainable," "participatory," "nonviolent" and "compassionate"—are amplifications of what this spirit of respect and care can mean in practice. A "just" community is one in which people are treated fairly and given full access to health care, education, work opportunities, and recreation, and in which there are no large gaps between rich and poor, precisely because people share in one another's destinies. A "sustainable" community is one that respects the capacities of the earth to supply renewable and nonrenewable resources and to absorb pollution, and one that develops in ways that are sensitive to the needs of other, nonhuman communities. A "participatory" society is one in which people have a voice in the decisions that affect their lives and feel "listened to" even when decisions run counter to their preferences. A "nonviolent" community is one in which people feel—indeed, are—safe to travel and speak freely; where differences are appreciated rather than feared; and where people settle disputes through arbitration and negotiation, not force. A "compassionate" community is one that practices a caring approach to other people, including those who might otherwise feel marginalized, and that advocates the humane treatment of animals, whether pets, animals used for research, or livestock, all of which have suffered profoundly the callousness of modern industrialized societies. Whether Buddhist or Muslim or Christian, ecotheologians feel called to help develop these kinds of communities.

Accordingly ecotheologians find themselves at odds with the idea that the health of a society is to be measured in strictly economic terms, as though economic growth were the sole criterion for evaluating progress. For ecotheologians progress is measured by how much people care for each other and share in one another's destinies; by how much people respect diversity—differences in culture, religion, ethnicity, and gender; by whether it is possible for people to recreate and enjoy their lives regardless of income; by how well the plants and animals that join us in the web of life are treated. A community can be rich in money but poor in community—this is not progress. Ecotheologians thus put forward the kind of question that is rarely asked in consumer society but is nonetheless critical to our human future: what would economic theories, policies, and institutions look like if their primary aim was the promotion of human community in an ecologically responsible context, rather than ever-increasing production and consumption? Ecotheologians do not reject market economies, but they do resist the idea that communities ought to serve the market.[6]

A COMMON SPIRITUALITY: LET US BE SENSITIVE TO THE
HORIZONTAL SACRED

In addition to this hope for just and sustainable communities, most eco-theologians also share a common spiritual intuition: that the sacred can be in the world—that is, be in it "horizontally" as well as "vertically." Some explanation is perhaps in order.

The vertical sacred is God or the Godhead. In the world religions it can be experienced as a higher power that is above us in some way, to which we feel attuned in faith and hope; or as a deeper yet bottomless source from which all things continuously emanate, moment by moment, of which we ourselves are expressions. The personal God of Abrahamic religions is often experienced as a higher power of this sort; and the transpersonal Brahman of Hinduism is often experienced as the deeper source of this kind. The first is above us, yet also around us and within us. The other is within us, yet around us and deeper. Many ecotheologians speak of a vertical sacred in one or both of these ways. They speak of God or Brahman, the Lord or the Abyss, the Goddess or the bottomless Ground of Being. But what is equally important to them, and often more important, is the way the Lord or the Abyss can be discerned in the world itself. What is most important is the horizontal sacred.

By the "horizontal" sacred I mean two things: first, the intrinsic value of each and every living being on earth, understood as a subject of its own life and not simply an object for others; second, the joy of mutually enhancing relationships, in which humans dwell in harmony with one another and with other living beings. The horizontal sacred is complementary to the vertical sacred. In the Abrahamic religions, for example, people say that the Spirit of the creator God is in the earth and within each living being as its animating breath. They emphasize that the Spirit is truly God, which means that God is truly in the earth and its many forms of life. They find the vertical in the horizontal.

Nevertheless, the horizontal sacred also can be experienced without reference to a vertical sacred. Indeed, there are many people in the world today who have a sense of the horizontal sacred even as they do not believe in something "higher" or "deeper," and there are certain religious traditions that are primarily concerned with the horizontal sacred.

Confucianism would be one example, with its emphasis on reciprocal relations between human beings as the place where the sacred is found. In the West since the Enlightenment, when some intellectuals have experienced what they might call a "death of God," they often turn to the horizontal sacred as their frame of reference. The horizontal sacred does not require worship, but it does indeed elicit awe and respect. From the perspective of these nontheistic ecotheologians, then, we do not need to look up beyond the universe or deep within ourselves to find the sacred. We can look at the stars and galaxies above us, and we can look into the eyes of another living being. This is why nontheistic ecotheologians are yet theologians. Of course they are theologians in a rather unique sense. Even if they may not believe in God or Brahman, the Lord or the Abyss, they have a sense of the sacred in the things of this world and in their potentially beautiful connections with one another.

THEOLOGY FROM THE OUTSIDE IN AND THE INSIDE OUT

Ecotheologians are thus people who trust in the sacred: vertical, horizontal, or both. Often we think of theologians as people who write books, or at least spend a lot of time thinking about religious questions. I myself am a theologian in this sense, specializing in a form of contemporary theology called "process theology." But those of us who are theologians in this narrow sense often meet people who do not think of themselves as theologians at all, but who exemplify the spirit of what we are talking about. Their lives are their sermons; their attitudes are their teachings. They exemplify living theology. Thus when I say that ecotheologians are theologians, I have in mind this living theology.

All of us have a living theology of one kind or another. It is simply a way of living that is guided by a set of concerns; and it includes our attitudes toward life, feelings about others, capacities for paying attention, motivations, and intentions. These more subjective and affective aspects of our living theology do not arise in our lives simply through our believing particular things about the world or by holding particular worldviews. Rather, they emerge from experience itself, from the outside in and from the inside out. "From the outside in" in this context refers to how the world shapes our own internal perspectives, consciously or unconsciously, healthily or unhealthily. For example, if we grow up in a culture where people think unlimited consumption is good, then our own embodied theology may be shaped by their attitudes, such that we ourselves live with the hope of consuming our way to happiness—our feelings are shaped by their feelings. "From the inside out" here refers to how we respond to the world that shapes us, from within our own subjective depths. For example, if we try to resist the call of unlimited consumption, seeking to live in a more humble and less acquisitive way, then our living theology includes this inner response—our feelings arise out of our own inner depths. Thus a living theology may include a struggle—an internal jihad—between two opposing forces, both of which lie inside us, but which might also impinge upon us from the outside. Most ecotheologians experience such an internal struggle.

FUNDAMENTALISM AND CONSUMERISM

Some theologians struggle with "fundamentalism," which espouses the possession of finite things as if they were infinite and at the expense of an openness to fresh possibilities from the future, which, if actualized, might challenge or change inherited ways. In our time, "religious" fundamentalism is an obvious reality, but there are other kinds of fundamentalism as well, among them "political" and "economic" fundamentalisms. Whenever a person holds to something finite as if it were infinite—a sacred text, an inherited ideology—he or she has become a fundamentalist, at least for that moment. Ecotheologians may sometimes hold onto their hopes and intuitions in this way, thinking that truth is exhausted by their experience of it. When this happens they become ecofundamentalists. They cannot laugh at themselves or laugh with others. They think that they are "pure" and other people are not, that they are "right" while others are "wrong." They become, in a word, insufferable.

But the more common struggle is with consumerism. Many ecotheologians struggle with the voice of consumerism and more particularly with the tension they inherit from being a part of consumer culture between two discrete ways of living, both of which call for response—a life of ambition and a life of integrity. If we are driven by ambition, our desire is to acquire prestige through achievements in the marketplace and conspicuous consumption. We want to "be successful" and "get ahead of others" so that we might make our mark in the world. On the other hand, if we feel called to integrity, our desire is to live in solidarity with others, without having to get ahead of, or be more successful than, others. These two forces—ambition and integrity—live within most ecotheologians. The ecotheology movement, however, is a way of living that prioritizes one, integrity, over the other, ambition. It finds value in being with others, not surpassing them. And it redefines "ambition" in qualitative terms; that is to say, once basic needs are met, the appropriate ambition in life is to grow in wisdom and compassion, not in wealth and prestige. Thus ecotheology is in profound tension with the culture of consumerism.

Consumerism, too, is a living theology. It is a way of living that takes appearance, affluence, and marketable achievement as the highest goods in life and that envisions happiness as an ongoing process of consuming ever more without end or limit. It tends to reduce the earth and other living beings to their cash value in the marketplace. It does the same to human beings, reducing them to the role they play in a growing economy. People are valued for their work but not their hearts, for their productivity but not their tenderness. Nonmonetary achievements in life, such as being a good parent or taking care of one's parents, are not valued in consumerism. The emphasis is always on marketable achievement.

Ecotheology seeks an alternative and counterconsumer way of living in the world. Its goal is the development of human communities that live with respect and care for the greater community of life. It relies on a sense of the sacred as something that lies inside and between living beings when they dwell in mutually enhancing relationships. It looks to provide a meaningful alternative to both consumerism and fundamentalism.[7] In what follows I offer a portrait of two living ecotheologians in the Christian tradition.

CHRISTIAN ECOTHEOLOGY

Imagine two women—one a Protestant, the other a Catholic—sharing a meal together in a restaurant in an urban setting somewhere in the United States: Omaha, Nebraska, for example. And imagine that they are, as it were, living ecotheologians. They would never call themselves "ecotheologians." They would simply call themselves people. For the sake of concreteness, I will call them Joan and Sarah.

True to the spirit of ecotheology, Joan and Sarah know that life is not all about "ethics" or "doing good." They laugh and play, sing and sleep, enjoy times with friends and family, and cherish being alone, too. They are moral but not moralistic. If asked which of three Platonic values they prefer—truth, goodness, or beauty—they might be inclined to say "beauty" because they enjoy the beauty of life itself.

Precisely because they are struck by life's beauty, though, they are also pained by the unnecessary sorrows and tragedies from which many people and other animals suffer. Accordingly they are active in their communities trying to make a constructive difference in the world, or at least to brighten their small corners of it. They are committed to what ecotheologians call "ecojustice."

Joan, the Protestant, practices her ecojustice as a social worker. She spends several hours a week visiting with poor families who struggle to make ends meet. Recently she has been lobbying city officials to offer more support for a local community center that provides opportunities for teenagers to be tutored after school. People find her good-humored, but also at times irritating. Like Jesus, she comforts the afflicted but also afflicts the comfortable.

Sarah, the Catholic, is of a quieter nature. She has always loved animals and the natural world, and she now finds herself spending many hours monitoring the effects on wildlife of the building of gated communities and suburbs for the wealthy. She has been lobbying officials to prevent developers from building a golf course on habitat best left preserved for nonhuman inhabitants of that region. People tell her that she is against "progress" and "development." She says that she is "pro-life" in the deepest of senses: she seeks the well-being of all life.

To an outsider it might seem the social agendas of Joan and Sarah are very different. One is concerned with people and one with other living beings. But in fact they share a common hope. They want to help build communities where unfair gaps between rich and poor are eliminated; where people have a sense of shared destinies and take care of one another; and where people have a sense of affinity with other living beings, such that they are motivated to live with respect and care for the community of life. They are committed to the common good: a good that includes the good of other living beings and of humans, especially the poor and powerless. This is why Joan and Sarah are, in their own unique ways, un-American. They do not want to get ahead of people. If "upward mobility" names a process of getting ahead of people, they are interested in "horizontal mobility": that is, in being with people and other living beings. Like Buddhist *bodhisattvas*, they do not want to be saved unless everyone else can be saved, too.

GRACE BOTH RED AND GREEN

In addition to sharing a common concern for just and sustainable communities, Joan and Sarah share a kind of spiritual kinship. For one thing, both have had significant tastes in their lives of what we might call "green grace." Green grace is the satisfaction—the harmony and intensity—that comes through enjoying rich bonds with other people, plants and animals, and the earth itself. Green grace includes healthy relations between people, as that between Joan and Sarah. But green grace also comprises healthy relations with the more-than-human world. People experience green grace when they have healing and healthy relations with their own bodies; when they gaze into the heavens and feel small but included in a much larger whole; when they look into the eyes of their companion animals and realize that these animals have lives of their own, filled with intelligence and beauty; when they learn about the flora and fauna of the communities in which they live, delighting in the birds and trees; when they find themselves amazed at the sheer beauty of the elements—earth, wind, fire, water. We might say, then, that green grace is multicolored. It is the brown of the house cat, the blue of the ocean, the yellow of fire, the green of the grass—each one glows in the eyes of one who can feel their presence with appreciation and wonder. As Joan and Sarah have lunch together, they bring with them their capacities for green grace. This is one reason they care about other people and other living beings: they are struck by their beauty.

But let us imagine further that Joan and Sarah also know the reality of "red grace." Red is the color of wine as shared in the Eucharist, through which Christians share in the living-by-dying that Jesus experienced on the cross. As ecofeminists Joan and Sarah are suspicious of understandings of the cross that lead women and others to be more self-sacrificing than they ought to be. They know that the cross has sometimes functioned to tell women that they are, or ought to be, mere doormats for others. Still, they find value in the cross as a symbol for shared suffering and personal accountability. They find value in red grace. This is the wholeness that comes through being honest about our own suffering and sharing in the suffering of others; accepting our own responsibility for whatever harm we have inflicted on others or ourselves; and recognizing that, even amidst our own harmful actions and suffering, new life is possible.

Joan and Sarah know red grace from the inside out. Joan has had a bout with cancer, and she has been shepherded through the process by Sarah's deep listening. In their shared suffering, a certain kind of bond has emerged that cannot emerge otherwise. This is one dimension of red grace. It takes heed of the first noble truth of Buddhism. However else we might differ from other people, we have a common bond in dis-ease, in suffering, in *dukkha*. Our *dukkha* need not overwhelm us, if we have others—a *sangha*—to understand it and somehow share it with us. Sarah is part of Joan's *sangha*.

Another dimension of red grace is its honesty and humility about sin. As middle-class Christians, Joan and Sarah have both struggled with the fact that their lifestyles too often illustrate the overly consumptive habits they critique. Only in being honest about their own complicity in sins against the earth can they be humble in their attempts to help the earth become a healthier place for all. In their humility, there is a kind of wholeness, too. It helps them overcome the illusion and burden of moral purity. Thus they know that in human life both kinds of grace are important. Red grace without green grace is morbid, and green without red is shallow. The two forms of grace complete one another, giving life a certain spiritual depth. Joan and Sarah seek this depth.

Of course the word "grace" suggests a grace-giver, namely God. Joan conceives God in personal and traditional terms. God is a higher power to whom she prays, who calls her and others to responsible stewardship of the planet and to care for others, and who is affected by the joys and sufferings of the world. But she does not simply think of God as external to the world. She also thinks of God as present in the world and, for that matter, as present in all living beings. She speaks of this presence as God's Spirit, and believes that the Spirit can be found not only inside each person, animal, and plant as their breath of life, but also in their relationships with one another. God, then, is in her relationship with Sarah, and God is also in her relationship with her companion animals. In short, she believes in the horizontal sacred as well as the vertical sacred, and she finds the vertical in, but not exhausted by, the horizontal. For her, God is One-embracing-many and One-inside-each and One-between-many.

Sarah conceives God differently from Joan. She speaks of herself as a nontheistic Catholic. For her God is inside each living being and between living beings when they enjoy healthy relations with one another; but God is not a transcendent and embracing reality who loves the world from afar. She believes in the Spirit, but not in the Father: in One-between-many but not in One-above-many. Part of her distrust of height imagery is that it has so often functioned to support a monarchical understanding of deity, whereby the divine is envisioned on the analogy of a holy warrior or conquering dictator. She does not have faith in dictators, divine or human. Another reason for her distrust is that she cannot reconcile belief in a transcendent God with the suffering in the world. "If God is all-loving and all-powerful," she asks, "then why is there so much suffering?"

Joan has a response. She says that God is all-loving but not all-powerful. In her words: "God embraces the world but does not and cannot force the world into love; God's power is inside us, and it lies in persuasion not coercion, in beckoning not force. Things happen in the world that even God cannot prevent. God helps us pick up the pieces whatever happens, and take a next step." For Sarah, however, it does not make much sense to believe in a transcendent power who cannot and could not control all things, at least in principle. So she chooses not to believe in it at all. She experiences grace both red and green, but not a separate grace-giver.

These different ways of conceiving God do not get in the way of their friendship. Both trust in grace red and green; both are committed to ecojustice and to lives of solidarity with the poor and with the earth; both agitate city hall; both choose integrity over ambition; both are more concerned with immersion in the world than with getting ahead of others. And both enjoy a good meal together. In their humanity, they exemplify living ecotheology.

THE BIGGER PICTURE OF THE WORLD RELIGIONS

The example above illustrates ecotheology as it might unfold in the lives of two middle-class women in the United States who identify as Christian. In concluding this essay, I turn to how ecotheology might be relevant to the larger context of the many world religions and to the better hopes of the world as a whole.[8] In so doing I will draw more explicitly on the monotheistic perspective of process theology as represented by Joan simply because it makes the most sense to me.

From a process perspective, religions are verbs (actions) rather than nouns (subjects). They are not settled paths on which people walk, but rather paths that are continuously being created by the people who walk them. Some elements of a path are given to each generation of walkers: the creeds, codes, ritual practices, role models, and memories that they inherit from predecessor generations. These elements help them get their bearings and gain a general sense of direction. Nevertheless, in response to contemporary challenges and opportunities, the actions of the present generation are forever adding new chapters to a religion's history. This means that a world religion, when understood as a social and historical movement in time, is slightly different in every age. The footsteps of each generation help create the path.

No one knows where the paths will lead. Of course, some monotheists might disagree with this suggestion. Some might say that the footsteps of a given generation are decided in advance by God, who creates and wills all things—the future is foreordained. Others will deny that God thus determines the future, but nevertheless insist that all footsteps are known by God in advance, even though the walkers make the decisions. But the tradition from which Joan spoke above—process theology— recommends an alternative point of view. It proposes that the thoughts, feelings, and actions of people are decided by people themselves as they creatively respond to the circumstances of their lives, and that these deci- sions are not knowable before they are made because prior to their oc- currence, there is nothing to be known. This emphasis on an open future does not mean that God lacks omniscience. It means that God knows all that can be known, but that the future, because not yet decided, cannot yet be known. Nor does the reality of an open future mean that God lacks power. But it does mean that the power of God is influential rather than imposing, invitational rather than manipulative, calling rather than coercing. Process theologians propose that God is inside each of us as a gentle calling to take steps that are wise and compassionate, for our sake and for the world's sake. Our task, as human beings, is to hear the call and respond. The precise nature of the call changes from moment to moment because situations change from moment to moment. Thus, like a Buddhist *bodhisattva*, God is omniadaptive even as God is omnipresent, forever beckoning toward wholeness for each and all.

In the Christian tradition the process of hearing and responding to divine beckoning is called "discernment." Individuals, small groups, even large groups can be discerning. Indeed, a contemporary generation, even a community of faith—indeed, even a religion—can be discerning, responding to a collective calling such that it takes footsteps in a certain direction. Thus certain questions emerge: at this stage in world history, toward what does God call the religions of the world? What would be the most constructive way in which the many world religions might evolve, given the situation at hand? That "situation" is globalization, in which the many peoples of the world now affect one another for good or ill through mass media, economic interchange, and international travel, and in which urban communities are emerging throughout the world that are ever more deeply diverse in their religious expressions. The Catholic writer Hans Küng says that there can be no peace in the world unless there is peace among religions.[9] What, then, does God call the religions to do?

This question is dangerous if it suggests that there is one—and only one—direction in which all religions should develop. From a process perspective the many religions can and should evolve in different directions relative to their needs and circumstances. After all, they start in different places and they can lead to different but complementary forms of liberation or salvation. Nirvana can be one kind of salvation, for example, and feeling enfolded in God's love still another. Both are saving in their own ways. There can be many salvations.

For that matter, even in a given tradition there are multiple directions in which evolution can and should occur. In some ways a religion is like an ecosystem; it needs diversity, and perhaps even competition, in order to flourish. Nevertheless, given inevitable and desirable diversity, people of the many religions now face common problems that require cooperative and collective response. Globalization makes this inevitable. The common problems include the disparities between rich and poor, the tragedies of war and violence, and the gradual depletion of the earth's nonrenewable resources caused by poverty, on the one hand, and the hyperconsumptive habits of developed nations, on the other. Somehow, with our ancient philosophical and religious traditions, and with our newly developed sciences and technologies, we humans have not yet learned to live gently with each other or lightly on the earth. The world needs its religions to play their constructive role. If we are religiously identified, it may not matter to future generations whether we are Christian or Muslim, Jewish or Buddhist, Confucian or Taoist, Navaho or Cherokee. What will matter is that we have responded to five distinct historical challenges.

The first: to live compassionately—that is, to identify resources for respect and care for the community of life within our traditions and to live from them, thus helping to build multireligious communities that are just, sustainable, participatory, and nonviolent. In order to respond constructively to this challenge, it can help to recognize that sometimes we have been overly preoccupied with questions of doctrinal truth and ritual purity at the expense of being compassionate. In order to respond constructively, we must allow our preoccupations with "right" doctrine and "right" ritual to be subordinated to a desire for "right" love.

The second: to live self-critically—to acknowledge tendencies within our traditions that lend themselves to arrogance, prejudice, violence, and ignorance, to repent for them and add new chapters to our given religion's history. In order to respond constructively to this challenge, we acknowledge that our religious traditions are indeed verbs rather than nouns: that is, social and historical movements to which new chapters can indeed be added by contemporary generations.

The third: to live simply—to present a viable and joyful alternative to the dominant religion of our age, namely consumerism, by living prudently and frugally, thereby avoiding the tragedies of poverty and the arrogance of affluence. In meeting this challenge, it can help to recognize, as I propose shortly, that consumerism now functions as the dominant world religion of our age, and that it stands in tension with the core visions of almost all the other traditions. For their many differences, almost all religions emphasize a kind of abundant living that transcends material affluence but brings deeper and more lasting joy.

The fourth: to live ecologically—to recognize that we humans are creatures among creatures on a small but magnificent planet who have ethical responsibilities to other living beings and to the whole of life. To address this challenge, it might be useful to acknowledge that human beings are kin to other creatures both biologically and spiritually, and also that healthy spirituality includes awakening to rich bonds between people, animals, and the earth.

The fifth: to welcome diversity—to promote peace between religions by befriending people of other religions, trustful that the truths of the world religions are manifold, all making the whole richer. Helpful in this regard is the observation that our own religious traditions need not be perfect or all-inclusive in order to be good, and that their differing insights can be complementary rather than contradictory.

To the degree that religiously affiliated people respond to these five challenges, there will be hope for the world, and religion will be part of the solution. And to the degree that we do not, there will be tragedy, and our religion will be part of the problem. Of course, these challenges will be more difficult for some religious people than for others. It may be easier for indigenous peoples to recognize kinship with other creatures than for more anthropocentric Muslims, Jews, and Christians. Equally important, not all of the challenges are relevant to all people. Consider the challenge to live simply. This challenge is most relevant to the one-fifth of the world's population who consume more than half the world's resources, depleting its nonrenewable resources and living in a way the wider world could not possibly emulate, much less sustain. It is estimated that, if the whole world lived as the average American does, it would take six planets the size of the earth to supply all the resources. It is thus the affluent who are called to live more simply so that the poor can simply live. When considered in relation to concrete circumstances, the five challenges may be seen to occupy a sliding scale of objective urgency and perceived need.

Nevertheless, for many middle-class, religiously minded people—and I count myself among them—the five challenges are relevant and complicated. We are called to live compassionately even if compassion makes us feel vulnerable; to live self-critically even if we are afraid of change and find it easier to criticize others than ourselves; to reject a lifestyle based on appearance, affluence, and marketable achievement even if we are deeply absorbed in it and gain from it; to recognize that we are kin to other creatures even if we prefer to think of ourselves as set apart and special; and to welcome religious diversity even if we are initially fearful of strangers and what they might teach us.

The best hope for us, it seems to me, is that we adopt a countercultural approach to life of the kind embodied in ecotheology. It helps meet the first challenge by reminding us that compassion is rightly extended to other animals and to the earth itself; the second, by inviting us to attend to those aspects of our own traditions that have been harmful to the earth; the third, by welcoming us into a post–consumer-driven way of living, in which integrity is more important than material ambition; the fourth, by encouraging us to recognize that we are part of, not apart from, a larger creation; and the fifth, by suggesting that we be sensitive to biological as well as cultural diversity, seeing religious diversity itself as part of a larger, beautifully complex whole. However, ecotheology itself can not fulfill this promise unless it is seen as a living theology—that is, a way of living. It is a movement of the body, heart, and mind.

SUMMARY: A MOVEMENT OF THE BODY, HEART, AND MIND

By *a movement of the body* I here mean physical activities such as living simply and frugally in one's personal life, without pretense or conspicuous consumption, and using one's time and energy to help build communities that are socially just, ecologically sustainable, and spiritually satisfying for all. These actions are not bodily in the sense that they necessarily involve strenuous effort. Writing a poem, for example, is a bodily act, but it is not physically strenuous. What makes them bodily is that they are observable acts whose effects can be seen by others. The bodily side of ecotheology can be manifest in how one runs a business, or raises a child, or writes a poem, or rattles the cages of city hall, if these are done with the aim of helping build a just and sustainable community.

Many of these bodily practices are cross-generational. This is important to emphasize because often when we think of theology we think of something that only adults can do. As a way of living, ecotheology is different. Children and teenagers can embody ecotheology by purchasing products whose manufacture is socially responsible, eating healthy food, using public transportation, and offering their time and energy to help others. And senior citizens can embody ecotheology by communicating with government officials, urging them to enact responsible public policies.

Equally important, the practice of ecotheology transcends class divisions, albeit in different ways. People who have too much money or overconsume can practice ecotheology by relinquishing possessions and living more simply so that others may have the resources to simply live. They can say: "I have too much and I want to live with less. I don't want to get ahead of others; I want to be with others." This practice typically involves a conversion on the part of the practitioners. They must realize that there is more to life than appearance, affluence, and marketable achievement, and that there is a conflict between integrity and ambition. In solidarity with the earth and other living beings, they may then choose integrity over ambition.

On the other hand, people who have too little can practice their ecotheology through grassroots community efforts that help lift them from poverty in ways that are ecologically sustainable as well as socially just. Thus they can help the world see that it is possible to live comfortably and humanely without the trappings of affluence or the tragedy of poverty. This, too, requires a conversion. It is a conversion from feeling like a victim or feeling helpless to feeling like an agent of one's own destiny. It involves learning to say, "I truly count in life," rather than "I don't amount to anything." For many who are accustomed to feeling unimportant, this conversion is as difficult in its way as is the conversion to integrity for the wealthy. Both conversions can be helped by others who provide models of humility and self-respect. In the language of Buddhism, both require a community of kindred practitioners: that is, a *sangha*. In the language of Christianity, both require a community of spiritual friends: that is, a church.

Of course the two conversions lead in slightly different directions. For the rich, ecotheology involves downward mobility—learning to live with less and immersing oneself in the lives of others where integrity, not ambition, is the guiding ideal. And for the poor, ecotheology involves upward mobility—struggling and succeeding in building a sustainable life that is comfortable but not affluent. The hope is that both groups, the rich and poor, rightly meet in the middle, living in ways in which, in principle, all others might have enough to live.

The emphasis on conversion points to the second side of ecotheology; it is a movement of the heart. By *movement of the heart* I have in mind certain attitudes and motivations that are subjective rather than objective, but that nevertheless inspire the kinds of actions named above. These attitudes include a sense of respect and care for the community of life; a love of animals; a delight in beauty; empathy with all who suffer; and a desire to be a healer, a person who helps reduce unnecessary suffering in the world and who seeks the happiness of all. The heartfelt side of ecotheology also includes what Buddhists call "mindfulness," a willingness to listen to others, to pay attention, to hear what they are saying and also, and importantly, what they cannot say. In the case of other human beings, this listening includes a respect for silences as well as sound: a realization that people do not always have to be talking to be together. In that of other animals, the listening side of ecotheology includes a sense that other beings speak to the world, not through human languages, but through actions of their own.[10] As a path of deep listening, then, ecotheology involves a willingness to be touched by others, sharing in their suffering and delighting in their joys.

This willingness to be touched is part of the existential power of ecotheology. Sometimes we think of power as a capacity to effect change in others through an act of intervention, but there is also a power in being able to receive influences from others and allow oneself to be changed by them. This capacity to receive is itself powerful, yet not interventionist. It is spacious and strong, like an ocean that can receive many streams without being overwhelmed. This is the power of deep listening. It affects others, not by forcing its will upon them, but by helping them discover their own capacities for creativity. In relation to other people and other living beings, it is the power simply *to let be*. Ecotheology involves and requires this kind of power, especially in relation to other living beings. They do not need to be managed; they need to be loved. We love them by letting them be themselves, and not manipulating them to our own ends.

By a *movement of the mind* I mean to suggest an outlook on life—a way of looking at things—that complements the more bodily and heartfelt dimensions of ecotheology. This mental aspect may or may not be formalized in terms of a system of belief. Moreover, it may or may not be spoken in words by the person who has it because ideas are always more than their verbal articulations. But the outlook is a way of seeing things, of interpreting the world. Like the bodily and emotional dimensions of ecotheology, the mental side is historically and socially conditioned. It does not drop from heaven; rather it arises in our world today as a reaction to more mechanistic, patriarchal, authoritarian, and consumer-driven ways of being in the world. It involves an intuition that all things are connected to one another, such that nothing can be self-contained or isolated; that all living beings have value for themselves that is worthy of respect by human beings; and that the universe as a whole is a sacrament of sorts, a visible sign of an invisible grace.

One important task of the more formal ecotheologian—the one who writes books—is to give words to the whole of ecotheology, to its bodily, heartfelt, and mental dimensions. Written and spoken texts are one way of communicating and exploring this way of living in the world. The way can also be communicated through images, sound, and dance. Sometimes it is better to dance the ideas of ecotheology, or to sing them, than to speak them in prosaic terms. And often it is more important to act them out in concrete acts of love and service. The hope of the future is not that the world embrace the term "ecotheology," but rather that "ecotheology" become an irrelevant and unused term whose attitudes have been fully integrated into the lives and hearts of ordinary people who have no special interest in theology, but who find beauty in the world and thus live with respect and care for the community of life. Only when ecotheology becomes irrelevant will it have succeeded.

✢ Talking the Walk: A Practice-Based Environmental Ethic as Grounds for Hope

ANNA L. PETERSON

INTRODUCTION

Environmental issues consistently rank near the top of people's concerns. Surveys indicate that as many as four out of five Americans consider themselves to be environmentalists. Environmental values have become mainstream, permeating politics, education, religion, and popular culture in myriad ways.[1] This awareness and concern, however, do not automatically correspond to changes in behavior. While 80 percent of Americans may identify as environmentalists, fewer than one in five regularly participate in environmentally responsible activities such as recycling, reducing personal consumption, supporting green businesses, eliminating waste and pollution, or engaging in environmental activism. As many "green" practices have stagnated or even declined since the onset of the mass environmental movement in the 1970s, ecologically destructive behavior has climbed exponentially in the same period, as measured in terms of fossil-fuel consumption, raw-resource consumption, expanding home-size, and population shifts to the suburbs.[2]

Political behavior follows a similar pattern of expressed concern tied to little or no practical action. More than 70 percent of U.S. survey respondents say they have never voted for or against a political candidate based on his or her environmental views or record. An October 2005 poll conducted by Duke University, for example, found that while 79 percent of respondents favored "stronger national standards to protect our land, air, and water," only 22 percent said environmental concerns have played a major role in determining whom they voted for in recent federal, state, or local elections.[3] Candidates' personal qualities and their positions on "moral" issues such as abortion, gay rights, and gun control appear to influence voters much more than environmental issues. (Especially since the November 2004 elections in the United States, a number of progressives have called for a "reframing" of environmental issues as matters of "morality" and "values," although the success of these efforts remains to be seen.)

The gap between environmental ideas and practices also appears in religious life and organization, identified by both scholars and activists as a crucial arena in the struggle for more ecologically sustainable ways of living. Many spiritual and environmental leaders have asserted that ecological destruction is a spiritual crisis, the resolution of which demands a new understanding and valuation of nonhuman nature.[4] This volume itself, which reflects more than three decades of theological work in articulating ecological theology, marks a significant contribution toward that goal. The call for a "greening of religion" has been met with a proliferation of "green" religious statements, conferences, and documents in the past decade or two, from a wide range of traditions, including Roman Catholicism and Tibetan Buddhism, historic Protestant denominations as well as "new age" spiritual groups. Despite an apparent consensus that religion must play a central role in building a more environmentally sustainable society, religious organizations and individuals have achieved few tangible results. In some cases, to be sure, principles have been implemented, usually by particular congregations or action groups. One of the most sustained and large-scale efforts has been the Environmental Justice Working Group (EJWG) of the National Council of Churches (NCC) (see Laurel Kearns's contribution to the present volume). Despite the time, energy, and resources that have been dedicated to initiatives such as this, however, it is far from clear that significant numbers of individuals and congregations affiliated with the NCC have reduced their energy consumption, or indeed changed their environmental practices in significant ways. In other words, even when religious groups make environmental issues a priority and devote resources to developing environmental values, these values do not automatically generate a significant transformation of members' behavior.

In sum, there is strong evidence in favor of two apparently contradictory truths. First, most people want a clean and safe environment, with abundant habitat for nonhuman species and wild places in addition to livable human settlements. Second, this valuation of the natural world is not always, and maybe not usually, reflected in people's personal consumption practices or political choices. There seems to be, in other words, little if any causal relationship between environmental value orientations, awareness, and concern, on the one hand, and behavior, on the other.[5] In this essay I explore these questions, both as a social ethicist interested in the ways that values translate into action and as an environmentalist concerned with the gap between our theories (our "talk") and our practices (our "walk"). These concerns are relevant to my own thinking and practice; I do not consider myself exempt from the dilemmas that vex environmentalists and ethicists. The challenge of understanding and addressing the gap between ideas and action requires the contributions of ecological theologians and ethicists, just as it challenges them—us—to reflect and act differently. While we will probably never close the gap entirely, we must acknowledge that today the divide between the value we say we place on nonhuman nature and the way we act threatens the survival of nonhuman and human worlds alike.

VALUING NATURE

If we do not always live by our explicitly held values, then what—if any—values are embodied in our actions? Backing up, we might ask whether our actions always (or most of the time) reflect values. Are we always walking a talk, or are we sometimes "just walking"? I propose that simply walking—entirely random behavior—is relatively rare, and that most times actions that contradict people's stated values in fact reflect other values—usually unsystematized, often unacknowledged, yet powerful. Rather than dividing behavior into that which is motivated by professed values and that which is random, then, I suggest we understand most actions as grounded in values of some sort or other. Sometimes (more often for some people and in some circumstances) these values are clearly understood and deliberately professed, while other times they are implicit and unacknowledged. We might describe the latter as "background" values, sometimes inherited from religious traditions, often reinforced by cultural patterns and social institutions. These background values kick in, so to speak, when it becomes too difficult or costly to follow values we may consciously prefer.

This might parallel the laboratory rats or mice that follow their accustomed route, even when it changes; that is, they will still jump over an obstacle en route to their food dish even when the obstacle has been removed. This practice reflects not stupidity but rather a good rule for self-preservation: if you do not have to think about the way home, because your feet automatically follow a path that normally does not change from night to night, then your other senses can be devoted to detecting predators and other threats. (This may help explain why people sometimes head in the direction of work or home even when they are supposed to go someplace out of the ordinary.) Maybe our nonconscious values are like a default path home—something to follow without having to pay much attention to it.

Another possibility is that self- or class-interest shapes our background or default values, and explicit moralities represent precisely the extent to which we have been able to transcend or at least hold in check our natural inclinations to favor ourselves. (This echoes points made by Immanuel Kant and Aldo Leopold, among others.) Sometimes we act "on principle" and give someone else the larger piece of pie, vote to raise our own taxes, or spend time and energy walking to work instead of driving. Other times—the evidence suggests most times—we do not.

What background values shape our ecological practices? They might best be described as an uncritical anthropocentrism, a valuation of the human over the nonhuman that is so deeply reinforced by our religious and cultural traditions, our economic and political institutions, and the patterns of everyday life that it is virtually impossible for even the most biocentric persons to avoid it completely. We might just accept this as inevitable human complexity and fallibility; we might even celebrate, as Walt Whitman did in "Song of Myself": "I contradict myself? Very well then, I contradict myself!" The consequences, however, are not always trivial or deserving of celebration. Few of us desire a world without polar bears or penguins, but that will be one of the many terrible consequences of our addiction to fossil fuel.

IDEAS AND PRACTICE

The question we now face is how our practices, including both individual behavior and that of collective institutions, might better reflect our professed concern for the preservation of the natural world. Educated and concerned people disagree, of course, on which actions are most effective—some emphasize working for structural change through electoral politics, while others prefer helping nature directly through ecological restoration activities, while still others might highlight personal actions, such as using alternative transportation, buying "green" products, and eating in more sustainable ways. Many also dedicate themselves to teaching others about the problems and possible solutions. Despite differing priorities and emphases, most environmentally concerned people know about at least some of the activities that can reduce our negative impact on the earth and thus reflect the value we place on nonhuman nature. And because of differing talents, life situations, home places, jobs, and incomes, environmentally concerned people can and must fight different battles.

This does not explain, however, why and how concerned and educated people choose so often not to act in environmentally responsible ways. What motivates us to save time or money instead of the planet—to skip a meeting, drive when we could walk, or buy the cheaper nonorganic produce when we have the option to do otherwise? These answers may seem obvious: we don't have the time or money. However, this obvious answer cannot be the whole answer, and the question is not as simple as it seems. People frequently spend more money than necessary on items they claim to care about much less than they do about the environment, and many people find time to watch television for hours every day without claiming that television is a "very important" value in their lives. There seems to be no clear correlation, in other words, between the importance of a value and the time or resources that people devote to it. This disjuncture deserves much more attention from environmental ethicists and activists than it has received. Scholars in particular end up spending very little time thinking about the obvious gaps between values and our practices, and instead devote themselves primarily to intellectual puzzles such as the distinctions between intrinsic and instrumental value or the ecological implications of St. Paul's vision of the End Time, and so forth. I appreciate that much theological and ethical reflection is motivated by a recognition that our value system has been implicated in our ecologically destructive behavior and that one set of values are still used to justify policy and action. I suggest, however, that at this historical moment, perhaps the most vital task of ecotheologians and environmental ethicists should be not to come up with better knowledge or values, but rather to figure out how to get people to live according to the good ideas we already have.

This raises a whole set of complex and interrelated questions: How might our practices help ground new forms of environmental knowledge and ethics? How, why, and under what circumstances do good ideas contribute to different practices? How might we communicate about what we value in order to increase the real impact of our arguments on behalf of nonhuman nature by changing the ways people use and consume nature? How, in other words, might we make our values effective in the world?

I find myself, then, asking why our ideas—those of environmental ethics and ecotheology, of biocentrism and stewardship and intrinsic value—have been so singularly ineffective. There are a variety of causes, of course, including many economic, political, and social factors far beyond the scope of this essay and of our individual capacity to change. However, our ineffectiveness also stems from causes internal to our own work, chiefly the fact that environmental ethics, my own field (leaving theological discussions to others, including many of the contributors to this volume), has been dominated by the same idealist logic that has shaped Western ethics generally.

According to this idealism, the primary task of philosophy and theology in general, and of ethics in particular, is to develop better theoretical models. If actual practices enter into the equation at all, they do so in relation to the "application" of ethics to "real-life" situations. Once established, a theoretical model need not, and in fact should not, change as a result of its application to any given situation. In many idealist models, including secular ethics based on rights and rules as well as religious moralities based on divine revelation, good moral principles are by definition abstract, unchanging, and universal. Thus, for example, if it is wrong to lie (or take a life, or use drugs, et cetera) in any circumstance, then it must be wrong to do so under all circumstances.

The privileging of the ideal over the practical might be viewed as the common thread that runs through Western ethics, both secular and religious. Amidst all the differences and disagreements within and among schools of thought, two shared convictions emerge: first, that a good action is one done with the right understanding or intentionality; and, second, that this understanding must be in place before action is taken. This understanding, further, is abstract and general: love of a universal God or good, not homegrown knowledge rooted in everyday life. We see these assumptions in the fathers of the Western moral tradition: St. Augustine believed that the will, or what one loves most, determines the city to which one belongs. Thomas Aquinas, of course, insisted that all good emanated from the divine law. While Martin Luther rejected much of Thomas's thought, he agreed that belief comes first, which is why justification can never come through works. And for Kant, of course, the individual's intentionality or will determines the morality of an act. Despite their differing foundational assumptions, sources of legitimacy, and social aims, these different approaches agree that the idea is always prior to the act, chronologically, symbolically, and ontologically.

The presumption in favor of ideas remains almost universal among contemporary Western thinkers of diverse ideological persuasions, including environmental philosophers and theologians. Well-meaning and deeply concerned people devote themselves to the details of argumentation and language, intrinsic versus instrumental value, rights versus interests, foundationalism and the grounds for various moral claims, and so forth. Others concentrate on the pitfalls of anthropocentric approaches or the recovery of ecologically friendly values from earlier traditions.

There certainly is a place for talking about religion and theology. Relative to secular environmental philosophy, religious traditions and communities hold greater potential for engaging the values of large numbers of people. Religion is the framework in which many people think about important moral and intellectual issues, so talking about ecological values in relation to religion may draw in people who are not engaged by secular philosophy. Further, attention to past and present communities of faith can illuminate the ways people live out environmental values, something more difficult for secular philosophy to address. Religiously based ethics may also have greater capacity to influence people through institutional and personal loyalties, transcendentally grounded values, and salvific hopes. Most ecotheology, however, has focused on statements of principle and theoretical critiques, not diverging significantly in this regard from secular philosophical counterparts. Again, this is not to deny the importance of articulating principles, especially for centralized institutions such as the Roman Catholic Church and other religious bodies. For many of these, a formal statement of principle opens the door to support for activist campaigns. Still, the potential of environmental ethics, and especially those of religious origin, has not yet been realized very fully.

This failure stems, in part, from the widely shared, rarely examined, and often related assumptions that theorizing by itself is adequate practice and/or that the right theories will lead automatically to effective forms of practice. Despite the evident lack of empirical and historical data supporting these assumptions, they remain largely unquestioned—precisely because of their own idealist logic. The role of philosophers, including ethicists, stops with the elaboration of the proper ideas; what happens to these ideas out in the world is too often someone else's concern.[6]

I do not dismiss the value of refining good theories and retrieving good traditions. These are important and necessary projects, which lay the groundwork and clarify our thinking as part of a much longer process, but they are not ends or foundations—let alone effective practice—all by themselves. The "demand to interpret reality in another way" does not, by itself, change reality. Like the Young Hegelians whom Marx criticized in *The German Ideology*, environmental philosophers and theologians often forget that "they are in no way combating the real existing world when they are merely combating the phrases of this world."[7] This does not mean there is no relationship between changing practices and changing knowledge, language, and values. This relationship, however, entails complex, unpredictable, and open-ended mutual interactions, rather than the straightforward linear influencing we would desire.

Most of us know that simply clarifying and articulating green ideas will not green the world. So why do we so often content ourselves with talking the talk? It is not just laziness or mistaken ideas of what makes a difference in the world. Rather, I think, a root cause is the difficulty of even knowing where to start making a real difference. Fear of failure—not just our own failure but failure of the project of averting ecological disaster—also plays a role. We may be confident that we can succeed in crafting a good argument or writing an elegant paper, but we have no assurance that even our most dedicated efforts will transform people's relationship to nature. These obstacles to praxis constitute something resembling what anthropologist Michael Taussig calls a "public secret," defined as "that which is generally known but cannot be acknowledged."[8] The "public secret" of environmental ethics is that we spend most of our time on tasks that do not lead to the ends we desire. We all know this, but we do not call each other or ourselves to task, both because we realize we all have good intentions and because we do not know how we might do things differently. (And, not insignificantly, professional prestige and rewards are rarely based on practical impact.)

Perhaps the biggest problem we skirt around is the undeniable fact that good ideas do not guarantee good practices, and environmental damage can occur even when dominant value systems do not legitimize it. Buddhism and Taoism are often held up as examples of environmentally friendly systems of thought, emphasizing the interdependence of humans and the natural world, the intrinsic value of nonhuman nature, the need to coexist peacefully and avoid gratuitous harm to other beings, and so forth. However, Asian societies dominated by Buddhist or Taoist traditions, among others, have well-documented histories of deforestation, water contamination, unsustainable agriculture, cruelty to nonhuman animals, and much else. Even Native American cultures, whose worldviews are seen by many environmental philosophers as the greenest of all, have participated in overhunting and other abuses of nature. Many of these cultures have not damaged the nonhuman world nearly as much, or as profoundly, as have Western industrial societies, of course; my purpose here is not to rank cultural worldviews in order of greenness but rather to underline the fact that apparently green worldviews do not prevent the destruction of nonhuman nature. A host of other factors, including unacknowledged assumptions and values as well as self-interest and political and economic pressures, shape people's uses and abuses of nature.

Once again, people's actions do not follow, in any simple or straightforward way, from their values. The relationship between values and action is extremely complicated, and no one lives in perfect accord with a set of principles. We all contradict ourselves, and, as Walt Whitman celebrated, each of us contains multitudes. What this means for environmental ethics is that regardless of the values they espouse, even people who value nature cause unnecessary harm to nature. Even the most principled among us live by our values only some of the time. Our principles are not thereby meaningless, nor do we forfeit the right to call ourselves "environmentalists" (or Christians, Buddhists, or deep ecologists) because our actions do not always accord with the principles we profess (of stewardship, *ahimsa*, or biocentrism). Yet the lack of agreement between actions and principles, and also the difficulty of resolving tensions between conflicting value systems, raises some very important questions and deserves more attention than it has gotten, particularly from ethicists.

While a host of factors are involved, one way environmental philosophy and ethics might contribute is by clarifying the relationships among ideas (or knowledge), attitudes/feelings, and action. Rather than just assuming that our job stops with ideas and knowledge, we need to go further and learn about what makes ideas effective in the world or, more broadly, what motivates people's behavior, especially changes in behavior, and when and how ideas and values enter into these motivations. One direction for research by environmental philosophers, then, should be ethnographies of the reception of ideas about the value of nature. How are environmental philosophies and theologies lived or not lived? We should not leave these tasks to others; they must become central to our own understanding of our methodologies and goals.

TALKING THE WALK?: TOWARD A PRACTICE-BASED ENVIRONMENTAL ETHIC

In short, we need to change our talk and our walk together. We also need to understand which "talk" we are walking in the first place. The flip side of rethinking the practical impact of our ideas in the world is recognizing the ways that our practical experiences in the world shape our ideas. The way we walk, the forms of transport and traveling partners available to us, the paths we can take and the destinations we can pursue all shape how we understand the world and what we value in it. This suggests that our abuse of nature and our feelings of separation from it are grounded not only, or primarily, in theories of human exceptionalism but in the ways we have lived, separately from, and with power over, the rest of nature. As Anthony Weston puts it, anthropocentric views of human nature, including value-laden ideas about human exceptionalism, are "in large part a consequence of the growing isolation of humans from any open-ended encounter with other lifeforms, an isolation that for many of us is now nearly complete. That isolation in turn is a consequence of the relentless humanization of nearly all environments and the increasing concentration of humans in the mostly relentlessly humanized of them."[9]

This suggests that if we want to experience communion we have to create it in and through our practices. We need not to identify more credible and constructive understandings of human nature but to *create conditions in which we experience* the interdependence, fragility, and humility taught by evolution and ecology and by green religions and ecocentric philosophies. This can be done not in theory but only in and through practice. The way to a morally, not just intellectually, *better* understanding of humanness might be by opening ourselves to the nonhuman world, reaching out, making room and conditions for new possibilities in terms of both thought and practice. We can do something similar by reaching out to the other human worlds around us. Here phenomenology offers an important resource, especially in its emphasis on experience rather than rationality as the grounds for understanding and valuing nature. As David Abram writes in *The Spell of the Sensuous*, the goal for environmental philosophy

> is a way of thinking that strives for rigor without forfeiting our animal kinship with the world around us—an attempt to think in accordance with the senses, to ponder and reflect without severing our sensorial bond with the owls and the wind. It is a style of thinking, then, that associates *truth* not with static fact, but with a quality of relationship. Ecologically considered, it is not primarily our verbal statements that are "true" or "false," but rather the kind of relations that we sustain with the rest of nature.[10]

An environmental ethic, in this perspective, must be grounded in true or right relationships with the more-than-human world. It is impossible, further, to construct an adequate ethic prior to the restoration of nonmediated and reciprocal human experiences of nature. This key insight of ecophenomenology helps create the possibility for grounding environmental philosophy on concrete situations and experiences. Despite its promise, however, phenomenology also contains pitfalls, in particular its tendency toward individualism and subjectivism. Additional resources are needed to move beyond a consideration of individual relations to nature and account for and shape the ways larger collectives understand, value, and treat the more-than-human.

Here the Marxian tradition offers a pair of helpful correctives. First, it suggests that the divisions among humans both affect and mirror divisions between the social and natural worlds. Reducing the experiences of human exceptionalism must entail reducing not only the exceptional experiences of humans as a species among other species, but also the exceptional experiences of privileged humans among other humans. Second, it insists that the transformation of practice and of ethics, in social or environmental terms, cannot be done on an individual or purely local level. Ecological destruction and social injustice are political problems that need political solutions, which means structural, institutional changes. We cannot have societies of good people in conditions that make it hard to be good. We need to have good choices and for those choices not to be dauntingly difficult for ordinary people. And those of us with more options need to make all the good choices we can to provide models and to expand possibilities. So, as much as possible, we should live *as though* the structures were good. This, I think, is what Wes Jackson means when he says that as long as we are living on the wrong side of what he calls "the brick wall of capitalism," we need to create alternatives and models. That enables us to "imagine life on the other side of the wall [and] . . . plan for the other side."[11]

This is the kind of restorative practice that Bill Jordan has spoken of. In *The Sunflower Forest*, Jordan writes that ecological restoration does not just reflect or express already held values but in fact is a work of *value-creation*.[12] This is a collective, not a private, process. In restoration we participate in and create community both among humans and between humans and nonhuman nature. We begin to act locally, yet with the broader vision that any "militant particularism," as Raymond Williams termed it, must have if it is to make an impact beyond its own backyard.[13] We engage in a reflective practice—what Marx understood as genuine *praxis*—self-conscious about its foundations and goals, open to change and surprise, and always seeking to connect at different levels and in different ways to other struggles. Ecological restoration, Jordan argues, is a ritual process, through which we clarify meanings and articulate and sometimes resolve conflicts. In and through the practice of ecological restoration we redefine the value of nonhuman nature and clarify our own role in it.

CONCLUSIONS

Environmental ethicists need to abandon the idealist assumption of a simple and unidirectional relationship between ideas and practice, in which practice is always derivative or secondary to ideas and which believes that if we get the ideas right, then the practices will follow. This does not mean that we should stop trying to get our ideas right, but we do need to stop thinking that this is an adequate intellectual and moral response to environmental crisis. Decades of work in environmental philosophy, ecotheology, and related fields have provided us with many good ideas (and hopefully taught us that there are many "right" ideas in our pluralistic world). Now the problem is how to live by them. This requires, I have suggested, not just an extension of the usual way of doing ethics but also a new way to think about this task.

Environmental philosophy has the potential to spark this deep rethinking of ethics, insofar as it calls into question the relationship between epistemology and ethics and thus basic assumptions about ethics itself. Jim Cheney and Anthony Weston have elaborated this possibility in relation especially to epistemology. Most traditional philosophy, they note, begins with knowledge and assumes that ethical action comes next, as a response to knowledge. This carries further assumptions, including that the world is readily knowable and that the task of ethics is to sort out the world in ethical terms.[14] As an alternative, Cheney and Weston propose that ethical action is or should be first and foremost an attempt to open up possibilities. With regard to nonhuman animals, for example, they point out that we have no idea of the capacities of other animals until we have approached them ethically. An open and loving approach enables us to see what others cannot see: "Love in this sense," according to Cheney and Weston, "is already an ethical relationship. It thus stands at the beginning, at the core, of ethics itself . . . a risk, an attitude that may . . . lead in time to more knowledge of someone or something."[15] Here the usual order is reversed. Action, rather than knowledge, is the first step.

This echoes Latin American liberation theology, which, as Gustavo Gutiérrez has pointed out, is not a new topic for theology but rather a new way of doing theology. Theology is a "second step" after engaged practice.[16] These insights have been systematized in the "hermeneutic circle" of progressive Catholicism, described by Juan Luis Segundo as "the continuing change in our interpretation of the Bible, which is dictated by continuing changes in our present-day reality, both individual and societal."[17] A similar process occurs in environmental science's model of adaptive management, which uses a "feedback loop" so that environmental-management approaches change as people learn from their experiences. This seems to me to be another version of the same basic idea: that our real-life experiences and practices should constitute the starting point, not the repository, of theory. What we deeply know from our experience opens us to better theoretical knowledge. Ethics, then, becomes less pedagogical and more dialogical—and the dialogue occurs not only with human partners.

Cheney and Weston argue for this approach in environmental ethics precisely because nature is continually surprising and rife with hidden possibilities. In such a world, ethics may proceed not by increments and logical extensions but by leaps. And these leaps may take us to unexpected places and conclusions. But that is acceptable, even desirable, if we see the task of ethics as not to explain the world but rather to explore and enrich it.[18] Cheney and Weston call this an "etiquette-based ethics" or an "ethics-based epistemology." This is very close to what I am thinking of as a materialist ethic. Practice, they write, "is no longer some application of ethical knowledge: it is now *constitutive of ethics itself*, our very mode of access to the world's possibilities."[19] Our ethical practice is not a result of our knowledge of the world but rather is the way we learn about, and in, the world. To return to the earlier example, we cannot know what nonhuman animals can do until we approach them. Rather than assuming that only humans do or feel something, then, we might first explore the possibilities.

I want to ask what happens to ethics when we begin with experience. Here we would not start with an ethical framework and ask how it gets worked out in a particular situation. Instead, we would begin by looking at economic and political institutions and everyday practices and ask which values and ways of moral reasoning these embody and legitimize.

I do not suggest that we just identify and legitimize whatever ethics are currently operative. As Wes Jackson argues, we cannot simply meet people where they are—or we will meet them "at Wal-Mart, where things are cheap and don't last."[20] This does not imply a condemnation of all people who shop at Wal-Mart. Wal-Mart is not a good choice—not only are its products largely short-lived and of poor quality, but in its advertising and stores it encourages unnecessary and excessive consumption, it treats its employees poorly, and it harms local ecosystems, economies, and communities. For some people, however, it is, or appears to be, the only choice for needed items like shoes, clothing, and groceries. The conclusion environmentalists ought to draw from this is, as I have already argued, that people need better choices. We should not meet people "at Wal-Mart" not because the people there are bad but because Wal-Mart is not a good place to be, physically or metaphorically, and certainly not environmentally. Meeting people there does not advance the conversation or move us forward along the path we need to be on—a path that does not cross-cut the Wal-Mart parking lot. Jackson's point, as I read it, is that meeting people "where they are" does not get them to where we all, as a society, need to be.

That said, we do need to know where people are, at least as a starting point. This means learning through ethnographic and empirical research what moves people, what choices they make and why, what values they live by regardless of the values they profess, how these lived-out ethics relate to professed ethics, and why gaps exist between professed and practiced values. Is it too hard to live by certain values? Too convenient and fun not to? What are the carrots and sticks that move people to follow a value, or not? What structural conditions and cultural dynamics drive people toward choices they may in fact not like? These questions need to be asked and answered in practice, not in abstract theory, and the answers will be different across cultures, classes, and places.

If the ground of ethics is practice and experience in the world, then *in order to construct a new ethic we must transform the world*. As Weston puts it, "We need to de-anthropocentrize the world rather than, first and foremost, to develop and systematize non-anthropocentrism—for world and thought co-evolve. We can only create an appropriate non-anthropocentrism as we begin to build a progressively less anthropocentric world."[21] This suggests a certain ambivalence regarding timing: do we change the practical world "first" or as we are also changing ideas? While there will never be perfect synchronization, on the whole, the changes have to be occurring simultaneously and dialogically. "World and thought co-evolve." This is the real hermeneutic circle. There is no set linear or chronological relationship between world and thought. Conditions of life make certain narratives, worldviews, and values possible, but these forms of thought and discourse, in turn, shape the world, and seeing, judging, and acting continually transform each other. In this light, the actions we undertake must be open-ended. We have to "reach out," as Weston suggests, "opening the possibility of being touched rather than touching, whatever might eventuate."

So we move once again away from any simple determinism. A practice-based ethic, which I think must be a materialist ethic—socially and ecologically grounded—necessarily sees the future as open. This challenges the notion that materialism is necessarily reductive and deterministic. Idealism is what turns out to be reductive, in its incapacity to imagine an outcome not foretold by its theory. As we begin with practical work in the world, the possibilities for new ethics will emerge.

This calls to mind Marx's famous eleventh thesis on Feuerbach: "The philosophers have only interpreted the world. The task, however, is to change it."[22] Marx mostly dismissed ethics, at least as usually done in his time and place, because he believed that moralizing would alter nothing. Values and theories would change when social structures and relations of power changed. This was not a simple dismissal of ideas as mere "superstructure" emerging directly from "real" structures. Marx recognized that the connections between theories and practices were always mutually dependent. In the third thesis on Feuerbach, for example, he writes that

the materialist doctrine that men are products of circumstances and upbringing, and that, therefore, changed men are products of other circumstances and changed upbringing, forgets that it is men who change circumstances and that it is essential to educate the educator himself. . . . The coincidence of the changing of circumstances and of human activity can be conceived and rationally understood only as revolutionising practice.[23]

In this revolutionizing practice, world and thought coevolve.

Marx also argued that ethical theories are useless, at least by themselves. His point is that we cannot ask people to do good in a bad system. More specifically: if the modes of production in a capitalist society encourage, even require, people to be alienated from and compete with each other, among other things, moral exhortations to benevolence are meaningless. Or, to use Weston's example, if our relations to nature are anthropocentric and utilitarian, then exhortations to ecocentrism are useless, or worse. Ethics must help make it possible for people to relate morally to each other and to nature. Simply telling them to do better focuses on symptoms while avoiding the deep causes of presently immoral ways of acting and relating. This is Marx's point when he calls religion the "opiate of the people." This phrase is often misunderstood. The point Marx was making was not that religion is bad in itself but rather that it is a symptom or, more precisely, an "expression of real suffering." The answer to this suffering lies not in the realm of religion but rather in the material and social conditions that give rise to it. Therefore "the abolition of religion as the illusory happiness of men is a demand for their real happiness. The call to abandon their illusions about their condition is a call to abandon a condition which requires illusions."[24]

I do not accept Marx's view of religion wholesale, certainly not his conviction that it always reflected pathology (and this in fact contradicts Marx's own emphasis on creative praxis). Rather I hope to draw a parallel that can help us see new dimensions of the environmental problems of anthropocentrism and human exceptionalism. We might say that the call to abandon people's illusions about their exceptionalness is a call to abandon a condition which requires and makes possible these illusions. Or we could rewrite another phrase from Marx. He writes that "it is the task of history . . . once the other-world of truth has vanished, to establish the truth of this world."[25] It is *our* task, perhaps, to establish the truth of interdependence, vulnerability, and humility: not to talk about this truth but to establish it.

This is, in embryonic form, what I mean by a "materialist" environmental ethic. Again, this is not a crude or deterministic materialism. Merely changing material conditions will not change ideas, as Marx consistently acknowledged. The point is that structures and ideas are always in mutually transformative relation, and that pure idealism—of the Hegelian or ecotheological variety—is ultimately sterile. We need instead a materialist environmental ethic. However, there are many materialisms (just as there are many versions of idealism). Which kind of materialism is most appropriate for environmental ethics, here and now? I hope we can develop not a reductive materialism for which ideas do not matter at all, but a nuanced materialism in which ideas do not exist in a vacuum. Then we face questions such as these: Whose ideas? Why are certain ones emphasized? Approaching these questions in environmental ethics, we can draw on materialist (and modified materialist) models such as the mutual determination of ideas and structures in humanistic Marxism and the "lived religion" critique of textualism.[26]

This sort of practice is a ground for hope. To ecological restoration we might add other potentially restorative daily practices such as work in a community garden or bicycling to work. In these rituals we live *as though* we were on the other side of the wall. These actions provide grounds for hope first because they entail living, as Christians might say, the reign of God in our midst. We experience hope in and through the experience of living in right relationships, or what Jordan describes as communion with nature and people (and what Marx might call the reduction of alienation within and among persons as well as between persons and nature).

Grounds for hope come not only from what is in our midst, here and now, but also from the future possibilities these practices create. Here is the hope that our ideas can matter in a new way and help shape future ways of living on the earth. We can help create the grounds, though not the inevitability, of a new relationship between ideas and practice, a new "walk" and a new "talk." This hope is related to the knowledge that if we can change in this way, walk in this direction, so might others, and another kind of life—not only of thought—may be possible.

✦ Talking Dirty: Ground Is Not Foundation

CATHERINE KELLER

On Christ the solid rock I stand;
All other ground is sinking sand.
—EDWARD MOTE, C. 1834

He explained that the Earth—the Deterritorialized, the Glacial, the giant
Molecule—is a body without organs. This body without organs is permeated by
unformed, unstable matters, by flows in all directions, by free intensities or
nomadic singularities, by mad or transitory particles.
—GILLES DELEUZE AND FELIX GUATTARI, *A Thousand Plateaus*

The North American continent may be more nomadic than most of its
inhabitants.
—*Scientific American*

BACKGROUND NOISE

Common ground has become uncommonly hard to find. Even those of
us who don't think much about our common ecology have been worrying about a base for democratic politics: without shareable ground, we
have failed to make common cause, even when the stakes were intolera-
bly high. These are not unrelated problems. Ecology and politics criss-
cross perilously within the transdisciplinary terrain of what is called
"ecological theology." For many of us, a feminist interrogation of the
common sources of anthropocentrism and androcentrism first lent this
terrain its allure, its transgressive width, and indeed its promise of a

shared method. But gender analysis breaks up as much common ground as it effects—thankfully. Yet I suspect that without some vibrant sense of collective ground—and not only shared goals—a ground that precedes and exceeds academic interdisciplinarity, the *trans*disciplinarity signified by ecofeminist theology cannot twist theory into earth-saving activity. At the mere hint of ground, however, let alone of the common, a high-pitched staccato chorale in my mind strikes up: *Decenter! Denaturalize! Deterritorialize! Differ! Destabilize the common! Deconstruct ground!*

Slower-moving voices answer from a low register: *And common earth, Gaia—mother of all the mothers, taken for granted, treated like dirt, dull dark mat(t)er, trodden underfoot, left behind, like all the earth-bound creatures, the darker peoples? Too base for their theopatriarchy, too common for your atheology. You want destabilization—you got it!*

The chorale, its dissonance sharpening, retorts: *Don't bore us with those mommy earth gender essentialisms, housewifely ecovirtues, naturalisms swinging between their romanticism and their objectivism. Your eco cancels out your feminism: get hip to the groundless.*

As my voices fade into the background, I hope their caricatures of ecofeminism and constructivism can mark for our discussion the poles of an obstacle to the transdisciplinary work of ecological renewal. It is not that the environmental movements need an academic *foundation*. They do, however, need institutional and intellectual *ground* to foster a broad-based leadership. For example, Mary Evelyn Tucker's account of the Earth Charter articulates "shared values for the larger Earth community as it seeks to build common ground for a sustainable future."[1] Any thinking that participates in the immense project that Rosemary Radford Ruether has named "conversion of thought to the earth" will assert *ground*—literally and metaphorically. Yet as eco*feminism*, such thought will also destabilize any cultural fabrications packaged as unquestionably "natural," and especially as "human nature." But no ecological model can honestly deploy a rhetoric of mere "denaturalization"—even with Judith Butlerian sexiness. For in this pure form of deconstruction nonhumans simply do not appear among the "bodies that matter."[2]

Is this to say we cannot serve both ecotheology and deconstruction? William Cronon and his interlocutors insisted in *UnCommon Ground* that the environmental movement is tempted toward a homogenized, dehistoricized notion of nature, "nature as naïve reality."[3] No doubt a naïve naturalism tends—quite understandably—to underlie much environmentalism. And surely we would want to disentangle ecotheology from a naturalized foundation that renders the earth a set of exploitable resources, a background for the human drama, and a rock upon which progressive politics inevitably runs aground. Bruno Latour puts it punchily: "Political ecology has nothing whatsoever to do with nature, this jumble of Greek philosophy, French Cartesianism, and American parks."[4] Nature, he warns, conjures the Platonic Truth of an uncontestable reality accessible now only to the Scientist (though once only to the philosopher or the priest). "[U]nder the pretext of protecting nature, the ecology movements have also retained the conception of nature that makes their political struggle hopeless."[5] This uncontestable nature trumps politics as such, defined as "the progressive composition of the common world."[6] Nature in this paradigm silences the negotiations by which the human and nonhuman agents of this earth collective might *collect* themselves. In this Platonized form, which surely haunts ecological "natures," despite or because of their all-inclusive intentions, nature keeps settling into a foundational rigidity.

In this essay, I'd like to sidestep the debate on Nature, however, and contend with its dirtier associate, ground. I'd like to experiment with a crudely down-to-earth hypothesis: *Ground is not foundation.* The deconstruction of foundationalism has shaken up any imaginary terra firma—of Nature, Being, or God. Yet deconstruction, or constructivism, has also had the unintended side effect of undermining ecological thought in the transdisciplinary academy, including theology. Yet perhaps it need not. It may actually (and also unintentionally) support a more inviting way to ground our thinking in the earth: a way of nesting gratefully, rovingly, and carefully among the *relative* stabilities of the earth. Of course, those stabilities are stressing out: we find ourselves on terra *infirma*. But long before we began to strain those stabilities catastrophically, theology and its philosophies were fleeing the shifty surface of the earth, upward or downward, in search of firmer foundations—which turn out, paradoxically, to be abstract, immaterial grounds, unchanging reasons. So what if we now declare: *let the earth itself be the ground*; let every grounding metaphor acknowledge its place, its earth, and its planetary context. Let it disclose its clay feet. Such an elemental ground will even soil the Christian trinity. Now we're talking dirty!

SOLID ROCK, SINKING SAND

Of the four symbolic elements, you'd think that earth, for earthlings, would be the element most present. Sculpted into the flesh of our bodies, lying faithfully underfoot (well, somewhere down there)—yet we can hardly hold it in thought. The very presumption of its "presence," of a stable, self-identical ground lying open to scrutiny, seems to put earth under deconstructive erasure. Besides, next to the moody mobilities of fire, water, and air, earth seems stodgy.

For the biblical heritage, the problem is oddly parallel. There are holy grounds; the earth rejoices; Adam is of *adamah*, humanity from "humus." The earth, or heavens-and-earth, brims with creaturely goodness; it "brings forth" in Genesis 1, very unlike the earth Neoplatonically inscribed as the heavy, base, light-trapping and sin-effecting element. Nonetheless, despite the ground-affirmative nondualism of the Hebrew worldview, earth remains, of the four elements, the furthest from the celestial God. The divine appears in the form of tongues of fire and burning bush, living waters, or spirit, *ruach* blowing where it will, breathing divinely—but never as *earth*. While the other three elements flicker, flow, and blow at the metaphoric edge between divinity and materiality, the earth remains simply creature, not the matter of a divine epiphany.

Of course Christians still seek their eternal foundation in Jesus. *"On Christ the solid rock I stand / all other ground is sinking sand,"* according to an old hymn of my tradition. The notion of grounding theology in the earth would still seem groundless to most Christians—like building a cathedral on quicksand. In the meantime, postmodern theory sinks pleasurably into the groundless.

This question of ground as elemental matrix rather than unshakeable foundation may not be directly relevant for environmental strategists, activists, and religious leaders. But it surely does matter for intellectuals whose cultural practice is anchored in the academies upon which, willy-nilly, the leadership and language of the ecological movements depend. It also matters for me and my students, who find ourselves irreversibly influenced by varieties of French-based theory, crucial both for our deconstruction of the theopatriarchal empires of our own foundations, and for the postcolonial complexity that encourages, beyond our identity politics, more emancipatory coalitions.[7] Yet at the very base of deconstruction (if it has one) is the identification of the metaphor of "ground" as a notion "belonging essentially to the history of onto-theology, to the system functioning as the effacing of difference."[8] It isn't that deconstructive thinkers are antienvironmentalists.[9] It isn't that they never write about ecology (especially in science studies);[10] nor is it that poststructuralist thought lacks animal and green potential.[11] Yet this potential has barely been actualized. Instead, an eerie, ecological silence follows in the generational wake of deconstruction—like a numbness or a boredom with the problems of that vastly nonverbal matrix of creaturely existence in which we live and move and have our texts. Only the human and speaking comes into play, with its "body" (an attractively cultured, naughtily sexed, and all-too-human body) as close as it comes to our material embeddedness. But even there, what fascinates turns out to be a writing of the body, the pillow-book of an inscribed surface; the body as text, indeed, in Butler, as "boundary." The depth of its incarnational intensity seems dissociated from its surface. This skin-thin "body-that-matters," however delightful, may separate us from our socio-ecological matter as surely as the dogmatic version of the resurrection body has done for Christians. So there has been an academic extinction, inadvertent like most extinctions, of the vocabulary by which ecological witness can be carried on.[12]

The disciplines of theology and religious studies—laudably seeking on their progressive side to avoid imperialist notions of common ground, let alone delusions of certainty and nature modeled on scientific objectivity—have taken in both the gift and the poison of deconstruction. So much theology gets stuck in a double groundlessness, a double denial of earth-ground, premodern and postmodern. The establishment Protestant antagonism toward any cosmological theology, especially process and ecofeminist variants, marks this double denial. Ecotheology gets caught between the rock—and the quicksand.

GROUNDLESS DECONSTRUCTION?

Let us then touch down lightly on an originative site of the deconstruction of ground. "The ontico-ontological difference and its ground [*Grund*] in the 'transcendence of Dasein' [*Vom Wesen des Grundes*] are not absolutely originary," wrote Derrida in *Of Grammatology*. "Difference by itself would be more 'originary,' but one would no longer be able to call it 'origin' or 'ground,' those notions belonging essentially to the history of onto-theology, to the system functioning as the effacing of difference."[13] He is here contending with Heidegger's *Grund* as importing the very onto-theology it had itself meant to deconstruct. He wants to liberate from this surreptitious "theology" his own first principle of *différance*. But is it *essential* thus to reduce ground to the notion of "origin" as absolute foundation? Or is it instead possible—precisely in the interest and the justice of a Derridean difference—to liberate ground to its "free play of signification," and first of all *from* ontotheological "origin," from a rigidly anthropocentric *logos*?[14] One hopeful sign: David Wood, a philosopher in this lineage, finds deconstruction "precisely adapted to our current situation, one in which the whole privilege of the human as a well-meaning but often toxic terrestrial, is quite properly being put in question."[15] He argues for the greening rather than the shunning of deconstruction.

Luce Irigaray also deconstructs the *Grund*. In *The Forgetting of Air in Martin Heidegger,* she liberates from the ground an elemental play of sexual difference. "The elementality of *physis*—air, water, earth, fire—is always already reduced to nothingness in and by his own element: his language."[16] Given her radical symbolism of embodiment, flesh, sex, and occasionally earth, it is noteworthy that her strategy is to pit the air, with support from the fluid, *against* this abyssal ground, rather than to deny its adequacy as ground itself. Would it not be more consonant with her generally ecofriendly project to let the air, water, and fire of earth come to enliven—indeed to *aerate*—the notion of ground? She does propose an alternative "groundless ground" comprising a "relation between," read as woman's "touching herself within herself and of each [female] other touching each other."[17] This delicately materialist gynomorphism "grounds" her ethics of eros in an interhuman interval or envelope. But it hardly touches upon the extrahuman. Might we not infuse the discourse of ground with elemental passion? The dirty ground, earth itself, does not compete with other elements but rather, at least from the perspective of any terrestrial, grounds them all.

Difference as the poststructuralist transcendental—whether deconstructive or sensible—opens an interval or space of irreducible alterity, freed from the naturalized foundation set in its empty "*Grund*."[18] But deconstruction might force us to choose between a purified transcendent

Grund and a purified transcendental *Différance*. However, doesn't the "originarity" of difference, as it fissures the foundations that had buried it, suggest a more elemental metonymy? Might it not expose to the air (in a fully Irigarayan sense) a dirtier ground, in which the groundless is not nihilistically triumphant but mystically irreducible? Precisely as earth-dirt-element might this ground perform an aporetic *grounding of deconstruction*? This would be a step toward the delightful possibility Wood calls "econstruction." Might an econstructive theology be at hand?

Deleuze and Guattari may provide support—oddly, given their militant antitheism. Yet they are preoccupied with the earth itself, recurrently attempting to think in its rhythms. Their joint work on capitalism and schizophrenia announces a "Geology of Morals." It solicits the multidirectional flows of the earth itself, its own "free intensities or nomadic singularities." This politically charged geology is an outcropping of their idea of the "rhizome." Its democratizing, deterritorializing and fractal flow, like the horizontal expansion of grass or water lilies, is contrasted to the image of the "arboreal," with its vertical system of roots and branches. "The wisdom of the plants: even when they have roots, there is always an outside where they form a rhizome with something else— with the wind, an animal, human beings (and there is also an aspect under which animals themselves form rhizomes, as do people, etc.)." With the caveat that trees also form rhizomes (except when lining Parisian boulevards), we may form a rhizome between constructivism and ecofeminist theology. Deleuze and Guattari, bent on deconstructing any hierarchy, exalt a "body without organs," meaning a body emancipated from its organic hierarchy. They privilege *de*territorialization. Still, when thinking with the earth, they make a meaningful concession. "Deterritorialization must be thought of as a perfectly positive power that has degrees and thresholds (*epistata*), is always relative, and has reterritorialization as its flipside or complement." And yet earth remains, they insist, *"the absolutely deterritorialized."*[19]

Earth itself does the dirty work of destabilization. The material ground *is* the effect of unimaginable intensities of flow and shift. Here, quite apart from human artifice, earth performs its own denaturalization. If I may risk reterritorializing these *differing* French strata: the sensory trace left in the Derridean fissures of language, the elemental passion folded in the Irigarayan intervals, the earth that "constantly," in Deleuze, "flees and becomes destratified, decoded, deterritorialized"—these all suggest the possibility of a deconstructive ground, or a soiled deconstruction. But then dirt itself, before we draft it as common ground for a transdisciplinary theological econstruction, bears a bit of narration. Ground *ad litteram*.

DIRT *AN SICH*

"The truth is," hyperbolizes my favorite dirt-writer, William Bryant Logan, "we don't know the first thing about dirt. All we can say is it doesn't come from here. Our own sun is too young and cool to manufacture any element heavier than helium."[20] The 112 elements naturally occurring on the Earth were not produced in our solar system; and the heaviest, uranium and plutonium, can only be forged in a supernova. "We are all stardust" is an understatement, he writes: "In fact, everything is stardust." So how does it get here?

Our cosmos is dirtier than we thought. For instance, the dark line across the center of the Milky Way comprises dirt that measures 65,200 light-years across. The "countless megatons of unknown dirt," the "ejecta, dejecta, rejecta, and detritus of ruined stars" float around the cosmos until it enters a field of force, shaped in fractal forms of spider webs, spirals, or labyrinths.

Imagine that the dust from a thousand different exploding stars has gathered along the weak lines of magnetic force in a spidery red nebula. A few light-years away another star explodes. The force of the bang sends out a shock wave that perturbs and twists the magnetic lines of the nebula, creating eddies and whirlpools, exactly like those in a river. In those eddies, the dust begins to gather. As it forms lumps, gravity for the first time becomes stronger than electrical forces. The process feeds upon itself, until spherical masses are formed.[21]

Order gathers at the very edge of chaos, where turbulent iterations in absolute unpredictability create a "phase transition" into a complex—and relatively stable—self-organization. This emergence of ordered complexity at the edge of chaos will provide whatever explanation is possible for the inexplicable phenomenon of life—but biological life has this pattern of feedback and iteration in common with the formation of the great geological bodies.[22] The earth "without form and void," *tohu va bohu?*

Earth: a huge mass, not so big as to catch nuclear fire and become a sun, not so small as to catch cold and remain a moon or comet. In the course of its evolution, "this compacted mass of interplanetary dirt called the Earth has two primary products: soil and atmosphere."[23] This location suggests a dirty translation of the Hebrew phrase for what Elohim created in the chaos of beginnings: *ha-shamayim et ha-aretz*, the so-called "heaven and earth." But how can we read the "earth" as *producing* "heaven and earth"? By solving the genesis puzzle of the *ha-aretz tohu va bohu*, which signifies, in an ancient intuition that may not be so purely mythical after all, that the earth in its chaotic ("formless and void") state precedes the subsequent heavens-and-earth![24] In the compact code of Genesis, this production is a process: "as the beings that make up organic life continue to exist, evolve, and cover the Earth, they create a rich, stable atmosphere and rich, deep soils. Only here on Earth does stardust engage in this extraordinary array of self-organizing behaviors. Only here on Earth does it perform the ceremony of continually creating an atmosphere."[25]

The volcanic chaos that sculpts the earth from inside out not only disgorged minerals from inside the crust; it also evolved the ground-crust upon which we stand through plate-tectonic activity. Geophysicists are now "eavesdropping on the conversations between earthquakes," finding that earthquakes interact in ways never before imagined—"even tiny stress changes can have momentous effects, both calming and catastrophic."[26] The so-called butterfly effect now moves mountains. In the interaction of fire from star and core, with water raining for twelve thousand years, it formed the sea—the primal waters (*mayim*)—as the "proto-soil, where Earth, air, water, and the solar fire met for the first time."[27] But it is the black, organic stuff of soil that gives birth to all life on the land. And what is intriguing is that "radical disorder is the key to the functions of humus." At "the molecular level, it may indeed be the most disordered material on Earth," exhibiting fractal self-similarity with no self-sameness.[28]

These stabilities of the heavens and the earth: how hard-won from the cosmic wilderness, the earth *tohu va bohu* and darkness over the sea of chaos. Earth, the "absolutely deterritorialized," hosts the self-organizing complexity of its territorial creatures. Its richness of life-sustaining inhabitability evolved over millions of years: for instance, the atmosphere in its yin-yang exchange with the dirt achieved the temperature range that has held stable for the past ten thousand years. It is this stability that we are wrecking: through the boundless exploitation of the elements of Earth, the use of the atmosphere as a free sewer; the overmining of fossil fuels, and the release of too much CO_2 and methane.[29]

So if the geology of morals has an ecoethic, it will now destabilize the elite academic disregard for *stability*. (*Oh how eco-schoolmarmish*, mocks my dissonant chorale. But I persist . . .) If stability gets situated automatically on the side of the conservative, then *destabilization* will seem subversive by definition, the work of difference and dissemination. But geophysics narrates a more complex, more radical story. Out of elemental turbulence emerge barely imaginable stabilities: constancies of a continental mass, a climate, a life-sustaining atmosphere, and the delicate skins of bodies encapsulating their fluids—able on land first to stay still, and later, again, and now at will, to move. At every scale, life is the story of complex emergence within a ground shot through with quantum indeterminacies and chaotic nonlinearities, composed of cosmic instabilities braided into tensely persistent orders.

These orders can be called open-system stabilities. They provide stability precisely *by* deterritorializing internal tendencies to rigid orders. We might need to reconsider the question of social stability, or place, as ecorelational constructs vis-à-vis gender, sexuality, and constitutional rights, not to mention religion and its institutions.[30] If we become more honest about our own needs for multidimensional, open-system stabilities, who knows, we might even find more common ground with a few of those unstable, stability-seeking swing voters.

Freed from the deadly binaries of destabilization versus stability, groundlessness versus ground, perhaps even constructivism versus realism and antinaturalism versus nature, ecological religion might teach our species to celebrate and cherish the brilliant constancies of our shared atmosphere and dirt. Precisely as we deconstruct the authoritarian, changeless or heterosexist Nature, we rejoin the dirty slow-dance of planetary life.

GROUND GOD AND THE GENESIS COLLECTIVE

The desire to ground theology in the earth comes down, perhaps, to a humbly exorbitant hope: that we who inhabit the progressive margins of the traditional religions might set off some earth-saving, far-from-equilibrium butterfly effect; that we might provoke some species-wide fit of elemental spirit at the edge of environmental chaos. But the neon Christian imagery of height and light seems to overwhelm any trope of divine dirt: how can dust shine brightly enough to emit God's glory? When dirt shines, as in José Saramago's *The Gospel According to Jesus Christ*, it is a gift of the devil![31] This absence of earth-epiphany poses a profound pretheoretical problem for ecotheology. The dominant forms of Christianity, until recently prone to apocalypticism or to indifference regarding the health of the planet, echo this absence. But other forms are evolving.

Mark Wallace declares an "Earth God," manifest biblically in the pneumatological iconography of the dove: "an earth creature vibrating with flesh and feather, bearing a bit of olive tree in its mouth." Far from being the immaterial substance demanded by the canonical theological lexicon, the Spirit is imagined in the Bible as a material, earthen life-form who "mediates God's power to other earth creatures through her physical presence."[32] Having observed all too closely the Manhattanite dirty lives of those doves referred to as "pigeons," I've decided that this trope works. The dove forms a rhizome with the olive branch, emerging from the raging deterritorialization of the *tehom*. Wind, water, and the greening power of the solar fire above all interact turbulently within the earth; an earthen spirit is no more inert than the planet, productive of dirt and atmosphere. Even the trinitarian theologian Jürgen Moltmann insists that the Spirit as *ruach*, breath, and wind is never biblically understood as immaterial but as force field.[33] Force fields are invisible but physical, like those that gather dirt across the universe. But the point of the present earth theology—we might as well call it an *eartheology*—is not to reduce the other symbolic elements, let alone spirit, to a single element. It means precisely to resist any reductionist account of the earth and its geobiological bodies; and by the same token to smudge every metaphorical body—social, textual or theological—with its terrestrial ground. This would be the eartheological analogue to the Derridean "trace."

Ecospirit might inspire a soil-ful reinscription of the whole trinity, that grand metaphor of interrelation so early alienated from its own ground, not to mention its ecofeminist potential. The constitutive interdependence of trinitarian multiplicity forms its own rhizome. The first person is infinite and therefore apophatically groundless ground. While Rosemary Radford Ruether wisely never *identifies* Gaia and God, she lets the Divine Matrix ground an elemental sense and practice of the divine,[34] and Anne Primavesi offers the trope of "Gaia's Gift" in engagement both of the Christian language of God's grace and the poststructuralist aporia of "the gift." To receive life and its elements as gift is the very opposite of taking earth for *granted*.[35] Further, the universe as the body of God—in the metaphor absorbed from Charles Hartshorne via Grace Jantzen and Sallie McFague—holds promise. What if we transcode God the first persona into the groundless ground—that is, the matrix, depth, or *tehom* of the universe, which is Her moving body? The body of God as "body without organs"?

This would be *pan*carnationalism. What of the singular incarnation, the becoming-flesh of the *logos* in that concrete Jewish male creature because of whom trinitarian language was invented? As Karen Baker-Fletcher writes, in almost Nicene orthodoxy: "Jesus is fully spirit and fully dust. Jesus as God incarnate is spirit embodied in dust."[36] Thus Christians may still insist that God got down and dirty in Jesus. So we could all get down. Some of us will ground ourselves in the body of Christ, even as we ground that body in its elemental planetarity. Other ecotheologians will be grounded in Christ more evangelically, as the full page ad in the *New York Times* proclaimed recently: "Our commitment to Jesus Christ compels us to solve the global warming crisis."[37] Ecumenical common ground *within* and *between* religions may increasingly signify the *oikos* of the shared and shaky heaven-and-earth.[38]

God the groundless ground, the dusty incarnation, the breathy spirit: this rhizomatic trinity, like every trinity, models and mediates all creaturely interdependence. And it recycles the eschatological hope for the New Creation into this originally soiled and now nearly spoiled *old* creation. If, when beginning, Elohim created the atmosphere and the dirt (the dirt having been already *tohu va bohu*), we creatures end up looking "very good" only in our collective planetary portrait. The creation of the humans, male and female, in the image of the divine collective, the "we" of Elohim, does not in any way extricate us from the whole messy collective of Creation. Not, at least, before the soilless Christian foundationalism gets hold of the text. The text of Genesis will go on *grounding* Western power one way or the other: it will run environmentalism aground with its foundational myth of origin; or it will graft current ecology onto a nourishing ancient rhizome.[39] The Genesis "dominion" will no doubt continue to reterritorialize Gaia as grant rather than gift. So we will need to keep deterritorializing Genesis, translating it as *emergence*: the becoming of the *genesis collective*. We can then read it as an anticipation of Latour's attempt to collect us in a "collective of humans and nonhumans";[40] or, in a different genre, of the vision of the Earth Charter, which has no squeamishness in using a strong collectivist language. "The choice is ours: form a global partnership to care for Earth and one another or risk the destruction of ourselves and the diversity of life." Or: "The spirit of human solidarity and kinship with all life is strengthened when we live with reverence for the mystery of being, gratitude for the gift of life, and humility regarding the human place in nature."[41]

The radical ecumenism of "the spirit of human solidarity and kinship with all life" is of course irreducible to the Christian "Holy Spirit," which is language for an ecclesial context. But when Moltmann identified the Holy Spirit, through the Spirit of Creation and its biblical trajectory, as the Spirit of Life, he had the human destruction of planetary life-systems in mind.[42] In the interest now of sharing planetary ground, we will become nimble as well as humble in our rhetorical substitutions. For if the resonances of Christian language in the Earth Charter can reinforce Christian ecotheology, the latter also celebrates the mysterious ecumenacy of a spirit that blows where it will—as, for instance, into the vast collective negotiation that produced the Earth Charter. I see no reason to shrink away from such bold evolutionary declarations of our terrestrial interdependence. What about when it claims, as activist rhetoric is wont to do, that "we urgently need a shared vision of basic values to provide an ethical foundation for the emerging world community"? Does this vocabulary not thwart my own distinction between ground and foundation? Not at all. Before those high-pitched voices get warmed up again, let me clarify. The metaphor of foundation in itself is neither oppressive nor ontotheological. If we understand a foundation, as most Christian and metaphysical foundationalisms do not, as just what humans make, "provide"—indeed, as what is always (by definition) grounded in the prior depth of the earth, to which it therefore owes careful fidelity—the metaphor does not totalize its claim. It, too, can help to deconstruct the groundless abstractions of origin. I will continue in my own contexts to avoid the language of foundations, to deterritorialize that of "nature" and "human community." Yet teamwork such as that embodied in the Earth Charter is not what needs to be deconstructed. It may, however, benefit from the emergence of common ground with deconstruction—impossible, perhaps, apart from the econstruction this volume begins to practice.

This essay has been an experiment in pushing up some of that shareable ground as a specifically theopoetic project. In order to renegotiate prior notions of the common, of nature—indeed, of the agents and subjects of representation—strategies and vocabularies must proliferate and collect, deterritorializing the transcendentals of theory and enriching the spiritual resources for political ecology. Among these proliferating forms of representation may be some econstructive theology, its own rhizomatic trinity in play. If we perch on the solidity of a rock for a time (even a *petri*fied Christ), the very stones will testify to the erupting depths of a self-deterritorializing earth. The genesis collective stands neither on solid rock nor on sinking sand. It moves, shakes, and self-organizes within a flowing ground. One way or another, the ground we share is always already shifting beneath our feet.[43]

❖ Ecofeminist Philosophy, Theology, and Ethics: A Comparative View

ROSEMARY RADFORD RUETHER

Ecofeminism has emerged in the late twentieth century as a major school of philosophical and theological thought and social analysis. The word "ecofeminism" was coined in 1972 by Francoise d'Eaubonne, who developed the "Ecologie-Féminisme" group, arguing that "the destruction of the planet is due to the profit motif inherent in male power." Her 1974 book *Le Féminisme ou la mort (Feminism or Death)* saw women as central to bringing about an ecological revolution.[1]

Ecofeminism sees an interconnection between the domination of women and the domination of nature. This interconnection is typically made on two levels: ideological-cultural and socioeconomic. On the ideological-cultural level, women are said to be "closer to nature" than men, more aligned with body, matter, emotions, and the animal world. On the socioeconomic level, women are located in the sphere of reproduction, child raising, food preparation, spinning and weaving, cleaning of clothes and houses, roles that are devalued in relation to those of the public sphere of male power and culture. My assumption is that the first level is the ideological superstructure for the second. In other words, claiming that women are "naturally" closer to the material world and lack the capacity for intellectual and leadership roles justifies locating them in the devalued sphere of material work and excluding them from higher education and public leadership.

Many ecofeminist thinkers extend this analysis to include class, race, and ethnic hierarchies. That is, devalued classes and races of men and women are said to lack capacity for intellect and leadership, denied higher education, and located socially in the spheres of physical labor as serfs, servants, and slaves in households, farms, and workshops. The fruits of this labor, like that of wives in the family, are appropriated by the male elites as the base for their wealth and freedom to exercise roles of power and culture. These male elites are the master class who define themselves as owning the dependent classes of people.

The ruling class inscribes in the systems of law, philosophy, and theology a "master narrative" or "logic of domination" that defines the normative human in terms of this male ruling group. For Plato, and even more for Aristotle, the free Greek male is the normative human. Descartes, a major philosopher for early modern European thought, deepened the Greek dualism between mind and body, seeing all bodily reality as mere "dead matter" pushed and pulled by mechanical force. The mind stands outside matter contemplating and controlling it from beyond.

In modern liberal thought essential humanity corresponds to rationality and moral will. Humans are seen as autonomous egos "maximizing their self-interests" who form social contracts to protect their property and in which their individual pursuit of profit can be guided by an "invisible hand" to the benefit of the larger society. Although such views of the self claim to define the generic "human," what is assumed here is the male educated and propertied classes. Dependent people—women, slaves, workers, peasants, and colonized peoples—are made invisible. They are de facto lumped with instrumentalized nature.[2]

This master narrative, with its logic of domination, has structured Mediterranean and Western societies for thousands of years. Since the sixteenth century it has been extending its control throughout the globe, eliminating smaller indigenous societies with alternative, more egalitarian and nature-sustaining social and cultural patterns. Most other urban civilizations and religions, such as Hinduism in India and Confucianism in China, also developed patriarchal ideologies with similar social expressions. But even these earlier patriarchal worldviews, which retained some sense of the sacrality of nature, are being subordinated in the twentieth and twenty-first centuries by the one triumphant master narrative of Western science and market economics.

How do ecofeminists envision a transformation of this deeply rooted and powerful ideology and social system? Some feminists have objected to any link between the domination of women and that of nature, seeing this as reduplicating the basic patriarchal fallacy that women are closer to and more like nonhuman nature than men. They believe that women need to claim their equal humanity with male humans, their parallel capacity for rationality and leadership.[3] They too, like males, are separate from and called to rule over nature. But this solution to women's subordination ends with assimilating a few elite women into the male master class, without changing the basic hierarchies of the ruling class over dominated humans and nonhumans.

Most women remain subordinated in the home and in low-paying, menial jobs, even as a few elite women make it into the cabinets and boardrooms of the powerful. The same can be said of racial, ethnic men. Token inclusion of women, black or white, and racial ethnic men buttresses the claim that American society is completely inclusive and is open to talent from whatever group. The many who do not "make it" have no one to blame but themselves. This show of "equality" thus masks the reality of a system in which the super wealth and power of a few depends on the exploitation of the many.

Some ecofeminists do claim that there is some truth in the ideology that women are "closer to nature." They see this closeness as having been distorted by patriarchy to dominate both women and nature as inferior to male humans. But this distortion is rooted in an essential truth that women, by virtue of their child-bearing functions, are more attuned to the rhythms of nature, more in touch with their own bodies, more holistic. Women need to claim this affinity with nature and take the lead in creating a new earth-based spirituality and practice of care for the earth.[4]

Most ecofeminists, however, reject an essentializing of women as more in tune with nature by virtue of their female body and maternity. They see this concept of affinity between women and nature as a social construct that both naturalizes women and feminizes nonhuman nature, making them appear more "alike." At the same time, by socially locating women in the sphere of bodily and material support for society, women may also suffer more due to the abuse of the natural world, and hence also become more aware of this abuse. But this is a matter of their experience in their particular social location, not due to a different "nature" than males. Such experiences would vary greatly by class and cultural

location. An elite Western woman living within the technological comforts of affluent, urban society may be oblivious to the stripping of forests, and the poisoning of water, while a peasant woman who has to
struggle for the livelihood of her family in immediate relation to these
realities is acutely aware of them. This awareness, of course, does not
translate directly into mobilization for change. For that, one needs a
conscious recognition of these connections and a critical analysis of the
larger forces that are bringing them about, together with the rise of leadership that can translate this into organized resistance to dominant powers and efforts to shape alternatives to them.

One must also question the universality of the cultural ideology that
places culture over nature as male over female. Preurban people who
depend primarily on hunting-gathering and small-scale agriculture often
have very different patterns of thought. Often males are associated with
either wild nature (the sphere of hunting) or the fields that men control,
while women are associated with the domestic realm. Men may see their
activities as superior to those of women, but this is a matter of opinion,
with women seeing their work as equal or better. A hierarchical sphere
of male elites controlling culture and politics has not yet subsumed these
earlier patterns that relate the whole society more directly to the fields
and forests. But these earlier peoples have today been largely subordinated to patriarchal societies that identify themselves with a culture transcendent to nature and regard tribal and peasant peoples as inferior.[5]

Ecofeminist hope for an alternative society calls for a double conversion or transformation. Social hierarchies of men over women, white
elites over subordinated classes and races, need to be transformed into
egalitarian societies that recognize the full humanity of each human person. But if greater racial and gender equality is not to be mere tokenism
that does not change the deep hierarchies of wealth and power of the
few over the many, there must be both a major restructuring of
the relations of human groups to each other and a transformation of the
relation between humans and the nonhuman world. Humans need to
recognize that they are one species among others within the ecosystems
of earth, to embed their systems of production, consumption, and waste
within the ways that nature sustains itself in a way that recognizes their
intimate partnership with nonhuman communities.[6]

In this essay I will briefly survey several ecofeminist perspectives that are emerging from a number of religious and cultural contexts—those of Vandana Shiva from India, Ivone Gebara from Brazil, and Carolyn Merchant, a North American historian of science. I will conclude with some questions about the utility of this effort to interconnect the domination of women and of nature, social justice, and ecological health.

* * *

Vandana Shiva, of Hindu background, was trained as a physicist but abandoned her career in nuclear energy to become an environmental activist who writes and organizes against the Western systems of "development," which she sees as destructive of the ecology and economy of India as well as responsible for the debasement and impoverishment of humanity and the earth, generally. She seeks to propose sustainable alternatives to the Western model of "development" that are rooted in traditional Indian peasant agriculture.

In her first major book, *Staying Alive: Women, Ecology and Development*, published in 1989, she enunciated the major lines of her ecofeminist critique and vision. She has continued to elaborate these views in her subsequent volumes: *The Violence of the Green Revolution: Third World Agriculture, Ecology and Politics* (1991); *Monocultures of the Mind: Biodiversity, Biotechnology and the Third World* (1993); *Biopiracy: The Plunder of Nature and Knowledge* (1997); and *Stolen Harvest: The Hijacking of the Global Food Supply* (1999). With Maria Mies, leading German feminist socialist, she wrote *Ecofeminism*, published in 1993.[7]

Shiva speaks of both Western science and development as "projects of patriarchy."[8] Western developmentalism picks up in the post–World War II period where colonialism left off and is a continuation of colonialism (neocolonialism). The British colonialists in India had stripped the forests for timber to build their ships and railroads and organized the land for expropriation into wealth that supported their empire. In the postcolonial period it becomes the national elites of India who continue the same model of exploitation and plunder in the name of modernization and development.

For Shiva this model of development is built on a false assumption that nature and women are mere passive objects that are unproductive in themselves. Both nature and human labor become productive only when taken into a system that uses them for profit within the dominant system of accumulation. This model of development, "lifting all boats," is claimed to produce wealth for all. But it actually creates only a short-term extraction of wealth for a global elite, while impoverishing women, poor people, and nature itself, destroying the very base on which it is founded.

The destructive results of this model of development are not accidental, but are themselves rooted in the distorted epistemology of Western science. Drawing on Western feminist critics of science, such as Susan Harding, Evelyn Keller and Carolyn Merchant,[9] Shiva views Western science as being based on an epistemology of male domination over women and nature. This epistemology abstracts the male knower into a transcendent space outside of nature and reduces nature itself to dead "matter" pushed and pulled by mechanical force. Western science thus "kills" nature, denying its possession of self-generating organic life. It also imagines nature as a dangerous female that must be tormented and forced to submit in the laboratory, even as witches were forced to submit to inquisitors in torture chambers.[10]

India, like other developing countries, has been traditionally a rural society, based largely on subsistence agriculture, in which women have played a predominant role. For Shiva, Indian rural women have been the base of a sustainable system of subsistence agriculture because they have understood the interconnections of the cycles of life in the land and animals that they have tended. Western science and developmentalism, by contrast, sees this work of women as completely unproductive and the forests they tend as mere "wasteland" because they have not been incorporated into a system of commodities for profit in the market of international exchange.

Shiva mounts a severe criticism against the Western-style imposition of agricultural development on India in the form of the "green revolution."[11] The revolution promised abundance for poor third-world farmers through an increased grain supply, but in practice it has been a disaster. Women are eliminated from agriculture—so, too, are their roles as the maintainers of sustainable soils, forests, and food for humans and animals. Petroleum-based fertilizers, pesticides, and fuel for tractors poison the soils and waters, and the water table itself is depleted through overuse for irrigation. Rivers and wells dry up, and regions that had had

sufficient water suffer water-famine. Farmers must buy their seeds, fertilizers, pesticides, and fuel and sell their harvest to companies that respectively charge them high prices and pay them poorly for their labor. The
result is impoverishment of the people and the land. Yet this calamitous
model continues to be promoted by Western global corporations, development banks, and national elites as the epitome of "modernization."

Shiva's solution to the miseries of modernization and development is
to turn back to traditional sustainable agriculture. There needs to be a
recovery of the ecological knowledge of how forests, water, plants, animals, and humans are maintained in a renewable system of interaction
with nature, a partnership that based itself on nature's own cycles of
renewal. Women and tribal people are the privileged repositories of this
ecological knowledge. Thus instead of seeing them as ignorant, primitive,
passive, and unproductive, one must learn from them how to maintain
genuine life within nature.

The Western model of knowledge and economic value has proven to
be delusory. We must shift to a different understanding of knowledge
and economic value. The epistemological model we need is not one of
dominating mind over passive body, but how to think within nature's
own interrelationships. The economic system that produces true value,
that maintains life, is decidedly not one that destroys nature, but rather
one that cooperates with it and fits human life within its cycles of self-
maintenance.

Shiva turns to traditional Hindu cosmology to express the worldview
that is needed for the recovery of ecological knowledge and life-sustaining practice. She speaks of this as the recovery of the "Feminine Principle": not in the Western sense of a dichotomizing of masculinity and
femininity as binaries of aggressive activity and dormant passivity, but as
a dynamic interaction of creative energy, female *Shakti* (activating energy), together with male form (*Purusha*), which together produce nature
(*Prakriti*). In Hinduism, both *Shakti* and *Prakriti* have been understood as
feminine and as goddesses. Shiva thus suggests that traditional Hindu
culture, both in its high philosophy and in its popular spirituality, has
seen women as the active principle for the maintenance of life.[12]

To reclaim *Shakti* is, for Shiva, a reclaiming of human interdependency
with, and immersion in, the organic vitality of the natural world. Shiva

argues that this veneration of the Feminine Principle as self-creative nature is not a gender ideology that makes women different from, or better than, men. Rather it is a rejection of the Western gender ideology that defines males via a masculinity of disconnection from the body, women, and nature, violent domination over women and nature, and the subsequent distortion of women and nature into passive objects of this violence. Men need to overcome their alienation and violence, and women their passivity and acceptance of denigration. Both men and women must see themselves as active participants in nurturing life in partnership with nature's own vitality.

Some Indian ecofeminists have been critical of Shiva for her use of this Indian tradition of *Shakti* and *Prakriti* as feminine cosmological principles for ecological life. They see her as ignoring the negative aspects of this feminine cosmology, which they view as world-negating and instrumental in the subordination of women to male control.[13] Some critics also see her as ignoring the caste structure intrinsic to Hinduism that has traditionally marginalized tribal people and *Dalits* (untouchables.)[14] Shiva often writes as though patriarchy was invented in the seventeenth century and imported entirely from the West, rather than having been a part of Hindu society for millennia.

These are valid criticisms, but they perhaps ignore Shiva's primary purpose, which is to use popular symbols in Indian culture to honor rural women as the base of knowledge and practice of a sustainable subsistence economy. She also wishes to point to a vision of nature itself as vital and dynamic, not as dead matter to be dominated by knowers disconnected from it. Perhaps it is not accidental that in her subsequent books she has ceased to speak of a feminine cosmological principle, of *Shakti* and *Prakriti*. But she has grown even more devastating in her critique of the model of development coming from Western neocolonialism that is impoverishing rural people in India and throughout the world. She sees rural women and tribal people as centers of resistance to this model of development and preservers of ecological knowledge of sustainable life. This is a precious resource for an alternative that we all—men and women, East and West, urban and rural people—need to relearn if we are to survive on the planet earth.

* * *

Ivone Gebara is a Brazilian and a member of the Sisters of Notre Dame. In her book, *Intuiciones Ecofeministas (Ecofeminist Intuitions)*, Gebara talks about how she came to adopt an ecofeminist perspective.[15] She acknowledges the criticism of some Latin American feminists that ecofeminism perpetuates the stereotype that women are "closer to nature." But she believes that her own viewpoint has nothing to do with essentialist anthropology. Rather it springs from her concrete experience in her impoverished neighborhood. There she observes that poor women are the ones who primarily have to cope with the problems of air pollution, poverty, poor quality of food, and lack of clean drinking water. This creates health problems for themselves and their children, for whom they are primarily responsible.[16]

Gebara speaks of doing her theology "between noise and garbage." The noise is that of a crowded neighborhood with machinery, trucks, and cars that lack mufflers, and also the shouts and loud music of the people as they find ways to survive each day. The garbage is the waste of society disproportionately discarded where the poor live, with little organized clean-up. To do one's theology amidst noise and garbage is to do it in daily awareness of the oppression of the poor and the degradation of their environment. It is also to do theology inspired by the vitality of the poor, who manage somehow to keep going and sometimes even to celebrate despite these challenges.[17]

Gebara sees her ecofeminist theology as a third stage of feminist theological work in Latin America. The first stage, in the 1970s, recognized that women "are oppressed as historical subjects." In the second stage, in the 1980s, women began to question the dominance of masculine theological symbols and to search for feminine symbols for God, such as Wisdom.[18] For Gebara both of these phases are still expressions of a "patriarchal feminism," a feminism that has not deeply examined the androcentric model of theology, of God and the cosmos, but rather has simply sought to include women in it. Gebara sees ecofeminism as moving to a more radical stage that calls for a deconstruction of patriarchal thinking, with its hierarchical structure and methodology of thought. Ecofeminism seeks to dismantle the whole paradigm of male over female, mind over body, heaven over earth, transcendent over immanent, the male God outside of and ruling over the created world—and to imagine an alternative to it.

Changing the patriarchal paradigm for an ecofeminist one starts with epistemology, with transforming the way one thinks. Patriarchal epistemology bases itself on eternal unchangeable "truths" that are the presuppositions for knowing what truly "is." In the Platonic-Aristotelian epistemology that shaped Catholic Christianity, this epistemology takes the form of eternal ideas that exist a priori, of which physical things are pale and partial expressions. Catholicism added to this the hierarchy of revelation over reason; revealed ideas come directly from God and thus are unchangeable and unquestionable in comparison to ideas derived from reason.

Gebara, by contrast, wishes to start with experience, especially the embodied experiences of women in daily life. Experiences cannot be translated into thought finally and definitively. They are always in context, in a particular network of relationships. This interdependence and contextuality includes not only other humans, but the nonhuman world, ultimately the whole body of the cosmos in which we are embedded in our particular location. Theological ideas are not exempt from this questioning from the point of view of embodied, contextual experience. In addressing ideas, such as God as Trinity, she asks, "to what experience is this idea related?" What in our embodied daily life is the basis for thinking about reality as trinitarian and hence ultimately of God as Trinity?[19]

Such an effort to dismantle patriarchal epistemology for ecofeminist thinking includes the nature of the human person. How do we move from a patriarchal to an ecofeminist understanding of the self? Patriarchal theology and philosophy start with a disembodied self that is presumed to exist prior to all relationships. The body is seen primarily as an impediment to the soul to be controlled, not an integral part of the self. The ideal self is the autonomous self, that which has extricated itself from all dependencies on others and stands outside and independent of relationships as a "free subject."

In this view of the human, only elite men are fully selves; women and subjugated people are by definition dependent. The apparent credibility of such a view depends on making invisible this whole structure of support on which the apparent male freedom is itself "dependent." This notion of autonomy and independence is translated into global corporations. Such corporations dominate and control all else, turning them into things and making invisible their dependency on them.

An ecofeminist understanding of the human person starts with the person in a network of relationships. The person does not exist first and then assume relationships; the person is constituted in and by relationships. One does not seek to extricate oneself from relationships in order to become "autonomous." Such autonomy is a delusion based on denial of the others on whom one depends. Rather, one seeks to become ever more deeply aware of the interconnections on which one's own life depends, ultimately the network of relations of the whole cosmos. One seeks to shape those relations in ways that are more life-giving and reciprocal, to respect the integrity of the other beings to whom one is related, even as one is respected by them and respects oneself. *To be is to be related*; shaping the quality of those relations is the critical ethical task.[20]

This reflection on the network of relationality reaches from the most intimate relation with one's own body-self, to interpersonal relations, to intergroup relations of one human to another and of humans to the earth, generally. It culminates finally in recognizing our interrelations with, and dependency on, the whole cosmos. It is on this understanding of interrelationality that Gebara bases her reflection on the meaning of God as Trinity. For Gebara, God as Trinity is not a revelation from on high that one imposes on people as eternal and unchangeable truth outside of, and incomprehensible to, daily experience. Rather, the idea of God as Trinity is itself an extrapolation from our daily experience of interrelationships. The Trinity is a way of expressing the dynamics of life as interrelational creativity. Creativity by its nature ramifies into diversity while at the same time interconnecting in community, leading to new diversification. This process of dialectical diversification and intercommunion can be seen on every level of reality.

This concept of trinitarian dialectics as the process of creation of life on earth raises the issue of good and evil. If whatever develops is part of a natural process, from whence come systems of violence and oppression? Gebara insists that there is good and evil in natural life itself. Natural life exists in a dynamic tension of life and death, creativity and vulnerability. Death is an integral part of life, not foreign to it as traditional Christian cosmology had claimed. But the very vulnerability and fragility of life provides the impulse for possible distortion of this dialectical process. Each being in its species context seeks to protect and expand its own life against others that compete with it. Nature limits the extent to which some species can expand at the expense of others. When some exceed their life-support niche by destroying others on whom they depend, this precipitates the collapse of the dominant group.[21]

But humans have developed an ability to stand out somewhat from these limits. They have been able to organize their own species power in relation to land and animals to monopolize the means of life. This takes place in the context of some humans seizing power and organizing relations to other humans so these subjugated people do the brute labor. Those in power extract this into means of wealth, dominating power, and leisure for themselves at the expense of others, while claiming to represent the well-being of "all."

This pattern of exploitation of some humans over others and over the nonhuman world has been endlessly repeated through human history. These systems of exploitative distortion are always based on denying the interconnection of the powerful with the powerless, men with women, ruling class with slaves, workers, and peasants. Those on top imagine themselves as "autonomous" and naturally superior, while the inferiority of those they rule over demands their subjugation. Thus the systems of exploitation "naturalize" themselves by shaping ideologies that pretend that these systems simply represent the "order of creation" and the will of God or the gods.

We are now living in the nadir of this system of distortion that has grown increasingly centralized worldwide while impoverishing the majority of humans and destroying the earth. Yet this system continues to claim that the privations it imposes on others are necessary for all to eventually prosper and attain comfort and leisure equivalent to those of the affluent. If the poor but "tighten their belts" a bit more, the wealth generated at the top will "trickle down." But this is a fallacious ideology belied by reality. This system of distortion, violence, impoverishment, and oppression is immoral or "unnatural" evil, built on the denial of the interconnection of all beings with one another.

For Gebara there is no original paradise of blessedness without finitude or death at the beginning of human history, nor is it possible to construct a paradise of deathless goodness in some future millenium. Rather humans need to accept our limits, our fragility, our partial joys and sorrows within finite life. We need to lessen the patterns of distortion that allow some few humans to flourish inordinately at the expense of most other humans and the earth. We must shape more egalitarian societies where joys and sorrows, flourishing within finite limitations, are shared more equally and more justly between humans and between humans and the other earth beings with whom we share this planet. This is the very real but limited utopia that Gebara allows herself, recognizing that within our lives today we can expect only momentary glimpses of this more justly shared life in interconnected mutuality.[22]

Carolyn Merchant is a historian of science whose 1980 work *The Death of Nature: Women, Ecology and the Scientific Revolution* has been formative for ecofeminist thinkers worldwide.[23] The book opens with a dramatic declaration: "The world we have lost was organic."[24] She goes on to detail the organic view of nature that the medieval world inherited from Mediterranean antiquity. In this worldview, nature was typically conceived as female, either as a virgin or nurturing mother, or as a witch—a disorderly, demonic woman. The whole universe was seen as organic and alive. The *anima mundi* (world soul), imagined as a woman, animated the universe.

In the seventeenth century, with the rise of Cartesian philosophy and Newtonian science, this model of the "world-as-organism" was converted into a view of the "world-as-mechanism." The clock and other machinery became its image. Metaphors based on the persecution of women as witches were brought into scientific thought by writers such as Francis Bacon. Nature was described as needing to be "unveiled," stripped of her concealing clothing, dragged by her hair into the laboratory, "vexed" (tortured) and forced to "yield her secrets."[25] Despite these personalized metaphors, nature came to be seen as a mechanical order composed of tiny, dead balls of matter (atoms) that are pushed or pulled by external mechanical force. All intelligence, soul, or life was taken out of the material world and lodged in a transcendent mind (God) and made manifest in the human (white Western male) mind. This intellect knows reality from the outside, objectively, in a value-free and context-free fashion based on mathematics.[26] Scientific knowledge was identified with the power to control nature, as fallen humanity was seen as having lost dominion over nature. Both humans and nature thereby fell into disorder. Through scientific knowledge, dominion is restored to humanity and nature, and humanity thereby redeemed.

Merchant sees the organic holistic view of nature being reclaimed in ecology. She calls for the redevelopment of "communities based on the integration of human and natural ecosystems." Her *Reinventing Eden: The Fate of Nature in Western Culture* (2003) picks up these themes from a different perspective.[27] Merchant here posits that Western cultures have been shaped by two opposing narratives, both aimed at the "recovery" of Eden, the primal paradise of the biblical narrative where humans, male and female, nature, and God were in harmony. One narrative, which has dominated Western thought since the seventeenth century, secularizes the Christian story of redemption through Christ that pointed to a transcendent heaven as the ultimate locus of this redemption. In the secular progressive redemption story, paradise is reestablished on earth.

Based on Bacon's view, human dominion over nature, given by God at the beginning, was seen as impaired by the fall. Disorder and savagery have reigned since this early collapse. But science and technology are restoring human dominion and thus transforming primitive disorderly nature into civilization. This task of civilizing nature is the "white man's burden." The white Western male is subduing the whole world, first Europe and then the colonized areas of the Americas, Asia, and Africa and elevating them to this higher order. Women, Africans, the indigenous peoples of the Americas are all to be subdued, domesticated, cleared out of the way, or transported as slaves to be the work force for civilization, although denied its full benefits.

Merchant sees the North American shopping mall as the ultimate image of the reinvented Eden.[28] Surrounded by a concrete desert of parking lots, the shopping mall presents an artificially constructed total world of commodified pleasure. Shops and restaurants line the walkways that are set in artificial gardens with waterfalls, flowers and trees, even with modeled animals and birds and live fish. Against this narrative of recreating Eden as a fabricated world freed from natural constraints is what Merchant calls the "declensionist" narrative. This narrative has been adopted by some feminists, ecologists, and ecofeminists. This narrative looks back to an original Eden that subsisted in human history for hundreds of thousands of years—before the rise of plow agriculture, urbanization, slavery, and war—some time in the eighth to fifth millennium BCE in the Ancient Near East. All these ills are often spoken of collectively as "patriarchy," the rise of societies dominated by a male elite who subjugated women, turned the majority of humans into slaves, and redefined all these humans, as well as nature, as property.[29]

Thus ecologists and ecofeminists call for an urgent revolutionary transformation of the world order that has been shaped by a "5000- year" process of domination, and the recreation of "Eden," as small self-governing communities that integrate democratic relations and economies of natural renewal. This is what David Korten, leading critic of the neoliberal corporate global economy, has called in his new work, "The End of Empire and the Step to Earth Community."[30]

Merchant concludes *Reinventing Eden* with the call for a "partnership ethic," which would integrate the narratives of both progress and decline but in a new way. The basis for this ethic draws upon new scientific and philosophical developments of quantum mechanics, as well as chaos and complexity theories, all of which have come to recognize that nature is not passive or mechanical, much less composed of "dead" matter. Rather nature is alive, holistic, and interconnected. Nature has its own self-organizing patterns of life. Humans must come to view nature not as comprising dead objects to be exploited, but rather as a totality of active subjects with which they must learn to partner.[31]

This comparison of three ecofeminist thinkers from North America, India, and Brazil reveals significant commonalities. There is in each a critique of Western epistemology that posits an isolated knower outside of, and unrelated to, the reality that is known and whose knowledge is a means of control over others. Each questions the model of the self based on the premise of an isolated individual disconnected from relationships and which ignores the actual support services that other humans and nature are providing to create the privileged appearance of this "autonomous" self.

They all also reject a view of nature as "dead matter" to be dominated in favor of an understanding of nature as living beings in dynamic communities of life. They call for democratic relationships between humans, men and women, ethnic groups, and those presently divided by class and culture. Ultimately, they each seek a new sense of partnership between humans and nature. The keynotes of interrelationship, interdependency, and mutuality echo across all three perspectives.

The circulation of ecofeminist ideas across cultures resonates with deep conflicts, struggles, and changes of consciousness that are happening worldwide. The destructive impact of a pattern of "dominology," based on a top-down epistemology and a destructive concept of the self and its relation to other humans and nature, is widely seen as the root, not only of sexism and racism, but also of imperialism, with its ongoing expressions in neocolonial exploitation of third-world societies and their natural resources. Groups of people around the world are working to change these patterns. Similar ideas of the needed alternatives are emerging in many contexts and linking up with one another.

It is widely assumed that there is a need to refound local community, in democratic face-to-face relations with the variety of people—across genders, classes, and ethnic groups—living in a given community. There is a need for renewed regional communities to redevelop their relation to the land, agriculture, and water such that they might be utilized in a sustainable way; such changes will need to be based on democratic decision making that takes all parties, including nonhuman nature, into consideration. This also means withdrawing from the centralized systems of control that have been forged by colonialism and neocolonialism. By banding together in communities of accountability, it is hoped that this system of domination can be undermined and changed to new ways of networking local communities across regions and across the globe.

Visions of humans in interrelation with one another and with nature express this longing for an alternative way of situating people in relation to society and the world. To see nature itself as a living matrix of interconnection provides the cosmological basis for this alternative vision of relationship. This common ecofeminist worldview shares some of the following characteristics. There is a rejection of a splitting of the divine from the earth and its communities of life to project "God" as a personified entity located in some supercelestial realm outside the universe and ruling over it. The concept of God is deconstructed. The divine is understood as a matrix of life-giving energy that is in, through, and under all things. To use the language of Paul in the book of Acts, God is the "one in whom we live, and move and have our being."(17:28) This life-giving matrix cannot be reduced to "what is" but has a transformative edge. It both sustains the constant renewal of the natural cycles of life and also empowers us to struggle against the hierarchies of dominance and to create renewed relations of mutual affirmation.

This divine energy for life and renewal of life is neither male, female, nor anthropomorphic in any literal or exclusive sense. It can be imagined in many ways that celebrate our diverse bodies and spirits. What is excluded are metaphors that reinforce gender stereotypes and relations of dominance. It can be called "divine Wisdom," the font of life that wells up to create and recreate anew all living things in what Thomas Berry calls "ecozoic" community.[32] "Ecosophia" calls us into life-giving community across many strands of tradition, culture, and history, and also empowers us to stand shoulder to shoulder against the systems of economic, military, and ecological violence that are threatening the very fabric of planetary life.

❧ Econatures: Science, Faith, Philosophy

⌖ Cooking the Truth: Faith, Science, the Market, and Global Warming

LAUREL KEARNS

"Have you heard the one about the rabbi, the priest, the pastor and the Toyota Prius? No, it's not a joke. And neither is global warming."

So reads the introduction to an action-alert email on Faith and Fuel Economy from the Interfaith Climate Change Network (ICCN). Only I might change it to a rabbi, pastor, priest, and preacher to more accurately imply the four constituent groups of the National Religious Partnership on the Environment (NRPE),[1] the group supporting the ICCN, who chose global warming in the 1990s as the one topic upon which they could all agree.[2] This particular action alert encouraged individuals to bring their faith to bear on deliberations in the U.S. Congress on increasing fuel efficiency. The alert was also related to the concurrent visit to carmakers in Detroit by Jewish, Catholic, and Protestant leaders in a fleet of hybrid cars driven by Catholic nuns. The Web site goes on to ask:

If God is With Me All the Time, Does that Include the Auto Dealership?
As people of faith, we use religious convictions to determine the movies we see, music we listen to, and activities we participate in. If we bring God to the movies, why do we leave God behind at the Auto Mall? There is no reason to drive gas-guzzling, climate-changing cars when there are options that give us freedom and reduce the impact on our environment. Because it's not just about vehicles, it's about values.[3]

This is one of many religious campaigns related to climate change in the United States that I have studied in my sociological research on religiously motivated environmentalism. These efforts involve a variety of approaches, as will be discussed below. But not all religious efforts on the topic are aimed at increasing awareness of the threat of global warming; some are aimed at discrediting global warming by referring to it as a religion, a theology, or an object of belief. These voices do not just single out global warming to attack; rather, their approaches range from a rejection of most scientific theory (a result of creationism campaigns), to a dismissal of climate change as a creation of the left. Many climate skeptics aver that perhaps it will even create new positive opportunities.[4] All base their objections, whether explicitly or not, in their concern that any action to combat global warming will be a threat to private property rights, free enterprise, and capitalism. The differences between the religious "pro–" and "anti–global warming" campaigns (as I somewhat facetiously call them) are too easily painted in terms of the dichotomy of religious versus scientific authority. But since actors on both sides are religious, the difference is more nuanced than that.

On the surface, it *does* seem like a battle between the claims of religion (arguing that the science is so speculative that it amounts to a religion or theology) and science (arguing that the science is so strong that it demands a religious response). Both sides, however, in fact want to make global warming a religious issue, and not just a scientific one. The "anti–global warming" forces want to make it so in order to undermine the authority of the science, and thus to discredit the existence of global warming, particularly insofar as it is human-caused. The "pro–global warming" activists want to make it a religious issue because they believe that basing any response on just the science is inadequate, for the issue is also about values. How each side frames the issue in terms of both religion and science is a major theme of this essay. I argue, however, that just examining responses to global warming in terms of religious views concerning science can be misleading. First, following Bruno Latour's work, the current framing of religion and science is very misleading. Secondly, the faith-versus-science framing can hide perhaps the real framework involved: the belief in the economy and the market that influence the stances on both sides of the issue.

BACKGROUND TO RESEARCH

I have long been interested in how theological ideas and religious concerns are translated into action by grassroots groups, particularly on ecological-environmental issues.[5] I approach this topic by looking at how people understand their lived religious traditions and what they select within these traditions as a basis of authority for action, for reinterpreted or reconstructed ritual, and for motivation, hope, and inspiration (especially in fighting off despair). As a student of social movements, I am also interested in what strategies and actions are chosen, and why, and the relative success of these strategies.[6] These two concerns—claims to authority and strategies of action—also help frame this chapter.

I became interested in how groups approached the question of global warming in particular because I was observing the crossing of boundaries, such as between liberal and evangelical Protestantism, Judaism and Christianity, and religious and secular environmentalism.[7] These boundary-crossings have provoked innovative strategies and cooperative efforts. Furthermore, it seemed an excellent chance to explore whether religious framing and grounding could motivate concern and response over a topic that the U.S. populace and government on the whole choose to downplay or ignore.[8] In other words, could grounding a scientific claim of the enormity of global warming in religious authority work to motivate a response?

My intent here, however, is not to cover all of the various campaigns around the globe related to the topic of global warming. Instead, I focus on the range of religious responses to growing concern over global warming in the United States, in part because the United States has refused to be a part of the Kyoto Protocol on Climate Change and in part because the United States disproportionately contributes to CO_2 emissions.[9] In a country where the official government response is to deny, or discredit, the reality of global warming, the response by religious groups can be an important element in supporting or challenging the authority of the government's position. After briefly introducing both movements—one portrait stemming from years of participant observation, and the other, admittedly less grounded, from an examination of group statements, press coverage, and media campaigns that arose from my research on the "wise-use" movement—the latter part of this chapter explores what the two responses then reveal about certain religious constructions of environmental concern, science, and economics.

"PRO–GLOBAL WARMING" ACTIVISM

I provocatively use the term "pro–global warming" to characterize the movement to enlist U.S. citizens to put pressure on the government, at all levels, to legislate a variety of measures aimed at reducing the rate of global warming, primarily through the reduction of energy and petroleum consumption and CO_2 emissions.[10] At the same time, many groups also strive to get religious institutions and individuals to change their energy-consumption habits. Thus they have a two-pronged strategy of action to bring about both structural change that would then enable individual change, and to change individual and institutional consumption habits that drive the marketplace. But I primarily use the term "pro–global warming" (which on the face of it seems absurd, for few are in favor of the changes that will be wrought by global warming to the earth's atmosphere and oceans) because what is largely at stake in the religious battles over the topic is whether or not global warming even "exists" and is an issue worthy of religious concern. Thus "anti–global warming" activists seek to deny or downplay the existence of global warming, for a variety of reasons to be explored below, and the "pro–global warming" movement is premised on the scientific consensus that global warming is happening, that human actions are in part responsible, and that immediate changes in human activity can affect the rate and consequences of global climate change.

Since the late 1980s the NRPE coalition has engaged in a variety of campaigns focused on climate change. Although they agree on the importance of the topic, each of the constituent members—the Evangelical Environmental Network (EEN), the U.S. Catholic Conference of Bishops (USCCB), the Coalition on Environment and Jewish Life (COEJL), and the National Council of Churches Eco-Justice Working Group (the NCC-EJWG includes mainline Protestant, Orthodox, and Historic Black churches)—obviously has different theological underpinnings and institutional constraints for their work on the environment. For example, they are divided on what the right approach should be: individual conversion versus political action and lobbying; getting people to appreciate

"nature" more or to stop their overconsumptive habits; or moving beyond an individual focus to changing congregational lifestyles by fostering a greener spirituality and worship. They are likewise divided with
respect to theological differences over the place and role of humans in
the larger creation and God's relation to creation.[11]

Consider the National Council of Churches Eco-Justice Working
Group (NCC-EJWG). Founded in 1984, the group has been working on
climate change since 1990 by promoting public awareness and working
for political change. Although their campaign to persuade the United
States to sign the 1997 Kyoto Protocol was ultimately unsuccessful, they
do feel that the campaign of television ads and petitions containing hundreds of thousands of signatures leading up to the Kyoto meeting influenced then Vice-President Al Gore to modify the U.S. stance.[12] They
have continued to work on the topic, organizing the third biannual NCC-
EJWG conference in 2001 around the theme "Let There Be Light," coordinating denominational and statewide efforts, providing training for
youth, and compiling and distributing educational and worship materials.

The focus of the NCC-EJWG, in partnership with the Coalition on
Environment and Jewish Life (COEJL), also shifted to trying to influence
key congressional members by establishing eighteen state Interfaith Climate Change campaigns under the umbrella of the Interfaith Climate
Change Network (ICCN).[13] Initially these were funded by a grant from
the Pew Charitable Trusts channeled through the NRPE to the NCC-
EJWG and COEJL (the EEN ran a separate campaign built around the
slogan "What Would Jesus Drive?"). Later the statewide campaigns were
more directly run by the ICCN and NRPE, and now have returned to
the constituent members. These groups focus on influencing key state
governmental officials all the way up to potential swing senators, as well
as on educating the public and stimulating action around energy issues.
Participants are trained in media relations, media appearances, political
lobbying, and so forth.[14] That work continues as both U.S. President
George Bush and Oklahoma Senator James Inhofe, former chairman of
the Senate Committee on Environment and Public Works, deny that
there is any need to respond to global warming.[15]

Encouraging people to buy green electricity (produced from non-pol-
luting, renewable resources), especially when combined with energy-
conservation measures, has been a tactic of many of the groups. Many

states have green power or Interfaith Power and Light (IPL) state organizations that often are an outgrowth of the ICCN campaigns. The largest, not surprisingly, is in California, where the movement started and where there is a heightened awareness about energy costs and consumption. Maine's interfaith climate change group worked to help found the Maine Green Power Connection. In 2003 the group had one thousand customers, including a pledge from the Maine state government to commit to buying half of its energy from renewable sources. The Regeneration Project, the group that coordinates various Interfaith Power campaigns, lists twenty-three states with active campaigns or organizations.

Other aspects of the NRPE global warming campaigns give a sense of the diversity of strategies. For example, each state group is encouraged to issue an energy charter for the state, and, more recently, to encourage congregations to host viewings of the films *The Great Warming* or *An Inconvenient Truth*. Several letters or statements have been circulated to gain endorsement from a wide array of religious leaders. A letter as part of the "Let There Be Light" campaign to President Bush from forty-one senior religious leaders read in part: "At stake are: the future of God's creation on earth; the nature and durability of our economy; our public health and public lands; the environment and quality of life we bequest our children and grandchildren. We are being called to consider national purpose not just policy."[16] In another letter on energy conservation signed by 1,200 religious leaders, a commitment was made to send out educational materials to 100,000 congregations.

The EEN's 2003 "What Would Jesus Drive?" (WWJD) campaign garnered a great deal of national and international publicity.[17] It depended upon a New Testament "love your neighbor," personal-ethics approach and stressed air-quality issues as part of a larger evangelical focus on health. Jim Ball, the head of EEN, toured the country in his hybrid car.[18] This campaign was connected with the larger NCC-EJWG/COEJL effort to pressure U.S. automakers mentioned earlier.[19] The WWJD campaign obviously could not be an NRPE-wide campaign because of its exclusively Christian focus, and Jewish groups responded with campaigns (or joked in discussions) based on the messages "What Should the Rabbi Drive?" and "What Would Moses drive?"

More recently, the EEN has been working to get key evangelical leaders to issue a statement on climate change.[20] Although the goal of a unanimous statement from the National Association of Evangelicals failed, it is far more significant that the February 2006 "Climate Change: An Evangelical Call to Action" statement was signed by eighty-seven key evangelical leaders, ranging from the popular author Reverend Rick Warren,[21] to the heads of many important evangelical colleges and seminaries, and to several of the staff at *Christianity Today*.[22]

Operating outside of the NRPE-related efforts, Religious Witness is another global warming activist group that was initially founded to protest drilling in Alaska's Arctic National Wildlife Refuge (ANWR), whose tactics included various forms of nonviolent resistance as well as intentional arrest. On November 12, 2003, they also hosted an event marking the fifth anniversary of the Kyoto Protocol outside the United Nations Building in New York City, at which 150 interfaith "witnesses" made personal and educational visits to various UN member offices, from Russia and the United States to many of the small island states who will be the most affected by rising water. They then gathered together on a street corner for a stirring service and sermon from an African American Unitarian-Universalist minister.[23]

Many denominations had separate campaigns; for example, the Presbyterian Church of the United States (PCUSA) had an eye-catching handout that was a cut-out picture of burnt toast that vividly brought home the point of "global scorching."[24] As denominations have faced staff cuts, the importance of somewhat local or regional special-purpose groups working on the issue, such as EarthMinistry, GreenFaith, or Eco-justice Ministries, has grown.[25] Although not an official ICCN or IPL state group, the interfaith group in New Jersey, GreenFaith, has taken a similar approach.[26] GreenFaith (then known as Partners for Environmental Quality) first worked to convince congregations to sign up for Green Mountain Energy. In return for either the congregation or individual congregational member signing up, Green Mountain donated $25 to the congregation, in the hope that it would be used toward energy-efficiency measures. Part of the sales pitch was that the cost of green electricity was no more than the cost of a single large pizza per month (a very familiar price unit in a state that abounds with mom-and-pop pizza restaurants), but even that slight cost increase was a hard sell in most congregations. Now GreenFaith is promoting the purchase of wind energy

from a wind farm off the New Jersey coast and purchasing green electricity through New Jersey's own clean-energy program. More successfully,
with the aid of state grants, GreenFaith has also worked to provide energy audits and contractors to encourage religious congregations to take
energy-conservation measures, and to encourage religious institutions to
commit to solar energy installations.[27] Genesis Farm, associated with Sister Miriam MacGillis and Thomas Berry, was one of the first beneficiaries
of this program. For this work, GreenFaith received an Energy Star
Award from the United States Environmental Protection Agency (EPA).

This somewhat cursory discussion does not give a full picture of all the
activity that has been stimulated by religious global warming campaigns.
Across the country, youth groups raise money by selling compact fluorescent light bulbs, and congregations work to reduce their energy consumption, employ green building design, and green nuns overhaul their
properties and build energy-efficient demonstration buildings (including
straw-bale housing). Across the country, over 4,000 religious congregations hosted screenings of the movie *An Inconvenient Truth*.

Consensus and Cooperation

It is worth noting that global warming was the topic that the four members of the NRPE, which formed in 1993, could agree upon as a top
priority.[28] This alone makes it an interesting case study in the development of the religious-environmental movement. Why such agreement
on this topic? First, because it was so all-encompassing that each tradition
had room to approach it in a variety of ways and still be faithful to their
own tenets. Second, there was obviously a biblical basis in the *scriptural*
authority of the Hebrew Scriptures/Christian Old Testament. The short
video produced for the initial campaign, "God's Creation and Global
Warming," starts by quoting Genesis 1: "In the beginning . . . God said
that it was good." By linking global warming to the very beginning of
the Scriptures shared by all constituent groups and with the divine assertion of the concepts of the integrity, sacredness, and care of all creation,
each group claimed a clear and unambiguous authority, a biblical mandate, to speak on the issue. Third, the potential effects of global warming
invoked a shared ethic of justice, of protecting the "least of those" who
will be the most affected—the poor, the more-than-human world, and
the future generations who will "inherit the earth." Viewed thus, the

amelioration of global warming is a clear issue of ecojustice. Fourth, the importance of the topic rested on an unusually strong scientific consensus that allowed each group to appeal to *scientific* authority while at the same time turning a scientific issue into a moral issue, as seen in the Colorado Interfaith Climate Change Campaign statement by religious leaders: "We call on the religious community to use its voice and actions to undergird the scientific consensus with a moral consensus." Finally, strategies of action such as energy conservation and the development of alternative sources of energy related to countering global warming could be "sold" in economic terms (as the activists of various action groups have commented), thus "making sense" to actors whose everyday thinking is premised upon the dominant economic paradigm. In other words, the topic could be conveyed with scriptural, scientific, and moral authority, and as we shall see, both challenge and still appeal to aspects of the dominant cultural economic ethos. It was a topic that could be constructed in various ways, supported by multiple claims to authority, with a variety of strategies of action that would be "acceptable" to diverse constituents.

It is not hard to imagine the range of issues that the four groups composing the NRPE might not agree upon—anything related to population being first and foremost. But one can also imagine other environmental issues upon which there would be agreement, such as endangered species.[29] It is the human *justice* implications of global warming that clearly stand out as the predominant framing of the issue for religious activists, in part because it is seen as the strongest response to the economic framing of this issue by corporate-political elites.[30] Bill McKibben, a high-profile environmental activist and author of *The End of Nature*, exemplifies the merging of religious and environmental concern on this issue.[31] (In 1999, when he spoke at the "Greening the Church" conference at Drew University,[32] McKibben was just starting to become a *religious* prophet for the environment, in addition to being a secular one.) He now speaks commandingly from the biblical prophetic tradition about the justice issues inherent in global warming, as in this sermon entitled "The Comforting Whirlwind" included in an Interfaith Study Guide to Global Warming prepared by the Seattle-based grassroots group Earth Ministry for the NRPE campaigns:

I am unwilling to pussyfoot around (global climate change) much longer. These things are happening in large measure because of us. We in this country create 25 percent of the world's carbon dioxide. It is the affluent lifestyles that we lead that overwhelmingly contribute to this problem. And to call it a problem is to understate what it really is: it is a crime. A crime against the poorest and most marginalized people on this planet. We've never figured out, though God knows we've tried, a more effective way to destroy their lives.[33]

The video presentation "God's Creation and Global Warming" mentioned earlier similarly drives home the unjust impact of global warming, giving the example of Pacific Islanders who will lose not just their land and homes, but their very existence as independent nations and peoples. One more example can be found in a statement on energy conservation signed by 1,200 religious leaders based on the *"long-standing principles of faith and values concerning all of creation—stewardship, covenant, justice, prudence, solidarity, and intergenerational equity"* as precepts for action.[34] All of these principles relate to biblical notions of justice, with the exception of stewardship of creation, which some activists consider to be an "Eleventh Commandment" that stands on its own.[35] And in most theological interpretations, Christian stewardship has a justice component.

The consensus of the scientific community is also a key grounding for religious action on the issue. Scientists have repeatedly called for a religious response on the issue, most notably in the 1991 Joint Appeal in Religion and Science that led to the formation of the NRPE, which grew out of a letter from thirty-four prominent scientists calling for religious involvement, stating that "efforts to safeguard and cherish the environment need to be infused with a vision of the sacred." Evangelical environmentalists in particular spend time strategizing and coordinating efforts to counter the "junk science" so easily accepted within conservative Christianity.

Yet economic framing often is just as much a key to the success of various "pro–global warming" campaigns, even though the concept of ecojustice strongly critiques the inequalities of the current economic system. The Interfaith Coalition on Energy in Philadelphia has "promoted" energy audits to inner-city churches precisely by appealing to economics in their handout: "reduce operating costs to create money for social programs." An article in *Church Business* entitled "Conserving Energy on Your Campus: Doing What You Can to Save the Earth's—and Your Church's—Resources Is an Issue of Ethics, Stewardship" also illustrates this point:

> "Energy costs represent a substantial proportion of most churches' budgets," Higgins-Freese (director of the EPA's Energy Star congregations program) explains. "If just 2,000 churches made basic conservation improvements, the savings would amount to about $2.4 million a year." Related statistics estimate that if the nation's 307,000 congregations shaved energy use by 25 percent, it would free up nearly $500 million to spend on other priorities. In terms of *environmental* stewardship, this translates to a surplus of about 13.5 trillion kilowatt hours of electricity—with no new electric plants—and the prevention of more than 5 million tons of carbon monoxide released into the atmosphere. This, in turn, is the equivalent of removing about 1 million cars from the road or planting 1.4 million acres of trees.[36]

In other words, although groups would like to convince people on purely ethical or scientific terms, it is often much easier to convince people to take the right and moral action when it makes good economic sense. Additionally, as in the case with the efforts of GreenFaith in New Jersey, the fact that congregations themselves act as model consumers—taking the perceived economic "risks" with new technology, such as solar power, compact fluorescent light bulbs, and LED lighting, and subsequently reaping the financial rewards—can help encourage individuals who might not consider such actions on their own. Several of the New Jersey congregations and synagogues with solar installations inspired members to use solar power for their houses or businesses.

Thus religious "pro–global warming" activists are able to offer a view of a religiously grounded economic ethics of sacrifice and asceticism with a consumer's dream of cost saving, and thus counter another pervasive religious paradigm, which posits that health and wealth are outward manifestations of religious faithfulness but negates concerns for the common good.[37] By framing a highly scientific issue with large-scale economic implications in terms of religious understandings, "pro–global warming" activists can attempt to combat a counterreligious message that dismisses global warming or sees it as a threat to the trinity of U.S. nationalism, salvation theology, and a belief in the market as God, (to use Harvey Cox's locution) that is dominant in much of contemporary conservative Christianity.[38]

This fivefold grounding (theological room, scriptural base, central justice implications, scientific authority and an economics of cost-savings) provides a clear authoritative stance for religious groups, who are often embattled with their constituency over concern that in their linking of ecological concern and religious faith, they are becoming pagan, or are worshipping the creation and not the creator.[39] The justice framing helps move the discussion beyond these debates, as does the economic appeal of the solutions: save energy, save money. There are plenty of groups, however, ready to challenge the particular linkage of environmental and religious concern represented by the NRPE and other groups, and to do so in the name of a religious environmentalism of a very different stripe.

"ANTI–GLOBAL WARMING" ACTIVISM

There is a significant amount of religiously based "anti–global warming" activism in the United States. Perhaps the strength and energy of this movement can be taken as an indication of the success and perceived threat of the activist campaigns demanding action on global climate change, or as an indication of how much is perceived to be at stake. The usual factors in the conservative Christian dismissal of the need for Christians to respond to environmental concerns are certainly at work here: the central focus of Christianity should be on salvation and saving souls; the fear of worshipping creation and not the Creator; a concern that religious environmentalism is "New Age"; and an apocalyptic focus on the End Times that negates the need to worry about the current state of the planet because a new earth will replace it. But it is more illuminating to understand how "anti–global warming" activism fits into an even broader picture of a) the "secularization," or diminishment, of the authority of science over religion (related in part to creationism and the anti-evolution debate;[40] b) the religious right's appropriation of the postmodern critique of science as socially constructed; c) the perceived threat of religious environmentalism; and d) the connection between certain strands of conservative Christianity, individualism, concepts of freedom, and market ideology.

Science As Religion
The "secularization" of science can be seen in the following two quotes, both recent: one from *Christianity Today* and the other from the *Wall Street Journal*. I'll let you guess which is which for the moment.

Perhaps no result of the creation-evolution stalemate is as potentially disastrous as the way it has stymied courageous action on climate change. . . . All science is ultimately a matter of trust. The tools, methods, and mathematical skills scientists acquire over years of training are beyond the reach of the rest of us, even of scientists in different fields. Thanks to the creation-evolution debate, mistrust between scientists and conservative Christians runs deep.[41]

It was Michael Crichton who pointed out in his Commonwealth Club lecture some years ago that environmentalism had become the religion of Western elites. Indeed it has. Most notably, the burning of fossil fuels (a concomitant of economic growth and rising living standards) is the secular counterpart of man's Original Sin. If only we would repent and sin no more, mankind's actions could end the threat of further global warming. . . . Much has been made of the assertion, repeated regularly in the media, that "the science is settled," based upon a supposed "scientific consensus." Yet some years ago in the "Oregon Petition" between 17,000 and 18,000 signatories, almost all scientists, made manifest that the science was not settled, declaring: "There is no convincing scientific evidence that human release of carbon dioxide, methane, or other greenhouse gases is causing or will, in the foreseeable future, cause catastrophic heating of the Earth's atmosphere and disruption of the Earth's climate.

One of the biggest hurdles for any religious campaign in responding to the reality of global warming is the long-simmering tension between certain varieties of Christianity and science. Although reverberations of the creationism-versus-evolution debate play a role, as the first quote— from a 2005 editorial by Andy Crouch in *Christianity Today*—illustrates, it is increasingly apparent that many Christians beyond creationists choose which theories they want to "believe" in science (gravity yes, certain medical research yes, evolution no, global warming no).[42] In other words, just as believers in a secularized and multifaith world are aware that they choose to believe whatever Christian worldview that they hold,[43] they also feel they can similarly choose to believe whatever science most fits within their worldview. Crouch intimates this with his comment that "all science is ultimately a matter of trust."

The science behind global warming is thus portrayed as just another belief system, another . . . religion. This is certainly evident in the second quote—from a 2005 *Wall Street Journal* editorial by James Schlesinger, former Secretary of Defense and Director of the CIA under Richard Nixon and Secretary of Energy under Jimmy Carter, on the G-8 Summit in Scotland. The editorial referred to "the theology of global warming" and questioned the idea of scientific consensus on global warming by citing the 17,000–18,000 scientists who signed the so-called Oregon Petition, which states that "research data on climate change do not show that human use of hydrocarbons is harmful. To the contrary, there is good evidence that increased atmospheric carbon dioxide is environmentally helpful." This petition can be found on the Web site of the Oregon Institute of Science and Medicine,[44] which explains that "science and mathematics consist of certain truths that people have discovered about the world and universe which the Lord created—simple truths that are within the limited abilities of the human mind to comprehend"; the institute, it might here be noted, also offers a home-schooling curriculum that addresses the problem of science.[45]

The petition was started with a mass mailing to thousands of scientists that included what looked to be a "reprint" of a scientific paper entitled "Environmental Effects of Increased Atmospheric Carbon Dioxide," which in turn had been formatted to look as though it were from the official *Proceedings of the National Academy of Sciences* and was accompanied by a cover letter signed by "Frederick Seitz, Past President, National Academy of Sciences, U.S.A., President Emeritus, Rockefeller University." The National Academy of Sciences quickly denounced the paper, noting that it was not peer-reviewed, nor were any of its coauthors climate scientists. My own limited investigation and the investigation of others reveal that the 17,000 scientist signatories range from specialists in hydrology and mining to high-school biology teachers and pesticide-industry public-relations personnel to nonscientists, even nonpersons (Drs. Frank Burns, Benjamin Franklin Pierce, and B. J. Honeycutt of the television show *M.A.S.H.* are listed), but extremely few are recognized and respected climatologists.[46]

The "anti–global warming" movement is adept at drawing upon the notion that if something is religious it is not necessarily verifiably "true," nor does the believer need verifiable proof, while science demands verifiable truth and proof, which they claim is lacking. They thus seek to discredit global warming science by calling it a religion.[47] Schlesinger is not alone in his references to the "religion of global warming"; such comments are numerous in the writings of the political and religious right. It is not at all difficult to find references to global warming as religion, whether in the form of a headline describing it as the "new European Religion,"[48] or another about the chief executive officer of British Petroleum (BP) "getting religion" for deciding that the company needed to follow the precautionary principle and take into account the possibility of global warming,[49] or the widely publicized remark by MIT Professor Richard Lindzen made in a speech to the National Press Club: "Do you believe in global warming? That is a religious question."[50]

It is clear that the reference to global warming as a religion was a major "talking point" for conservatives. These talking points—in this case the linkage of global warming with religion, faith, and belief—become what feminist sociologist Dorothy Smith calls an "ideological code" that prompts the reader to a series of connected thoughts. As Smith comments, "An ideological code operates to structure text or talk. . . . Ideological codes don't appear directly; no one seems to be imposing anything on anybody else; people pick up an ideological code from reading, hearing, or watching, and replicate it in their own talk or writing."[51]

This move to construe science as a religion is part of the secularization, or loss of societal influence, of science. In other words, just as people pick and choose their religious beliefs in a secularized world where many religious beliefs compete, science, too, has been similarly relativized, as Steve Fuller comments, so that, like religion, "people continue to believe in science, but now also believe that they have a choice as to which science they believe."[52] This shift in viewing science can be seen in the conservative Christian emphasis on evolution and global warming as theory, or theology, just as the account of creation, or intelligent design, is a "theory." The implication is that theories are unproven, and thus are matters of belief. Of course, gravitational force is also a theory, but the religious and political right's insistence that global warming is a theory (thus needing more research) or a religion (and thus belief is a matter of choice), is meant to counter the significant scientific consensus that it is happening. This understanding of theory, related to the dismissal of evolution as a mere theory, is illustrated in the quote below made by H. Sterling Burnett in his July 27, 2005, commentary "Global Warming: Religion or Science?":

> The "theory" of global warming posits that human activities . . . are causing an increase in the amount of heat-trapping greenhouse gases in the atmosphere, enhancing the natural greenhouse effect. . . . I placed the word "theory" in quotes because I am reluctantly coming to the conclusion that the idea that humans are causing global warming is really more akin to a religious belief—a revealed truth about human sins (fossil fuel use) and their consequences (all manner of calamities)—rather than a testable scientific explanation.[53]

There are several aspects undergirding this view of science. Believers in a God who intervenes in history have long attacked what they term "atheistic" or "secular" science, for it challenges their beliefs in miracles, scriptural literalism, divine intervention, and divine providence. To many, evolutionary science clearly challenges these beliefs, but so does global warming, for it implies that humans have caused changes in the planet's systems of such magnitude that little can be done except to try to reverse the process and diminish its effects twenty to thirty years hence. To take global warming science seriously enough to change individual patterns and restrain the consumption of fossil fuel that currently drives economic expansion is to challenge the notion of a personal and omnipotent God who knows and has preordained the future (similar arguments, often grounded in belief in the apocalypse and the approaching End Times are made against other environmental concerns), who can perform miracles that defy the laws of nature, and who is held to have blessed the current economic order.

Further contributing to the ability of "anti–global warming" spokespeople to portray science as a belief system is their use of the postmodern critique of science's objectivity.[54] Advanced not just by scholars in various fields, but also by nuclear activists, environmentalists, feminists, and postcolonial critics, this critique takes on the inherent subjectivity and belief systems or paradigms that shape scientific research and the interpretation of its results.[55] Frequently, as Sandra Harding, and before her, Thomas Kuhn, have noted, scientists make assumptions that reflect the dominant cultural assumptions, such that research on only white men is taken to provide a definitive picture of universal, or "human," health.[56] Schlesinger draws upon this critique of science to undermine the authority of climate change science by citing the examples of Galileo and Copernicus to prove that the dominant scientific view is not always "right."[57]

Just as the postmodern critique of science as socially constructed is appropriated, the postmodern "spiritualizing" of ecology and cosmology by green thinkers such as Charlene Spretnak, Thomas Berry, and Brian Swimme (and perhaps some of us in this volume) is used to lend credence to the critique from the right that environmentalism is a religion, and thus all connected to it is suspected as being part and parcel of an alternative religion. This makes any religious endorsement of science all the more suspect and threatening as a competing belief system. The degree of this threat from religious environmentalism can be seen in a speech delivered by former U.S. Representative Helen Chenoweth (R-ID) to the U.S. House of Representatives on January 31, 1996, which attacked "environmental policies [that] are driven by a kind of emotional spiritualism that threatens the very foundation of our society." She added, "There is increasing evidence of a government-sponsored religion in America. This religion, a cloudy mixture of New Age mysticism, Native American folklore, and primitive earth worship (pantheism), is being promoted and enforced . . . in violation of our rights and freedoms."[58] More recently, in 2006 the Southern Baptist Convention passed a resolution stating that "some in our culture have completely rejected God the Father in favor of deifying 'Mother Earth'" and "have made environmentalism into a neo-pagan religion."[59] This view of environmentalism-as-religion is often accompanied by the portrayal, by those in the "wise-use" movement, of environmental groups as radical.[60] Phillip DeVous, former staffmember of the Acton Institute, which has funded both "anti–global warming"

publications and probusiness, pro-"wise-use" interfaith environmentalism, called the collaboration to try to prevent oil drilling in the Arctic
National Wildlife Refuge between the Sierra Club—a "radical group"—
and the National Council of Churches an "unholy alliance."[61] Acton was
also instrumental in the founding of the Interfaith Coalition on Environmental Stewardship (ICES; now the Interfaith Stewardship Alliance [ISA],
launched at the White House in spring 2006),[62] as well as the Cornwall
Declaration on Environmental Stewardship that states: "Some unfounded or undue concerns include fears of destructive man-made global
warming, overpopulation, and rampant species loss."[63] According to the
Web site, this statement was signed by over one thousand religious leaders, including James Dobson of Focus on the Family, Bill Bright of Campus Crusade for Christ, and Richard John Neuhaus.[64] These are many of
the same evangelicals who opposed the Evangelical Declaration on Climate Change issued in February 2006.[65]

The differences in the two declarations are instructive of perhaps the
most significant factor in the opposition to global warming: economics
and the support of free enterprise and capitalism.[66] The Cornwall Declaration, written in response to the Evangelical Declaration on the Care
of Creation[67] affirms private property and market economies, while the
Evangelical Declaration discusses "lifestyle choices that express humility,
forbearance, self-restraint, and frugality" and "godly, just, and sustainable
choices."[68] The Cornwall Declaration stresses the needs of humans over
nature, arguing that free-market forces can resolve environmental problems, and denounces the environmental movement for embracing faulty
science and a gloom-and-doom approach. In contrast, the Evangelical
Declaration encourages Christians to become ecologically aware caretak-
ers of creation. They disagree on the place and privileges of humans
relative to nature, issues of biblical interpretation, the definition of stewardship, God's sovereignty, and the concern that saving the earth is replacing the central Christian emphasis on saving souls. These
disagreements reflect more than internal Christian conflict; the Cornwall
Declaration reinforces the "wise-use" emphasis on the continuing improvement of the environment through human technology, the abundance of resources given by God for human use, the privileged place of
humans, and disagreement with seeing more-than-human nature as an
idyllic, harmonious state that must be preserved (the latter picks up on

the postmodern criticism of the social construction of nature, for which William Cronon received much grief for aiding and abetting the enemy).[69]

In general, it is the conflict of differing "belief systems" regarding the current economic system that arouses some of the strongest global warming-as-religion comments, as illustrated in this quote from Tom De Weese in *Capitalism* magazine (February 16, 2005).

Global warming is nothing more than a euphemism for redistribution of wealth from the rich, development nations to jealous dictatorships who refuse to allow their citizens the right to gain their own wealth through free markets. It's about political redistribution from strong, independent, sovereign nations into the hands of a power-hungry global elite cowering in the United Nations. The truth is there is no man-made global warming. There's only the scam of an empty global religion designed to condemn human progress and sucker the feeble-minded into worldwide human misery. I rest my case. Amen.

Or this, from the Canadian *Financial Post* (December 26, 1998), an editorial by Terrence Corcoran aimed at discrediting the Canadian Environment Minister, Christine Stewart:

In another statement quoted by the *Herald*, Ms. Stewart gave another reason for adopting the religion of global warming. "Climate change [provides] the greatest chance to bring about justice and equality in the world." Here she gets closer to the core motivation of some of the leading global warming activists. Where socialism's attempt at a global redistribution of wealth ended in economic catastrophe, global warming is being wheeled in as the next new economic crusade.

In general, despite significant diversity within the "anti–global warming" movement, which includes both avowedly secular and explicitly religious voices, participants in the movement see environmentalists as ultimately threatening the sacred values of free enterprise and globalized capitalism, and seek to undermine the claims of scientific authority as ground for any significant response by relegating it to a matter of belief.[70]

DOES RELIGIOUS FRAMING MAKE A DIFFERENCE?

Yes and no. It is clear that the religious framing and organizational activity on both sides of this issue have elicited many responses. As previously discussed, there have been well-organized "pro–global warming" campaigns by religious groups working at all levels to influence the president of the United States, corporate CEOs, institutional (congregational) and individual consumer behavior, and these efforts are producing some clear and moderate changes (of course, the same could be said of religious anti–global warming campaigns, which have received a warm welcome at the White House). There are also well-orchestrated campaigns specifically aimed at conservative Christians, on the one side, by linking the issue to Jesus' command to "love thy neighbor," and, on the other, to denounce the science of global warming by appealing both to conservative Christian distrust of science and trust in neoconservative economics.

As elsewhere in the religious landscape of the United States, special-purpose groups such as those discussed in the first section are the most prominent and successful because they can focus the efforts of people with similar concerns on a specific set of issues and cross societal and interfaith boundaries in ways that secular groups often cannot. This can be seen in the recent efforts of the EEN to get global warming on the agenda of the National Association of Evangelicals, and thereby attract the attention of both supporters and politicians of the Republican Party, while aligning evangelicals with the global warming stance articulated by other members of the NRPE.[71] Special-interest groups can draw a constituency across religious and political lines while often sidestepping other theological and ideological issues. This is clear in the interfaith efforts of both GreenFaith and the ICES/ISA, which cross religious boundaries in order to mobilize "pro" and "anti" religious voices on the issue. This ability to pull together a constituency from various locales within the religious sphere is a significant part of their success,[72] for the results of all my research on religious environmental activism indicates that most individuals feel isolated in their local contexts.

Second, these groups have the ability to use religious framing to pitch the issue in a different frame than do scientists, secular environmentalists, economists, and industry, and one different from that held by the U.S. population generally. As one Louisiana activist commented with regard to the strategy of holding contentious hearings about water issues in churches: "It just brought about an element of dignity to the discussions that aren't necessarily found in other avenues of public hearings."[73] So the ability of religious "pro–global warming" activist groups to use moral and theological language and religious contexts to articulate the issue within an already shared justice frame of reference is essential, as is the denunciation from the public pulpit of global warming as a religion for the "anti–global warming" forces. Implicit in "pro–global warming" activism is the need to change personal and corporate consumption of fossil fuels, and this can mean major lifestyle and economic changes, depending on the point from which one is starting. By framing scientific-economic issues in religious frames, and grounding action in faithfulness and morality on both sides of the issue, these groups are more successful with their constituency than the so-called doom-and-gloom scenario imputed to much environmentalism by its opponents (it is notable that antienvironmentalism might well be accused of a similarly dour outlook given its reliance upon threats of job losses and economic collapse to drum up support for its efforts against environmental regulation). Religious framing enables activists to counter the frightening message of global warming with messages of hope, faith, and concrete corporate action, or, contrarily, with messages legitimating denial or dismissal and reassurances that no response is needed.

Third, as sociologist Christian Smith has pointed out, religious groups and their constituent congregational and individual members act as movement midwives, providing already existent resources in terms of staff, buildings, and symbolic capital (as in the case of the civil rights movement), and, perhaps most importantly, contact lists, ready-made audiences, and the technical apparatus for mobilizing large audiences, as demonstrated in the various petition drives and "call Congress" campaigns, or in the recent success of Republican political campaigns.[74] In the Louisiana example cited above, it was clear to the activist that holding all of the hearings in religious settings was important to their success.

So why would I also say no, religious framing does not make a difference? First, it is clear that neither side of the issue is successful in convincing everyone. Second, however, is that framing the issue in economic terms may be just as vital to success. As already pointed out, "pro–global warming" groups "sell" conservation and alternative energy sources in terms of economic savings. Many religious congregations "buy" into energy conservation because of the economic cost savings and the satisfaction of being good "stewards" as much of financial resources as of being stewards of God's creation. GreenFaith's more recent campaigns offering solar energy installations, energy audits, and the technological means to help the congregation use less energy are all "marketed" to congregations in a "win-win" sense—that of making sense for the financial bottom line and being good for the earth. "Anti–global warming" forces are perhaps more up-front about the centrality of the appeal to the economic interests of their constituency. As Calvin Beisner comments, "Man was created to rule over the earth, not to be its slave, and any economic system that puts nature above humanity, as do some modern environmentalist movements, is sub-biblical."[75] In most U.S. official discussions of the issue of global warming,[76] the current market and economic system appear to be more important than any future threat to the planet's inhabitants. Ironically, it is the insurance companies, especially after the recent powerful hurricanes that have hit Florida and Louisiana repeatedly, that are taking global warming seriously as they see already the threat it poses to their bottom line.

Obviously, the answer concerning religious framing is mixed. There is much evidence that although the hope of campaigns to get religious groups involved is that a relatively easy change toward environmentally friendly behavior might open the way to stronger commitments, economics is still the bottom line for many churches. Energy conservation that saves money is fine. Churches, such as the Energy Star Award–winner Hebron Baptist Church in Dacula, Georgia, can be enthusiastic about energy conservation, and the related $170,000 annual cost savings, without buying into global warming. They are currently building a huge new sanctuary, and nothing in their Web site mentions any ministry, sermon, committee, and the like related to environmental issues. Given the profile of Southern Baptists, I am sure that at Hebron there are many members who agree with President George W. Bush and his current energy and global warming policies. It is clear that for many the economic frame is far more persuasive than any moral or religious ecoactivism. This indicates that while the smaller message of energy conservation is heard, especially as natural gas and oil prices rise, the larger message concerning the need for significant worldview and lifestyle changes that include the current value system of economics has not been as widely accepted as activists might hope.

The results of a survey conducted by Cassandra Carmichael, Rebecca Gould, and me of participants at the 2003 conference of the NCC-EJWG indicate similarly mixed results. Only roughly 33 percent, or 35 of the 108 respondents, listed global warming as a central concern. If you add in the number who listed energy as a major concern—and I think that combining the two issues is justified since the consumption of fossilized fuels to produce energy is central to the problem of global warming—the total is only 45 percent. While this is hardly a significant representative sample, however, to even be at a NCC-EJWG conference, one has to already be fairly knowledgeable and involved, and the previous conference in 2001 was explicitly focused on global warming. Yet only 45 percent of survey respondents listed global warming as a central concern for themselves as individuals (the question did not ask for the most important issue, but was designed to solicit multiple-issue answers), and only 18 percent listed it as a concern of their congregation. Despite firm theological and moral grounding, significant campaigns, and economic incentives, the importance of religious action to combat global warming is still a hard sell, although foundations and environmental organizations have long recognized its importance. Religious framing of "anti–global warming," however, is seen as a good bargain, as major companies, such as Exxon-Mobil, have taken to funding groups like the Acton Institute.[77]

CONCLUSION: REVISITING SCIENCE VERSUS RELIGION

The fact that the United States continues to refuse to meet any global warming–related emissions-reduction goals, to sign onto the Kyoto Protocol, or to meet European standards for CO_2 emissions demonstrates that the global warming activists first profiled in this paper, and their secular counterparts, have a long road to success ahead of them. It also demonstrates that the arguments and framing made by global warming opponents have succeeded in challenging the authority of the scientific consensus in the minds of voters, and that they have allied themselves with politicians and economic interests that prefer not to address the issue, thus giving one more illustration of the daunting task of religious environmentalism. If the authority of the science about global warming can be challenged, it is hard to make other arguments based on science, or suggest strategies of response, and vice versa. On the contrary, one can also view the variety of efforts to dismiss global warming science as evidence that the messages that global warming is real and in need of amelioration are getting through, and that "anti–global warming" forces have been forced to scramble so as to forestall the United States from reaching the conclusions that a majority of the globe has already reached.

Global warming is happening, and actions must be taken now. By sacralizing science and nature, by grounding responses to scientific knowledge in religious values of justice, or by challenging scientific knowledge by reducing any response to a matter of religious belief, the two sides of the debate crystallize a long-standing divide in the United States. It seems tempting to understand the complex picture of religious responses to global warming as simply a battle over the authority and relationship of religion to science, and indeed, such an understanding does prove illuminating in part. James Proctor, in research sponsored by the National Science Foundation, discovered that there are strong correlations between trust in the authority of religion and in government, or trust in the authority of science and of nature.[78] His respondents linked these items in terms of such clusters, choosing one cluster over the other, with obvious ramifications for the issue of global warming. In this scenario, trust in science is pitted against trust in the government and in religion, and the heightened focus on perceived threats of terrorism has reinforced a general sense of the need to trust (a trust seemingly misplaced) the executive branch of the government, no matter how much their actions are secretive or untrusting of U.S. citizens. By rallying certain sectors of conservative Christianity in support of the government's stance on global warming and a wide range of unrelated issues, it becomes a matter of "who do you trust?" By claiming that global warming science is a religion, anti-global-warming activists further undermine trust in the authority of science.

Bruno Latour sheds an interesting light on this phenomenon in his article "Thou Shalt Not Freeze-Frame," which challenges the very way we view science and religion:

By ignoring the flowing character of science and religion we have turned the question of their relations into an opposition between "knowledge" and "belief," opposition that we then deem necessary either to overcome, to politely resolve, or to widen violently. What I have argued . . . is very different: belief is a caricature of religion exactly as knowledge is a caricature of science. Belief is patterned after a false idea of science, as if it was possible to raise the question "Do you believe in God?" along the same pattern as "Do you believe in global warming?"[79]

This observation is part of his larger argument about the false juxtaposition of the claims of religion versus those of science, for we are accustomed to viewing science as about "the visible, the direct, the immediate, the tangible," and religion as about the invisible. First, he argues that in most ways quite the contrary is true. Religion is about experience and the mundane, and science is about that which is far away and accessible only through mediated chains of experiments, theories, and so forth. This in itself explains some of the reaction to global warming science—for global warming is the hard-to-see (melting ice caps, however, is making it more visible). Yet Latour also points out that both science and religion are about the immediate, the visible, and the far-away in that they are chains of mediations, and that we make the mistake of "freeze-framing" the flowing character of each.

Latour's comments also point to, in my analysis, what Proctor's research leaves out: trust in the authority of government is bound up with trust in the authority of the market. By making it seem to be an issue of science versus religion, "anti–global warming" activists have succeeded in diverting attention from perhaps the real "faith" at stake: that in the economic system. As Harvey Cox points out in his well-known essay "The Market as God," it is the workings of the market that cannot be challenged or managed.[80] Replacing the maxim of the separation of church and state (so that the state supposedly cannot interfere with the practice of religion), already in flux and turmoil in U.S. society, is that of the separation of market and state, in which any state should not negatively interfere with the practice of global free-market capitalism. Anything other than voluntary individual responses to global warming is seen to threaten the unimpeded progress of the market, to disrupt the "invisible" laws that current economic theory posits make the market function properly and thus correct itself in the face of all problems. Neither science, nor science linked with religion, should be allowed to challenge the dominant economic model, which does not allow for externalities like natural limits.

As has been noted, a more chastened reliance on the workings of the market can be seen in the religious campaigns to bring attention to global warming, for it is the appeal to economics that sometimes seems just as strong as the appeal to justice. In both instances, this hidden "faith" is not always the first line of argument: we are far too used to assuming the inevitability of capitalism as the economic system, while arguing with each other over the place of science in religion.[81]

As stated in the beginning of this essay, my interest in this topic is directed toward the success of religiously motivated environmentalism in responding to global warming. In this I have stepped away from the supposed "neutrality" of social scientists. I have tried to give both a sense of the breadth of religious "pro–global warming" activism, and, by profiling "anti–global warming" activism, to show the many challenges the former faces. The movement is very dependent on key figures/activists/leaders, many of whom "burn out." The opposition is well organized, well funded, and has easy access to the halls of government. Religious environmental activists are struggling with the question of how to sell the message, as the examples in this essay have shown, and are caught between their critique of consumerism and an appeal to the consumer value of saving money that motivates much consumer behavior (as, for example, at WalMart). They are also subject to larger cultural, religious, political, and economic forces that constrain their efforts and shape the interpretation of scientific findings and challenge the authority of science as a basis for individual or societal action. The question of belief in global warming would seem to point up another pertinent fact: that for most U.S. citizens, the science of global warming, and much science generally, is inaccessible in its complexity, and thus perhaps more easily dismissed.

Religious "pro–global warming" activists, like religious environmental activists in general, must struggle with the tension between a focus on "guilt- and despair-inducing" topics or on "greening theology" and spirituality (as Fletcher Harper's chapter on GreenFaith in the present volume nicely delineates), so that many prefer to respond by developing a greener spirituality, rather than by focusing too much on changing environmental and economic behaviors, in the hope that a changed worldview will help motivate the needed change. This is a struggle for all of us who have contributed to this volume: where to put our efforts for the most effect. Finally, the success of the movement may be limited if the selling point for many remains whether action makes good economic sense, not good ecosystem sense (although with high petroleum and natural gas prices, and a few more powerful hurricanes, it may not be such a hard sell). Not all ecologically friendly behavior is an easy sell, and responding to global warming will have economic costs (as will not responding, and those are perhaps greater, if not incalcuble).[82] In that sense, Anne Swidler's analysis in her classic article "Culture In Action" is very pertinent.[83] In unsettled times, she argues, the gap between action and ideas, such that people do one thing and say another, becomes very clear. In these times, as I would argue we are now in, new strategies of action and ideology are being worked out, yet the familiar pathways—or "mazeways," as anthropologist Anthony Wallace terms them—still feel most comfortable.[84] The early successes of the "anti–global warming" movement clearly illustrate this: by naming the threat not as global warming, but global economic restructuring, it creates a climate of fear concerning the actions and ideology that we are most familiar with, which hold that the growth and progress of the U.S. economy is "essential" for the "American way of life." Although there may be changing patterns of energy consumption for reasons of economic savings or for patriotic resistance to being dependent on foreign oil (for example, as the motivating reason for buying a hybrid car), these do not mean that one sees the interconnectedness of human action with the planetary system. So it makes sense that many old pathways, in this instance a certain dominant form of economic reasoning and consumer choice, are hard to change. It is always difficult to directly connect a different set of desired ideas and actions while still trying to ultimately challenge the worldview and social order of which those pathways are a part.

So, in the final analysis, "pro–global warming" religious ecological activism has to be about both theology and action, changing beliefs and worldviews, and patterning action that fits those changed beliefs. What is also clear is that neither one comes before the other; that is, action does not necessarily lead to changed worldviews, nor do changing one's beliefs necessarily lead to the desired action. This is a good Weberian conclusion; one must do both, for the two, action and belief, are in a dialectical relation, and what religious activists are looking for is the religious ecological ethos, or "switchman," as Max Weber termed it, to help us change tracks away from the dominant economic ethos. Yet Weber was not optimistic from his viewpoint in the first decade of the twentieth century concerning "that mighty cosmos of the modern economic order (which is bound to the technical and economic conditions of mechanical and machine production). "Today," he prophetically pronounced, "this mighty cosmos determines, with overwhelming coercion, the style of life *not only* of those directly involved in business but of every individual who is born into this mechanism, and may well continue to do so until the day that the last ton of fossil fuel has been consumed."[85]

Although Weber's pessimism may be well warranted, we need not stop there. What is promising is that global warming activists realize that it is the economic system that is the problem, and that solutions are not just about asserting scientific certitude or inspiring religious action. Rather, changes must come from deliberately linking a changed worldview with ethical economic practice that challenges the current belief in, and workings of, the market and that promotes a more equitable and just cohabitation of the planet in order to ensure a future for all of creation.

Ecospirituality and the Blurred Boundaries of Humans, Animals, and Machines

GLEN A. MAZIS

EMBODIMENT AS SELF/OTHER

When the call to an "ecospirituality" has been raised in the recent past in the West, it has usually been as a result of a critique of the overindustrialization of the landscape and its attendant damage to the earth. Ecospirituality has been associated with augmenting and deepening the appreciation of the human relationship to other creatures in the surround—or at least other features of the biosphere, such as water, wind, soil, and rock. The emerging ecological conscience of the wrongs resulting from self-enclosure in the project of "human progress" has been a spur to a transforming sense of the spiritual and religious, so that humanity may develop an awe and compassion for the rest of the planet. Like the Buddhist insistence on the human interdependence with all beings in the surround and the correlative sense that "Buddha-nature" is in all beings, Western religious and spiritual traditions have recently turned to the surround to expand the sense of divinity from its overly human center. Also, similarly to the Buddhist and Taoist ancient traditions of pointing to kinship with tigers and insects, trees and rivers, the biosphere has been the site of this expanded sense of sacrality. To note two famous examples: Aldo Leopold's articulation of a "land ethic" in his widely read *The Sand County Almanac*, in which he declared, "The land ethic simply enlarges the boundaries of the community to include soils, waters, plants, and animals, or collectively: the land";[1] or Theodore Rozak's *The Call of the Earth*, in which, after articulating an "ecopsychology" of humanity enmeshed in the planet, Rozak states we need a new sense of divinity that brings together the human and "the deep systems of nature" in a synthesis of "religious thought and natural philosophy."[2] There is an explosion in current decades of this sort of literature. Although the critical call for limiting toxic effects of the now-biotechnological and still-industrial overgrowth is imperative for the future health of all the creatures of the planet, humans included, to exclude dimensions of the built environment from a sense of divinity or sacrality is equally destructive to the planet and to the eventual meaningfulness of articulated spiritual paths. Including the biosphere in spiritual community with humanity is insufficient without also including the ever-expanding and transforming realm of machines. They are too large a part of our landscape and reality to exclude from a deeper relation of meaning and respect. Spirituality at its most vital should be affirming of the widest possible sacrality of experience, and so should not exclude vital parts of existence.

In particular, the machine, and particularly its leading-edge "information machines" or "artificially intelligent" machines, seems to be cast in the same demonic role as that which the creatureliness of the "natural world" had assumed for centuries, embodying the power of "antispirit," from which it was humanity's task to draw a clear boundary and extirpate from within our very being. The new machine seems to have become the postmodern animal. The machine currently represents an external threat to humanity in a way analogous to how "the wilderness" was seen as threatening to civilization as both its basis and yet hostile "outside." Similarly, as the animality within us was seen as a force to overcome and not to embrace, so now the mechanism inherent in humanity's biology or even mental capacity is seen as a submerged force within us that, if allowed to become unduly augmented through human practice, could destroy human uniqueness. The perceived threat of animality at the heart of our biology and psyche in previous centuries was a central justification for the increasing control offered by rational human "progress." These animal forces were to be "civilized out of us."

Currently, many acknowledge a spiritual kinship with animals and nature, recognizing, for example, that animals recognize their mortality, or sacrifice themselves for the good of fellow animals, or have a sense of beauty, and even of wonder, as articulated in books, such as Gary Kowalski's popular *Soul of Animals*.[3] In detailing this kinship, there is also a proliferation of literature on animal consciousness that details the many ways in which we overlap with them in capacities we had thought uniquely human, not least language and abstract thought. Paradoxically, now that this kinship is being acknowledged and even celebrated, the dimension of our being that had been associated with the irrational "natural world" and perceived as a threat to our rational self-legislation is now seen as that which is uniquely human. Currently, the cry is heard that our unique human emotions and sensibility separate us from other beings, especially machines. In an age in which science and other enterprises can only advance by computers performing calculations, engineering possible schemata, or projecting outcomes—all using logical, mathematical, or other rational abstract principles, we can no longer claim reason, logic, and intellectual capacity as solely ours, what elevates humanity above the rest of creation. Now, we are afraid "progress" will go too far in reducing the world to a merely mechanical realm where the visceral is lost. A host of recent movies show popular culture's concern with this imagined threat, from the clumsy adaptation of Isaac Asimov's *I Robot* to the immensely popular *Matrix* series (both of which portray machines attempting to take control of the planet) to *Equilibrium* (where experiencing human emotion is a crime punishable by death) to *Gattaca* (where passion has been genetically bred out of humans, who have become "perfect").

To resort to an embrace of the overlap with animals, creatures, and the biosphere while maintaining a hostility against the developing capacities of machines is a new kind of romanticism that seeks to keep intact the culture-nature divide. Currently, however, certain parts of cultural production are seen to further some aspects of teleology inherent in nature while yet remaining "natural" despite their technological overlay, while other advances that call for a shift in how humans define themselves are labeled "artificial" and thought to violate some prior sense of "nature" and humanity. It is no longer thought unusual to grant that animals may have souls, be kin, and deserving of inclusion in the divinity of the planet as direct participants in a larger community; machines, however, are still widely considered infernal. This dynamic has emerged into political discourse and the "culture wars:" one side can claim that the use of technology to keep a catastrophically ill person alive is "natural"—in keeping with the sacrality of a creation that has natural rhythms humans may not disturb; the other side can claim that the refusal to use any technological intervention for the terminally ill allows for human self-determination and permits a dignity beyond nature, beyond the Darwinian impulse for sheer survival. Obviously, at this point the terms used to consider humanity's relationship to animals and "the natural world," as well as to machines and the built environment, are often ambiguous to the point of being unhelpful. These divisions between human and machine are asserted at the same moment in which more and more of our physiology, brain chemistry, sensory apparatus, neurological development, and genetic makeup are understood in certain senses to be machines, though the acknowledgment is not thought to compromise our humanity. Our relationship with animals is far warmer than in centuries past, when the animal in us and around us was often reviled and violated, but this unease with the animal and abuse of myriad animal lives in objectifying and destructive ways is far from past. This essay will assert that to make a simple division between humans and machines in regard to their sacrality is unwarranted, and overlooks new dimensions in machines' evolution—that it misunderstands dimensions of our relationship to what we build, and fails to fathom that the overlap among human, animal, and machine has a spiritual significance potentially as expansive and liberating as has been the earlier inclusion of animals and the biosphere in sacrality. I will also contend that discussion of any one of these three modes of being requires the inclusion of the other two in order to facilitate an understanding of their mutual participation in a spirituality that embraces the depths of meaning in materiality.

In meditating upon the strange relation among these three realms and the paradoxes this enmeshment raises for seeing the divinity of the human within the field of the planet, I would like to offer as a concrete beginning for meditation a poem I wrote about six years ago when starting to puzzle over the interweaving of human, animal, and machine. At the time, I was starting to address how these relationships might mark the beginning for a new spirituality responsive to Nietzsche's (too often misunderstood) call for one that "remained faithful to the earth." The poem is an imaginative reconstruction of an event—actually, course of events—that had bothered me for more than a decade and a half. In 1982, a University of Utah surgical team replaced the diseased heart of 61-year-old dentist Barney Clark with a device called the Jarvik-7. Air cables ran from a washing-machine-sized battery and ventilator source that powered the heart. The pneumatic pump operated the mechanical heart and sustained Clark for 112 days. In response to pictures disseminated through the media of the patient enclosed in tubes, pipes, and machinery there was a considerable outcry that this was a horrible violation of humanity. A second patient, William Schroder, soon after also received the Jarvik-7 and lived for 620 days, but for most of that time suffered infections and four strokes as a result of blood clots; he had to be tube-fed for a year.

My poem imagines the thoughts and confusions of doctors, scientists, and the patients, as well as what these events might have meant for all of us observing what was difficult to fathom at the time. The medical team seemed intent on extending the quantity of human life, but to critics like the *New York Times*, whose editorial page called the device "the Dracula of medical technology," there seemed to be scant understanding of the quality of human life. Though little may have been explicitly discussed then about the meaningfulness of embodiment in relation to the surround, my poem hints at implicit understandings that may have been overlooked. Doctors and biologists, as well as philosophers, have come to understand both humans and machines in differing ways in the past two decades, and this is significant for considering the spiritual overlap of these realms. The old-fashioned, simplistic sense of the machine as a self-contained unit designed to achieve a certain physical task may have plagued this particular biotechnical breakthrough:

The Famous Artificial Heart Experiment

There were no boundaries for him
once they removed his heart.
The bedside lamp broiled his liver
with its scalding light, the chill
draft running across the hospital floor
went straight to his tongue—
it froze in mid-sentence waggle—
and the nurse's hips
crushed his squeaking lobes,
surrounding them with their fleshy elegance.

Tied by throbbing hoses
to the clanking Jarvik Seven,
he was propelled relentlessly into space,
pushed assertively further
since his blood no longer
danced with the moon,
but was driven by metallic fanaticism.
The beats hammered at his mind
until holes opened
and he leaked away.

The doctors had feared
his senses would become too mechanical
paced by a cast iron pump,
after the pulpy center was removed.
How his thoughts got dispersed
was a puzzle, since clever with gears
and tubes, they never felt how we think
from the body's rhythmic kinship
with all that beats around it,
in time with clouds scudding skyward,
the bobbing of flowers on their stalks,
and even the ebb and flow of traffic:
the way the beating of the world
beats out its time in our hearts,
and so in our heads.

There is an ancient ability
of cells to listen
while pumping, to receive
in the midst of pushing,
that allows the world
to herd our feelings,
like wayward cattle,
into the corral of who we are.[4]

The 120 days Barney Clark lived tied to this throbbing prototype heart machine were days with hoses coming in and out of his body, tethered to what seemed to many a menacing presence. It was a very early attempt to physically create the fusion of mechanical and biological parts in a hybrid cyborg being. The Jarvik-7 was initially intended to be a lifelong fusion with the patient. However, as we will see, in attempting to artificially combine the realms of humanity with machinery, what was not obvious at first are the ways in which the realms work and could work together more fruitfully in augmenting meaning in existence. To have human and machine cooperate to deepen the meaning of lives would seem to be at the heart of spiritual life. This concern has since emerged from the initial focus on the mere physical assistance of machines.

The poem focuses on the way in which embodiment is not merely a matter of functioning physiological organs. The openness of embodiment is also about a sense of one's life, a fluidity of boundary, in which the human takes into its heart the sense of the world and people around it, and also takes in the sense of the surround as it would appear as though perceived through the things in the surround. The body does this through a rhythmic enmeshment with its "around world," the way in which embodying is an ongoing dynamic process of returning to oneself through the interconnections within the surround. This is to abandon the understanding of the materiality of human existence as atomistic, opaque, or set into a space of separation. Rather, the materiality of human existence is interstitial in being materially enjambed in a space of relationship with all sorts of other beings of differing qualities, which in turn betokens myriad sources of meaning. The scientists in the poem are surprised that Mr. Clark has lost the sense of where he ends and the world begins, because they imagined that tinkering with the biomechanics of pumping blood could only affect other biomechanical phenomena and not transform the patient's sense of kinship with the world. If there were to be any adverse effect, they thought (even as they doubted it), it would be a mechanization of the man's sense of his experience, since his blood would now be driven "mechanically." They did not reckon upon the "give and take" of the patient and his world in which humans are partially also the sky, the flowers waving on their stems, and even the rhythm of the traffic going by. The body's perceptions are not entertained on an input platform distinct or at a distance from the perceiver; rather, as Maurice Merleau-Ponty put it, "to look at an object is to plunge oneself into it."⁵ That means that as embodied perceivers, we are "in" what we perceive in some way. We even have a "reverse" sense of how the world would be seen from the vantage point of what we enter in perception—how these things would look back at us—since "to look at an object is to inhabit it."⁶ When we concentrate on a stone, we live the world through the stone for the moment. Thus stately rocks are placed in a Zen garden to engender the unmoving, serene, stable sense they seem to lend to us when we gaze upon them. Within that gaze, if we open ourselves to its power through concentration, the stillness of the rock becomes part of who we are at that moment. It is this cooperation of human and world, this enmeshment of embodiment and surround, that spirals back and forth and is overlooked by the doctors as imagined in the poem.

Yet insofar as human biological structures are mechanisms, the doctors may be said to know what they are doing. There is no meaning without the mechanical interaction within the human and with the surround. It would be quite plausible for them to believe that if they could restore the functioning of the organism through mechanical intervention, the dimension of sense and the quality of life might take care of themselves in a realm that is beyond their expertise. Yet the poem muses within a space of interconnection, where the mechanical gains significance from the human demand for meaning, while the sense of experience is insepa-rable from the mechanistic functioning at its base. The doctors are bewil-dered that their patient's sense of being both one and coming to himself from the many is dispersed, gone—lost in the cause-and-effect march of events. They have not yet come to the idea that a sense of each person comes to them in "reverberation," to use a Bachelardian term, in the resonance with the surround, the way the outside spirals around into the inside like a Möbius strip. These insights have occurred to many scientists working at the boundary of humans and machines in the decades since this surgery, which I'll discuss below.

The poem speaks of how the biological, as taken up in attunement over the course of a lifetime, is part of a context of material interrelations that has been ripped asunder by the abrupt insertion of the heart ma-chine. The nurse's sensual presence is not now a co-presence of the pa-tient's own body, the way others move within us while still being other, but instead overcomes him, as does the chill or the heat. The patient as imagined in the poem has become overwhelmed by the lack of boundaries, since the material interrelatedness of ongoing embodiment is that thickness of flow that is equally about the overlap and extension of being among beings, and also about distance or boundedness within the same rhythm. Again following Merleau-Ponty, the world is something our bodies are woven into over time. This means open embodiment is also at the same stroke a distance of communication across boundaries that have gradually been criss-crossed through interaction, rather than achieved by a sudden fusion of human, machine, and world, which has cost this patient his sense of self in relation to others and the world.

The poem refers to the rhythmic kinship with objects that are to be identified as part of the biosphere, but it also points to the reverberations with the mechanical, the technological, and the cultural. As Donna Haraway says in *The Companion Species Manifesto*, we can only speak of "naturecultures," in which we find ourselves in a terrain where "postcyborg, what counts as biological kind troubles previous categories of organism. The textual and the machinic are internal to the organic and vice versa in irreversible ways."[7] More important than cleaving to these obsolete distinctions for understanding the plight of Barney Clark is to recognize that his embodiment has been ripped away from its "world," its context, or the sense in which embodiment is only an ongoing dynamic process with those things to which it is related in its surround. Embodiment is the ongoing dynamic process of unfolding with those things to which it is sufficiently related. The poem suggests that the physical, rhythmic beating of the heart is the underside of the rhythmic beating of existence, with all that to which it is related in forging a sense, a meaning, and a sense of identity. These other beings are included in an "extendedness" of the body to differing degrees, given the centrality of the contribution of each at the moment. The isolated body could not beat out a life or an identity or take in and express meaningfulness; only a body in context, only an embodiment made up of "the flesh of the world" (Merleau-Ponty) or of the communal effort among beings, thrives and finds meaning in a larger circulation of material energies and sense. The poem's implication is that this "co-constitution" might be the case for other life-actions we take as "ours," such as thinking, communicating, caring, imagining, and in this meditation perhaps importantly, praying.

The point is similarly made by a Zen archery master to Eugen Herrigal when, after seven years of trying to hit the target, the master corrects him for thinking he finally made the correct shot. The master tells him that for once, he did *not* shoot the arrow, and that is why it was the correct shot. The Master explains, when a maximum of attunement, concentration, and ceremonial embrace of the surround, allowing a flow of energy and meaning, has been established, "It shoots."[8] I think that little phrase—"It shoots"—is good shorthand for an idea that seems complex for an individualistic, dualistic culture, as it conveys how each being is to be appreciated in its uniqueness, as other to all others, and yet as inextricably meshed in the moment of action and interaction, in the rush of creative expression to which the human can, with care, have access.

The kind of unity that is articulated in "It shoots" or "the flesh of the world" or "shepherding" of a sense of identity named in the poem through the interplay with all sufficiently related to it (again, Nietzsche and "perspectivism") is not one of cause and effect (external relations) or of subsumption under a more fundamental, monistic being, but rather a fluid identity emerging from the interplay with all that is in the surround to which it is related. One way of attempting to find the overlap between humans, animals, and machines is to be reductive and seek a common ground outside each of these distinctive dimensions. Traditionally, there has been a reduction of the distinctiveness of humans, animals, and machines, both from "above" and from "below." From both ends of the spirit-versus-matter binary, humans, animals, and machines have been seen as "actually" one sort of being in a way that makes it impossible to savor and appreciate what each realm contributes to existence and to each other. From "below," it has been declared that we are merely a concatenation of chemical, electrical, and mechanical processes of the very same sort that powers the most distant star to shine or makes the rain fall from the sky. Then, in some very real sense, being of the same stuff, there are no longer humans, animals, and machines as distinctive realms of being, for everything in the cosmos is reducible to a common denominator, a universally graspable foundation in the material universe. From "above," it is thought that all beings are just instances of a life force or a divine creation of goals and purposes that these beings strive to realize together, whether apparent or not. In this perspective, however, everything *is* just the divinity or this life force—instances of another sort of being. This first approach is the denial of a felt kinship in that word's deeper ontological connotations of compassion, wonder, and gratitude, all of which are key to spirituality; and the latter gives rise to a possible spirituality and attendant theology that leaves the particular and contingent behind for a spirit that no longer "remains faithful to the earth."

For those of us who seek an "ecotheology," however, one that takes seriously the "earth household"—recalling thereby the title of Gary Snyder's lovely collection of poems, as well as the etymology of "ecology"—we need concepts that give their due to both the material and the contingently unique meaning of each being. To have an ecospirituality that is most inclusive of this planetary community, every being on this planet must have a potential role to play in participating in the creation of that overarching meaning. That is not to deny these same beings can also be disruptive or become destructive forces, but to affirm their potential positive significance in the interrelatedness of material beings. To cite Catherine Keller's warning in the *Face of the Deep*, spirit can be taken as something that floats away from the planet and is seen as an immateriality at the price of losing its weightiness, which seeks to affirm the chaotic earth. Spirituality can envision a "more perfect place" or it can seek to affirm this planet. In seeking an ecospirituality, we seek to ground spirituality in the ongoing becoming of the household of the planet. To do this, Keller states that spirit must "resist its own disembodiment" and come down into "the chaos of materiality."[9] Without this sinking down into the earth and into all its beings, the relations of humans, animals, and machines become either just another issue of the day, irrelevant to the destiny of the spiritual. The relations among humans, animals, and machines may be seen as just a scientific and empirical matter, but ecospirituality views these earthly interweavings as the currently emerging site of self-discovery, revelation of otherness, and a key opportunity for the evolution of a spiritual sense of the larger community of the planet.

Given the need to abandon human-centeredness for a more open sense of community, to point to "human-like" behaviors, "human-like" feelings, and "human-like" capacities in animals (as many of us have articulated in the past), or to search for such similarities with the inanimate and machines, is a misguided way to bring these realms together in a sacrality. The idea is not that either animals or machines have to achieve "our level" to be recognized as joining humanity within a community.

To recognize otherness and yet be inclusive is to set up a dialogue that is affirming and reverent. Kinship may be found by more indirect comparisons that recognize parallels among capacities in quite different ways than we have traditionally assessed humans. Part of the shift away from anthropocentricism is toward a more complex materialism that considers these realms not in terms of "mental capacities," but rather in terms of embodiment. A nonreductive approach to embodiment is not merely about comparative biology or anatomies of varied creatures. Nor does it center upon the physiology of humans; as humans and animals have evolved, so, too, have machines when considered in light of their development through the centuries. In addition, if we are to see the bodies of humans, animals, and machines in their overlap and kinship, it is only as dynamic beings that these ties will emerge. We need to see them in the flow of time and interaction, as we would a cell of the body as it lives in community and exchange with other cells, blood, nerves, and so forth, rather than as fixed within a glass slide under a microscope.

For the purpose of this essay, I am focusing on the kinship of these realms of being that must be at the heart of an ecospirituality and how they might be creatively forged into a new sensibility. However, it must be noted that the blurred boundaries among humans, animals, and machines is at the heart of much suffering and distress for the planet. As Donna Haraway articulates the current situation in her *Modest_Witness @Second_Millennium.FemaleMan©_Meets_OncoMouse*™: "My hyper-text nodes and links to totipotent stem cells are very unlike the air-pump [of the original cyborg] because they are all part of a material technology for tearing down the Berlin Wall between the world of subjects and objects, and the world of the political and the technical. They all attest, witness, to the implosion of nature and culture in the embodied entities of the world and their explosion in to contestations for possible, may be even livable, worlds in globalized science."[10] This trouble has been with us for some time and in many forms: in the eighteenth century, we started cutting up humans more routinely to make repairs, and now do so often to replace parts (utilizing the machine dimension of humans); in the nineteenth and early twentieth centuries, when disabilities now seen as organic were then thought to be based on the failure to achieve certain sorts of culturally established language and communicative norms, such that persons were disenfranchised and/or imprisoned in institutions (defining the proper "machine function" of humans by cultural ideals); since the 1950s, when agricultural cultivation began utilizing animals as unfeeling, replaceable cogs (confusing animals with machines). Yet, as Haraway rightly contends, we are now entering new global territories of technoscience with far greater destructive potential. The list of confused, destructive conflations of human, animal, and machine has become overwhelming, and need not be rehearsed in this essay. Such an enumeration could, however, help elucidate that we must first define these differing realms in *both* their inseparability *and* difference, in their capacity for mutual destructiveness and their capacities to aid each other's realm achieve distinctive excellence.

THE SURROUND AND THE OPENNESS OF HUMAN AND ANIMAL EMBODIMENT

Looking within the flow of life-activity, it has been the cornerstone of ethology that the animal lives and makes sense of its world through its *Umwelt* (around-world). The context of relations with myriad other material beings are what give it direction, place, and a way to function. Its embodiment as a vitality reaching out for nurturance, community, and continuance can only occur as emplaced in a surround. This insight was the result of the work of the nineteenth-century founder of animal behavior studies, Jakob von Uexküll. For von Uexküll, the very idea of *Umwelt* within which each animal was embedded—forerunner of ecological discourse's "environment," in which each creature is what it is and thrives within a web of relationships with what surrounds it—meant that each creature was confined within its separate "world." For von Uexküll, animals were not reducible to a common physicalistic denominator (reduction from "below"). He distinguished the environment-world (*Umwelt*) of a context of significance for each creature from the objective world (*Umgebung*). Yet he saw the body as a separate, bounded zone of isolated creatures. The power of the *Umwelt* concept was that it broke the strict lines of cause and effect, and indicated that the animal was free to respond to its environment from within its sphere of needs in relation to the world, instead of being merely a function of surrounding physical factors.

Where von Uexküll and his followers held the *Umwelt* to be an isolating force for individual creatures, many current ethologists, breaking with tradition, have demonstrated how, on the contrary, the behavior of certain birds, bees, turtles, chameleons, chimpanzees, and other animals reveals an immediate, felt connection, orientation, and organization with the surround and its creatures. For them, this suggests a kind of apprehension that must be valued as a different kind of "knowing"—one that is open to the world and its inhabitants. A striking case is that of Clark's nutcracker, a bird of the American Southwest, which searches out pine seeds and hides them throughout the landscape. The bird will harvest pine seeds during a period of just a few weeks in the late summer and early fall, and then fly about the forest and hide them in up to two thousand different locations. The usual nutcracker cache holds two to five seeds, and a single nutcracker may store up to a total of 23,000 seeds when the cone crop is heavy. It is estimated the nutcracker has to retrieve at least 3,000 seeds to make it through the winter.[11] Experiments with the nutcracker have demonstrated that it remembers these caches within its surround to an extent far beyond human capacities. Unfortunately, ethologists call such phenomena "mental mapping," a phrase that might more accurately be rendered "embodied mapping" were the rationalist habit of privileging the disembodied mind overcome. In considering the Clark's nutcracker and similar cases, leading ethologist Donald Griffin declares, "Awareness of the cognitive abilities of these animals forever changes our perception of them and their place in nature, and ours."[12] Griffin suggests we must reconsider animals as possessors of consciousness and feelings, which, even if different than ours, nonetheless reveals the world.

Heidegger was greatly impressed by the work of von Uexküll and one can only wonder how much it contributed to Heidegger's description of those structures unique to human existence in *Being and Time*. Therein, he articulated how in lieu of finding ourselves in a rationally laid out space, we rather are enclosed in a space made up of the places to which we are related in an immediate way, because they mark those aspects of the surround with which we are concerned for our nurturance and survival. He states that instead of a centrally conceived space, "in each case the place is a definite 'there' or 'yonder' of an item of equipment which belongs somewhere."[13] In other words, there are only arrayed around us distinct places that are the sites of that which is useful or needful for us to continue our lives and get done what we have to get done. "Equipment," as Heidegger uses the term here, indicates such things as have a practical role to play in getting something of necessity accomplished for that being. This "equipment" has a place within our surround, or context of work and concerns. This is why Heidegger says that we are inseparable from the world as context of all these practical engagements. There are worlds nested within a larger horizon, toward which we are implicitly directed by an "understanding" (*verstehen*) that is not reflective, but rather "lived" as a sense of belonging and feeling immediately vectors of direction around us. He calls this kind of human being "being-in-a world" and makes it one phrase (*innerweltsein*) in order to indicate that we are inseparable from internal relations to all these spots around us. By articulating this sense of human space, Heidegger has given us a different sense of how building relationships with what is around us may be our most basic sense of "understanding" our world. This idea of a "lived space" and an "understanding of implicit relationship" allows us to see the human overlap with the being and understanding of animals.

To clarify what Heidegger missed is to indicate that this analogous spatiality and mode of understanding that accompanies it presents a parallel that can be further articulated among creatures—human and animal—who inhabit the world differently, yet are similarly caught up in relations of use and investment of varied energies. Animals have a sense of place in which food, prey, rest, refreshment, play, dangers of varied sorts, and so forth are all arrayed around them at certain directions and distances. These places are registered by them in an array of feelings such as hunger, fear, rest, joyfulness, et cetera. The situation of animals is thus parallel to that of humans, who are also inseparable from their particular "world" or context of affective engagements within the surround.

For those who take the human identification with self-reflection and rational manipulation as a paradigm for "understanding," this level of human apprehension is passed over as insignificant, and animals' parallel abilities are similarly dismissed. In animal studies, the classic example is that of "Clever Hans," the horse whose trainer claimed it could do mathematics. When shown cards that called for addition, subtraction, multiplication, or division of numbers, Hans would tap out the correct answer with his forefoot, such as eight taps in response to being shown a card reading "2 x 4." The world of ethology was scandalized when it was revealed that Hans could not do abstract mathematical operations, but rather—no matter who displayed the cards—was able to detect either slight bodily gestures or facial expressions attendant upon whether or not he would stop at the right number of taps. Rather than being damning evidence that horses are "dumb" and can't perform human mathematics, it was a startling display of how sensitive the horse was to the people around him and his desire to please.[14] This classic example can serve as a warning that we must readjust our expectations from human criteria of intelligence and mind and increase our sensitivity to others' ways of knowing.

Yet Heidegger refuses to grant animals a "world" or even any analogous apprehension of the surround as do other philosophers, ethologists, biologists, and authorities. Many of us who read Heidegger's work want to read it sympathetically and claim his analyses of existential space in an affective attunement implies shifting the locus of subjectivity from an abstracted consciousness to embodiment. Heidegger, however, does not bring embodiment into his exhaustive descriptions in *Being and Time*, and does so infrequently elsewhere. Heidegger's lack of development of the idea of embodiment shows in his dismissal of animals as "captivated" by the surroundings to which they are drawn in relation. By coining this concept, Heidegger agrees with von Uexküll that the animal is inseparable from what is around it, but that this being caught up in an *Umwelt* is "impoverished" to the extent that "beings are not manifest to the behavior of the animal in its captivation." He adds: "Neither its so-called environment nor the animal itself are manifest as beings."[15] Since the animal is "driven directness," it is suspended between itself and the environment, but without being able to encounter either. Instead, the animal is "possessed" by things in such a way that "it precisely does not stand alongside man and precisely has no world."[16] For Heidegger, there is a "ring" of these instinctual drives encircling the animal that makes it "incapable of ever properly attending to something as such."[17] The animal is inseparable from all the entities within the ring of the environment that surrounds it, but "in a fundamental sense the animal does not have perception" as it is "self-encircled" by the things around it, which it does not encounter.[18] For Heidegger, animals are unable to encounter other beings: "the animal in its captivation finds itself essentially exposed to something other than itself, something that can never indeed be manifest to the animal as either a being or a non-being."[19] Only humans can encounter beings. Heidegger declares finally, "the animal is separated from man by an abyss."[20]

In holding to the assertion that "the animal is poor in world," Heidegger not only denies that animals can encounter objects, other animals, or even light, he sees all reactions to light and other objects according to models of tropism, which are cause-and-effect reactions that involve no "encounter" between the animal and another being.[21] Heidegger holds to this view despite the evidence of experiments—of which Heidegger was aware—demonstrating, for instance, that bees use the angle of the sun in navigating a path to their hive. Heidegger also denies that animals can have any sense of death. He states that animals can "only come to an end . . . but cannot die in the sense in which dying is ascribed to human beings," insofar as humans are aware of their impending end and of the nothingness that invades life through death.[22] For Heidegger, this is what makes humans distinctive.

Heidegger's claims seem increasingly dubious as ethologists show us again and again how animals are deeply affected by the death of other animals and may well have a sense of the fragility of their own lives. They certainly are capable of grief, whether it is a group of elephants nursing a dying member of their herd, then attempting to resuscitate her before ultimately burying her and keeping vigil until the next night, or Koko, the gorilla who was taught American Sign Language and cried for a week after her kitten companion was killed. Koko not only howled at the time of the accident that killed the kitten, but he responded both then and later when asked to describe the nature of death—it is "sleep forever," he said.[23]

Despite Heidegger's intention to part with Descartes' reductive perspective that viewed animals as machines, and to find a distinctive sense of animal being, the former falls back upon a mechanistic sense of animals because he retains a mechanistic sense of embodiment. Heidegger sees animal embodiment as a site of drivenness by instinct in a straightforward cause-and-effect manner. Acting from this center, the animal is rendered a machine in the most reductive, unknowing sense. Heidegger's interpretive stance becomes most obvious not in the theoretical part of his lectures, but when commenting on specific scientific experiments.

Within the confines of this essay, there is not space to examine all of Heidegger's interpretations, but enough only to cite the most egregious example. In commenting on an experiment where a bee's abdomen has been cut in half while it continues to suck honey, Heidegger declares, "This shows conclusively that the bee by no means recognizes the presence of too much honey."[24] Heidegger sees the bee solely as motivated, driven, "being taken by its food," which impels it, rather than recognizing there is a bodily awareness that has been destroyed by the catastrophic damage to the bee's embodiment. The utter maiming of the bee's body is inconsequential for Heidegger, as it was for the scientist in charge of the experiment who doubtless assumed "intelligence" is disembodied.

To recognize the indirect parallels of humans and animals, a turn is needed toward a sense of understanding through embodied perception, as has been well articulated by Merleau-Ponty. As stated before, for Merleau-Ponty, unlike Heidegger, understanding occurs in a primary fashion through perception, in the body inhabiting what it perceives and having a sense of the world arising not only from its own perspective but those of other humans, animals, and objects.

When we plunge into an object or commune with an object in perceptual consciousness, we are entering a linkage among objects, events, and beings through our shared material being. Merleau-Ponty states that "every object is the mirror of all others."[25] With this image of ubiquitous mirroring, Merleau-Ponty means, for example, that the lake over there in being seen by us is simultaneously seen by us as if it had been seen by the sky above it and by the fish within it and by the trees beside it. This permeation of myriad points of view results from the fact that our body is not "here" where one physically stands, but rather is dispersed in feeling and meaning throughout the sensible landscape. We don't have to consciously think about these other aspects of the lake; we "see" them from within whatever perspective we "literally" see the lake, because to have a body is to inhabit the sky above me, the birds overhead, the bushes beside me, in a circulation of sense that comprises "perceptual understanding." Of course, the top of the tree is given to me with a different degree of determinacy—that is, the view as seen from the sky above or from the vantage of the bird flying over me—but it is still given to me nevertheless as one of the layers or levels of my perception. Not everything about me is perceptually given with the same degree of deter-minacy and certainty, which is the power of reflection and intellection to isolate, examine, and detail, but there is still rich meaning and under-standing all about me. This is the power of having a body that flows through the "perceptual field" about me. When it sees or touches some-thing, the body "synchronizes with it" in order to come to a greater sense of what had been an instant ago "nothing but a vague beckoning." In one example, Merleau-Ponty talks of turning toward the blue of the sky not as set over and against it, but rather as "I abandon myself to it and plunge into this mystery, it thinks itself within me; I am the sky itself as it is drawn together and unified."[26] My embodiment allows me to enter the things of the field and pull them together.

Merleau-Ponty claims that insofar as it is kinesthetic, the body is pro-jected into all those movements that it may never literally make—such as soaring in the sky as the bird or cloud above—but that nevertheless inform the sense of the scene for one and reverberate through one's body. Merleau-Ponty shows in his detailed study that "movement is not a particular case of knowledge," but "has to be recognized as original and perhaps as primary." This means that, pace traditional conceptions of knowing, "my body has its world, or understands its world, without having to make use of my 'symbolic' or 'objectifying' function."[27] A later articulation of Merleau-Ponty's is that of "the flesh of the world": "that means that my body is made of the same flesh as the world (it is per-ceived), and moreover that this flesh of my body is shared by the world, the world reflects it, encroaches upon it and it encroaches upon the world."[28] We feel the heat of the hundred-degree afternoon infiltrate us and we tend to become hot-tempered—this is not just an empty figure of speech, unless we can modulate its presence within us. The flesh means we sense, because we are *of* the sensible itself. I can see because I am seen by others, whether in the vision of other humans, the dog at my feet, the fly buzzing about me, or as if the tree had a certain vantage over me sitting at this table, which it does, that would reveal my thinning hair seen from above and my squat figure like a trunk.

Touching is an even better paradigm for reversibility, since to touch is necessarily to be touched back. Artists have always been particularly aware of this dimension of perception and our bodily relationship with things and others. Cézanne spent ten years painting different canvases of

Mount St. Victoire and claimed that the mountain was painting itself through him. Michelangelo said the same of a piece of marble: the statue that he was to carve was already in the stone, and he had only to follow its sense with his body. Paul Valéry spoke of writing poems that were given to him by the voices of the forests. The list could continue of artists and others who said of the objects of their surround that there was a communication back to them from the object's vantage to which they were to give expression. This is not to say that the tree was literally looking back at Stieglitz as he photographed it in his backyard for five decades, but that he could feel the vision of the tree by paying greater attention to his perception of it.

Merleau-Ponty notes, however, that we never become one with the perceived; the hand touches and is touched back, but does not fuse with the touched. Not only is there no overarching oneness, but the perceiver never loses the priority of his or her perspective, and always has a better sense of its registration of the world than how it might appear to other beings. As Merleau-Ponty put it, "I am always on the same side of my body" and not on "the side of" trees, mountains, other people, or birds. Yet one's hold on what is known is always indeterminate, open, and one does somehow, to some degree, escape oneself through the body into the sense of these other beings—but only hauntingly so, with rich feelings and suggestions, but not any determinate grasping. Merleau-Ponty designated this never-to-be-overcome distance the "écart" of perception, the "gap" between perceiver and perceived. Over time, as sense fluctuates back and forth between perceiver and perceived, this "criss-crossing" overcomes some sense of the distance, blurs its sharper edges, but can't overcome it. As Merleau-Ponty put it, we are "in one another" as having a body that resonates, reverberates, extends, and infiltrates, but "which is not a group soul either."[29] It is the relation and interconnection of specific aspects perceived that gain greater depth and reverberation, something made up of concrete perceptions that both the scientist and the artist can bring to attention.

Instead of looking at humans and animals "side by side" in direct overlap, so to speak, the sense of the unity, for both perceiver and perceived, with the surround suggests we look for the overlap by looking into the surround of criss-crossing spaces and energies. Thinking in a similar vein, right before his death, Merleau-Ponty wrote, "Why not the synergy existing among different organisms, if it is possible within each? Their landscapes interweave, their actions and passions fit together exactly: this is possible as soon as we no longer make belongingness to one same 'consciousness' the primordial definition of sensibility, as soon as we understand it as the return of the visible upon itself."[30]

If embodiment of humans and animals is in part the body of the flesh of the things around them to which they are related, there is an openness and interchange among us through the surround. It is into this openness of embodiment that machines have also become woven. Haraway has detailed how we are caught up in biotechnological market mechanisms that have restructured culture; Lyotard has demonstrated how we are caught up within media, communication, and informational economies that have "become" our dislocating bodies; and Deleuze and Guattari have catalogued the varied ways we are machine-beings in larger systems, the most frightening being the war machine. Yet, if we are cyborgs and have always been to some extent,[31] if the biotechnological and information revolutions not only drive globalizing oppressive economies, but also allow humanity to fight disease, allow blind people to see, deaf people to hear, calculate solutions to scientific problems with computers, and create more food, to take but a few examples—then we need to think more seriously of Heidegger's claim that "there is then in all technical processes a meaning . . . which lays claim to what humanity does and leaves undone."[32] Despite being a vigorous critic of technology's destructive power over the planet, Heidegger challenges humanity in his "Memorial Address" in *Discourse On Thinking* to meditate more deeply on the "hidden mystery" and meaning within technology. It is also the only way for ecological concerns to avoid a dead-end of romanticism, and instead, as Bruno Latour has articulated in *Politics of Nature*, to enter into a dialogue in which the voices of varied sciences and technologies are heard in an assembly of inclusive collectivity. Latour warns those of us who identify with spiritual concerns, with ethical and political concerns, who follow the path of humanities or social science, that "we need to count much more on the sciences."[33] We have the tools today needed on "both sides" of the aisle, whether within the varied humanities approaches (philosophical, theological, psychological) and within the varied sciences (biology, cybernetics/robotics, physics), to come together.

THE "WORLDLINESS" OF MACHINES AS EITHER EMBODIED OR OF "THE FLESH OF THE WORLD"

By philosophically thinking of embodiment, whether human or animal, as that which comprises "the flesh of the world" and takes in the surround, this way of thinking encourages us to look at machines differently and to make sense of developments in science and technology. The difference is useful in any effort to find "common ground" with machines and with those aspects of animals and ourselves that are machine-like or caught up in machines. In order to expand upon this idea, let us look briefly at two different avenues for exemplifying what this might mean—to "Artificial-Intelligence" (AI) research and to research in so-called human-centered technologies. As described in Anne Foerst's *God in the Machine*, a work that also aims to expand humanity's sense of sacrality through the inclusion of machines in a spiritual community, she describes the recent shift in artificial-intelligence research that instantiates Merleau-Ponty's earlier insights about embodiment. MIT's AI research took a turn away from the model of intelligence that envisioned the human grasp of the world as emanating from a central mind that imposed abstract rules and categories upon it (to make sense of a confusing sensory input through judgment) toward the idea that human learning takes place through a process of "embodied understanding."[34] Unlike the traditional approach to AI, which conceived of intelligence in terms of a central computer processor controlling the peripheral movements of the several parts—so that "when an arm movement is needed for the execution of a specific command, the controller makes a plan about what motors, sensors, etc., it should move, sends the specific commands to specific parts, and they move, hopefully as quickly as our printer starts to print when we give it the command"—these robots "interact because through experience they 'learn' motor control and coordination of several motors in reaction to the demands of a specific task."[35] The analogy is made to infants who first wave their arms about aimlessly until they learn to focus on the object desired; the robot also waves aimlessly at first and learns through trial and error how to grasp something. As Foerst summarizes: "It has no abstract understanding of space. Instead it learns to coordinate its arms though feedback loops that register the weight the joints have to carry in a specific arm position." The sensors and actuators that react to the environment are connected in such a way that the system can react flexibly, instead of following some set internal plan. The overall sense of what occurs is that "the systems body is situated in the environment, interacts with its environment, and creates new and complex behaviors out of simple interactions. This philosophy gives this AI direction its name, 'Embodied AI.'"[36] Foerst here implies that machines are developing a "situatedness" within the surround in such a way as to have some sort of equivalent of embodied cognition—the body as a way of knowing. Movements of the body are ways of inquiring of, and then dialoguing with, the environment.

Much more might be said about the ways in which machines are becoming able to interact with their environment. To cite one example, visitors to the MIT lab have been greatly surprised by how the "intelligent robots" have responded to the visitor's mood by asking them what was making them sad; Foerst herself was caught off-guard when Kismet, the robot, responded directly to her frustration: "I became quite frustrated . . . and so complained in a sad and disappointed voice. I accused Kismet of ignoring me and said it didn't like me anymore and many more such things, and suddenly Kismet started to concentrate on me, soothing and calming me."[37] Visitors have been startled to discover how the intelligent robots can respond to various "melodies" of language spoken to them in order to respond to their visitor's mood (among four possible affective "melodies").[38] Visitors to the lab have also been taken aback by being followed about the room by the robot's gaze and having the robot come forward to shake their outstretched hand.[39]

In a similar fashion, those working on the problem of how technological advancements can be "human-centered" have focused upon the nature of human embodiment and how this entails seeing human cognition differently, which in turn revisions the human relationship to its surround. In his *Natural-Born Cyborgs*, Andy Clark articulates how all the most daring advances in technology, from the expanding telepresence of humans to the use and development of various complex prosthetics, focus not on human-robot hybridization, but rather on new kinds of embodiment. Clark summarizes: "The larger lesson then is that embodiment is essential but negotiable. Humans are never disembodied intelligences; work in telespace, virtual reality, and telerobotics, far from bolstering the mistaken [notion] of detached bodiless intelligence, simply underlines the crucial importance of touch, motion, and intervention."[40] Clark continues by claiming that what matters are all the complex feedback loops between body and surround, however mediated by technology, and ultimately concludes, "it is the flow that counts."[41] Human cognition subsists only "in a hybrid, extended architecture" and humans "repeatedly create and exploit various species of cognitive technology so as to expand and reshape human reason.[42] "We—more than any other creature on the planet—deploy nonbiological elements (instruments, media, notations) to complement our basic biological modes of processing, creating extended cognitive systems."[43] In other words, embodied knowing happens through the surround of built components and technology must enter this circulation to be effective. Clark's book is a dazzling array of technological advances, from the cyber-performance artist, Stelarc, who paints and dances with a third mechanical arm attached to his body which he controls as fluidly and "naturally" as his two other arms (via electrodes and processors attached to four muscle sites on his body), to the example of an owl monkey at Duke University whose brain signals are used to operate a robot six hundred miles away in an MIT lab.[44]

I will focus briefly, however, on another bit of technological wizardry that appears in Clark's book—the interactive installation set up in Nexus Contemporary Art Center in Atlanta in 1996 by Eduardo Kac. In that installation, the viewer was confronted with an aviary with about thirty flying birds and one robot bird. The viewer was able to put on a virtual reality headset that was getting a video feed from two digital cameras that were the robot bird's "eyes."[45] The sense that the viewer had was of seeing himself or herself from above by the birds and also of being aloft with the bird. This example startled me since it is the technological equivalent of the example of embodiment I have used for decades drawing upon both Merleau-Ponty's work and the ending of Susan Griffin's *Women and Nature*. In its concluding pages, Griffin writes about the sense of soaring "through" the bird flying above and experiencing her body as being in "flow" with it. She and Merleau-Ponty describe how our embodiment gives us this dual sense of being on the ground and above simultaneously through emplacement within the surround. Technology has now literalized and made explicit this implicit bodily experience. This articulation of experience has been delivered to the viewers of this art/

media exhibition through technology, demonstrating the power of the machine to insert itself into the circulation of embodiment's meaning. What had been implicit by inhabiting the surround or having this "external scaffolding" of the body is now made explicit and intensified in its palpability. This is a clue to how machines are becoming part of a dialogue with us about the sense of our experience.

Presently perhaps, most of us are only in an extended sense cyborgs, but many are becoming cyborgs in much more literal ways. It is illuminating to look at Michael Chorost's story of his adaptation to his cochlear implant in *Rebuilt: How Becoming Part Computer Made Me More Human*. Chorost was thirty-six years old, a recent PhD graduate in educational computing, when suddenly he went from being hard of hearing, though able to hear fairly well with hearing aids, to stone deaf. He was offered the chance to hear again by having a cochlear implant put in his skull.[46] Chorost found himself with a new kind of hearing, one that it was hard to translate into what he had experienced before with his hearing aids. He had to give up on the sense that there was one way "to be in the world," one way to experience the world, for his hearing through the implant would be different: "It would not be hearing. It would be the equivalent of hearing to hearing."[47] Again, there is not enough space to develop this example, but it is important to point out that Chorost had three startling breakthroughs in learning to hear with his implant. The first was the realization that his body adapted to the technology, even his brain, through "neural plasiticity": "Thanks to neural plasticity, the neurons in my auditory cortex were slowly reorganizing themselves to handle the bewildering new input from the implant."[48] Disturbing sounds and aberrations changed in their significance or disappeared because, as Chorost relates, "over the weeks and months my auditory cortex obediently refined its topography, making physical distinctions where none had existed before. The implant was literally reprogramming me." In this instance, the computer, the machine, is intimately joined with the human body, is in dialogue with the physical makeup of that body, and especially its neural network, since they are plastic, adapting to what is around them.

Even more surprising, however, were the other two breakthroughs in his growing capacity to hear through the technology. At one point, Chorost realizes that he is able to hear better through the technology when he has learned to care more about his interlocutors and come to have a different relationship with them. He realizes that in order to become able to hear with the potential that the machine has provided him, "I would have to become emotionally open to what I heard."[49] The machine is part of his context, both internally and inescapably. It allows him to be woven into what is going on around him in an auditory way, but for the machine to work, it needs him to develop a closer relational bond to people and events around him. The other surprise is that Chorost cannot hear through the implant when he concentrates too hard on what he hopes to hear. The key to his really hearing seemed to be that "I had to pay attention. Just not too much attention." He tested this out further and found that, indeed, the radio drifted in and out of focus as he "played with different levels of attention."[50] He thought it similar to finding the "the sweet spot" on a tennis racket, which like listening requires some sort of attunement, being in a flow with things around one: not willing, not trying forcefully, but becoming alert in an unfocused way.

What each of Chorost's experiences demonstrate is that the machine, too, has to become woven into the surround in a flow of embodiment. Only in becoming so interwoven is there a give and take, a reversibility and dialogue, in which all come to be altered within the experience. The machine has to be taken up into embodiment via emotional attunement and an openness to give and take or it remains outside the spiral of deepening meaning. The machine, too, as reactive and interactive, has also a fluidity within the flow of the surround into which it can be inserted.

Michael Chorost is a philosophical person who had tried to read Haraway at an earlier point and found her meaningless. Now, after four years with his implant, he realizes that there is an indeterminacy of embodiment, perception, and sense of the self—and he comes to understand some of what Haraway is elucidating. He has come to see reality as something that alters within the interaction of machine, surround, and person. Michael Chorost has experienced that the key to adjusting creatively to having his auditory access to the world mediated by a computer (and its often upgraded software) is to realize that he doesn't have a determinate biological base as a body, but rather his embodiment is an openness to inputs of all sorts that evolve and mutate:

Haraway's essay now struck me as a straightforward description of my life. I experienced joint kinship with animals and machines, feeling oddly affectionate toward my robot vacuum cleaner yet also reveling in the smells and lusts of my animal body. I was permanently and pleasurably adrift in eternal uncertainty about teapots and microwaves. Cyborgs "have no truck" with master narratives because there is no single story running through their bodies. My sensory universe is now constructed by squadrons of programmers, not the garden fields of my ear. Unitary identity? Not anymore, if ever; there are two minds in my skull, one built by genes, the other by a corporation. I am a walking collective, a community of at least two.[51]

It is in the openness of embodiment as running through the surround and its human, animal, and machine interlocutors that there can be sensitive equilibration that augments a deeper significance, of which they are all partners. Within this spiraling of communicated sense, the machine embodies a collectivity of human expression, as well as echoes of the biosphere. Human, animal, and machine can enter an ongoing dialogue for creative enrichment.

It may seem more obvious that whatever is alive in the *Umwelt* of humans and animals is in a "reversible" relationship. It is easier to envision such beings "speaking back" in an indirect way that can be taken in on some level of human apprehension. If, however, machines have become a part of our surround that actively infiltrates into our capacities to perceive and to act, or at least enter into the flow of our actions as "facilitators," it is easy to see that they have become interlocutors with us and with our ability to create, apprehend, and express ourselves within the world. In the early phases of the industrial revolution, many were intoxicated with the potential for machines to greatly increase productivity, for machines to accomplish tasks that could not be done without them, and to open up new areas of human endeavor. At first, these were mainly physical tasks, such as moving large masses; later feats, like flying or exploring outer space, were merely fantastical. At first, especially in Europe after the scientific revolution, given the way of looking at the world that predominated in the cultures developing machines at an accelerated rate—that of the distanced subject versus objects rationalized mathematically and logically—it seemed as though the proliferation of machines was something that did not alter the being of humans, animals, or the rest of the surround. Machines were "there," in the environment, but could not get "inside us" psychically—they were mere tools. Of course, cause-and-effect relationships between humans and machines were seen, but not deeper ones. In the nineteenth century, however, writers and thinkers articulated how machines could alter the way we experienced human reality, created values, and made sense of things, with Marx being perhaps the most famous theoretician and Mary Shelley the most famous novelist. Marx sounded the alarm bell that who we are might become produced in ourselves by the machines we had originally produced: the same reversal of creator and created that Shelley saw in how humans might become machines on account of their lack of fellow feeling.

What is ultimately important for this meditation about the spiritual relatedness of humans, animals, and machines is to realize that we can no longer distance ourselves utterly from machines and say we have affirmed who we are in relation to some deeper meaning or spiritual beckoning. Machines are increasingly the heart of the surround that comprises our bodies. It is not just a matter of pacemakers or cochlear implants, of computers embedded in our biological flesh, but of our bodies within the flesh of this new world, of having machines "under our skin" in ways that have great potential for both violation and augmentation, for "taking in" the world knowingly and creatively. There is a continuum from the experience of someone with a computer implant within their skull; to the fortunate blind individuals who wear a camera and have implants that give them some sight; to a patient within a scanner for hours hoping the machine will make visible the flaws in the functioning of his or her biological machinery; to those of us whose continued functioning depends on chemicals machined for us in labs as a result of the read-outs of other machines used in exploring our makeup; to the tremendous computer power that gave rise to gene maps; to those of us whose tissue makeup is partially DDT and a host of other chemicals produced by machines; to all of us who see the world and make life-and-death (and other less earth-shaking) decisions on the basis of a view of reality provided to us by machines. Machines tell us of ozone holes, the earth's molten center, the age of dinosaurs, the dangers of radioactive materials, the composition of cells, and much else. If, as has been said, we see the world through the way the world sees itself—so that its soaring quality is transmitted to our bodies by birds, for example, or even rocks—think of how much more staggeringly true this is of how we see faraway galaxies, the electromagnetic spin of particles, the weather patterns sweeping the globe, the fractal geometry of chemical particles, the surface of the planet Venus—all of which has been seen and apprehended by humans only because it has first been "seen" through the "eyes" of machines. So much of what we worship or appreciate with a sense of the holy and profound has come to us through our kinship with machines.

A SPIRITUALITY OF WORLD AFFIRMATION

I venture to suggest two aspects of an ecospirituality that are resonant with the human situation I have described: the community of those beings who are to be considered part of a spiritual community, and the sense of spiritual reality that informs this community. I have already tried to describe the way in which humanity emerges from a fleshly dialogue with the creatures and machines of the surround. To take this dialogue further, I would like to cite two of Anne Foerst's themes in *God in the Machine* as being particularly helpful in trying to increasing our kinship with both animals and machines. First, she reminds us that the heart of human being is *"homo faber,"* "the being that makes things." In our consumerist culture, "making" often becomes degraded to a relationship between mass production and frenzied materialist purchasing. Foerst reminds us that at a deeper level, making things is about channeling the spirit within all things as well as our own spirit, and in that sense, "every act of creativity is prayer."[52] She asserts in looking at the evolution of machines: "the more complex things we build, the more we praise God."[53] If we are building machines that have been inserted into the surround in ways that augment the capacities for embodied knowledge and expand who we are, and could be used to let the surround become more of a place of thriving for animals, too, then machines, insofar as they function in this way, are members of a spiritual community. As Anne Primavesi expresses in *Gaia's Gift*, each part of the natural surround—from animals we know to the prehistoric and others currently invisible without research, to every other aspect of the planet—for eons has given the gift of life and its subsequent sources of meaning, thriving, and spirit.

And we have reason, too, to hope that in their creative possibilities machines are also contributing to this flourishing. Human, animal, machine—all parts of the surround seem to be an integral part of this gesture of prayer insofar as they make recognition, reverence, and ongoing creativity possible. If what inspires us and what we cherish about existence on this Earth—from the vastness of the cosmos to the intricacies of the most minute structures—is in part manifest through human and machine interaction, then machines, like animals, have a place in a community of spirit. They, too, belong within our intuition of beings of value; they, too, help to make life sacred. If there is a community that witnesses what is wondrous—from black holes to spider webs to human acts of self-sacrifice, kindness, or beauty—then machines, animals, and humans all belong to it.

My poem that appeared earlier in this meditation was a chronicle about the fate of the early heart machines. The initial outcry about the violation of humanity and the seeming demonic character of these machines put a halt to their development for quite some time. Now, however, newer heart machines have been developed in ways that are not intrusive and work more cooperatively with the patient for his or her level of required assistance. We are learning to make machines fit more reverently into the fabric of our shared surround. These creations may indeed seem like prayers to many of us, especially to those of us who would not be hearing, moving, or indeed even living without them.

Foerst notes also that spirit only thrives in combination with a guiding mythos. To cultivate reverence, gratitude, and love requires rich narratives that detail the unique attributes of others and of all experienced aspects of the greater cosmos.[54] To be included within these stories is to be part of a community. Humans, animals, and machines are part of a shared story of coming to thrive and coming to witness the extensiveness of the sacred place of Earth within its universe. Insofar as this planet continues to birth that which has intrinsic value, is worthy of reverence and love, it is because humans, animals, and machines are part of a cooperative venture, a larger "community" of diverse voices. With this sensibility and this idea, it is to be understood that our "household" is comprised of a moving flesh that streams within the surround by way of human, animal, and machine interrelatedness—this is the movement of spirit. An ecological approach to the community of inspirited beings cannot leave out the machine world.

In concluding I would like to turn to the theologian and philosopher Louis Dupré, who felt that theology was called upon to articulate "a particular version of the transcendent that matches the age." In explaining what this meant for him, Dupré casts a vision in which embodied beings recognize that they are caught up in a world of material beings that are not adversaries to their spiritual nature; rather, both spirit and matter are inextricably woven of the same worldly fabric. These day-to-day events composed of objects, creatures, and other humans are the focus of human aspiration and insight. This weaving is what we are and yet also that which moves us beyond ourselves. These concerns might be mundane, but they have necessarily a depth of meaning and spirit that can be uncovered. Dupré: "Instead of the traditional distinction between sacred objects, persons and events, and profane ones, spiritual men and women in the future will regard existence increasingly as an indivisible unity, wholly worldly and self-sufficient, yet at the same time will be aware of a depth dimension that demands attention and that they allow to direct their basic attitudes to their life."[55]

Dupré calls this discovery of the divine within all worldly things and events "a spirituality of world affirmation." Such a spirituality, I believe, recognizes that there is no separate spiritual realm. Otherwise the interweaving of humans with other humans, animals, machines, as well as other parts of the surround could not work cooperatively, in continual dialogue and mutual transformation, to open up insight, knowledge, wonder, love, thriving, and transformation. Dupré considers this a time in which humanity sees there is a mystery of ongoing creation "that discovers a transcendent dimension in a fundamental engagement to a world and a human community." If, as Merleau-Ponty articulated, each particular perception has inexhaustible depths of meaning, should we wish to explore it, and therefore each perception is "a birth and a death" because there is in each moment of this planetary existence such a surplus of meaning that it is enough for a lifetime and is a lifetime in each of its moments of unfolding—then we dwell in an ecospiritual reality, a divinity of the earth. The surround, of which we are part, is the source of spiritual inspiration and the object of reverence, the gift-giver to all living and nonliving beings and the necessary focus of our care.

๏ Getting Over "Nature": Modern Bifurcations, Postmodern Possibilities

BARBARA MURACA

Far from "getting beyond" the dichotomies of man and nature, subject and object, modes of production and the environment, in order to find remedies for the crisis as quickly as possible, what political ecologists should have done was slow down the movement, take their time, then burrow down beneath the dichotomies like the proverbial old mole.
—BRUNO LATOUR, *Politics of Nature: How to Bring the Sciences into Democracy*

INTRODUCTION

In *Politics of Nature,* Bruno Latour questions the value of the term "nature" in advancing the ecological agenda, commenting that "under the pretext of protecting nature, the ecology movements have also retained the conception of nature that makes their political struggle hopeless."[1] He then proceeds to outline an intriguing journey toward a postmodern political ecology pointing beyond any reference to nature, while at the same time recommending a slow process that instead of "cutting the Gordian knot" aims at untying "a few of its strands in order to knot them back together differently."[2]

However, the insistence that we drop the term "nature" altogether surely invokes the very strategy of cutting the "Gordian knot" that he himself advises against. If we were to start a process of untying some strands in order to open possibilities for interweaving them differently, we might more productively hold on to the rich texture that is offered to us and start looking into its historical complexity, rather than dismissing it too early.

As Latour himself would acknowledge, *"the* conception of nature" shows itself to be more multifaceted than it might at first seem. While referring broadly in the first part of this paper to his analysis of the establishment of the concept "nature" in the contested field between science and politics, I question his very attempt to overcome the use of the term. I then draw on the skepticism that Gianni Vattimo with reference to modern metaphysics addresses toward any form of "overcoming." Vattimo proposes instead a form of recovery that, while remaining inscribed within the same tradition and its diseases, assumes this very tradition as an efficacious and inescapable inheritance, a sickness we cannot simply get rid of, but which we can heal by taking upon ourselves.[3] With Vattimo we are invited to "get over" nature rather than beyond it, by drawing back hidden threads into the foreground of a more complex and dynamic picture.

In this fashion I try briefly to retrace some of the ways "nature" has both entered and escaped philosophical concern. I propose a reconstruction of nature as a complex field of meanings, drawing on Heidegger's analysis of the *physis/techné* dichotomy, its reference to a premodern concept of nature and its consequences for modern technology. I then focus on Whitehead's notion of a "bifurcated nature," which supports and grounds the development of modern science. Finally I present Vattimo's concept of "getting over" as an alternative to an overhasty overcoming that risks repressing its own long shadows.

In the following paragraphs I will refer to the term "nature" as a provisional open field, in which the different definitions offered by different actors in different (historical) settings are invited to participate. I prefer using the singular term instead of the plural "natures" because this is the way it has been used throughout the traditions I reconstruct. Referring to society in comparison to nature, Latour asserts that "like nature, and for the same reason, society finds itself at the end of collective experimentation, not at the beginning, not all ready-made, not already there."[4] Similarly I see this reconstruction as an "experimentation," in which neither a beginning definition nor a clear boundary already set are at stake, but the journey explicitly moves within the vague region where boundaries have been built, shifted, demolished, and reestablished in the course of time.

NATURE: ALLY OR THREAT FOR ECOLOGICAL CLAIMS?

Nature has always lingered as a misty presence at the boundary of our field of concern. It has long supplied the object of scientific research and the hidden background of our economic system. What we call "nature" has been deployed in philosophical thought, by science, and in political discourses in many ways—as a mute partner, at our disposal, infinitely elastic and serving multiple purposes.

Nature can be locally managed within urban fences, and it can be an object of analysis and controlled experiments. But for the most part it is something to be kept at bay beyond the boundaries of our human and cultural world. Nature as an almost inaudibly buzzing background is located outside the city walls: *hic sunt leones* (out there the lions), as the Romans used to say of what extended beyond the borders of empire.

This view does not necessarily lead to the elimination of nature from the whole horizon, but—to borrow a popular concept from mainstream economic theories—does at least engender its *externalization*. Nature, both as source and as sink with respect to our production cycle, is currently considered by neoclassical economics to be a mere *externality*.[5] It is located outside the human world of activity and production and, in the absence of dramatic scarcity, is excluded from any economic concern.[6] As the economist Robert Solow once declared, "The world can, in effect, get along without natural resources, so exhaustion is just an event, not a catastrophe."[7]

Nonetheless for us, in the light of the ecological disasters we are facing, nature acquires a dramatic new importance. Ecologists, environmental activists, "green parties" all over the world demand respect for nature, they raise political and ethical concerns about nature, and even claim legal rights for "natural" beings (animals, species, ecosystems). New or renewed naturocentric paradigms meet with wide approval within activist groups. A new metaphysics of nature is said to be needed, in order to tear down the walls of separation between the human and nonhuman world and to grant nature a new citizenship within human concern.[8] Once we acknowledge that nature is endangered, and that this threatens our very survival on earth, we cannot avoid standing up for it, reclaiming its place within the political discourse as the chief issue on the agenda.

However, as Latour clearly shows, the concept of "nature," to which ecological activists generally refer, is already enmeshed with strong political interests. Setting nature outside of the city walls reflects the political need of constructing an indisputable ground for truth to refer to as an ultimate matter of fact in order to cut off discussion about disturbing topics. Scientists were given the role as keepers of the secret that nature would reveal only through their instruments of analysis.[9] In the name of what was considered "against nature" or "in conformity with nature," certain categories of people and models of thought have been excluded from, or included in, political discourse at different times. The alleged objectivity of the concept "nature" thus obtained has operated for centuries as a legitimating basis for social values as well as for the allocation of rights and privileges among different groups of people. As long as nature plays the role of a background of unquestionable references within an essentialist framework, it leads to the institution of a fixed order of things and meanings supporting given structures of power. Women, especially, are aware of the consequences of being thrown back unto nature in order to exclude them from political and social life. Nature as the ultimate ground of reference carries the burden of exclusion, reduction, and silencing in the name of ultimate definitions of being.

This seems to be true independently of the specific concept of "nature" established as the reference background and would just as much affect alternative accounts, different from the modern framework—for example, the ecofeminist revolt against reductionism, as Vandana Shiva describes it: "reductionist science is at the root of the growing ecological crisis, because it entails a transformation of nature such that its organic processes and regularities and regenerative capacities are destroyed."[10] When "nature" is assumed to be speaking for itself, it gains the power to silence any other voices from within the walls, since it then sets standards that are considered independent from any of the different perspectives arising in the urban space. This is true also if it happens in order to claim a place for issues related to nature within both political discourse and moral consideration. When, for example, some environmentalists argue for nature as an independent realm of self-generating and regenerating processes that ought to be left to flow without human intervention, they hypostatize a truth, which does not allow for contra-diction (that is, no-body can say anything against this) from within the walls.

In this context, the very idea of "harmonizing" with nature as an entity haunted by the shadow of some unquestionable ground of all beings and thinking can be a scary prospect. Going back to "nature" as such runs the risk of subordination to metanarrative and the consequent reduction of multiple voices to what some natural philosophy may have decided about ultimate truth. Therefore, nature has been excluded from some emancipatory discourses, to the extent that emancipation is intimately connected to liberation from the cage of a "nature" already complicit with racial, cultural, and gender discrimination. In this regard, poststruct-uralist feminists, by reacting against essentialism and its matrix of domi-nation, have with good reason challenged any understanding of nature as something given independently from cultural attributions. Untying nature and gender—"denaturing gender," as Judith Butler puts it—has become a major battlefield for emancipatory struggle.

This denaturing process, however, can lead to the complete exclusion of references to nature from our political landscape. Ecological issues concerning nonhuman beings risk neglect within the antinaturalistic ten-dency of those postmodern approaches that perpetuate the modern alienation of human from nature.[11] The pressing necessity of unmasking the patterns of domination derived by a metaphysically constructed con-cept of nature risks eliminating the ecological issue altogether. As Cather-ine Keller points out, "Poststructuralism and most of the feminisms and theologies adapting its 'discourse' maintain an urbane and systematic si-lence, a kind of hostile boredom, toward all things nonhuman."[12]

It is the aim of this paper to explore various paths through these issues and questions, well aware that throughout modernity the concept of nature has been haunting our thought. Can we today still speak about nature while taking seriously the postmodern criticism of its roots caught within the metanarratives of modern metaphysics? Can we avoid speak-ing of nature in a way that simply closes the gates and constructs a self-deceiving refuge within the city walls? Do we really need the term "na-ture," or can we dismiss it without dismissing the claims that are attached to it, as Latour seems to endorse? Yet, if we do need it, which concept of "nature" are we then trying to hold on to, and which reject?

NATURE AS *PHYSIS*: APPROACHING MODERNITY

As both Whitehead and Latour clearly show, it was difficult to speak about "nature" as something distinct and separated from the largely human world before the seventeenth-century scientific revolution. Rather, the very concept of "nature" itself—as we know it now—seems to be the historical outcome of the modern Western tradition of thought.

Looking back at the ancient Greeks' use of an analogical term for nature, *physis*, it is clear that the latter refers to a completely different framework. The long list of works titled *"perí physeos"* or *"de rerum natura,"* from the time of Thales to that of Lucretius, concerns subject matter that is much closer to what we would today term "cosmology" rather than to natural science. We can usefully locate a split concerning nature in ancient Greek philosophy by following Heidegger's treatment of the dichotomy *physis / techné*, which alludes to the difference between the man-made and the "natural." While *techné* refers primarily to human ends, natural things are conceived as having their ends in themselves.

Physis does not require human shaping activity to generate; rather, it is a kind of *poiesis*, a bringing-forth from concealment to disclosure;[13] it is said to be *"poiesis"* in the highest sense of the term, the bringing-forth process in itself. Since the characteristic of nature as *physis* is its creative activity independent of man's hand, it is not present-at-hand—that is, it cannot be simply objectified and reduced to passivity.

On the other hand, *techné* implies the use of means to achieve certain ends and therefore encompasses modern technology. The use of means does not, however, necessarily involve a reduction of *techné* to a mere instrument employed to achieve pragmatic goals, as could be said for modern technology. According to ancient Greek culture, *techné* was intimately connected to *physis* and was therefore not limited to technology, the handling of a mere tool. *Techné* once included art: "There was a time when it was not technology alone that bore the name *techné*. Once that revealing (*Entbergen*) that brings forth truth into the splendor of radiant appearing also was called *techné* (. . .). And the *poiesis* of the fine arts also was called *techné*."[14] As seems obvious, even without adopting Heidegger's concept of art, art is not a matter of simply utilizing available means in order to shape formless stuff according to distinctly preformulated goals; it involves creativity, inspiration (the Muses!), and even divine ecstasy,[15] that is, a complex self-transcending process of relation with more than mere means. *Techné*, both as art and technology in the ancient sense of the term, discloses, reveals, and brings from concealment to unconcealment. As such, it is different from, and yet intimately connected to, *physis*.

If we follow Heidegger this far, we must acknowledge that a straight opposition and reciprocal exclusion between man-made and natural world was not present in ancient times. Rather, *physis* and *techné* were thought of as flowing into each other, implying each other, linking to each other in an ongoing dynamic of activity/passivity and creative process: ancient craftsmen and artists used to work in partnership with nature. The shape given to natural materials did not completely hide nature itself—as *physis*, nature would always "shine through" as the source of the continuous generation of forms. The natural material would still be recognizable through the artifact or, to put it more precisely, the artifact would not wipe away its inner process of becoming. As David Tabachnick, rephrasing Heidegger, rightly notes: "A carpenter imposes the form of a chair onto wood but once the chair is finished that wood still maintains its natural characteristics to rot and decompose in the same way a fallen tree rots and decomposes on the forest floor. In other words, the craftsman's chair is a site of openness for the revealing of nature."[16] Considering things as means does not automatically lead to a reduction of them to mere tools with no reference to the horizon of their emerging. Rather, their distinct coming into the foreground is intimately linked to the dynamic process of emerging from a complex reference to inaccessible backgrounds.[17] In this sense we can talk of a process of disclosing (*Entbergen*).

Modern technology, however, in spite of its also being a sort of disclosing or opening of the world, does not acknowledge the *poietic* element. The passage from *techné*, as cooperation with the self-creative power of *physis*, to technology as a sheer instrumentalization of "nature" implies a paradigm shift, which, according to Whitehead as well, took place with the development of modern natural sciences and resulted in the definition of a modern concept of "nature" much better adapted to a technological approach. It is not until the bifurcation of nature, as it has been accomplished by modern science and philosophy, that *techné* turns into technology and starts referring to an objective, passive world at our disposal. Instead of revealing nature, *techné* as technology reduces it to a raw material delivered over to technical transformation.

As Heidegger points out, modern technology, rather than "bringing forth" in cooperation with nature, "challenges-forth" nature itself.[18] Nature is thus reduced to a passive "*Bestand*," a standing reserve of energy for different services according to human goals,[19] from which men can extract what they need. Since nature is now considered as a general storage for stocks of energy, it has to be measured, quantified, and predicted. Heidegger explains in this context the approach of modern experimental physics, which interrogates this enframed nature as to whether and how it responds.[20] The challenging-forth of nature as stock of energy is termed "*Ge-stell*," en-framing: "We now name that challenging claim which gathers man thither to order the self-revealing as standing-reserve: '*Ge-stell*' [enframing]."[21]

It is tempting to recall here Francis Bacon's torturing of nature to force her to reveal her secrets to the scientist.[22] Instead of opening a space for (self-)disclosure, nature is forced to respond to precise and narrow questions. The link to the persecution of witches during the same period is hard to ignore, a weird but not accidental coincidence! Evoking here Latour's intriguing concept of scientists as "spokespersons" for nature, who let nature speak for herself but are no more than translators, interpreters,[23] it is interesting that early modern scientists did allow nature to have her say, but only on the torture rack: she could speak as long as was required and must then be silent. As with witches, nature did not have the right to raise objections, to appeal against her judges. She had to sign her confession and die—to be dead and never speak again.

Heidegger's focus is not so much upon a possible restoring of nature for ecological purposes within the human world. Rather, it is a dramatic reflection on the reduction of humans to means and "beings at disposal," as an effect of the exclusion—oblivion—of nature as *physis*. He wonders: "[D]oes not man himself belong even more originally than nature within the standing-reserve?"[24] On Heidegger's account, the absence of *physis* as the never-given background—the abyss (*Abgrund*), the nonground of Being—is responsible for the reification of human beings and their social worlds. Apart from this "bringing-forth" there are but things, beings in their simple "presence-at-hand."[25]

BIFURCATION(S) OF NATURE AT THE DAWN OF MODERNITY

Although it is difficult to account for the shift that took place from the ancient meaning of *physis* to "nature" in a modern sense, it is possible to reconstruct briefly its main highlights by drawing on Whitehead's description of the different factors that have led to the formulation of the modern concept of nature. Interestingly enough, this is precisely the notion of "nature" still referred to by contemporary ecology, and indeed by some elements within the environmental movement. According to Latour, political ecology, too, has sustained a concept of nature that does not support its own efforts, since it has bound its shaping possibilities to an ultimate given set of descriptive truths grounded on scientific observation and systematization.[26] Every political ecology needs therefore to deal with the complex inheritance of the concepts on which it depends.

The modern concept of nature goes back to the slow and yet revolutionary emergence of modern natural sciences throughout seventeenth-century Europe.[27] Whitehead aims to show the dramatic impact of these new sciences while insisting at the same time on their continuity with the past so as to maintain a complex account of their contribution and consequences. While the nascent modern sciences needed to get rid of the narrow limitations connected to the obscurantist dogmatism of late medieval thought,[28] claiming a freedom of research in the face of a set of preformed value-laden questions, they also owed to those very same assumptions most of their methodological orientation as well as their intuitive belief in an ongoing order of nature. In fact, the modern sciences could be said to be characterized by the powerful combination of a medieval legacy of a passion for generalizing abstractions (and proclivity for systematic thought), on the one hand, and a quite new and extraordinary attention to the details of experience, on the other. Moreover, the development of new mathematic patterns connected with a wider use of algebra and the focus on periodicity offered a stable basis for their need for predictability and regularity.[29] Without this extraordinary mixture, it would have been impossible even to think of natural laws!

Modern sciences were driven both by a fervent interest in general principles and the urgent need to rely on what Whitehead calls "irreducible and stubborn facts" in order to face the challenge of the ecclesial authority.[30] The interrogation of nature á la Bacon aimed basically at the attainment of ultimate matters of fact, on which the whole scientific framework of induction could be built. Nature played the role of a no-longer-questionable horizon of truth, with its own *poietic* bringing-forth once and for all reduced to silence, and further silencing any attempt to cast doubt on its ultimate and absolute voice, especially by such authorities as the political establishment or the church (at a time in which churches were busy fighting each other and their truth-claims had never been so fragile).[31]

As Whitehead observes, the achievement of ultimate matters of fact regarding nature was the outcome of unreflective assumptions, which allowed the sciences to be guided in their further development by the fiction of well-supported empirical data. In fact, if we extend experience to its whole range of complexity instead of reducing it to mere sense perception limited within the narrow borders of laboratory experimentation, there is no immediate empirical givenness of bare matter.[32] By grounding their theories on the latter, however, the newly emergent natural sciences developed a concept of "nature" as a passive stuff—"brute matter"—endowed with certain particular characteristics, such as extension and external relations.[33] The modern sciences acquired these characteristics by a careful elimination of all the other aspects from the whole picture: "the entity bared of all characteristics except those of space and time, has acquired a physical status as the ultimate texture of nature; so that the course of nature is conceived as being merely the fortunes of matter in its adventure through space."[34]

To achieve such a clear and functional concept of "nature," the modern sciences needed to exclude all those aspects that did not fit the picture of bits of matter wandering through an empty universe of mechanic relations. In *Concept of Nature*, Whitehead speaks of "the theories of the bifurcation of nature" in order to address this complex phenomenon.[35] The modern sciences needed for their development a nature reduced to aspects one could easily grasp by measuring instruments. Therefore, qualities like weight, size, and speed were thought to belong to nature itself, while other aspects like color, smell, and taste had to be excluded from any scientific consideration since it was not possible to register them as degrees or numbers and thus locate them within a coherent theory of general laws. Hence, the modern sciences needed a distinction, which philosophy promptly delivered, between primary qualities, or those essentially belonging to nature itself, and secondary qualities, or those that arise only in the percipient subject and have no existence apart from it: "Primary qualities are the essential qualities of substances whose spatio-temporal relationships constitute nature (. . .). These sensations are projected by the mind so as to clothe appropriate bodies in external nature."[36] This separation between different kinds of characteristics opened the path to a purely mathematical consideration of the object "nature" as well as to its quantitative and systematic tabulation (for example, Bacon's classification schemes).

This elimination of any kind of nonmeasurable aspects like perceptions (colors, smells, sounds) and emotions, as well as teleological implications and activity, implied a complete exclusion of experience, which had to be cast out of nature in order to render it easily describable in quantitative terms. Accordingly, what was now relevant to natural sciences was "only that part of nature that can possibly be an object of sense perception, *leaving out* the experience itself, in which the object appears."[37] The poor vestige that remains, which we call "nature," contradicts our most immediate image of it. As Whitehead notes with scalding irony:

> Bodies are perceived as with qualities which in reality do not belong to them, qualities which in fact are purely the offspring of the mind. Thus nature gets credits which should in truth be reserved for ourselves: the rose for its scent: the nightingale for his song: and the sun for his radiance. The poets are entirely mistaken. They should address their lyrics to themselves, and should turn them into odes of self-congratulation on the excellency of the human mind. Nature is a dull affair, soundless, scentless, colourless.[38]

It turned out to be impossible, however, simply to dismiss these not-merely-mechanical aspects, since it would have meant the denial of any possibility for human freedom, and therefore even for scientific research itself. Thus they had to be somehow parked on a different ground and fenced off. Science needed the subject of modern metaphysics as an Archimedean standpoint to support its concept of a *reduced* nature. For this free field of scientific analysis to arise, these excluded secondary qualities had to be brought together on a different stage—the big scene of the representational subject. This brilliant move cleared the ground for scientific research and eliminated all the awkward elements that resisted integration into the theoretical scheme: "the enormous success of scientific abstraction, yielding on the one hand matter with its simple location in space and time, on the other hand mind, perceiving, suffering, reasoning, but not interfering, has foisted onto philosophy the task of accepting them as the most concrete rendering of fact."[39] As soon as we accept that nature *is* matter endowed only with primary qualities we automatically exclude ourselves as percipient subjects, with our experience of secondary qualities, from nature itself and posit ourselves in sheer opposition to it.[40] The necessary outcome of this bifurcation was both a fixed and objectified nature, on the one side, which could no longer "speak" except through the scientific channels, and on the other a free, percipient subject, defined at first only through the remains of that process of exclusion.

This subject thus emerged as the waste-disposal site, the dumping ground, of a purified nature coming into being as the place in which to dispose of all that might hinder scientific research. At the same time, the modern sciences externalized the question about this exclusion itself, letting it be relevant only for philosophical interrogation and relying on it for their very foundational grounding.[41] The modern sciences needed the subject in both respects: the ground as a field for the excluded elements, and the "ground" legitimating the very possibility of their existence within a bifurcated framework. While the subject did plainly fulfill the task of being the disposal site serving scientific explanations of nature, it also legitimated the construction of an utterly scientific, object-like nature and became therefore the foundational principle for the existence of the sciences as such. The bifurcation called for a justification that the sciences could not deliver and yet needed in order to define themselves. The major task that fell then upon the modern philosophy of the subject, was, specifically by means of a theory of knowledge as well as of epistemology, to support the bifurcation in order to keep the stage clear for science.[42]

In the light of its modest emerging, the appearing of the modern subject seems to be a less glorious event than might be supposed. Yet what began as a disposal site developed through the ages into a rich system of thought: the philosophy of the subject followed its own developmental paths along the history of metaphysics and transcended the limitations of its origin. Along with this process of refinement, it maintained and strengthened its foundational role and became the ontological ground itself.

Through this process of bifurcation, on the one side "nature" has been defined as devoid of any glimmer of freedom, experience, and internality, reduced to a matter of fact; on the other side, the subject has been reduced to its functional role for science as well as—to put it in Seyla Benhabib's words—to a "pure subject of knowledge, to consciousness, to mind."[43] As a consequence, whenever we think of a reunification of human and nonhuman, of nature and subjectivity, of what is "out there"—the lions—and the urban sphere of life, we must acknowledge that the modern concept of "nature" is based on a bifurcation and is essentially linked to, and deeply rooted in, the modern metaphysics of the subject. When we talk about finding a way back to nature or a desire to reunify the bifurcated poles, we should recall that each pole is what it is because of the bifurcation itself and each pole would not exist without the other one. A reencounter between *that* subject and *that* nature would imply a conflagration of both, the destruction of each side, since each can stand only because of, and within, the separation. It is therefore not a matter of recovery, of rolling time back, of building bridges over the walls and the rift separating the city from the wilderness.

Would a condition of pre-bifurcation offer a fertile path to political ecology today? Can such a possibility even be thought of, and if so, how? Can we just dismiss the modern concept of nature and, at the same time, the modern metaphysics of the subject with a gesture that wipes out the history of our provenance and destination? Would such a move not be a new foundation carried out by the same foundational subject in its self-constructed, powerful location? Are there other paths opening through the traces of the complex, multilayered history of modern thought and through the fractures in its nonlinear structure? The fundament itself seems to be already shaking on a groundwork of its own construction, and between these fractures its interwoven texture shines through.

THE FUNDAMENT IS SHAKING: POSTMODERN CRITICISM

Along the path (or one of the paths) of modern ontology and metaphysics, the subject has developed a growing awareness of its foundational character for science. Revealed finally as the ground for truth, the subject began to tremble on its own unsteady groundlessness. At the peak of modernity, the all-sustaining ground comes to recognize the missing soil under its feet. The re- and deconstruction of "the" modern subject has been one of the most important contributions of postmodern thought:

the foundational principle has been unveiled in its weakness, unsteadiness, and fragmentation.

The ground for truth collapsed; it fell apart into a great number of fragments without any continuity other than the provisional texture woven by language. Accordingly, modern ontology and metaphysics, revealed now as *metanarratives*, fell to pieces along with their sustaining principle. As Jane Flax points out, "postmodernists wish to destroy all essentialist conceptions of human being or nature," including the very idea of the subject as such; "the subject is merely another position in language."[44] In spite of Flax's radical claim, we maintain that, rather than literally destroying the claims of the modern subject, postmodernists seek to expose the alleged unitary and stable subject of modernity in its being specifically "another position in language" and no more. In doing so, postmodern criticism registers and uncovers the fragile consistency of the city walls and reveals them as a papier-mâché scenery for our self-representation. It recalls the *ungroundedness* of the subject's structure and the lack of legitimacy of its unquestioned foundational role. The well-guarded, sheltered subject of a bifurcated modern world is thus no longer secure within the walls. It spreads out of the imposed boundaries; it multiplies, fragments, falls apart, ventures beyond the secure limits of its self-constitution toward more dangerous possibilities.

Yet the metanarratives of modern metaphysics—including ontology as well as the ontology of nature—are not simply dead once and for all, letting us back in a disorder of localities with no other interconnection but the texture that we weave and unweave among them.[45] This position still seems to be caught within the bifurcation, since it registers the fragmentation of the subject and the outcome of differences without taking note of their complex interwovenness. The (textual) plot connecting local stories thus veils the weave, from which they emerge as identifiable stories.[46] No local story makes its appearance all of a sudden as an isolated fragment, broken and separated; rather, it emerges from a web of relational references with faint, poorly defined edges.

Referring to Whitehead, we might therefore conclude that both the concept of one given foundational totality and also sheer isolated fragments can be considered abstractions from the concrete ongoing relationality of beings. Both the "whole" and the "wholes" are to be considered dynamic, interconnected webs that can be, and are in fact, isolated only by means of abstracting processes. Therefore, local stories in their isolation can be as abstract as metanarratives; at the same time metanarratives can be unmasked in their being (provisional) broad interconnections of local stories. According to Whitehead, nothing "simply happens," nothing is simply located and isolable in its simple location as a fixed standpoint. Rather, each event, each entity of whatever kind arises from its constitutive relationships to all other events and entities—all standpoints are interconnected: "each part is something from the standpoint of each other part, and also from the same standpoint every other part is something in relation to it. . . . Every location involves an aspect of itself in every other location."[47] The universality-claim of metanarratives attains a slightly different, and less threatening, meaning if we link it up with the ongoing relationality of all beings: "Universality of truth arises from the universality of relativity."[48] If the fragments resulting from the dissolution of the subject as unifying principle remain separate (the only possible connection being given by the language or the text intended as texture) and are, no less than the modern subject itself, monads locked up without internal relations, differences then risk becoming absolute and therefore reciprocally indifferent. On the contrary, if we take account of the dynamic, ongoing relationality connecting all things, we must acknowledge that "the differences arising from diversities are not absolute. Analogies survive amid diversity."[49] By following Whitehead on relationality we can take seriously the plurality emerging from the exposure of the groundless ground without ending up floating among shattered fragments.

While the dissolution of the modern subject(s) advocated by some postmodern thinkers does not seem to open a path away from bifurcation as such—since it remains caught within the same separation by focusing almost exclusively on the "urban" realm of language—it nevertheless renders possible a different approach to the bifurcated poles by revealing them in their fractured heterogeneity. Although the city walls are not entirely reduced to dust, their secure grounding has been finally unmasked as an illusion. Hence, by naming the unsteadiness of the ground, postmodernism offers a chance for tracing different paths within the complex bifurcation story itself. One of these paths is that proposed by Vattimo of recovering (from) metaphysics, getting over it without dismissing it. This is the path I will follow in the next section.

Having lost the subject as the ultimate ground, some might be tempted to consider "nature" as the secure basis for science and philosophy, as well as for ecological claims. In the face of a pervasive suspension and suspicion, "nature" presents herself as the stable self-identical landmark while still allowing a multiplicity of opinions and relativistic standpoints on the other side of the fence. As Latour points out, mononaturalism seems to be the necessary ground for multiculturalism. It is because we have *one* speechless nature that we can say anything and more within the human world without getting lost.[50] The one unquestionable nature outside allows plurality within. The boundaries are neat: different voices can arise in the marketplace as long as they are contained within the city walls, defining their limits and enframing their humming.

Furthermore, if we bring to mind the fact that neoclassical economics rests on the presupposition of the modern natural sciences, we might realize the danger of this enclosure within the city walls.[51] Here plurality is at home, all preferences are welcome and never questioned as far as they take place within the walls and accept the presupposition of a nature infinitely at disposal as source and sink. The very fundament of our economic system is indeed nature as a presence-at-hand, nature as passive matter, nature as a silent source to be exploited.

We need to keep this in mind when we think of the necessity to render nature more present within the walls in order to support ecological claims. Nature is nowadays all but absent in our economic, scientific, and political discourses. It cannot be therefore a matter of simply bringing nature back into the political space, and certainly not *the* nature of modern natural sciences, since it is already all too present and pervasive.

It is also not a matter of simply substituting for nature *physis* or some holistic concept, aiming thereby to get over bifurcation and finding a more profound unity of the opposites. The need to return to nature expressed by some spiritual as well as ecological groups often takes the shape of defining *one* nature as the common background beyond cultural distinctions, while acknowledging at the same time cultural pluralism.[52]

TOWARD A POSTMODERN ONTOLOGY OF NATURE: A RECOVERING SHIFT

Bifurcation is a very complex phenomenon that cannot simply be reduced to one all-present fact to be overcome. "Nature"—or whatever we might call it—can be described as bifurcating into *physis* against *techné*, as passive matter in opposition to subject, as wildness in opposition to civilization, as the divine power of creation in opposition to the human reshaping of materials, as the female reproduction process in opposition to the male production system, and so on. Although all these bifurcations rest upon a similar approach, which is related to the tradition of Western modern thought, they are probably not all the same and cannot be simply brought back to one common archaeology.

While we need to acknowledge the provenance of our views of nature in continuity with modern metaphysics and therefore accept them as an inheritance that constitutes us, we may also have to recognize that they do not constitute *the* modern destiny of Western humanity par excellence as a single line of inheritance; rather, they represent a dominant set of self-imposing patterns, calling for replication. In Heidegger's terms the word "destiny" (*Geschick*) does not refer to a fate that we have to embrace with resignation, but to a destination, an efficacious stream of influences that constitute us, that send us forth. As such, these patterns, which have been established over centuries, are indeed strong enough to bear some *obligatory* requirement for conformation and repetition;[53] nevertheless, this call is neither an absolute claim nor an unbreakable chain. Taking seriously the destiny of bifurcation as well as of modern ontologies means to acknowledge that we are inscribed in its frames as the structure that sustains our stability and possibility of acting.

As Vattimo, following Heidegger, elucidates, we cannot get beyond modern metaphysical ontology simply by substituting a different perspective on reality. The illusion of a new general vision of the world bears the risk of old shadows striking back. Any attempt to replace metaphysics with something different can only lead to a new metaphysical scheme. Metaphysics—and, for our topic, modern ontology of nature as well—is not something that can be just cast aside like an old dress; rather, it is more like a sickness, a destining, something we come from and are delivered to, something we can recover from, and yet never completely.[54] Likewise, we cannot overcome (*überwinden*) bifurcation without reproducing it over and over again. The criticism against the enframing of nature (*Gestell*) cannot find its expression in the imposition of a different frame, a new frame.[55] According to Vattimo, the enframing itself can in fact shift (*verwinden*), or turn into a more original event,[56] opening the way to its own passing into new—and yet not completely new—possibilities. Vattimo proposes here Heidegger's use of the term *Verwindung*, transformative renewal, in order to sketch a path through the destiny of metaphysics, which might be nonetheless no longer "metaphysical."[57] This concept seems very fruitful for a possible postmodern ontology of nature that can support political ecology and reorient philosophical concern about nature without either remaining stuck within modern fixed categories of thought or giving up any ontological frame whatsoever.

Verwindung bears the same root as *überwinden*, overcoming, but instead of the prefix "over" it presents a prefix, which in German always has to do with a setting in motion, an activating and at the same time a twisting. Sometimes the "twisting" leads even to the wrong move, like dialing the wrong number,[58] or slipping into the wrong word by chance.[59] Therefore, it is an active move, although not necessarily a conscious one, something that "slips" as from the side, that occurs, like a *lapsus*, a suspension in the course of the regular habit of an expected and obvious continuity. Among its stratified meanings, *Verwindung* implies also a distortion, a twist, a setting into motion and shift in the direction of the movement while doing this. Hence it is possible to "live metaphysics as well as the *Gestell* as a chance, as a possibility for a change, by which both curl towards a direction that, while not being the one expected by their most proper essence, is still connected with it."[60]

Moreover, *verwinden* suggests in everyday German the healing process through a sickness or even the mourning after a loss.[61] A serious sickness cannot just be overcome as if it had never been present, and a loss cannot be simply excluded from the horizon of relevance and thus completely withdrawn. These processes always need time, imply a slow recovering that bears all the scars. If modern ontology is partly a sickness, if bifurcation of nature is an error—intended more as an erring, a wandering off—within the history of metaphysics in which we still find ourselves, healing might be a path, too, intended as facing the destiny, reading the traces, and slowly healing the wounds of bifurcation itself without fully dismissing it.[62] *Verwinden* means "to recover," intended as both getting better and getting back, regaining something lost or out of reach. It does not mean to drop bifurcation, but to regain it in its multilayered complexity, to keep it as a part of the story, although not as the whole story.

CONCLUSION: RE-SIGNING (TO) THE BOUNDARIES?

If we were to find an analogy with the complex structure of our bodies, we could think of slowly getting over old stratified patterns of standing and moving, which might very well have been useful in the past but imply pain and tension in the long run, without completely dismissing their necessary supportive function. We know we cannot achieve a complete "wiping out" of patterns once curdled into habits just by pressing the "reset" button of our bodily constitution. Even if this were possible (which it is not, given that we are the result of our own stories of postures and gestures), we would still remain entrapped within the illusion of an infinite power of the mind over the body, of the subject over nature in a bifurcated framework. Alexander Technique offers a powerful analogy for a transformative renewal (*Verwindung*) of ontologies of nature, as a path of healing painful postures without altogether detaching given frames.[63] According to Alexander Technique, to deal with established patterns is to slowly go through the immediate bodily experience of a possible liberation, of a possible feeling of easiness and wellness. It is not a matter of actively doing something against the old patterns (sudden replacing of old frames with imposed new ones), which would instead be reinforcing the pain and the tension in the muscles produced by the patterns themselves. As in this case the tension, the need of controlling, is part of the problem, it is very unlikely to offer the way of solution. Rather, it is a matter of exploring other feelings, letting them come through, sensing for short moments of suspension the lure of new possibilities of being, of new patterns or of a different composition for the old ones. As with *Verwindung*, it is a process of recovering—both as healing and as recovering by remembering one's own history—a slow process of becoming open to possibilities that had been excluded in order to preserve the tensed and established balance.[64] This requires that we follow the weak traces, the treads woven underneath, and those residues that the dominant patterns have kept hidden on the background for a long time. It is a slow process of dismissing mechanisms of compensation for pain and tension—as well as for ecological destruction. Starting anew without all compensations would destroy—dis-member—one's own very stability and steadiness. There are no blazed trails, no paved paths. As Heidegger would put it, it is like finding a clearing all of a sudden within the woods and resting there, going on with the experience of the broken paths, which probably do not lead to any final goal, but to other clearings, and other paths. It is not a maze, although there is some risk of getting lost.

In a similar way Latour proposes in his constitution for a new collective of human and nonhuman beings the cooperation between the two powers of representation. On the one side there would be a welcoming house with the power of taking into account and the requirements of perplexity (taking into account, without simplifying the number, the voices of the excluded elements, those left outside the walls, the traces hidden in the background) as well as of consultation (facing the necessity of some criterion of admission by means of relevance). On the other side there would be an ordering house, with the power of arranging in rank order and the requirements of hierarchy (verifying the compatibility of the new proposal within the established patterns, seeking integration through negotiation) as well as institution (ending provisionally the negotiation in order to settle reference points, stable patterns that can support the process itself).[65] Provisional stability is a necessary condition for all living beings, facing their unsteady balance on the edge of chaos: established patterns render more creative and efficacious actions possible, by allowing a withdrawal of attention from habitual gestures. If we were to rediscuss at each step of our political commitment the use we make of the term "nature"—its ambiguous and bifurcated history of development, its risks and burdens—we could never make a step toward political decisions and social as well as ecological actions. Yet if we simply took the concept for granted, we would risk remaining captured within its fatal contradictions and reproducing its devastations. Also, choosing other terms instead of it does not really solve the problem. As ecofeminist Alicia Puleo, among others, has shown, the term "environment" expresses a form of reductionism, by which nature appears like a simple scenario for human self-realization.[66] Similarly, the term "nonhuman" risks reproducing the traditional logic of noncontradiction along the Aristotelian path, in which the opposite term becomes simply the reversed mirror of the positive one. None of these alternatives escape the burden supposedly borne by the term "nature."

The dynamic of the process moving back and forth through Latour's houses—including the crucial need for a provisional end of negotiation,

Whitehead's swinging between fecund abstractions and their misplacement in a pervasively relational pluriverse, Vattimo's path of *Verwindung* as remembering, distortion, and healing—all these offer a possible field for doing justice to both critical awareness and political decidability.

We can follow here the other meaning of the term *Verwindung*—to wind, to spin, to move around—to give a kick to the static, nostalgic bifurcations not in order to remove them, but with the aim of getting them spinning around as in an opening spiral for new possibilities of thought and action. Returning to the image of the walls around the city, we would neither want to tear them down all of a sudden, and so let the lions jump in, nor to reinforce them. Rather, by working locally though connectedly on the walls, a process of transformation might set in, allowing the walls to grow into a labyrinth-like structure, creating passageways among and through them, as well as through the city and the "outside." Instead of resignation to the historical destiny of modern ontology of nature, we might move toward a positive definition of this development, re-signing (to) it—as Vattimo proposes on the same line of *Verwindung*—in the sense of marking the boundaries, where necessary, to build walls, to define. Yet those lines remain provisional, not deeply rooted into the ground, renegotiable, always located in relational settings, porous.[67] They operate more as gates through the walls that shelter the necessary stability of established patterns and structures while slowly allowing "externals" to enter and reshape spaces.

We might, for example, want to refer to ecology as a natural science belonging to biology and yet inhabit the rifts traversing its attempt to mediate between the scientific need for exact predictability and the nevertheless highly unpredictable complexity of its object. In fact, contemporary physics and ecology itself are each questioning their modern legacy. It is precisely because of the failures of practical ecology in predicting natural phenomena that even according to the scientific model the concept of nature needs to be renegotiated.[68]

By reconstructing some of the relevant stages in the development of the modern concept of "nature" and by outlining the bifurcations that took place at the very beginning of modern science, I have tried to sketch a path toward possible postmodern ontologies of nature, which acknowledge their being inscribed in the tradition in which they arise without remaining simply caught within its grave contradictions. We need to create spaces for (inter- and transdisciplinary) encountering, recovering, and rediscovering relations, letting the different ways of speaking about nature, and of nature's own speaking, find rooms and paths to emerge and meet. In this way we can allow for the use of the term "nature" as inviting a texture of different traditions, dynamic possibilities, and negotiable definitions. As Whitehead recalls: "Unless we produce the all-embracing relations, we are faced with a bifurcated nature."[69] The all-embracing, all-through relations—in the plural form—weave a complexity of patterns and structures, through which abstractions, fixed points, milestones can be provisionally traced and crossed again.[70] We are inscribed in the same destiny of what might be considered as outside the walls: a destiny that can be twisted into new creative realizations. The fight for ecology is not about "nature" tout court—neither the nature on the rack, silent and silencing, nor that lying beyond the city walls, nor yet the hidden irreducible ground of all beings. Rather, the battle takes place in an opening toward a dynamic relational field twisting across the gates, both within and outside the walls.

❧ Toward an Ethics of Biodiversity: Science and Theology in Environmentalist Dialogue

KEVIN J. O'BRIEN

INTRODUCTION

In an often told and possibly apocryphal story, a member of the Christian clergy anxious to engage in dialogue asked the Marxist physiologist John Haldane what his study of the natural world had taught him about its Creator. Haldane replied that if there is a God, it seems to be one with "an inordinate fondness for beetles," citing the hundreds of thousands of distinct species of the insects already catalogued and the uncountable others that human beings have never seen. Haldane was likely annoyed by the question, and trying to shock the clergymember out of a follow-up. But as an environmentalist and a Christian, I am delighted by the exchange—delighted enough, in fact, to take up the questioner's position myself in this essay. I hope to ask more careful and involved questions, and I hope to receive more detailed and expansive responses, but the fundamental goal of this essay is a dialogue between scientific ecologists and environmentalist moral theologians.[1]

More specifically, I want to know what role biodiversity plays in the world and how we should respond to it ethically. A more developed and scientific description of biodiversity will come below, but by way of introduction I define the concept broadly as the "variety of nature," signaling the vast diversity within all species, between them, and among the ecosystems in which they are organized.[2] Biodiversity is, in this sense, a way of talking about the world as a whole—in theological terms, the sum of living creation—while emphasizing the vast multiplicity of its expressions and patterns.

In addition to making an assertion about the *variety* of nature, biodiversity is also commonly used to reference a "crisis" of loss. Just as human beings have come to glimpse the vast diversity of the natural world, we have also begun to reduce biodiversity at a rate unprecedented for at least 65 million years. Indeed, although it is not as commonly referenced as global warming, this "biodiversity crisis" is frequently used to represent environmental degradation more broadly as a generic environmentalist rallying cry. To preserve biodiversity is, in some sense, to preserve life itself.[3]

So, inherent in the concept of biodiversity are two claims: that life on this planet is vastly diverse, and that this diversity is under threat from our species.[4] This essay asks how to respond to these facts, and I will offer two answers. The first is a broad, methodological argument that, however environmentalist moral theologians respond to the concept of biodiversity and the natural world it represents, we must do so in conversation with the scientists who have most carefully defined and monitored it. I make this point in part by articulating a methodology of naturalism, but also by modeling it, developing my own position in dialogue with scientific ecologists who study biodiversity.

The second answer is a constructive theological argument about how Christian thinkers should understand the world and humanity's role in it. My claim is that theologians and theological ethicists must take biodiversity seriously as a place where humanity encounters the rest of creation, and that this means becoming fully aware simultaneously of the immense power our species has established over all other forms of life and of the limitations in our ability to use that power responsibly.

The work of this essay is therefore to demonstrate the potential of a genuine dialogue between ecological science and Christian ethics by using such a dialogue to articulate, defend, and explore an ethics of biodiversity. Like the clergymember who nagged the physiologist, I want to learn more about my faith by entering into dialogue with the life sciences. I want to know if God is calling me to an inordinate fondness for biodiversity.

THE METHODOLOGICAL TURN TO NATURALISM

The most obvious starting places for a theologian thinking about biodiversity are the resources of the Bible and the Christian tradition, which can certainly provide substantial insight into the concept. The book of Genesis begins by dividing lifeforms into groups based on when they were created. Reflecting upon the order and interdependence of this creation, the divine character in the story observes: "indeed, it was very good."[5] This is a discussion of biodiversity and its value. Just as compelling to me are the words of a central theological figure in my own faith tradition, Thomas Aquinas, who asks in the *Summa Theologiae* whether or not the vast variety of life is the work of God. He answers that it must be, because God's "goodness could not be adequately represented by one creature alone" and therefore "the whole universe together participates in the divine goodness more perfectly, and represents it better than any single creature whatever."[6] Inherent in this claim is a theological affirmation of creation's diversity, implying a value for that diversity over any single species, even our own. Much work could be done with such Biblical and theological sources to offer a nuanced Christian ethical position on biodiversity.

The argument of this essay, however, is that these sources by themselves are *not* sufficient, and indeed should not be our starting place when developing an ethics of biodiversity. Instead, theological reflection on biodiversity should work from a deliberate and careful understanding of the world around us as a natural system. I adapt such a naturalistic methodology from the Christian ethicist James Gustafson, for whom theological truths must be "derived from observations about all life in the world that are backed by modern sciences."[7] For Gustafson, the first question of a naturalistic theological ethics is not "What should we do?" but rather "What is going on?"[8] Gustafson's own answers to this question of course differ according to the ethical issue he is examining, but a common theme throughout his work is that our best and only means of understanding what is going on in this world is to examine it as closely and carefully as possible. Human beings are natural creatures, and so our lives are circumscribed by the natural world; we cannot hope to live or to think by any but natural means.

Theological ethics, in Gustafson's view, attempts to understand the same world as all other academic pursuits, and is constrained by the same complexities and human limitations. Theology, therefore, has no special authority or privilege and it works from the same basic source as all other pursuits: human experiences and interpretations of the natural world.[9] The theo-ethical enterprise of asking how humanity should live and what meaning can be found in our lives is not somehow distinct from the empirical attempt to discern what is happening in the world and what rules and trends govern it. Both are answered by reflecting on the nature of reality and so both shed light on the natural world and the God who created it.

This does not mean that normative truth can be straightforwardly derived from the world. Naturalism asserts that morality grows out of natural contexts, but it does not assume that there is anything like a single set of moral truths self-evidently revealed by studying the natural world, or that nature is a static entity in which universal norms can be clearly discerned. Even after we understand our evolving world as best we can, we must still interpret it and discern any moral implications, as well as cope with the inevitably divergent interpretations and discernment between peoples with different social locations and interests. The point therefore is not that paying attention to the natural world is the *only* step of ethics, but that it must be a *first* step.

With this understanding, theological naturalism calls for a deeply interdisciplinary approach to ethics, and this is particularly clear in the case of environmentalist ethics. As Gustafson writes, the work of a theologian concerned with the nonhuman world ". . . intersects with the research of ecological scientists, economists' accounts of cost-benefit analyses, and both domestic and international political powers. Faced with different analyses of present and future circumstances, the theologian has to assess scientific explanations of environmental problems."[10] When thinking about a moral problem like the loss of biodiversity, Gustafson argues, we find ourselves at an intersection of multiple perspectives, each of which sheds light on different dimensions of the problem and possibilities for a solution. If, he argues, "there are multiple interacting causal factors" behind this problem, then scholarship and its proposals "have to be inclusive," not putting any single discipline in charge but rather remaining at, and working from, the intersection.[11]

I cannot pretend to deal with all the relevant sources for an ethics of biodiversity in this essay, but will attempt to take Gustafson's argument seriously by working at the intersection of scientific ecology and environmentalist moral theology. Again, this must be understood as *a* step toward an ethics of biodiversity rather than the *only* step. Naturalistic ethics cannot turn to scientific research with an expectation that scientists have all the answers, or are always right about the answers they do offer. Science is by its nature incomplete in its understanding of the world, and scientists are as limited and capable of failure as any other human beings. Thus, a turn to science should never be a means to undermine other ways of thinking. Furthermore, the complexity and partiality of scientific data must be respected, and ethicists must guard against the temptation to generalize scientific findings and too readily assume that they fit a pre-existent ethical program.[12]

A naturalistic approach to ethics is based on the assumption that we cannot ignore our world, nor can the limits of scientists or our limited understanding of their work be excuses to avoid dialogue. Thus, the methodological starting point of this essay is that environmentalist moral theologians have something important to learn about biodiversity from a more careful examination of what ecologists are saying about it.

ECOLOGY, UNCERTAINTY, AND BIODIVERSITY

The phrase "biological diversity" was first contracted into one word by Walter Rosen, a botanist who helped to organize a 1986 "National Forum on BioDiversity." This shortening was quickly popularized by the publication of the proceedings from the forum in a volume edited by E. O. Wilson.[13] As biodiversity became a concept of growing importance for ecological scientists, it began to spur research focused on chronicling the varieties of life and the role that diversity plays in the functioning of ecosystems.

As discussed above, biodiversity is as much an activist concept as a scientific one, and this has been true throughout its two-decade history. Nevertheless, it is overly simplistic to say that the popularization of the concept is *solely* a response to environmental degradation. Rather, to understand biodiversity properly as a scientific concept, it must also be thought of in the context of the worldview that pervades most contemporary ecology, where emphasis is placed upon complex, unpredictable ecosystems heavily influenced by humanity.

It has become common in the last two decades to talk of a paradigm shift in ecology away from an older model, frequently traced to the early-twentieth-century plant ecologist Frederic Clements, which was based on the idea that natural systems progress along a predictable and comprehensible succession of stages until a state of equilibrium, or "climax," is reached, at which point the system becomes stable and mature. According to this paradigm, systems are naturally balanced, or on their way toward such balance; the role of ecologists is therefore to decipher this balance and the means by which nature achieves it.[14] In contrast, many ecologists now work from a paradigm that asserts that natural systems are fundamentally unpredictable, prone to—and even dependent upon—disturbances, random changes, and broad shifts or "flips" between different states. Ecologists who embrace these ideas no longer seek to understand the typical succession of ecosystem types toward equilibrium, but rather tend to study particular systems and attend most carefully to the surprises and disturbances that influence each one. In this view, as one ecologist puts it, there is "no mother nature" and no "balance of nature."[15] As environmental historian Donald Worster points out, this new perspective calls many of ecology's formerly unifying concepts, such as "health" and "balance," into question.[16]

So, in light of this view of the world, many ecologists see the diversity of biological and ecological systems as increasingly central to their thinking. "Biodiversity" is a concept that stresses not a predictable pattern of ecosystems, but rather the wide, varied, and endlessly complex distribution of life across the planet. It is not based on the assumption that ecological systems reach comprehensible steady states, but rather on the unpredictability of natural systems and the inevitability of disturbances in their development. It therefore fits neatly into the basic worldview of contemporary ecology, which helps to explain why it is currently such a popular idea in the science.

Of course, the widespread use of the term "biodiversity" does not mean ecologists are univocal on every aspect of the concept. There are heated disagreements about what kind of definition biodiversity should have, about how it should be understood as a product and/or producer of functioning or stable ecosystems, and about what (if any) role scientists should play in its conservation. Given such disagreements, the enormous quantity of work being done in the science, and the simple fact that ecological knowledge is evolving, it is difficult and risky to characterize "ecological thinking" with any generalizations. Nevertheless, I think it is safe to say that there is a virtual consensus about the incredible complexity of life on earth and its vast, perhaps even unknowable, extent and character.

One of the central projects pursued under the banner of biodiversity research is the cataloguing of life on earth, and an early conclusion in that attempt has been that there is far more than can ever be known or observed by anyone. Estimates vary widely, but a common figure suggests that there are fourteen million distinct species on the planet today.[17] Furthermore, many ecologists note, a simple count of species only captures a tiny slice of biodiversity, which is also reflected in the genetic diversity within each species and the array of ecosystems in which species interact.[18]

The uncertainties and unexplored limits of biodiversity were emphasized recently when two marine zoologists excitedly published the first ever pictures of a giant squid in the wild.[19] These are animals that, when fully grown, range from twenty to eighty feet in length and wrestle with whales, and yet the first live pictures were taken in the year 2005. At this writing it is still the case that no human being has seen one alive in the water. Given that we have so much trouble tracking down and observing even an enormous squid, we can only imagine how much there is in the oceans and all around us that might be even harder to find.

Diverse expressions of life on earth are also very old. The earliest organisms on this planet are thought to have existed 3.5 to 4 billion years ago, with mutlicellular organisms taking shape about 1.4 billion years ago and animals appearing about 600 million years ago. Our own species is famously a latecomer in this history, with the fossil record showing the first modern human beings between 150,000 and 200,000 years ago.[20] Here, too, there is a lesson about how little we know: these figures are based on the fossils that have been found and on the best guesses of scientists about what might have existed that was not fossilized; virtually no scientists claim that anyone can accurately reconstruct all of the history of life on earth.

One of the most interesting conclusions these researchers are willing to propose, however, is that biodiversity seems to have been steadily increasing throughout virtually all its development. Despite a few notable extinction events and some periods of relative stasis, graphs showing genetic, species, or ecosystem diversity over history of life on earth tend to climb steadily upward. This is particularly clear over the last 200 million years, over which period diversity has apparently spiked quite substantially. Why this is the case, why life on this planet seems to have evolved steadily toward greater diversity, is so far not conclusively explained.[21]

There is, however, a widely agreed-upon explanation for the very recent evidence that global biodiversity is now declining, and it is human development. Researchers estimate that species extinctions are occurring up to a thousand times faster than they would without our influence.[22] This figure does not even take account of the other levels of biodiversity: species in which only certain genetic strains are thriving or ecosystem types that could disappear entirely. Our crisis is not yet historically unprecedented in scope: there have been five other massive extinction events, and biodiversity continued to climb afterward. But this is believed to be the first time that extinctions at such a scale have been caused by the activity of a single species. Biodiversity, ecologists have found, is a phenomenon increasingly at the mercy of humanity.

This last point fits, incidentally, with the contemporary paradigm in the science. While earlier versions of ecology tended to dismiss human activity as a disturbance that interrupted the course of natural systems, ecologists today usually consider disturbances inevitable in any subject of study. Humanity, as a growing source of such disturbances, is thus increasing in importance for ecological research. As David Tilman puts it, "If ecology, as a discipline, is just the study of 'nature,' with nature defined as species living in habitats that experience minimal human impact, it is a discipline headed toward extinction."[23]

In a widely cited 1997 article from the journal *Science*, Peter Vitousek and his colleagues discuss the "Human Domination of Earth's Ecosystems," and argue that "no ecosystem on Earth's surface is free of pervasive human influence."[24] A selection of their statistics is informative, if disturbing: 16–23 percent of the planet's land has been "wholly changed by human activity," with much more "substantially changed." Human beings have also caused an exponential increase in levels of atmospheric carbon dioxide since 1800, and have driven "as many as one-quarter of Earth's bird species" extinct in the Common Era.[25] This is but a sample of how our species is changing the variety of life that we can hardly understand.

ENRICHING THEOLOGY WITH ECOLOGY: CHALLENGING AND
ACCEPTING HUMAN DOMINION

To summarize this sprint through the science of ecology: biodiversity is a way to talk about the variety of life that captures its complexity, the limits of our capacity to understand it, and the historically unprecedented impact humanity is now having upon it. Switching to theological language, I might paraphrase that biodiversity is at once a sign of the richness and unknowability of God's creation and a signal of how influential our role in that creation has become. The lessons I gain from studying ecological research into biodiversity therefore impact two distinct but related theological themes: the centrality of humanity in human ethics and the proper relationship of our species to the rest of creation.

A prevalent theme in environmentalist moral theology has been the insistence that human beings are not the sole or central value in creation, and scientific evidence has frequently been marshaled in support of such arguments against anthropocentrism. For instance, the ecofeminist Sallie McFague argues that human beings must stop seeing the nonhuman world as raw material for development, and defends her argument with cosmological and ecological language. She insists that other beings "exist within the vast, intricate web of life in the cosmos, of which they are and we are all interdependent parts."[26] As this inclusion of human beings in cosmic interdependence demonstrates, a Christian environmentalism ought not become misanthropic, dismissing the value of human beings in order to emphasize the importance of the natural world. The value of human communities, however, comes from our connections to the rest of creation, rather than from our distinctness. As Dieter Hessel puts it, "we are called to appreciate the goodness and interdependence of all created beings, human and nonhuman."[27]

Based on the research discussed above, there seems to be considerable ecological support for such rejections of anthropocentrism. A world this incredibly full of life, which was so full of life for billions of years before our species evolved, does not seem to have been created for our benefit. If there are fourteen million species besides our own, and if the course of evolution took over three billion years to produce all of us, then naturalistic evidence does not straightforwardly suggest that we are the center and cause of all that is.[28] Furthermore, if biodiversity has steadily increased up to now, it seems odd to suppose that there could not be further diversification beyond us, and that we might not sooner or later be driven extinct by competitive species. To believe that God would not allow such a thing ignores the millions of species that have already come to the end of their time on Earth.

As I have discussed, however, biodiversity research emerges in ecology as part of a new model for the science, a model that increasingly incorporates humanity. In other words, as theologians and ethicists have been working to extend awareness and moral value beyond our species, ecologists have been working to attend to human communities and their impacts upon ecosystems. This offers an important reminder to environmentalist moral theologians that there remain legitimate reasons to attend to our own species and its uniqueness, to direct our ethical attention not just to biodiversity and life in all its variety, but also to the particular form of life we know best, which is exerting ever-greater influence upon God's creation. The task of theological ethics is not merely to emphasize human continuity with the rest of creation, but also to discern the proper place for humanity within creation's diversity.

Theologians addressing this question frequently turn to an analysis of Genesis 1:28, in which God tells the newly created human species: "Be fruitful and multiply, and fill the earth and subdue it; and have dominion over the fish of the sea and over the birds of the air and over every living thing that moves upon the earth." H. Paul Santmire offers a typical environmentalist interpretation, stressing that this passage affirms human connection to the rest of creation rather than emphasizing our distinction. He writes that this divine command must be understood "in the context of that all-pervading, harmonious world of *Shalom* that Genesis 1 presupposes, a world where humans and animals enjoy a marked commonality and where the Creator clearly has purposes for all of creation that transcend instrumental human needs." The biblical discussion of dominion is therefore understood as "an ecological construct" that "refers to humans assuming their divinely given niche in the earth alongside other creatures" rather than the permission for unchecked authority that has often been read into it.[29]

Many interpreters have argued that this divine mandate should therefore be understood not as a dominating lordship, but rather as stewardship over creation. Human beings do not own the planet, but are called to care for it responsibly, as stewards of God's earth.[30] Whether based in biblical interpretation or ethical commitments, such arguments aim to dismiss the problematic implications of dominion and instead offer a Christian theology that inspires responsible care for the nonhuman world.

A naturalistic approach to theological ethics must approach this topic somewhat differently. According to the methodology laid out above, the first response to the question of how human beings should relate to the nonhuman world is not an appeal to biblical imagery, but an analysis of "what is going on"? What relationship exists here and now? Turning to ecological research for an answer, we see that human beings throughout the world have become managers of our ecosystems. In the *Science* article cited above, Peter Vitousek and his coauthors make an implied biblical reference by identifying this status as "dominion over earth's systems." According to these and many other ecologists, human beings are currently acting as dominators. Furthermore, this is not a position that we have any feasible short-term option of abdicating: we are dominant to such an extent that we cannot easily remove ourselves from this role.

What distinguishes this notion of dominion from the biblical one is that it is a historical condition. Human dominance—or management—over earth's systems is something that has happened in recent evolution, and is therefore not necessarily a basic right inherent in the structure of the world. Indeed, according to the standards by which the *Science* article declares us dominant, we have only become so in the last two hundred years. This may be the most valuable theological lesson I derive from ecological science: our dominance over natural systems is a reality, but a reality human beings have created, and therefore one that we can choose to believe was *not* ordained from on high. Our dominance is not therefore a duty we must simply accept, it is a responsibility we inherit and one that within historical memory we can recall not having had.

Theologian Jay McDaniel makes a similar point, arguing that Christians must retain the idea of dominion "because 'dominion' names a historical fact." He goes on to argue that we must discern "an image of what might be called *right dominion*" or "dominion-as-stewardship" because, with current human population levels, abdicating our authority over the rest of creation is impossible: "In the best of scenarios, we are doomed to dominion."[31] This assessment that we have no immediate option but to manage the rest of creation seems to be supported by naturalistic evidence. As ecologists Reed Noss and Alan Cooperrider point out, even the decision to fence an area and leave it alone as a wilderness reserve is a managerial decision, particularly because in most areas of the world such an action requires considerable enforcement and monitoring to sustain. Furthermore, with pervasive pollutants, relocated organisms, and anthropogenic species extinctions across the globe, it is increasingly difficult to imagine that any system could be fully protected from our influence. "Letting things be" is therefore not an option.[32] Instead, ecologists teach us, our decisions about how to treat the planet's biodiversity must be intentional and careful, consistent with our societal goals for such systems and our own relationships to them.

If, however, we understand our dominion historically, in the sense that it has happened but is not necessarily something which we theologically believe *should* have happened, we must ask whether this is an authority that human beings deserve. After accepting that members of our species must deliberately manage Earth's ecosystems here and now, we can consider whether our most appropriate course of action in the long term might be to relinquish that authority. Human beings can ask, in other words, whether we should begin the inevitably long process of relinquishing our dominance, and perhaps even our stewardship, over the creation.

A relevant ecological lesson here is the extent of our uncertainties and ignorance about how creation works. The millions of species with whom we share this planet and the billions of years they have taken to evolve reveal a world of rich intricacies we cannot hope to fully master. Ecological research into the behavior of other species reveals a creation where giant squid and the whales they wrestle experience and participate in a reality we have no real capacity to understand. This seems to be a creation that, as a whole and in most of its parts, is utterly beyond human ability to make sense of or describe. Along those lines, it becomes reasonable to suppose that we might not be equipped to manage this incredibly diverse planet well. If this is the case, human beings—at least those of us in positions of privilege and power who have done most to create this condition of dominance and degradation—must consider making our authority over creation a temporary condition, learning instead to become participants in biodiversity, seeking to act like one species among a vast variety.

This proposal for modest human participation in creation may seem incredibly idealistic, or even impossible. It is certain that it would take centuries to deliberately remove human beings from positions of management while still preserving a commitment to just and free social structures. Nevertheless, my interpretation of reality is that human beings are not equipped to manage Earth's ecosystems, and this leads me to the ethical position that we must begin to find ways to coexist with the rest of creation without dominating it. My hope is that bringing this ethical claim into conversation with a careful and empirical analysis of our current status can help us to see how far we have to go to reach this goal, and that a further naturalistic investigation might suggest what we might do first to move in the right direction.

The ethics of biodiversity toward which I am working therefore sees human dominion as problematic, and calls us to begin relinquishing it, while accepting that contemporary management of earth's ecosystems is nevertheless necessary. One lesson of a theological ethics of biodiversity that takes ecological research seriously is that there is no way to live out a nonanthropocentric morality in an anthropocentric world, and we live in a world that human beings have made increasingly anthropocentric. To be precise about what I am claiming: a naturalistic perspective reveals that our species has become an incredibly influential force throughout this world, but it reveals no evidence that human beings *should* have such power. Indeed, it strains credulity even to argue that we are meant to understand most of this world. If we accept that creation was not made for our benefit, but that we nevertheless find ourselves the dominant species in the world; if we take on the management of God's creation on Earth as an inherited responsibility rather than a divinely ordained right; and if we seriously consider whether our best course of action might be to find a way to begin reducing our power and influence over the rest of the planet, we can begin to make a world in which human beings will participate rather than manage.

ENRICHING ECOLOGY WITH THEOLOGY: THE NATURE OF ETHICS

Clearly, a theological ethics of biodiversity must be carefully aware of the power human beings have over the rest of creation and the limitations of our ability to use that power wisely. In making this argument, I hope I have shown that environmentalist moral theology can learn productively from a dialogue with ecological science. To make this exchange truly dialogical, however, it is important also to move in the opposite direction, arguing that scientific ecologists have something to learn from environmentalist moral theologians as well.

We must first admit that a dialogue around biodiversity is not one in which theologians and scientists have equal footing. Ecological scientists developed the concept of biodiversity, and have spent decades trying to understand it and the implications of human activity upon it. On this matter they have an expertise that few, if any, theologians can claim. Furthermore, while scientific claims are meant to be rationally defensible and widely applicable, theology is informed by the particularities of believing communities reflecting on the natural world with a faith to which we cannot expect scientists to give credence. My frequent reference to biodiversity as part of "creation," for instance, is a theological interpretation of the natural world that I would never suggest should be accepted on scientific grounds.[33] Despite the limited applicability of some of my claims, however, as a moral theologian, I draw on a long ethical tradition while ecologists have only recently begun to question how their research might relate to moral norms and truths. It is along these lines that I believe theologians can and must contribute to this dialogue, offering a more holistic vision of an inspiring ethics of biodiversity.

This point can be well demonstrated by reviewing the moral claims of ecologists who publicly advocate conservation. Moved by the scope of human power over the nonhuman world and the pace at which that power is growing, many ecologists use their scientific authority to take positions of explicit advocacy on behalf of global biodiversity.[34] Generally, they do this by enumerating and explaining lists of all the ways biodiversity is valuable. Such lists vary, but they generally review its economic uses and potential, its necessity for human survival, and its intrinsic worth. Ecologist-advocates therefore argue that biodiversity is valuable to all economic actors as a resource, to human societies as a precondition for our survival, and in some cases also to caring people as a phenomenon worthy of preservation for its own sake.

Apparently assuming that economics is the most universally accepted and politically motivating justification for valuation, most ecologists-advocates have concentrated on the argument that biodiversity is important to the basic structure and continued prosperity of human economies. All economic interactions, the thinking goes, are predicated on resources and processes provided free of charge by the nonhuman world. These resources and processes, referred to as "ecosystem services," are therefore the most frequent justification for biodiversity's value. E. O. Wilson exuberantly notes that our current economic uses for the planet's biodiversity are only the beginning: pharmaceutically, we have discovered "but a fraction of the opportunities waiting," and there are "tens of thousands of unused plant species" available for use as food likely to be "demonstrably superior" to foods commonly eaten today.[35] For many, such arguments lead naturally to work with economists placing a monetary value on the services that healthy ecosystems with intact biodiversity provide or could potentially provide to human economic communities.[36]

Others argue that biodiversity's value is more fundamental than economic terminology can capture. Using some of the same evidence, these advocates argue that biodiversity is simply essential to the survival of our species, regardless of whether its importance is recognized by markets or quantified for consumers. To support this claim, David Tilman and Joel Cohen cite the Biosphere II project, an ecological laboratory in which researchers tried to create a closed system where human life could be supported with no contributions (after initial construction) from the earth's naturally occurring systems, which were dubbed Biosphere I for the purposes of the experiment. The attempt was to create an artificial, life-sustaining world. And it was a failure. Technicians were unable to stabilize the atmosphere or keep a majority of the nonhuman species alive during the two-year experiment. The lesson, according to Tilman and Cohen, is simple: "At present there is no demonstrated alternative to maintaining the viability of Earth. No one yet knows how to engineer systems that provide humans with the life-supporting services that natural ecosystems produce for free."[37] What we learn from Biosphere II, these advocates suggest, is that the thriving ecosystems which have evolved to be diverse and self-sustaining are the only known life-support systems for our species.

These arguments about economic and life-support value call on the public to care about biodiversity primarily from self-interest, defending the necessary conditions for humanity's continued health and prosperity. Such calls for the preservation of our own interests are important, but some advocates supplement them with a more selfless sense that biodiversity has value regardless of its importance to human life. Along these lines, Kent Redford and Brian Richter end their analysis of biodiversity conservation with the claim "that all biological entities and their environments have intrinsic value independent of their usefulness to humans . . . the preservation of biodiversity for its own sake, in its entirety and in its component parts, is a legitimate objective in itself."[38] Biodiversity, they argue, need not be valuable *for* humanity to be valued *by* human beings.

Even ecologists who argue for biodiversity's intrinsic value tend to make lists detailing its importance, providing careful, rational arguments about why it matters. Some recognize that this is not a complete or convincing ethics. Along these lines, James Miller argues that the most neglected task of conservation scientists is to "convey the importance, wonder and relevance of biodiversity to the general public," primarily by providing experiences of naturally occurring biodiversity that will inspire people to care about it.[39] Similarly, Paul Ehrlich suggests that the way to protect biodiversity is to inspire "a quasi-religious concern for our only known living companions in the universe."[40] Ehrlich's language explicitly opens the door to a dialogue with theological ethics, recognizing that people are motivated not merely by rational lists and arguments of valuation, but also by their fundamental belief systems, and by the images and ideas that capture the imagination.

I greatly value ecologists' enumerations of biodiversity's value, but as a moral theologian I find such arguments far from complete or persuasive. Missing is a sense of biodiversity as a phenomenon that can inspire rather than just require a better way of living, a touchstone from which a coherent and inspiring moral vision can emerge. Environmentalist moral theologians have long been engaged in the project of communicating about environmental degradation in ways that are not merely alarming on a rational level, but also convincing and powerful on a personal level. Anna Peterson, for instance, writes that while environmentalist ethics must be connected to a rational understanding of the world, we must not assume that this connection is straightforward; we must instead appreciate the complexity and richness of how moral decisions are made. "Ethics are lived and concrete," she writes, "encompassing histories and relationships as well as abstract principles."[41] Similarly, Larry Rasmussen writes that the search for an "Earth ethic" in religious traditions must begin from "the way a community *does* the truth'—its practices," and must work to understand and shape those practices toward justice and sustainability.[42]

The understanding that drives Peterson, Rasmussen, and many others like them is of theology and theological ethics as practical, not just in the sense of offering concrete suggestions for how believers are to live, but also in appealing to the whole lives of human beings, changing not just the ways they think about or understand the world, but also their basic predispositions and behaviors toward it. The upshot of this for ecologists is that most of us will not decide to conserve biodiversity based solely on a careful list of its values; we must also be inspired to appreciate biodiversity and trained to rethink our lives in light of that appreciation.

A theo-ethical approach to biodiversity creates an inspiring and practical ethics by offering a vision of what it would mean to treat God's diverse creation with moral consideration. I hope I have begun to demonstrate this in the previous section by framing my own ethics of biodiversity as the responsible participation in God's creation. I do not pretend that this image is a finished or complete theological ethics, but it attempts to offer both a broad vision of how humanity should relate to the nonhuman world and some insight into what steps might be taken to make this ideal more real. Such a theological vision stands in stark contrast to the list of reasons to value biodiversity so often provided by scientific ecologists.

I do not wish to argue that ecologists should simply accept the ethical proposals and imagery of theologians, which would deny the open and mutual dialogue I am trying to advocate. Rather, I am suggesting that a dialogue with theological ethics will give ecologists more understanding of how broad a project the development of ethics is, and how important it is that ethics not just persuade people of biodiversity's value, but also inspire them to act on its behalf.

CONCLUSION

The present volume is based on the assumption that a relevant and challenging language for environmentalist ethics and theology must be developed in conversation with a wide variety of other disciplinary languages and considerations. My contribution has attempted to demonstrate the potential of a dialogue between environmentalist moral theology and ecological science. Such dialogue produces a more synthetic and inspiring ethics than ecologists could develop without theologians, and a more grounded and careful perspective on biodiversity than environmentalist moral theologians could grasp without ecologists. When both perspectives are taken seriously and brought into dialogue, a far more nuanced and fruitful ethics emerges.

A scientific attention to biodiversity calls for recognition of the incredible complexity of the world and the species and structures that compose it, revealing organisms and systems that we can hardly imagine, much less understand. A theological attention to this fact calls us to see God's creation as diverse beyond our understanding and comprehension. It also calls us to see and accept the ambiguous status of human beings in creation: one species among fourteen million, which has established a power over all others without conclusively demonstrating the understanding or wisdom to use that power wisely.

The naturalistic methodology that this essay models leads to a dialogue with ecological science not merely because of the importance of ecologists' findings, but also because this science is one path toward understanding the world itself, toward a better view of God's creation, and therefore a better understanding of how we are to live in it. Informed by naturalism, a theological perspective on biodiversity simultaneously offers a picture of the creation of which we are a part and the ways we have drawn apart from the rest of creation. In so doing, it calls us to learn how to participate more carefully and more modestly in the world. It calls us to be aware of what is going on and what is possible. It also calls us to constantly struggle to improve our behavior, making it possible for God's diverse creation to thrive and for diverse human communities to thrive within it.

◆ Indigenous Knowing and Responsible Life in the World

JOHN GRIM

INTRODUCTION

The term "indigenous" refers to that which is native, original, and resident to a place. By introducing perspectives on time, however, "indigenous" becomes somewhat ambiguous. That is, evolutionary history presents a story of change over time among landforms, plants, animals, and peoples. While a consideration of time would seem to introduce simply a scientific agenda, it can lead to problematic political positions. Thus, in India, Malaysia, and Indonesia, for example, there is an objection to the use of "indigenous" and a preference for terms—*Adivasi* and *Orang Asli*—that refer to the "first peoples." This standpoint, it is argued, avoids an overly exclusive emphasis on tribal peoples as indigenous residents. In India, for example, the claim is made that Dravidian and Aryan peoples have been in South Asia for millennia. On the other hand, in South America "indigenous" has been claimed by a wide range of native organizations as an overarching and positive term for referring to diverse native peoples in international conferences and agreements.

The ambiguity of "indigenous" also arises from the diversity of peoples indicated by the term. While often thought of as remote minorities, indigenous peoples are a significant global population of over five hundred million peoples in Africa, South Asia, Southeast Asia, Central Asia, Australia, the Pacific region, Northern Eurasia, and the Americas.[1] No claim is made that such different cultural groups can be described by the word "indigenous." This term can, however, be used to refer to the family resemblances among these small-scale societies. Having been utilized in international settings, moreover, "indigenous" has been claimed by these diverse local, tribal, folk, native, and traditional peoples as they struggle for their right to exist.

Its appearance in the consultations and documents of international bodies affirms my usage here of "indigenous." For example, the United Nations declared a "Decade of Indigenous Peoples" from 1994 to 2004. "The Earth Charter," a United Nations–related document, identifies ecological integrity, socioeconomic justice, democracy, nonviolence, and peace with an appreciation of indigenous knowledge. Furthermore, article 1 of the International Labour Organization's Convention 169 (1989) regards people "as indigenous on account of their descent from the populations which inhabited the country, or a geographical region to which the country belongs, at the time of conquest or colonization or the establishment of present state boundaries and who, irrespective of their legal status, retain some or all of their own social, economic, cultural, and political institutions."

The social and political emphases in this definition are important dimensions of any discussion of indigenous communities, but other religious and philosophical features of indigenous lifeways associated with "ecospirit" are also significant. Topics such as homelands, cultures, traditions, histories, languages, institutions, rituals, and sacred symbols can be raised in this comparative context. Indigenous spokespeople have described these themes as not only inseparable from knowledge but interwoven into the fabric of their existence, their cultural totality or "lifeway," as a people. Indigenous knowing among small-scale societies around the planet is embedded in particular languages, rituals and story-cycles, kinship systems, worldview dispositions, and integrated relationships with the land on which they live.[2] The interaction of these distinct aspects of indigenous lifeways provide pathways into understanding indigenous ecospirit.

Obviously, more is intended by "indigenous" than simply historical time in a locale, or political expediency. While this "more" is difficult to articulate, one way to consider it is in terms of an abiding cultural and spiritual identity established by a people in relation to a particular place. The term "ecospirit" is suggestive of the religious relationships established in diverse indigenous traditions between a people and their homeland. While the word "spirit" carries a strong transcendent orientation, the suffix "eco-" grounds this emphasis on spirit as immanent. That is, a strong family resemblance among indigenous peoples is a reverence for nature and local biodiversity. This veneration, of course, does not eliminate the necessity for hunting and horticulture. Rather, indigenous reverence is embedded within subsistence practices that manifest responsible relationships with ecospirit.

Diverse forms of ecological responsibility, environmental ethics, and sustainable behavior are also broadly identifiable features of indigenous societies. Yet, the deeper implications of specific indigenous religious and conceptual orientations to nature are extremely varied and difficult to generalize. In fact, a standpoint that presumes to present an all-knowing perspective on such diverse peoples within, for example, South Asia, South America or Africa is inappropriate and continues essentialist misreadings. The question for this essay, then, is what helps a reader understand the religious and philosophical relationships with local bioregions that these small-scale societies have developed over centuries of habitation. Moreover, these religious and conceptual expressions may be quite different than the Christian sense of "ecospirit" presented in this volume. An exploration of an indigenous "ecospirit" is what I will explore in the pages to follow.

It has become something of a commonplace to assume that any study of the environmental practices of native peoples implies historical assumptions that "primal" peoples lived in uniformly sustainable ways and were consistently ecofriendly. This naïve standpoint corresponds to an Edenic reading of indigenous peoples as having never left the primal garden. Obviously this mythic reading of human-earth relations can undermine efforts to understand indigenous understandings of the bush, the wild, or nature. That is, interpretations from a nonindigenous mythic perspective tend to fluctuate between an exaggerated empathy for the "noble savage," in which environmental practices of native peoples are idealized, and an overly critical position that reduces indigenous peoples to exploiters of their bioregions in ways parallel to those of industrial societies. An extreme exploitative standpoint, for example, overemphasizes ecological conflicts experienced by native peoples. Thus, the reduction of biodiversity, deforestation, and soil depletion become a basis for wholesale condemnation of any consideration of indigenous environmental awareness.

A more balanced understanding of indigenous cultures requires that their survival as distinct peoples not be tied to romantic readings of their environmental practices. The historical treatment of indigenous traditions should be based on moral considerations apart from our evaluations of indigenous environmental practices. Still, any ethical evaluation of indigenous peoples positions them in some relationship with land, biodiversity, or the dominant societies that typically surround them. Some evaluative perspectives on indigenous peoples recast them by using terms that are subtly quite negative.

In some development settings, terms such as "marginalized" or "poor" have been used to replace "indigenous." These terms introduce pejorative views of native peoples that undermine their cultural heritage. For example, who wants to accept traditional environmental knowledge from "poor" or "marginalized" peoples? Recently, more neutral terms such as "rural," "local," or "folk" have replaced "indigenous" in some contexts. Yet, even these terms tend to erase their subjects' diverse forms of cultural and ecological knowledge. They strike this author as similar to the religious and political negritude attached to many indigenous peoples by colonialists in the nineteenth and twentieth centuries to subvert and convert native peoples.[3] The questions thus arise: Is there an academic standpoint that allows for investigation of the religious orientations of indigenous communities in relation to bioregions? Would such a standpoint be helpful to indigenous peoples themselves? Furthermore, is such an investigation useful for environmental studies?

LIFEWAYS AND INDIGENOUS KNOWLEDGE AS DIMENSIONS OF ECOSPIRIT

This essay contends that indigenous ways of knowing must be set in the context of their diverse lifeways; that is, the obvious differences both within indigenous communities and between different native cultures should not blind us to cultural "wholes." Such a cultural whole among an indigenous people is evident in their shared language, kinship terms, mythic narratives, rituals, subsistence practices, and modes of environmental awareness. Certainly, cultural wholes are lived realities that change. Moreover, that process of flourishing and diminishing in the indigenous context involves individuals in community relationships with biodiversity and place. One example from First Peoples in Australia illustrates this process relationship using the metaphor of ownership in a different key.

April Bright of the MakMak (Sea Eagle) peoples, of the Marranunggu language group in Northern Territory, Australia, gives expression to the lifeway concept when discussing what "traditional ownership" meant to her mother. She comments: "Traditional ownership to country for my Mum was everything—everything. It was the songs, the ceremony, the land, themselves, their family—everything that life was all about. This place here was her heart. That's what she lived for, and that's what she died for."[4] Embedded within "ownership," then, for this MakMak elder, as transmitted to her daughter, was not simply individual control of private property or separate knowledge systems of collected data. Rather, ownership appears to be interwoven with diverse forms of indigenous knowledge. The land and knowledge of the land are centered in one's personal body and interwoven with the surrounding social body, ecological body, and cosmological body. Here "body" presents a metaphor for the embodied knowledge evident among indigenous peoples.

A lifeway perspective obviously draws on historical, social, and economic questions raised by the study of religion and ecology. Thus, lifeway as cultural whole does not eliminate cultural change, hybridity, and interactions with other peoples. Understanding the relationships of distinct practices to cultural wholes engages the attention not only of academics but also, and primarily, of indigenous thinkers themselves. Need we acknowledge that indigenous thinkers have always provided outsiders with the basic information and understanding of indigenous societies.

Lifeway describes the close connections between territory and society, religion and politics, cultural and economic life whereby indigenous peoples have maintained their knowledge systems. Indigenous lifeways as ways of knowing the world are presented as both descriptive of enduring modes of sustainable livelihood and prescriptive of what Peet and Watts call "ecological imaginaries."[5] These are deep, attractor relationships between place and people that activate the affective, cognitive, and creative forces at the heart of cultural life. When homelands of indigenous peoples are literally cut down or mined away the whole possibility of imaging oneself and one's community is torn asunder.

Obviously, the single most demanding challenge faced by all indigenous lifeways is their survival as distinct peoples in relation to their homelands. All indigenous peoples face relentless pressure to accommodate the ideologies of dominant nation-states, as well as to resist and respond to the demands of extractive industries seeking to log, mine, and often ruthlessly exploit the remote regions into which native peoples are squeezed. Moreover, even major conservation groups have at times been seriously challenged to understand the dynamics of the science of biodiversity and the ecopolitics of indigenous sovereignty.[6]

It is in the context of these overwhelming difficulties that preservation of indigenous knowledge (IR) becomes a cardinal issue for indigenous people. Retention of IR is basic to their very identity as distinct peoples with languages, kinship systems, shared stories, and creative relations with ecosystems.[7] The intention here is not to present a detailed study that separates indigenous knowledge from Western usages of the term "science."

It can be remarked, however, that even Western analytical science has its own roots in local, embedded, experiential indigenous and Asian knowledge.[8] Moreover, the use of scientific categories to describe indigenous knowledge tends to manipulate it as a global, universal, commercial commodity that is, like science, supposedly neutral. Thus, attempts have been made to protect IR as "intellectual property" but the demands of state legal systems have the effect of making IR vulnerable to being adapted into dominant modes of economic exchange.[9]

Typically, indigenous ways of knowing are framed by such Western intellectual templates as monotheism, social contract theory, private property, individual rights, scientific views of the objectivity of reality, and views of democratic governance. This has had the effect of decontextualizing indigenous knowledge such that some aspects are adapted into scientific categories, while other domains, logics, and epistemologies are rejected as unassimilable. Still, as Arun Agrawal points out, it is worthwhile to remember that constructing a "sterile dichotomy between indigenous and Western" may simply obscure ideas and practices that unnecessarily constrict peoples' considerations of potential knowledge transfers.[10] Thus, "indigenous knowledge" remains a helpful term in development, political, and academic settings to acknowledge and authenticate these diverse ways of knowing.[11] Foremost among these relationships is that between the wide-ranging ways of indigenous knowing and local place as expressed in indigenous lifeways.

INDIGENOUS KNOWLEDGE AND LOCAL PLACE

Indigenous knowledge is inextricably tied to engagement with bioregions in a local setting. The relationships are often actively pursued in dreams, rituals, and symbol systems entangled with biodiversity. Indigenous ways of knowing are not simply expressions of instrumental rationality, or a functional knowledge for accomplishing specific tasks. Indigenous, or native, science knows the world as something other than as inanimate objects, yet indigenous knowledge is certainly capable of objective understanding and of recognizing utilitarian relationships. More importantly, indigenous worldviews and epistemology are not primarily characterized by techniques of quantification and measurement, or through the experimental method. Rather, while indigenous knowledge can be labeled a *scientia*, or knowing, it is quite different than Western science and is closer to Western ideas of *philosophia*, or a love of wisdom.

Many indigenous peoples transmit their knowledge in songs, oral narratives, and symbolic actions that relate to specific places and events. For example, the Duna people of the Lake Kopiago district in the Southern Highlands province of Papua New Guinea assert social identity through a form of narratives called *malu*. The Duna recognize the rights of individuals by means of *malu* narratives to claim precedence for garden locations or for the gathering of medicinal plants. Prerogatives with regard to narrating *malu* are believed to have been transmitted from the time of the ancestor-animals. *Malu* for the Duna operate at the intellectual level of explanation of the world, at the affective level of ceremonies that keep powerful ancestors benevolent, and at the level of contemporary policy formation. That is, the performance of *malu* assist Duna people in framing reactions and responses to the current development of their land. Referring to or reciting *malu* for the Duna simultaneously generates social stability in mythic-ceremonial contexts as well as symbol-making creativity to meet the Duna peoples' needs for adaptive change.[12] This way of knowing is a behavior, then, that does not simply filter out pragmatic time in favor of a mythic time, but seems to bring both to bear upon each other.

Indigenous knowing is thus a way of thought that looks to place and the life given in place as storied being, or ecospirit, rather than abstract principle. This inquiry is not an attempt to romanticize indigenous knowing as a purely spiritual realm, in which nature is divinized in a turn toward pantheism. Rather, this is an inquiry into indigenous knowledge as a form of responsible life in the world.

Just as "responsible" holds within it a sense of one's response as something deeply felt, unconscious, and gratuitous, so also it suggests a more conscious ability to act. Indigenous knowing, or science, manifests this twofold affective and reflective characteristic in its intuitive connections with natural processes. As the indigenous educator Gregory Cajete, from Santa Clara Pueblo in the Southwest of North America, says, native science is "a metaphor for a wide range of tribal processes of perceiving, thinking, acting, and 'coming to know' that have evolved through human experience with the natural world. Native science is born of a lived and storied participation with the natural landscape. . . . In its core experience, native science is based on the perception gained from using the entire body of our senses in direct participation with the natural world."[13] Following Cajete's lead, this study regards the "lived and storied participation" of indigenous knowledge as individual participation in, and communal reflection on, the process of giving in the cosmos. Behavior resulting from this form of life is oriented and responsive to different values than revelation from, and faith in, a monotheistic God. It is also different than meditative abiding in ceaseless change, or an ethical imperative as an intuitive principle guiding behavior.

Religion, in this context, does not simply function in indigenous life as one among a range of alternative views of the world. Rather, the religious commitments of indigenous societies emerge from the deep interweaving of a people in sustained and sustaining relationships with ecosystems experienced as giving. Thus, knowing is not separated out as the action of human reason, intention, or will—though these modes of psychological action can be identified in indigenous thought—but as an exchange whereby indigenous peoples know life as a whole cosmological giving.

Indigenous modes of knowing are primarily sense-based and their conceptual expressions are integrally tied to these affective relationships with the nonhuman world. In these ways of knowing, knowledge is not a grid system of analytic exactness but rather a feeling for the place and what it provides for the people. One writer described the relationships of the Miwok, Ohlone, and Yokuts peoples with what would eventually come to be known as the San Francisco Bay region of California:

> From the perspective of the first peoples of this land, every plant community had its own requirements for human interaction. Three of the most important were grasslands, oak woodlands, and riparian corridors. While people along the margins of the Bay and ocean obtained an incredible amount of shellfish and other kinds of sea-food, grasslands likely contributed the greatest abundance of plant and animal foods. Local tribal peoples expertly applied two specific management techniques to grasslands—judicious but wide-ranging use of fire, and cultivation through digging.[14]

Here we see indigenous knowledge interacting with the uncertainty and indeterminacy of natural forces perceived in the Bay Area grasslands. Fire and cultivation provided interactions in which indigenous knowledge was expressed. Moreover, gender, song, and status responsibilities in many indigenous settings continue to emerge from out of these harmonious interactions with component forces in these place-based knowledge fields. Rather than grid-like mapping for control of natural processes, this harmony flows from careful observation and behavior, prayerful thoughts, actions and offerings, and participation in a cycle of rituals that opens individuals and communities to the larger cosmos.

This place-based knowledge of indigenous peoples often results in cosmologies, or meaningful stories and song-cycles of the larger reality, that emphasize personal, social, and contextual responsibilities. The affirmation of indigenous knowledge need not be understood as promoting a "myth of primitive environmental wisdom."[15] Rather, it can be said that indigenous knowledge affirms local peoples and ecologies while meaningfully orienting them in ceremonies and oral narratives to larger cosmic realities. Therefore, in the wider context of the environmental crises facing humanity, indigenous knowledge provides viable alternative visions of human-earth relations. This knowledge is primarily for indigenous communities themselves, but it also bears significantly on an emerging environmental awareness in any multiform planetary civilization that promotes sustainable life.

EXAMPLES FROM THE DENE LIFEWAY

As a way into understanding the pervasive and mutual triangulation of indigenous knowledge, lifeway, and land, consider the statement by the Gitksan and Wets'uwetén elders, Gisday Wa and Delgam Uukw, before the Supreme Court of British Columbia, Canada:

> Each Gitksan house is the proud heir and owner of an *adáox*. This is a body of orally transmitted songs and stories that act as the house's sacred archives and as its living, millennia-long memory of important events of the past. This irreplaceable verbal repository of knowledge consists in part of sacred songs believed to have arisen literally from the breath of the ancestors. Far more than musical representations of history, these songs serve as vital time-traversing vehicles. They can transport members across the immediate reaches of space and time into the dim mythic past of Gitksan creation by the very quality of their music and the emotions they convey.
>
> Taken together, these sacred possessions—the stories, the crests, the songs—provide a solid foundation for each Gitksan house and for the larger clan of which it is a part. According to living Gitksan elders, each house's holdings confirm its ancient title to its territory and the legitimacy of its authority over it.
>
> In fact, so vital is the relationship between each house and the lands allotted to it for fishing, hunting, and food-gathering that the *daxgyet*, or spirit-power, of each house and the land that sustain it are one.[16]

Here the song cycles, sounded as *adáox*, are presented as a body of Gitksan and Wets'uwetén knowledge. Rather than an abstract body of knowledge, however, they are described as living archives connected to social organization, political leadership, and subsistence practices. This lifeway, here in the sense of "social body," is a way of knowing the world that activates the "breaths of the ancestors."

Lifeway can be a limiting concept if it leads to images of societies that are frozen, timeless, and ahistorical. This standpoint leads to an erroneous view that indigenous peoples must preserve cultural norms outside of the world of change. Similarly, the lifeway concept can be misleading if understood as a unit of analysis that separates out fundamentals when, in fact, indigenous knowledge is oriented toward a field woven from a warp of experiential encounters with the world and a woof of normative social coherence. In this sense, ceremonies, economic transactions, and kinship systems are singular expressions of lifeways at the interface of indigenous knowledge and land. Here, the Gitksan and Wets'uwetén social body of songs meets the personal-somatic intentions of a practitioner in the larger field of knowledge. As oral expressions these transmissions can be adjusted to the maturity of the learner, his or her level of related-ness to the land and experiential knowledge of the holistic order. Thus, the song cycles are not fixed as in a literate system but accommodate personal learning and teaching styles as well as existential, spiritual accomplishment.

In a comparative aside, it is interesting to observe that social science methods can accommodate and understand objective forms of organiza-tion and political structures. The affectivity and mutual embodiment of this knowledge that achieves connection through ancestral breath is, however, beyond the purview of scientific method. Realization gained through personal experience is not reducible to exact units of analysis, nor is it falsifiable by means of the experimental method. Rather, the millennia-long transmission of experienced traditions as indigenous ways of knowing are grounded in personal accomplishment and respon-sibility, community awareness and approval, and ecological response and sustainability.

The Gitksan and Wets'uwetén elders state that the ceremonial settings at which the songs are performed involve crests and stories of the lineage ancestors. These serve to substantiate and affirm individual and group claims to subsistence rights, traditional sanctions, and relationships with land and spirit-powers. Moreover, the songs transmit epistemological in-sights into the nature of time, space, authority, and spiritual presence. In this path of knowledge these are not simply objective, reified, abstract topics. On the contrary, the passage above strongly suggests sensory par-ticipation in ancestral knowledge by means of synesthetic experiences of auditory, visual, emotional, and intellectual ways of knowing. This interweaving of sensing with mental cognition appears to be a recurring mode of indigenous knowing that has scant parallel in Western natural sciences, social sciences, or humanities. It is in these ways that Dene indigenous knowledge simultaneously creates and functions within the ecospiritual dimensions of their lifeway.

In the quote above, the Gitksan and Wets'uwetén elders suggest that the synesthetic character of the songs transport members into mythic time. This leads to reflection on the ways in which multivalent modes of indigenous knowledge not only express particular dimensions of an indigenous lifeway but actually serve to structure occasions for the active creation of that lifeway. While an extensive overview of indigenous views of time and space is not possible here, even a brief consideration of these topics suggests facets of the interface of indigenous knowledge, lifeways, and land.

Strikingly different notions of time are evident in indigenous ways of knowing than those familiar to Westerners through the images of a linear "arrow" of time or "flow" of reality, or those that imply a rigidly demarcated past, present, and future. Obviously, the knowledge embedded in the Gitksan and Wets'uwetén songs brings practitioners into deep participation with time. By "deep" is meant more than simply a mystical unknowable. Rather, indigenous knowledge probes an original depth by exploring the character of time and space as a pervasive whole that is interactive, process-punctuated, and marked by spiritual presences.

Awareness of place in the indigenous world may be undeveloped until punctuated by arresting experiences that afford their own original ways of knowing. For example, the metaphor of the darkness of night suggests a field that is at once unknown yet filled with potential, and is knowable in relationship to an illuminating, existential experience. Similarly, many indigenous ceremonials transform a limited individual understanding as microcosm into a larger macrocosm of self-reflection by means of symbols that activate existential awareness of place and biodiversity. Authority for Gitksan and Wets'uwetén leaders, for example, is established in a worldview linking land, house, affect, ancestors, and spirit-powers. How shall we interpret this type of knowledge that evokes powerful emotions just as it establishes social organization and political authority? Such a ceremonial opens practitioners to complex contemplations of time. Place-based analyses can suggest interpretive depth but the existential authority of indigenous knowledge arises from an original and experiential knowing of place.

Moreover, Dene knowledge, embedded within complex ceremonials, is not exclusively liminal—that is, an entry into extraordinary space and time outside of that which is ordinary. In fact, the opposite is the case; indigenous knowledge manifested at these peak ceremonial moments has deep and abiding connections with both extraordinary presences, such as ancestors and spirit-powers, as well as with ordinary events, such as canoe making, gardening, gender roles, and healing practices. Time is a key, it seems, to an interpretation of ways in which space, authority, and ecospiritual presences manifest one another in the ceremonial and sym-bol-making contexts of indigenous knowledge.

Another insightful interpretation is provided by Henry Sharp in his observations among the Dene peoples of North Central Canada. Acknowledging continuities with Western modes of time, he relates the special character of Dene reflections on *inkoze* as a form of indigenous knowledge treating time, space, and spiritual presences. *Inkoze* refers to the Dene sense of causality that comes from the power of dreams, and the ways that animals mediate these special relationships. He observed that

> Dene time usage, though it uses linear, directional, and cyclical time, includes usages that do not correspond to any of these. Animal/persons are not limited by our perceived constraints of the physical universe. Their license to suspend the restrictions of what the West considers physical reality is embedded in their *inkoze* and is particularly conspicuous in their dispensation with the restrictions imposed by time. Dene culture is not dominated by the idea of a now, and time is not seen as a flow between a no-longer-existing past and a not-yet-existing future. Communication and connection between past, now, and future are all possible. The connections between time and the three ordinary dimensions of physical reality are uncoupled. Time sometimes becomes a thing independent of motion within it. All places in time become equally accessible. It is possible for some beings to move anywhere in time rather than having to move in only one direction in time. The effect of this is to make it seem as if the Dene sometimes use time the way Western culture uses place.[17]

Here Sharp gives his understanding of the ways in which Dene indigenous knowledge responds to time and space as equally available to ecospirit. His final remark comparing time in Dene knowing with place in that of the West evokes the linkage forged by contemporary Western physics as *spacetime*. But in Western science this unifying concept holds authority absent of spiritual presences. For the Dene, the linkage of time and place weaves an experiential tapestry of deep, cosmological relationships of self with a world of ecospirit.

Peter Chipesia, a Dene elder of British Columbia, in describing the vision quest experience of his people, points toward an embodied knowledge that gives fuller experiential description to Henry Sharp's insights about the indeterminacy of indigenous knowledge. In this passage he describes the fasts from food and water that induce a visionary ecstasy and somatic knowledge that comes from intimate experiences of animals in the natural world.

> When you are close to something, an animal, you are just like drunk. You don't know anything. As soon as this happens you have trouble thinking straight, like being drunk. Everything is just like when you see this animal it is as if he were a person. If you take water then, everything will get away from you and you will be a person again. You won't see anything. That is why you can't drink water before you go out into the bush. When kids are about the same size as Joe (about five), when they are just old enough to think, to talk, to walk . . . when they are older, the animal shows them how to make a medicine bundle.[18]

Thus the young person undergoing the experience described above is not conceived as an isolated unit, as is so typical in many modes of philosophical individualism. Instead, the "person" is presented as an emergent being in relationship with dreams, ancestors, and animals. Just as medicine bundles are gradually assembled with guidance from the agents of ecospirit, communal personhood also is assembled with orientations from place, biodiversity, and spiritual presence. Ecospirit is one way of describing the relationship that is established both in the maturing individual and in the personal assemblage of sacred medicine materials.

This relationship is not a dogmatic teaching that can be stated or written, nor is it simply a singular individual's religious experience. The knowledge acquired by individuals directly relates to the mythic stories of the people that describe the ancient cosmological spirits. This social knowledge relates as well to specific animals in the environment. The work of implementing the vision is part and parcel of the lifepath of the individual person in relationship to spiritual forces that have "adopted" him or her. This work of acquiring personhood is an embodied knowledge accomplished over years of interaction with the life of local regions.

ACQUISITION AND TRANSMISSION OF INDIGENOUS KNOWLEDGE

Marlene Brant Castellano makes a trio of pertinent observations regarding the acquisition of indigenous knowledge that may serve as a conclusion to the present overview. First, *traditional knowledge* has been handed down by indigenous peoples more or less intact from previous generations. Second, *empirical knowledge* is gained through careful observation of the natural and built environments. Third, *revealed knowledge* is acquired through dreams, visions, and intuitions that are understood to be spiritual in origin.[19] These points provide a useful foundation upon which to note several characteristics of indigenous knowledge as ecospirit.

First, *traditional knowledge* is transmitted within an indigenous community primarily by elders. This rich concept of the "elder" is an indeterminate, spontaneous, community-generated role that may or may not settle on an older person; rather, an "elder" is that person whom an indigenous community recognizes as imbued with the knowledge, responsibility, and spiritual awareness for actively living the role requested of them. In the contemporary world indigenous elders, women or men, often embody knowledge of language, worldview, and ecology.

Transmission of language is widely recognized as central to the authentic acquisition of that traditional knowledge to which indigenous elders refer when they speak of their identity and meaning as an integral people. As H. C. Wolfart and F. Anenakew note in the title of their volume on the Cree, elders of that Northern sub-Arctic tribe are strongly identified with the substance of their traditional knowledge as expressed in their ancestral tongue: *"kinêhiyâwiwininaw nêhiyawêwin* (The Cree language is our identity)."[20] J. Y. Henderson comments on this, noting that "Aboriginal consciousness and language are structured according to Aboriginal peoples' understanding of the forces of the particular ecosystem in which they live. They derive most of the linguistic notions by which they describe the forces of an ecology from experience and from reflections on the forces of nature."[21] Thus, just as indigenous languages derive from experiences of the land, so also worldviews emerge, in indigenous lifeways, from visionary experiences of the depth of origins, the beginnings of life, and the unfolding of the universe. Just as elders embody an engaged knowledge involving language and worldview, so also they live their connection to ecosystems in ways of knowing that exemplify *empirical knowledge*.

Empirical knowledge emphasizes the strong orientations of indigenous peoples to the material world. Thus, the use of the term "spirit" can be misleading insofar as knowledge of the sacred often arises from observation of the land and biodiversity. This empirical sense in indigenous knowledge is pervasively charged with a spiritual potential. Yet it does not bring the one who holds this empirical knowledge into a transcendent, spiritual realm void of matter. Traditional environmental knowledge, for example, is now widely acknowledged as a seminal contribution of indigenous peoples to the world's knowledge of the earth.

There is no doubt that traditional environmental knowledge has captured the public imagination more than any other aspect of indigenous knowing. In entering the public sphere it also reminds us of ways that indigenous knowledge has been commodified. The allure of pharmaceutical panaceas gathered from indigenous knowledge sources, for example, tends to mask potentially exploitative activities such as genetic piracy. In some contemporary agendas regarding indigenous peoples, biological materials are extracted from native peoples and attempts made to patent the distinctive genetic heritage of indigenous groups.[22] Furthermore, efforts to protect particular expressions of indigenous knowledge using international intellectual property rights has proven ineffective at best. As suggested above, intellectual property arguments involve indigenous elders and spokespeople in the bureaucratic and legal procedures of multiple nation-states. More significantly, intellectual property situates the empirical knowledge of indigenous peoples in an epistemological context that is wholly inconsistent with the knowledge being protected. That is, the languages, ideas, and values supposedly protected by the concept of "intellectual property rights" are set within a neo-colonial mindset in which indigenous peoples are subservient clients rather than careful observers, users, and co-creators of their environments.

Finally, the character of *revealed knowledge* in dreams, visions, and spiritual intuitions among indigenous peoples is manifest in many of the examples above. Two of the most striking case-studies of indigenous visionary experience involve the making of *ayahuasca* in South America and maize agriculture in Mesoamerica. Several scholars have discussed the sophisticated processes involved in making the ecstasy-inducing drink, *ayahuasca*.[23] Its preparation involves not simply the vine *banisteriopsis* but delicate infusions of several other prepared substances. More important, however, is the widespread understanding among the native peoples who work with this substance that the *plants themselves* revealed in dreams and visions the complex processes required to produce *ayahuasca*.[24]

Similarly, the mythical origin of maize among Nahua peoples of central Mexico presents a major example of revealed knowledge in which the gods "robbed" maize for the people and transmitted the knowledge of its production and sustenance in the context of lifeway and land. Thus the profound teachings of indigenous knowledge open insight into the embodied character of ecospirit among these indigenous peoples. The empirical presences of the holistic order in which ecospirit manifests itself are the human body, the social body, the ecological body, and the cosmological body, as denoted by J. G. Silva.[25] This passage from the *Florentine Codex* encourages the reader to reflect on the fourfold manner in which maize is embodied in the life of the people:

> Listen: *Tonacayotl*, the maize, Our Sustenance, is for us, all-deserving. Who was it who called maize our flesh and our bones? For it is Our Sustenance, our life, and our being.
>
> It is to walk, move, enjoy, and rejoice. Because Our Sustenance is truly alive, it is correctly said that it is he who rules, governs and conquers. . . .
>
> Only for our sustenance, *Tonacayotl*, the maize, does our soil subsist, does the world live, and do we populate the world. The maize, *Tonacayotl*, is the true value of our existence.[26]

For these Mesoamerican Nahua peoples, maize not only gave life, but knowledge of its stories and cultivation enabled participants to touch the deeper dimensions of life through their own bodies as well as the bodies of their communities, the bodies of their ecosystems, and the larger cosmic body. While marked by cultural differences, this revealed knowledge among the Nahua is related to the distinctly different understandings of the neighboring Mayan peoples. The family resemblances among these Mesoamerican peoples regarding the origin, production, and deeper symbolic meanings of maize agriculture all manifest dimensions of their ecospirit.

CONCLUSION

I have sought above to associate ecospirit with indigenous knowledge in a plurality of forms. There is a family resemblance, but not a unity, among the lifeways of indigenous peoples in their different homelands. Indigenous scholars have observed that

> Perhaps the closest one can get to describing unity in Indigenous knowledge is that knowledge is the expression of the vibrant relationships between people, their ecosystems, and other living beings and spirits that share their lands. . . . All aspects of knowledge are interrelated and cannot be separated from the traditional territories of the people concerned. . . . To the Indigenous ways of knowing, the self exists within a world that is subject to flux. The purpose of these ways of knowing is to reunify the world or at least to reconcile the world to itself. Indigenous knowledge is *the way of living* within contexts of flux, paradox, and tension, respecting the pull of dualism and reconciling opposing forces. . . . Developing these ways of knowing leads to freedom of consciousness and to solidarity with the natural world.[27]

Indigenous ways of knowing attend to the flux and tensions of life that are embedded in, and resolved through, living responsibly in relation to one's local place. These are the dynamics of indigenous ecospirit.

These shared ways of knowing and attention to difference underlie the wisdom and the specificity of indigenous knowledge. Thus the Maori scholar Linda Tuhiwai Te Rina Smith observes, "*Whakapapa* is a [Maori] way of thinking, a way of learning, a way of storing knowledge, and a way of debating. . . . *Whakapapa* also relates us to all other things that exist in the world. We are linked through our *whakapapa* to insects, fishes, trees, stones, and other life forms."[28] The wisdom transmitted here seems to speak across its specific Maori context, but only the Maori can determine if that wisdom is transferable as more than environmental poetry.

What this sacred knowledge reveals today is not simply the ashes of an extinct, failed way of knowing, but rather the embers that indigenous elders are rekindling to confront their peoples with awareness of deeper purpose. The metaphor of fire evokes that renewed understanding of indigenous knowledge whose wisdom may not be available to other peoples. Still, it calls to mind the social and political vitality of peoples who have faced the challenges of life knowing what the Nahua refer to as "the true value of our existence."

❧ Econstructions: Theory and Theology

The Preoriginal Gift—and Our Response to It

ANNE PRIMAVESI

Freely you have received, freely give.
—MATT. 10:8

THE PREORIGINALITY OF GIFT

In *Sacred Gaia* (2000) and again in *Gaia's Gift* (2003), I explored the concept of "gift" while taking for granted that it involves more than simply an exchange of goods between two people. It is now time to spell out, as precisely as possible, what this "more than" refers to. Briefly, it presupposes that gift is essentially a community event constituted by diverse inputs over time from the environments—natural, biological, and social—of giver and receiver. Within this extended framework, such events are best understood as instances of symbolic behavior that mediate and disclose more of the basic and evolving relationships between people and their lifeworlds than can be discerned in the immediate context of the gift itself or in its content. Potentially, the framework extends as far as earth's limits: or, if desired, beyond them to include the birth of our solar system and the life-giving power of the sun.

In line with this expanded context the qualifying term "preoriginal" suggests that the origins and source of present gift events be traced back through evolutionary time to a point or place before the emergence of the human species. It is true that, as Levinas points out, such a linear regressive movement would never be able to reach "the absolutely diachronous pre-original which cannot be recovered by memory and history."[1] Nevertheless the effort to do so directly acknowledges the essential contributions to present gift events made by "more than" those participating in them now. They include antecedent generations of living beings: all those who, by their lives, their labor, their deaths, their vision, and their patient endeavors have made such events presently possible. We can, therefore, respond to their contribution with a gratitude that is itself, as Levinas remarks, grafted onto an antecedent gratitude for finding ourselves able to be thankful.[2]

At first glance there is nothing new about this expanded perspective. A seminal work on the phenomenon of gift by anthropologist Marcel Mauss unapologetically took societies as its context. Those studied belonged to an era before the institution of traders, of modern forms of contract and sale (Semitic, Hellenistic, and Roman) and before money was minted and inscribed. During that era, gift appeared as a type of behavior that necessarily emerged from the relationships between those who constitute a particular community and between neighboring communities.[3]

Now, however, modern forms of trading have taken us into a cyberspace of electronic exchange, far beyond not only the realm of market stalls piled with produce but also the environmental limits imposed by geographical borders and the physical reality of minted or printed currencies. As a result, the communal context for our transactions has all but disappeared from view and from our consciousness. Today the dominance of capitalist models of trading supports a now commonly held assumption: that one individual gives to another primarily in the expectation of at least an equivalent return in personal wealth, however defined.[4] On closer inspection, however, it is clear that these apparently isolated individuals can only give something to each other because they are in fact embedded in communities with access to, and input from, their own or others' resource systems.

This fact supports my proposal for a context for gift events spatially and temporally extended to correspond to the contemporary concept of the "more-than-human" lifeworld. David Abram coined the phrase "more-than-human" at a time, he says, when we appear to interact almost exclusively with other human and human-made technologies, forgetting that "we still *need* that which is other than ourselves and our own creations." For we are human "only in contact and in conviviality with what is not human."[5] This describes the lifeworld out of which we

emerged and within which we live, just as it describes that which preceded and followed from those antecedent societies studied by Mauss.

GIFT'S MORE-THAN-HUMAN CONTEXT

The origins of the "more-than-human" that is essential for our humanity
date back some 3.5 billion years ago. Then, many and varied forms of
single-celled organisms began evolving into the enormously complex systems of interacting, symbiotic multicellular entities we now broadly clas-
sify as plants and animals. The first plants to reproduce not by spores,
like the earlier fungi, but by seeds, appeared a little more than 360 million
years ago. By this time all the transformations required for life on earth
to manifest as plants (such as oaks and redwoods) had taken place. At
least 150 million years before the time of the first dinosaurs, the basics
were in place, although there were many refinements still to come. Bo-
tanically speaking, says Colin Tudge, it was the beginning of modernity.
Many lineages of seed plants have appeared since then, most long extinct.
Five are still with us. Two of them, the conifers and the flowering plants,
dominate our terrestrial ecosystems.[6]

The emergence of flowering trees and plants was a significant event in
the evolution of life on earth. The availability of flower seeds brought
about a flourishing of animal life and reciprocal, beneficial relationships
between birds, insects, animals, and plants. This contributed to the evolution of the planetary biosphere system and to the earth's surface itself
in ways that later gave emerging human communities the water, temperature range, and nutrients they needed to survive. In a very precise sense,
the gift event of the emergence of flowering plants, their relationships
with and within the earth itself and with its other-than-human life forms,
was preoriginal in the diachronous sense that it preceded our own species' origins and is vital to our continuing existence. Tudge simply says:
"The human debt to trees is absolute."

Until quite recently, however, the evolution, continuance, synchronic-
ity, and necessity of these life forms, environmental life support systems
and our continuing dependence on them were thought (by those who
did not attribute them directly to God) to be a "happy accident"—until,
that is, modern technologies enabled earth-system scientists like James
Lovelock to study their complex interactions and distinguish between the
physical and chemical conditions that first allowed life to emerge (such

as the presence of water) and those necessary for life's support *after* it had emerged. This entire system, on which we depend and in which we participate, together with its self-organizing principles, is what Lovelock calls "Gaia."

Other studies (such as Tudge's) have taken close account of the way in which the participation of living organisms, from diatoms to dog-woods to dinosaurs, has modified the soil and atmosphere of the earth. It is now clear that over the past two hundred years or so we, too, have modified it to an extraordinary degree. Our activities and technologies have so recast the global nitrogen and phosphorus cycles as to strongly favor species that thrive on heavy diets of these nutrients. In such ways we have now become equivalent to a natural force (such as volcanoes and glaciers) in environmental transformation.[7] Our present experience of the effects of climate change reminds us that we are not exempt from the consequences of exerting such force.[8]

For we are no less dependent on Gaia's self-regulatory processes than were our ancestors. They, like us, needed the natural resources "given" them freely by Gaia to supply what Mauss referred to as the exchange of total services and counterservices between communities. He saw this exchange among more settled groups within human societies as imposing on them an obligation to reciprocate for gifts received. This obligation has, however, become formalized, ceremonialized, and com-modified to the point where gift-giving is almost entirely structured in interpersonal exchanges and defined by the expectation of an equivalent commodity return calculable in monetary terms.

Unfortunately this has colored our thinking about the preoriginal, dia-chronous gift of Gaia to such an extent that we are hardly aware of its true character: that of unconditionality. The sun shines, the rain falls, waters flow, seeds germinate, and bees gather honey for rich and poor, just and unjust, the miserly and the generous alike. Losing any perception of this quality of unconditionality, of giving freely without expectation of a reward, has also meant the loss of any impulse, or of the imagination needed, to respond positively to what is freely given us. So today the gift of "total services" given by the more-than-human members of the earth community to our ancestors, and through them now to us, is generally accepted without heed, without question, and without any discernible impulse to make a return, even that of simple gratitude.

By contrast, before European traders invaded North America, indigenous communities living on the Pacific coast responded to the annual gift of the first salmon appearing in the rivers with rituals of thanksgiving for the unquantifiable increase of life that gift would bring to the tribe. They also ritualized the desire to give something to the sea in return; in their case, the carefully preserved salmon bones. Taken as a whole, the rituals expressed a coherent, if largely generalized, awareness of mutual, interdependent relationships between people, rivers, sea, and salmon.[9] Again, in the older written Vedic tradition, among the stated reasons for offering hospitality to strangers and giving them gifts of food and money is not that they may one day do the same for the givers, but because when human beings are born, they are born with debts to the gods, to the seer-teachers, to the ancestors, to other humans and, in some texts, "to all nonhuman living beings."[10] It is now a sad fact that to be born in some countries is to be born into debt to the World Bank.

GIFT AS SYMBOLIC

The comprehensiveness of such contrasting views, even when, as here, they are abstracted from the settings and modes in which they were and are expressed, points to a most important dimension of our evolutionary history, one that is crucial in understanding the character of gift and our response to it. This is the development within human communities of symbolic behavior: of the ability to use actions, words, images, and other signs in contexts where their literal meaning can also connote a figurative one. Fires "flare up." But when someone's anger flares, we know there is no physical flame. The literal meaning of "flare" has been exceeded, taken beyond its denotative scope by using it symbolically, that is, in a context where it cannot be taken literally. The context tells us what its sense is and how it is to be interpreted.

This shows that symbolic usage is relational. It only makes sense within a shared perspective on the world as a whole, a whole that nevertheless consists of clearly distinguishable units that preserve their identity and so set them apart from each other. But they remain related to, not severed from, each other. So the Pacific Ocean, or the food on a Hindu table, forms part of the undifferentiated whole of a pristine vision out of which definite characters, such as the salmon, the ancestors, or the sea themselves emerge.[11]

Expressing this vision, however inchoately, means symbolizing its parts and characters, encoding them so that every detail is charged with meaning, and images can be used to envisage and re-envisage a tradition or event in ways that make sense to us. This requires developing the ability within a community to organize, transfer, acquire, and impart a vast amount of information through words, actions, artifacts, ideograms, music, and other signifiers. It also requires laying down (and in turn adding to) a deposit of old, abstracted concepts that represent an ever-increasing treasure of metaphors. Or should I say a metaphorical treasure? The givenness of these coded forms can itself be seen (symbolically) as a preoriginal gift from our ancestors.

Cassirer describes the nature and function of this deposit as a "living web of thought." Taken in itself, he says, it is a whole and closed, but it is a living whole by virtue of the fact that here "a single step stirs a thousand threads and all these threads flow unseen." Viewed in this way, when language is used symbolically, it is not a product but an activity: an interaction between expression, representation, and signification. There is a force in this living process of speech that does not leave the world of linguistic forms in the same condition in which it was found. It is affected as a whole; is changed, however imperceptibly, and made receptive for "future, new formations."[12]

Through this process we can and do routinely symbolize important events by bringing together words or actions within the framework of a coherent whole embracing past and present. More information is conveyed from sender to receiver by this means than that signified by any one identifiable unit within that whole. Brought together in this way, different types of words and image-making serve to create, to mediate, or link us to stories, concepts, and relationships beyond those of historical fact or actually evident in our immediate surroundings. For in symbolization, words, gestures, and artifacts function as signs within a system in which their meaning is dependent *both* on the relations they have to the way objects and actions in the world are experienced by others *and* on the relations they have to other signs in the cultural system. A symbol cannot exist in isolation. It is always part of a network of references.[13]

Jablonka, Lamb, and Langer use our evolutionary history as a framework for the symbolic dimension of human behavior, showing how it flows from genetic, epigenetic, and behavioral variations in our biological inheritance. In my discussion in *Sacred Gaia* of *autopoiesis* (self-making) I described this symbolic dimension as the *PoieticScape*: the linguistic, poetic, intellectual, creative, imaginative, and expressive aspect of our autopoietic relationships.[14] But I stressed that it cannot be separated in any real sense from the other three dimensions, classified as our *SelfScape*, *SocialScape*, and *EarthScape*. However named, together they stand for the variations in our individual, social, biophysical experience of tight coupling with our environments. Through this we become what we are and, indeed, as we now see, contribute to how our environments become what they are. And interactions between the *Scapes* enable us to express, symbolically and otherwise, what they give to us and we to them.

The above analyses also make clear that the ability to see gift as symbolic is not only a gift in itself. It also adds "more" to whatever is given than a simple exchange of goods. This symbolic excess exists because words, sounds, actions, and artifacts function, often subliminally, as parts of the greater whole that we call "cultural inheritance." The gift event, whether structured as an elaborate annual ritual or a homely daily meal, can then be read as a form of discourse where its referential content communicates more than its physical, calculable, discrete ciphers ever could. Its network of references connects it, overtly and covertly, to concepts such as "past," "present," and "future"; to emotions such as gratitude, horror, joy, and reverence; and, within particular cultural systems, to traditional ways of expressing these concepts and emotions.

Therefore within a communal framework (and which of us does not dwell there?) a gift *means* more than it says, or can say directly. But *what* it says, the symbolic content of offering food to a stranger or of returning a salmon's bones to the sea, is only accessible to those within the community who intuit or learn the deeper levels of meaning attached to these expressive events. Consciously or not, they alone are able to take account of the networks of information, the tacit dimension in local knowledge of a landscape, and the latent emotional content, all of which constitute the lived historical tradition of their community. Without that interactive relational element, casting bones into the sea is nothing but a primitive form of waste disposal.

Where that element *is* present, however, we are in the realm of symbolic behavior and understanding. There the symbols belong within a shared discourse in which imagination, thought, emotions, meanings, and perception function coherently. When, for instance, Gaia is perceived as a primal earth deity, mythic images of her giving birth to life in the world can emerge from, and sometimes merge with, contemporary cultural and scientific consciousness. (Although in the latter case, scientists often reject this symbolic element so violently that Lovelock at one time considered dropping the name "Gaia" for his theory in favor of "geophysiology.") With this wider mythic understanding, however, comes the true perception that long before we emerged to do things *to* Gaia, she was doing things *for* us: providing us with all that is essential for our lives then and now. So a theory that for many today stirs only the rational, intellectual passion of our age has at the same time awakened in some a sense of gratitude to Gaia for the gifts that have nourished, and continue to nourish, mortal beings like ourselves.[15]

When the theory does inspire such gratitude, it does so by recovering the deeper sense of connectedness implicit in the scientific theory. Making gratitude explicit through exploring the symbolism of gift, especially in regard to food, makes living a gift event in itself. This occurs most obviously at meals, when we are clearly gifted with what we need to live. It occurs whenever we consciously realize that we are continuously gifted with life by other organisms and species that have not evaded or ignored the demands we make on them—even to their death. Ultimately, this does not allow us to evade or ignore *our* dependence on, and belonging within, the more-than-human community of life on earth; nor to ignore our responsibility to return its members, at the very least, the gift of gratitude. This impulse to give thanks for the life given us in and on "the planet on the table" has been the basis for religious rituals throughout the ages.[16]

RELEARNING THE SYMBOLISM OF GIFT

How, then, do we recover this sense of gratitude in a culture that appears to have lost the symbolic sense of relationality and community? Contemporary evolutionary theory reassures us that we can, in fact, relearn, regenerate, and reconfigure the symbolism of gift. This is the case because symbols are socially agreed-upon conventions, disclosing their meaning within socially agreed-upon systems. Therefore, as we have seen with the evolution of Gaia theory, they can be changed, translated, regenerated, and reconstructed into other corresponding conventions.

We can and do "translate" symbols from one form to another; we separate and combine different symbolic forms and levels by following general principles of coherence—as does every piece of music, every play or poem intended for an audience. What we call "the shock of the new" comes very often from seeing or hearing "the old" in a different place or form, at a different age or in altered circumstances. The fact that we experience it with the help of one or another agency points to an important factor in symbolization: it is transmitted through active instruction. Just as we are taught the symbolic system of literacy, of mathematical symbols and rules, "we are *taught* how to understand and participate in the rituals of our culture. The framework needed to interpret symbolic information has to be learned."[17]

As the reference to the Vedic Scriptures shows, religion and religious myths have provided such frameworks and such instruction throughout human history. In Western culture, however, there is now no overarching religious framework that gives us a socially agreed-upon convention, myth, or cultural ritual for understanding or learning the symbolism of gift. Take, once again, the sharing of food. In my own religious tradition, how many Christians now teach their children to "say Grace" before and after meals? How has the development of "fast" food affected our perception of it and of the more-than-human community's contribution to it?

Yet within the framework of our evolutionary history and religious histories we see that with the phenomenon of past gifts present among us there is also the phenomenon of a gratitude that is itself, as Levinas remarks, grafted onto an antecedent gratitude for finding ourselves able to be thankful. Gratitude, too, is latent in our biological inheritance. Paradoxically, we may also respond directly to the preoriginal gift potentially present in the future when, as Derrida suggests, we give of our time, energy, and material resources without taking account of thanks, of future recompense or, at the very least, without conditions attached.

This sequential and consequential logic, in which receiving is necessarily antecedent to giving and the manner in which we receive then influences how we give, is strikingly expressed in a biblical passage that appears in the Gospels literally without precedent or parallel, and which serves as the epigraph to my contribution to the present volume. As he sends them to preach and to teach, Jesus instructs his disciples: "Freely you have received; freely give" (Matt. 10:8). The verses following this injunction set out some of the more usual requirements governing the conduct of newly hatched disciples. They are to carry no gold, silver, or copper coins, no knapsack, second shirt, sandals, or staff, thereby expressing a posture of complete trust and reliance on providence.[18] The injunction to give "freely" or "without payment," just as they have received, calls for a heightened awareness on their part of the relationship between the modes of receiving and giving. It also emphasizes a distinctive attitude of unconditionality with which they are to give to others. All of which predates and underlines Levinas's and Derrida's insistence upon taking account of our own responses to what we are given and linking them with how we ourselves give.

Gift-giving is, then, a form of emergent behavior that is transmissible, that can be taught, learnt, and relearnt. It is also tied in to perceptions of the prehistory of that behavior and to our reactions to what we have been and are given. So an effective response to the preoriginal gift of life would be to work within a framework that combines present and future concerns.[19] This requires understanding how the gratuitousness of gift events lies beyond the range of current market economics, beyond the measure of any demonstrable human calculation. We can only learn this, however, by breaking through the closed mindset of exchange and return, production and consumption, profit and loss that operates within market economies.

SEMANTICS OF GIVING

This breakthrough is necessary because the most basic sense of giving today is that it brings (at least) two people together, and that the implied dichotomy of two basic orientations—self and other—governs debate about the generosity required to give a gift. Can a gift be for the other without being against the self? Can a gift that does not subtract from the self really add to the other? Part of the problem is vocabulary. Traditionally the issue is framed in psychological terms as the possibility of "altruism" and conversely, the limits to "egoism." Both terms, according to Stephen Webb, have theological roots. "Altruism" for him is the secular equivalent of the religious category of sacrifice, "egoism" another name for the sin of selfishness. He points out that modern Western culture has undertaken a massive and prolonged rehabilitation of "egoism" and "selfishness" (note the success of Richard Dawkins's *The Selfish Gene*) so that the very purity of the ideas of "altruism" and "sacrifice" has become the easy target of ridicule and rejection. Indeed, the very grammar of giving is threatened in this linguistic climate.[20]

Webb could be seen to threaten it further by appearing to confine the act of giving to something that occurs only between two people and at some cost to one of them. Limiting the ability to give in this way follows naturally from his analysis of gift in terms of the free-market principle that there is no such thing as a "free" gift. I give to you if, and only if, you can and will give to me. Webb calls this the *model of exchange*. Although the (presumed) circularity of this model (what is given always returns) can, he says, highlight the values of equality, responsibility, and reciprocity, it is more likely to permit the cunning of self-interest to dominate every social interaction.[21]

This rather gloomy view of human generosity takes routine exchanges of goods and money between individuals within a market economy as normative. By doing so, it excludes any perception of the contribution of the more-than-human community of life. It also directly contradicts what Calvin Schrag regards as an essential ingredient of gift: "the eternal requirement of unconditionality," as in the Gospel passage cited above. In the light of this requirement, Schrag examines Derrida's claim that "the insertion of gift giving into a network of exchange relations results in a virtual negation of the gift as gift." For it is indeed the case that within a capitalist market economy a gift inevitably incurs indebtedness. Then even the customary practice of thanking someone for the gift can be seen as a way of giving something back. On Derrida's reckoning, however, a genuine gift "would need to be forgotten in the very moment it is given and received."[22] For one might say that a gift that could be recognized as such in the light of day, a gift destined for recognition, would immediately annul itself.[23]

I take Derrida's rhetoric not as a rejection of the possibility of gift but as a conscious, formalized subversion of the prevailing capitalist logic of exchange relations. It assumes that they are based solely on investment for return and on the accompanying structure of competitive accumulation.[24] Scientist and artist Glynn Gorick recognized this subversion operating on an even wider scale when he read my account of gift in *Sacred Gaia* while illustrating the 2002 UNESCO introductory booklet for the

Encyclopaedia of Life-Support Systems. This latter work indicated differences between the human system of economic accounting and our computing of the natural flows of energy and materials within the associated cycles and systems that constitute the planetary biosphere. Gorick perceived "gift" as operating outside *all* such systems views. It is not, he said, measurable, but rather a quality item of very high power that plays a huge role in the evolution, stability, and security of all life communities. Echoing Webb, Gorick says gift may be the mystery "altruism" factor that puzzles evolutionary biologists: "Indeed the process of giving in its purest and most powerful sense makes standard economic nonsense. . . . It seems to be contrary to the laws of conservation of energy or the logic of genes being selected for advantage over other genes. There seems to be a connection here with our understanding of what love means."[25]

Gorick's perception of the process of giving as "economic nonsense" and as being somehow connected to love agrees, as we shall see, with some of Derrida's own conclusions. Schrag, too, goes on to note that the conditions that inform what for Derrida is an aporia have to do principally with the semantics of possession and dispossession, reserve and expenditure, surplus and squandering, having and giving. These catapult the gift into the market model of exchange relations existing within a matrix of social relations based on the accumulation of wealth and the institution of private ownership.

POSSESSING AND DISPOSSESSING OWNERSHIP

Within this semantic and social matrix, to be able to "dispossess" oneself of something presupposes prior possession or ownership; in a market economy, only those privileged by ownership of private property are able to give objects that need first to be produced, distributed, and owned before they can be given away. Schrag quotes Michael Walzer to draw attention to some possible effects of this view of gift on our self-perception: "If I can shape my identity through my possessions then I can do so through my dispossessions. And what I can't possess, I can't give away."

Entwined with this consumerist definition of "gift-giving" as gift exchange are a host of presuppositions regarding the extension of the metaphor of possession: not only to the human subject's relation to the goods of the earth (and therefore to the earth itself) but veritably to the human subject itself.[26] Schrag is rightly concerned not only about the effects of a consumerist view of gift on our *SelfScape*, but also on how we relate to the *EarthScape*. Similarly, in *Gaia's Gift* I look at how the very odd notion that the earth belongs to us (rather than the other way around) leads us to say that we "own" land and its resources and that we may, therefore, use and abuse them for our own ends. But how can we say we *own* something that was evolving and creating life forms for billions of years before we were born, and will continue to do so after we are gone? We *assume* possession of the earth, in every sense, and are then prepared to contest our claim to some part of it even, if necessary, by killing others.[27]

Schrag notes further that to analyze the gift back into the framework of possession and dispossession, reserve and expenditure, is at once to privilege the private over the public while failing to discern that objects given as gifts never issue from a zero-point origin of absolute ownership. The meaning of "ownership" itself, he says, requires inspection. The goods of the earth and the services of humankind can indeed be "privatized" by us. But such privatization needs to subordinate personal ownership and property rights to an acknowledgment of the earth as an abode shared by all its inhabitants. Rather than viewing the earth as belonging to us, parcelled out by us as our property, our relation to the earth is enhanced if we view it as that to which we belong. Such enhancement, says Schrag, belies Walzer's rather thin notion that one shapes one's identity through one's possessions and dispossessions. If that were the case, those unfortunates without possessions are destined to exist without knowledge of who they are!

So a consumerist perspective on the gift invites a consumerist theory of self-identity and a consumerist morality. The moral self is portrayed as a storehouse of virtues, a surplus of moral acquisitions that the self is obligated to give away in fulfilling its moral duties to others.[28] According to Derrida, this sense of obligation belongs to an ambiguous type of exchange economy that would renounce earthly, finite, accountable, visible wages only to capitalize on this by gaining a profit or surplus value that is infinite, heavenly, incalculable, and secret. But the moment the gift is infected with the slightest hint of calculation, its value is destroyed as if from the inside.[29]

A biblical analogue with a contemporary ecological reference occurs in the parable of the rich fool who, when his crops yield abundant grain, pulls down his barns, builds larger ones in which to store it, and, on the basis of having these ample provisions, looks forward to many years of taking his ease, eating, drinking, and being merry. But God says: "Fool! This night your soul is required of you and who then will own your possessions?" (Luke 12:16b–20). True, the emphasis here is on the size of the storehouse and the accumulation of its contents rather than on any obligation to give away the surplus. Nevertheless, the parable exposes and then subverts the consumerist perspective of identifying oneself and one's moral or material possessions with one's destiny.

Such an identification necessarily entails ignoring the effects on one's *SocialScape* and *EarthScape* of accumulating material possessions. So there is a strong link here to ecojustice issues, too. For if the fool in the parable represents those identifying with the aims of a capitalist economy or preaching the gospel of prosperity, oriented toward ever-increasing returns on any transaction, then the reckless accumulation of goods extracted from finite natural resources means that future generations will have none to inherit.[30]

THE SEMANTICS OF GIFT EVENT

So what is the difference between giving and exchanging? Or rather, what difference does it make if we perceive gift as an event rather than an exchange? Reflecting on Derrida's theory of giving and, in particular, on the (im)possibility of responding to Derrida's rhetoric, Webb concludes that the attempt to explain gift, to bring reason to bear on it or to apply logic to it is to deny the gift altogether, as thinking can only measure actions that are repeatable, calculable, and regular. Perhaps, he wonders, "giving is something we grasp intuitively but have trouble explaining." Indeed. For as Webb points out, Derrida himself characterizes the gift as *excessive*. "The problem of the gift has to do with its nature that is *excessive in advance, a priori exaggerated*."

The symbolic nature of a gift event already signalled its carrying an excess of meaning. The point here is the irrationality associated with this excess. A moderate, measured gift would not be a gift. The most modest gift must pass beyond measure, beyond reason. Thus "it means the extraordinary, the unusual, the strange, the extravagant, the absurd, the mad." Giving must, therefore, be at once voluntary, a free act, and at the same time involuntary, or unconscious. These two conditions, Derrida says, "must—miraculously, graciously—agree with each other."[31]

John Caputo throws some light on such descriptions of gift by reminding us of a well-known biblical story in which "a gift happens when I give not what I have but what I do not have." The widow in Mark's Gospel who gives two copper coins, not from her superabundance but from her lack and deprivation, gives what she herself needs to live and of which she does not have enough (Mark 12:41–44). This gift, says Caputo, shatters the horizon of expectation; is unplannable, unforeseeable, immoderate, and immeasurable: "[It is] a tear in the circle of time and in temple-giving practices, an *événement* [event] in which something other breaks through, breaks out, and shatters the regular flow of now-time; in which something inconceivable, hardly possible, *the* impossible, happens."[32]

His paradigmatic example of Derrida's understanding of "gift" illustrates not only the difference between giving and exchanging but also the primary condition on which the difference rests. It is that giving shall be excessive, and therefore beyond measure; beyond calculation by human reason or rational ideas of equivalence. Behind Gorick and Caputo's discernment of this condition lies a deeper discernment: that it is dangerous to take rationality or logic (in a narrow Cartesian sense) as the sole yardstick of credibility, value, or proper activity in a person's life or behavior. The relevant point here is made by Simone Weil: "We must suppose the rational in the Cartesian sense, that is to say mechanical rule or necessity in its humanly demonstrable form, to be everywhere it is possible to suppose it, in order to bring to light *that which lies outside* its range."[33]

The symbolic gift of Gaia can be conceived of (as it is by Lovelock and other scientists) in rational terms, in what Langer calls "discursive thought." But it can also be conceived in what exceeds, what lies beyond discursive thought: in imagination and dream, myth, and ritual. The parent stock of both conceptual types, she says, of both verbal and nonverbal formulations, is the basic human act of symbolic transformation: "The root is the same, only the flower is different."[34]

Once this transformation happens, the "excessive" nature of gift not only places it beyond rational accounts of economic systems, of measuring natural flows of energy and materials within the biosphere. It also,

Caputo notes, breaks through to what lies beyond the demonstrable range of human forms of accounting: the emotional, sensuous, imaginative, religious, empathic, and, occasionally, excessive content of our interactions with each other. In other words, to the mythic, religious dimension of our evolutionary, biological inheritance.

It is also important to note the language Derrida uses to describe the nature of gift—at once voluntary and involuntary, simultaneously a conscious and unconscious act. This makes it the "very site of the most decisive paradox, namely, 'the *gift that is not a present*,' the gift of something that remains inaccessible, unpresentable, and as a consequence, secret. The event of this gift would link the essence without essence of the gift to secrecy."[35] It also makes it impossible to describe gift in rational terms alone. The metaphorical form of paradox works within the bounds of reason and sequential logic by subverting them. Using it to describe gift, to locate gift, means crossing between linguistic codes and conceptual types and, by alluding to one in the context of the other (voluntary, involuntary: gift, unpresentable), transforming it symbolically in order to convey information about something more or other than the words themselves convey. In Weil's terms, it brings to light that which usually lies hidden, secreted beyond the rational use of language. The decisive paradox of gift is that it subverts any attempt to give logically possible explanations for the event of giving.[36]

It also lays bare the impossibility of confining the process of giving within the perceivable bounds of human interactions alone, and so encourages us toward gratitude for gift events manifest within the whole community of life on earth. The two basic orientations involved, of self and other, extend then to "more than" the human "other." By disclosing this "beyond," paradox reveals what is "unpresentable" by us: the excess within nature's economy that lies beyond and beneath the customary flow of goods between us.

❧ Prometheus Redeemed?
From Autoconstruction to
Ecopoetics

KATE RIGBY

Over the past decade or so, a lively and fruitful conversation with post-modernist and poststructuralist strains of thought has been taking place among theologians and biblical scholars. As indicated by the recent work of Catherine Keller and Mark Wallace, among others, this is a conversation that some ecologically inclined religious thinkers are also beginning to join, and in so doing to reframe: for everything looks different when viewed from the perspective of an endangered earth.[1] In general, though, a certain mutual distrust between postmodernist and ecological thinkers still tends to prevail across the humanities and social sciences, not least in my own home discipline of literary studies, with the postmoderns commonly suspecting the ecologists of naïve naturalism, and the latter accusing the former of persistent idealism—in either case, the other is deemed self-deluding.

There are doubtless several dimensions to this discordance; including, I suspect, underlying differences in style and milieu. Intellectually, however, and at the risk of oversimplification, the primary bone of contention can readily be identified as the question of constructivism. Is the human condition, or more dynamically, are the possibilities of human becoming predominantly self-made, determined in the main by social relations, cultural conventions, and (as some would insist) the very words that compose our mental world? Or is not human life, for all its ramifying complexity of diverse languages, cultures, and societies, crucially dependent upon and powerfully shaped by that which is other-than-human; from microscopic organisms through to planetary systems and ultimately (again, as some would insist) a divine creative agency beyond and/or within the created order? And what are the implications—religious, philosophical, ethical, and political—of prioritizing that which we make (and over which we therefore imagine we have some measure of control) over that which we are called upon to respect as given, and hence as limiting no less than enabling our chosen undertakings? There is, I think, a well-founded ecosophical concern that constructivism, at least in its more doctrinaire guise, threatens not only the life of faith (what is God if not a phantasm of our own making?) but also our ability to keep faith with life; to acknowledge the giftedness of our existence in kinship and in relationship with the multifarious other creatures with whom we have been called into being on this our earthly home, or *oikos*. At the same time, I believe that there is also clearly some justification in the converse anxiety that an ecosophical insistence on that which is held to be naturally and/or divinely pregiven threatens to stymie the unfolding of human creative potential and emancipatory endeavor, which is perhaps itself naturally and/or divinely gifted.

In this essay, I would like to address these concerns by initially taking a step back from the fray of current debate to consider one particular locus in the archaeology of constructivism that I have discovered in the work of a writer and amateur scientist whose complex contribution to the emergence of modern ecological thought is only now beginning to be more widely appreciated: Johann Wolfgang von Goethe. Starting with his extraordinary recasting of the figure of Prometheus, in whom both the emancipatory promise and the other-denying menace of modernity are fatefully entwined, I will then consider a number of contemporary philosophical responses to constructivism before returning to Goethe in order to contemplate whether, and how, in the postmodern present, Prometheus might in some sense be redeemed.

Penned in the same year that certain uppity colonists in the New World dumped a shipment of tea into Boston Harbor in a symbolic gesture of defiance against the sovereignty of the mother country (I dread to think what has ended up there since), Goethe's "Prometheus" summons an Old World figure of rebellion in order to issue a declaration of independence for humanity at large. This famous ode belongs to Goethe's early work and was written in the context of what was possibly Europe's first youth movement, Germany's decade of literary "Storm and Stress" (*Sturm und Drang*). Here, the young poet sets myth to work in the cause of demystification, effecting in the process a remarkable protoconstructivist critique of religion. In the course of railing against Big Daddy Zeus in emphatically free verse (the very form of the poem embodying a refusal of conventional constraints), Goethe's mythic rebel—and prototypical modern man—invalidates divine overlordship by disclosing it as a human, all-too-human projection, arising from a particular psychic economy and sustained by cultural convention:

> I know of nothing more wretched
> Under the sun than you gods!
> Meagrely you nourish
> Your majesty
> On dues of sacrifice
> And breath of prayer
> And would suffer want
> But for children and beggars,
> Poor hopeful fools. (ll. 12–20)[2]

Goethe protects himself from the then very serious charge of atheism by adopting a Grecian mask, but it is fairly obvious that one of his targets is conventional Christian piety (another being a naïve faith in more worldly lords and masters). The most radical reading of this second stanza is that it is not God who created the world, but humans who created the gods. More moderately, one might conclude that what is at stake here is not the existence of deity per se, but a particular model of the divine. In this reading, one might even hear an echo of the biblical tradition of prophetic zeal against idols. Significantly, the idol that Goethe's Prometheus is set on unmasking is cast in the opening stanza as a power freak who likes to throw his weight around in a manner that is seen as positively infantile:

> Cover your heavens, Zeus,
> With cloudy vapors
> And like a boy
> Beheading thistles
> Practice on oaks and mountain peaks—

Still you must leave
My earth intact
And my small hovel, which you did not build,
And this my hearth
Whose glowing heat
You envy me. (ll. 1–11)

This dismissal of a wrathful kind of God who would like to disfigure the earth's topography, such as Goethe might have encountered in another dimension of the prophetic tradition (for example, Isa. 40:4), does not open the way for an affirmation of a more merciful model of the divine, however. Indeed, belief in a deity who (recalling Matt. 11:28–29) succours the oppressed and lifts the burden of the heavy-laden is explicitly rejected as self-deluding:

Once too, a child
Not knowing where to turn,
I raised bewildered eyes
Up to the sun, as if above there were
An ear to hear my complaint,
A heart like mine
To take pity on the oppressed.
. . . .
I pay homage to you? For what?
Have you ever relieved
The burdened man's anguish?
Have you ever assuaged
The frightened man's tears? (ll. 20–26; 36–40)

By the end of this paradoxical hymn, not of praise and thanksgiving, but of defiance and self-assertion, the whole concept of godhead appears to have been discounted as an irrelevance to human existence:

Here I sit, forming men
In my image,
A race to resemble me:
To suffer, to weep,

To enjoy, to be glad,
And never to heed you,
Like me! (ll. 50–56)

The fuel that fires Prometheus's bid for freedom is said to lie not in rational reflection, as one might expect in the Age of Enlightenment, but in his "holy and glowing heart."[3] Goethe's sacralization of the passions marks this text as a protoromantic work of counterenlightenment. Yet his Promethean vision of human emancipation from blind faith presages the egress from self-imposed minority that Immanuel Kant would later posit as the answer to the question, "what is enlightenment?" in his essay of that name from 1786. *Sapere audi*, asserts Kant, is the name of the game: "Dare to think for yourself!"[4]

The philosopher (unlike the poet, incidentally), did not hold high hopes for the fair sex in this regard, assuming that women would rather play it safe and let their menfolk think for them. The likes of Olympe de Gouges and Mary Wollstonecraft would soon prove him wrong, however. Daring to think for themselves, many women and other oppressed groups have continued to follow the lead of Goethe's Prometheus in unmasking relations of domination once deemed to be natural and even divinely ordained as actually culturally constructed and open to change. Today, within the discourse of much feminist, queer, antiracist, and post-colonial theory, the constructivist critique is fiercely guarded as the necessary guarantor of the still unrealized promise of a particular kind of human liberty: the freedom, that is, to refuse heteronomic roles and identities, and to make and remake oneself on one's own terms.

While the demystificatory practice of constructivism continues to offer a valuable technique of resistance to reactionary reinscriptions of what is natural, good, and godly, its blessings have been mixed, to say the least. For one thing, since we have forfeited Goethe's youthful and protoromantic faith in individual self-fashioning (above all in the wake of the perceived failure of the student movement of 1968), the emancipatory promise of the Promethean project has tended to give way to a new determinism, albeit one of "culture" rather than "nature."[5] For those who wish to affirm the force of that which precedes and exceeds cultural inscription, whether in the guise of a god beyond or as the earthly *oikos* that sustains us, the dogma of constructivism is particularly problematic:

for theists no less than for ecologists (and thus doubly so, perhaps, for ecotheologians). In order to explore this tension further, it is helpful to return once again to "Prometheus."

Goethe's version of the Promethean rebellion culminates in the crafting of a new species of earthling. Goethe reminds us here that the emergence of humans qua humans, earthlings of a very special kind, was predicated upon the appropriation of fire: fire, by whose light and warmth we learnt to steal part of the night from sleep that we might tell those tales and sing those songs that spun a human world of words in whose warp and weft we would come ever more to dwell; fire, by means of which we learnt to alter local ecologies and to fashion those tools and weapons that would facilitate an ever-more-radical reshaping of the earth; fire, that transformed us from omnivorous predators among others into uniquely cultivated (if, for all that, no less predatory) cooks; fire, the primal human technology, by means of which we brought ourselves into being, namely, as *homo faber*, man the maker, master of the element of transformation.

It was no coincidence that Goethe should have chosen to privilege the figure of Prometheus at precisely this historical juncture, a few short years after the opening of the first cotton mill deploying Arkwright's "spinning jenny" inaugurated the Industrial Revolution: as ever, he had his finger on the pulse of the times. Dismissing divine sovereignty as an all-too-human projection, Promethean man lays claim to the earth as a place on which to build a new world of his own designing: "you must leave / *My earth* intact, / And my small hovel, *which you did not build.*" Two and a quarter centuries later, the ominous significance of the words that I have highlighted here comes more fully into view. In declaring his independence from the gods by disclosing their dominion as a cultural construction, *homo faber* is at risk of devaluing not only those old notions to which he was formerly enslaved, but also the entire realm of the given: all that he did not make; all that, could he but see it, stubbornly eludes his grasp; all that might yet resist his intentions and unsettle his expectations, even in the things of his own making. With the birth of Prometheus's new breed of men, *techné* (to recall the Greek distinction that Martin Heidegger helpfully brought back into philosophical currency) comes to lord it over *physis*. Divine hegemony, or hegemonic divinity, has been sent packing in order that man might claim the mantle of the Creator.

Historically, two main paths lead from here. One we have already traced: the cultural constructivist path that aims to enlarge the realm of human freedom by unmasking oppressive constructions in the hope of dismantling the social relations of domination that they support. The other is the technoscientific path that seeks to increase human self-determination by gaining mastery over the physical conditions of human existence; including, these days, our own genetic blueprint. To some extent this latter trajectory was already implicit in the beginnings of modern science, when Sir Francis Bacon presented his *Novum organon* as the means by which to dramatically expand man's rightful dominion over creation. But whereas Bacon still needed God as the guarantor of the righteousness of this venture, his Promethean descendents evidently share James Watson's cocky view that "if we don't play God, who will?"[6] Today's cultural constructivists might well resist being associated with this kind of biological reconstructionism. But as Goethe's prescient poem reveals, they have a common point of departure in the devaluation of the given. The skeptical gaze that his prototypical modern man casts toward the heavens has now fallen also upon the earthly: nature, no less than God, is under grave suspicion from cultural constructivists and genetic engineers alike, even though their terms of reference might differ markedly.

In seeking to counter an increasingly hegemonic constructivism, Australian ecofeminist philosopher Val Plumwood speculates that "the deep contemporary suspicion and scepticism about the term 'nature' may play some role in the contemporary indifference to the destruction and decline of the natural world around us."[7] The fundamental problem with constructivism, in her analysis, is that it perpetuates the long tradition of Western dualism in denying agency to the other-than-human, reducing nature to a passive screen upon which we project culturally relative and ideologically distorted views of reality. In mounting this argument, however, Plumwood has herself undertaken a constructivist critique, unmasking constructivism as a culturally relative and ecophilosophically suspect view of reality. Indeed, one of Plumwood's central arguments is that we cannot afford to do away with the critical tools of constructivism altogether. In the interests of overcoming its anthropocentric bias, however, it is equally essential that we find ways of affirming the independent interests and potentially resistant agency of other-than-human entities.

Another of my ecophilosophical compatriots, Freya Mathews, has sought to articulate a new metaphysics that is capable of doing just that. Within her renovated panpsychism, the entire realm of matter is accorded agency. In *For Love of Matter*, Mathews asks that we abandon our objectifying quest for power-knowledge in favor of an ethos of erotic encounter, whereby humanly made things, no less than natural entities, are seen to participate in the psychic dimension inherent in all matter and hence to possess their own vitality and capacity to surprise.[8] Paradoxically, Mathews argues most recently in *Reinhabiting Reality* that reductive materialism, by denying mind to matter, actually engenders "derealization," the currently dominant version of which within Western academia she takes to be cultural constructivism: "When matter is thought to appoint no ends or purposes for us, that same matter becomes increasingly irrelevant to the way we experience our lives: it is perceived merely as the inert backdrop to our meaning-making, a backdrop that can be epistemically constructed in countless ways."[9] Mathews nonetheless acknowledges that "we do, to a significant degree, construct the world that we inhabit, in ways biological and neurological, social and cultural, not to mention discursive and ideological, as poststructuralists and feminists, and before them Marxists, have so illuminatingly demonstrated." Yet she insists that our perceptions and understandings must also be seen as co-constituted by the more-than-human world itself: "It will reach out to us, regardless of the particular sensory forms it assumes for us and the discursive constructions we impose on it."[10]

Similarly, although in a more poststructuralist vein, and hence with a greater degree of epistemological caution, Jane Bennett recalls the hylozoism of the early Greek philosophers in the name of the liberation of what she calls "thing-power": "Flower Power, Black Power, Girl Power. Thing Power: the curious ability of inanimate things to animate, to act, to produce effects dramatic and subtle."[11] Drawing in particular on Lucretius (whom she brings into conversation with a somewhat unlikely crew that includes Baruch Spinoza, Henry David Thoreau, Franz Kafka, Henri Bergson, Gilles Deleuze, Felix Guattari, and Antonio Negri), Bennett allows herself a moment of (not so) "naïve realism" in order to allow "nonhumanity to appear on the ethical radar screen." While she concedes that "any thing-power discerned is an effect of culture," to the extent that it comes into view only in the frame of an "onto-story" that affirms material agency, she observes that "concentration on this insight alone also diminishes any potential we might possess to render more *manifest* the world of nonhuman vitality": a world that inheres in the human person too, in the depths of our very bones. "For a thing-power materialist," Bennett concludes, "humans are always in composition with nonhumanity, never outside of a sticky web of connections or an *ecology*."[12]

Bennett's metaphor of the "sticky web" points to the tangled imbrication of ideality and materiality that Catherine Hayles refers to as the "cusp," constituted, as Mark Wallace explains, "in the interface between the ebb and flow of natural processes and the positioned assumptions and contextual experience of the observer."[13] In her contribution to the volume *Reinventing Nature: Responses to Postmodern Deconstruction*, Hayles proposes a "constrained constructivism," which "points to the interplay between representation and constraints. Neither cut free from reality nor existing independent of human perception, the world as constrained constructivism sees it is the result of complex and active engagements between the unmediated flux and human beings."[14]

Retracing our steps to "Prometheus," it is worth pondering a moment in the text, in which Goethe, too, admits an element of constraint on *homo faber*'s heroic self-fashioning. Following the lines previously quoted that negate divine reassurance, Goethe's model modern does in fact acknowledge a certain limit to his self-determination:

Was it not omnipotent Time
That forged me into manhood,
And eternal Fate,
My masters and yours? (ll. 41–44)

This concession to the power of time and the role of fate (to be understood, I think, more in the Greek sense of *tyche*, contingency, than of divine providence) threatens to undo the constructivist project at its very outset in a way that resonates interestingly with the recent work of Elizabeth Grosz. In *The Nick of Time*, Grosz returns via Deleuze to Darwin,

Nietzsche, and Bergson in order to develop a philosophy of temporality, which unfolds a more open-ended and complex understanding of natural forces than that which predominates in most contemporary political, cultural, and social theory. In her chapter, "The Nature of Culture," in *Time Travels*, the companion book to *The Nick of Time*, she, too, contests constructivism, arguing that once nature is recognised as "dynamic, prodigious, excessive, differential—that is, to the degree that it is construed as temporal, historical, and unpredictable (a picture that some in the biological sciences have worked hard to elaborate since the writings of Darwin himself)—culture in all its permutations must be understood as the *gift* of nature, its increasing elaboration and complication through the efforts of life to transform itself."[15] Indeed, if we follow evolutionary biology, as Grosz urges, perceiving nature in terms of perpetual variation, "and life as the evolutionary playing out of maximal variation or difference . . . then culture rather than nature is what impoverishes nature's capacity for self-variation and becoming, by tying the natural to what culture can render controllable and what it sees as desirable."[16] Grosz appears reluctant to associate herself with ecophilosophy, but it is surely within the horizon of global ecological crisis, entailing a loss of biodiversity on a scale unknown since humans appeared on the scene, that the capacity of culture to *materially* impoverish nature comes most starkly into view. And it is precisely at this historical juncture that we are called to acknowledge the force of that which lies "outside" all cultural constructions: an other-than-human Outside, that is, with the capacity to disrupt, to intervene, to break down expectations, and thereby also to incite thought, innovation, and transformation in the realm of human culture.

Progressive naturalism, panpsychism, thing-power materialism, riding the cusp, transformative temporality: all offer valuable ways of contesting the mis-taking of materiality as necessarily subject to human remaking, without giving ground to those who would appeal to an unchanging nature in order to endorse oppressive social identities and relations. But what about that other target of the Promethean assault upon the given, that which is construed as the ultimate source of the given in the monotheistic tradition that is the masked object of Goethe's critique? In their endeavors to theorize an ontology (or, more modestly, "onto-story") that accords agency to the other-than-human, Plumwood, Mathews, Bennett, and Grosz all leave God out of the picture—understandably so. For, at least within the dominant western conception of creation, nature was always already construed as a work of *techné* rather than as a manifestation of *physis*, to the extent that it was held to have been conjured out of nothingness by magisterial decree. As Ilya Prigogine and Isabelle Stengers remark, the mechanistic reduction of matter to mute malleability was dependent upon nature's prior deanimation within mainstream Christianity, whereby it was "stripped of any property that permits man to identify himself with the ancient harmony of natural 'becoming.'"[17] From this perspective, it becomes evident that the Promethean model of man as *homo faber* is flawed precisely insofar as it replicates, in a secular key, what Catherine Keller in *Face of the Deep* terms the "dominological model" of God within the theology of *creatio ex nihilo*. Playing God, that is to say, is particularly problematic if God is taken to be the sovereign One, for whom precisely nothing is taken as given. Today's— and yesterday's—swashbuckling genetic engineers aspire to emulate this supreme Creator by recoding living entities as artifacts of human design. Meanwhile, those constructivists for whom it is language that creates the world would evidently also have humans play God: the God, that is, of Genesis 1, who is said to have brought all things into being by speech-act.

Clearly, then, appealing to a theocentric order beyond the human in order to counter Promethean hubris will not necessarily bring us any closer to respecting the autopoietic agency of our earth others, or of materiality in general. Even so, if we are to address the problem of Prometheanism at a deeper level, we cannot afford to overlook its theological background. Countering constructivism constructively (rather than simply critically) might then require nothing less than a radical reconceptualization of creation. This is, of course, precisely what Catherine Keller undertakes in *Face of the Deep*. Within her extraordinary close reading of Genesis 1, the dominological doctrine of *creatio ex nihilo* gives way to a counterdiscourse of *creatio ex profundis*, whereby the differentiations called forth by the Word of Elohim are understood to presuppose both a prior creative capacity figured biblically as the *tehom*, or deep, and the *ruach*, breath or spirit, which connects Elohim and *tehom* and holds all things in relation. Importantly, Keller stresses that Elohim does not "unilaterally order a world into existence, Elohim 'lets be'. . . . The earth is *agent* of Elohim's creation. The same again with the sea. . . . *Creation takes place as invitation and cooperation.*"[18] From this perspective, creation ceases to be an artifact of divine handicraft with a definitive origin and trajectory and reappears as an open-ended process of becoming, a perpetual incarnation or (delightfully) pancarnation of multitudinous interrelated entities, all of which participate in the ongoing activity of material co-creation in and with a plurisingular God.

The sources of Keller's theology of becoming are themselves multitudinous, but include contemporary scientific understandings of complex dynamic systems popularized as "chaos theory," as well as astronomy and ecology. As a flurry of new research on European romanticism has indicated, many of the shifts in thinking about the physical world that have been taking place in highly specialized corners of the sciences over the past century were to some extent prefigured by none other than Goethe and a number of his younger contemporaries in Germany and Britain around 1800. While romantic science and *Naturphilosophie* fell into disrepute in the later nineteenth and early twentieth centuries (while nonetheless crucially informing the process thought of Alfred North Whitehead that continues to inspire Keller's work), they are currently being revalued as a crucial locus of resistance to that nasty and persistent nexus of dualism, mechanicism, positivism, and instrumental rationality. In place of Descartes' mindless clock, the romantics posited nature as a self-organizing, dynamically unfolding, and ultimately theophanic multeity-in-unity of which humanity was integrally a part; thus nature could never fully and finally be known by us as a whole, but only ever partially and provisionally.[19] The question was, and remains, though, what *is* our part? How should we answer to the agency of the other-than-human? What is the vocation of humanity in the continuing becoming of a divinely graced creation?

The romantics had various and conflicting views on this, but none doubted that we were in some sense special. Unlike deep ecologists, for whom humans are simply one species among many, most ecotheologians view humans as distinctive, even though they are at pains to acknowledge, and break with, the violent legacy of the divine warrant for human domination construed as given in Genesis 1. So, too, Catherine Keller insists that, read in context, "'dominion' can only mean *caretaking* for that divinely loved whole, which only with *all* the species is called 'very good.'"[20] The role of caretaker or steward is the human vocation most commonly avowed within ecotheology, and it undoubtedly points to a more earth-friendly ethos for Christians than that proffered either by the Baconian discourse of mastery or by salvationist doctrines privileging redemption in an otherworldly beyond. To someone of my deep-green sensibility, however, these terms still seem too closely bound to a hierarchical conception of the chain-of-being, harboring connotations of a certain coolness and distance in our relations with the rest of creation, and perhaps even implying a level of overall influence that we almost certainly do not actually possess. At the same time, I have grown dissatisfied with radical ecocentric egalitarianism, to the extent that it risks eliding the opportunity that we might perhaps have been given in this grim moment of crisis to discover, or recover, a specifically human potential to enrich rather than impoverish the unfolding of life in the dynamic process of continuing creation.

In order to begin to consider how this might be configured, I would like to return once more to Goethe. The figure of Prometheus was to haunt him for much of his life, metamorphosing into the tragic hero of his greatest work, *Faust* (of which more anon). Long before his final reckoning with the Promethean subject of modernity in Part Two of that veritably cosmic drama, however, Goethe had drawn away from the defiant self-assertiveness of his *Sturm und Drang* days. In 1781, at the beginning of a new phase in his life coinciding with his early scientific studies, he wrote another ode, the very title of which implies a retraction of the Promethean rebellion: "Limits of Humanity." This work of self-correction, written in the same form as "Prometheus" but embodying a profoundly different ethos, opens with a prayer of thanksgiving to God as the source of the natural bounty upon which all human making is dependent:

When the ancient,
Holy Father,
With tranquil hand
From roiling clouds
Over the earth sows

The blessing of lightning,
I kiss the last
Hem of his gown,
Childlike awe throbbing
True in my breast. (ll. 1–10)[21]

To rejoice in the givenness of earthly existence with words and deeds of
thanks and praise is surely one way—one that has long been widely
cultivated across human cultures, however variously the giver has been
construed—in which humans can be seen to augment creation, precisely
by welcoming it as given. Here, Goethe allows himself a "naïve moment" in affirming that childlike gratitude to a paternalistically imaged
deity that he had formerly ridiculed. This movement of simple piety is
nonetheless complicated in the penultimate stanza, which calls upon the
image of the watery deep in order to locate the limits of humanity, not
in any kind of infantile dependency, but rather in the hard-won recognition of our own finitude:

What then distinguishes
Gods from men?
That many waves
Before them move,
An eternal stream.
Us the wave lifts up,
Us the wave swallows,
And we sink. (ll. 29–36)

For *homo faber*—the sovereign self of modernity who takes life not as
interdependently given in a way that, as Anne Primavesi emphasizes,
utterly defies reciprocation,[22] but as autonomously self-fashioned—death
acquires a whole new horror (is it any wonder that Goethe's renovation
of the figure of Prometheus coincides with the emergence of the genre
of the gothic?). Placing an untranscendable limit to human self-fashioning, while also being stripped of more-than-human meaning, death becomes obscene: it is either banished from view entirely or foregrounded
as a locus of obsessive fascination (sex, it might be noted, has had a

similar fate in modernity). To what extent, I wonder, have our ever-more-elaborate strategies for endeavoring to control our physical and social conditions of existence and to predetermine our future, both individually and collectively, been driven by repressed anxiety about our ultimately mortal condition? If so, is it not ironic (or perhaps entirely fitting) that our unconscious efforts to keep death at bay should be breeding a kind of "double death" in the mass extinction of species and imperilment of the generativity of whole areas of the earth, raising the specter of the death of death, and with it the death of the possibility of new life?[23] Keeping faith with life thus calls for a new relationship with death. As Tony Kelly has suggested, in keeping with the tenor of Goethe's "Limits of Humanity," this might be found in the recovery of a certain kind of "humility," recalling that this word derives from the Latin *humus*, earth, soil, dirt. By "'re-membering' ourselves in the ground of the universe," Kelly suggests that death might be allowed "to emerge from its subterranean place of influence, no longer able to sap our energies or drive us to the frenzy of illusory immortality projects, but to connect us with the whole, and to immerse us in a universal mystery."[24]

A similar reframing of death as serving life through the incorporation of the self into a greater nexus of becoming is also undertaken in Goethe's ode of newfound humility. Although our individual lives are finite, we are reassured in the final stanza that we participate in the infinite through our connectedness with others, those "many generations" of life, human and otherwise, who precede and follow us:

A little ring
Limits our life,
And many generations
Link up continuously
On the endless chain
Of their existence. (ll. 37–42)

This reassurance does little, though, to deprive death of its sting. Any hope that we might harbor that death is not, after all, the end (however that "not ending" is conceived) needs to be counterbalanced, as Kelly has Vergil remind us, by the acknowledgment that "there are tears at the heart of things, and we are deeply touched by all this dying."[25] A recognition of the anguish that attends our encounter with mortality, others' as well as our own, is rendered emblematically in the post horn that announces the arrival of the human in the fourth movement of Gustav Mahler's Third Symphony: the post horn, which, in the poem that inspired this moment in Mahler's remarkable symphony of the universe, was said to have been played by a friend's grave. With humanity, we might conclude, the community of life is enriched by the emergence of a being who is specially endowed with the capacity to give voice to grief, to make (an) art out of mourning. This is not to say that other beings might not sorrow and indeed rejoice after their own fashion. However that might be, facing our own finitude in fittingly human ways, we are also empowered to feel for others, human and nonhuman, in the brevity and often adversity of their lives. The epigraph to the final movement of Mahler's symphony, entitled "What Love Tells Me" reads: "Father, look upon my wounds, / Let no being be lost and gone" (*Vater, sieh die Wunden mein, lass kein Wesen verloren sein*).[26] Today, what love tells us is surely that we are called upon to protest not death itself but the dealing in double death that is today leading to the loss of ever more beings by breaking the chain of their generations forever.

Mahler was himself a great admirer of Goethe, not least of his *Faust*. In the final act of the second part of this epic drama, written shortly before the author's death in 1832, Promethean man appears in ultramodern guise as a ruthless developer, whose hostility to the given has reached pathological proportions. In a classic case of what Keller terms "tehomophobia," the aging Faust determines to drive back the sea and drain the swamp in order to win new land for some humanly meaningful purpose, assenting to the Mephistophelian eradication of all that stands in the path of the creation of his brave new world, including the native inhabitants of this place, the old couple Philemon and Baucis, who dwell by the shore. Earlier, in the "Classical Walpurgis Night" of Act 2, the ocean was hailed as the source of all life, the primary locus of the eros of biotic becoming; but the aging Faust, having lost his taste for love with the death of Helen and their phantasmic child Euphorion, comes to view the sea with revulsion:

Landward it streams, and countless inlets fill;
Barren itself, it spreads its barren will;

It swells and swirls, its rolling waves expand
Over the dreary waste of dismal sand;
Breaker on breaker, all their power upheaved
And then withdrawn, and not a thing achieved!
I watch dismayed, almost despairingly,
This useless elemental energy!
And so my spirit dares new wings to span:
This I would fight, and conquer if I can. (4.10212–21)[27]

With extraordinary prescience, Goethe's Faust here prefigures that sense of absurdity and alienation that would only take hold in Western modernity a century and a half and two horrific world wars later. Of that line of development that leads from Promethean rebellion to Sartrean nausea, Erazim Kohák has written:

If there is no God, then nature is not a creation, lovingly crafted and endowed with purpose and value by its Creator. It can only be a cosmic accident, dead matter contingently propelled by blind force, ordered by efficient causality. In such a context, a moral subject, living his life in terms of value and purpose, would indeed be an anomaly, precariously rising above it in a moment of Promethean defiance only to sink again into the absurdity from which he rose.

If God were dead, so would nature be—and humans could be no more than embattled strangers, doomed to defeat, as we have largely convinced ourselves we in fact are.[28]

With Mephistopheles at his side, Faust, however, is not yet ready to accept defeat. Driven by the Promethean impulse to transform the given into an artifact of his own making, thereby keeping nausea at bay, Faust cannot rest content while there is still a reminder of the world that he did not create. Thus, the premodern dwelling place of the shore's indigenous inhabitants with its line of linden trees, modest shack, and crumbling chapel, which had so far survived the "Faustian project of discipline and drain,"[29] must also give way to his bid for absolute environmental sovereignty:

I chose that linden clump as my
Retreat: those few trees not my own

Spoil the whole world that is my throne.
From branch to branch I planned to build
Great platforms, to look far afield,
From panoramic points to gaze
At all I've done; as one surveys
From an all-mastering elevation
A masterpiece of man's creation.
I'd see it all as I have planned:
Man's gain of habitable land. (5.11240–50)

Faust is nonetheless shocked and chastened when he hears of the death of the old couple. Blinded by the allegorical figure of Care, in an apt concretization of his metaphoric blindness to the price of his Promethean quest—that is, his failure to care and to take care—he dies dreaming of the "brave new world" of "free people" who he hopes will one day inhabit his "inland paradise" wrested from sea and swamp (5.11561–68). In Goethe's subversively avant-garde rewriting of the Faust legend, Mephistopheles is comically foiled in his attempt to make off with the errant hero's soul by his erotic attraction to the androgynous angels who cheat him of his prize, and we are left to ponder the mystery of Faust's redemption by means of something that Goethe refers to in his famous (and, to contemporary ears, infamously patriarchal) last line as the "eternal feminine" (5.12110).[30]

Here it would seem that Goethe once more retracts his earlier retraction of Promethean striving in "Limits of Humanity." This is no unconditional endorsement of the modernist project of autoconstruction, however. The human desire for transformation and self-transformation is affirmed, but disregard for the being of others and disrespect for the unfolding of a physical world that is not of human making, and potentially resistant to human makeover, is disclosed as tragically blind.[31] Meanwhile, nature's resistance to human mastery is signalled by Mephistopheles, who foretells the ultimate failure of Faust's engineering schemes in the face of the greater power of the elements: "Do what you will, my friend," he observes in an aside, "You all are doomed! They are in league with me, / The elements, and shall destroy you in the end" (5.11548–50). Yet it is not evil, but eros, that has the last word; and this, too, is figured as coming from beyond the self, indeed beyond the human. Faust might be redeemed on account of his continued striving, but he is said to be saved despite himself, by a "Love beyond / Time" (5.11964–65) that is quite definitely not of his own creation or at his disposal.

Within Goethe's *Naturphilosophie* of ceaseless becoming, the creative capacity that humans have evidently been endowed with in such glorious and perilous abundance is to be honored as enabling us to participate in the more-than-human process of ongoing creation. Human self-realization, he nonetheless came to believe, inevitably entailed an element of strife with the earthly conditions of our existence. In an essay on meteorology from 1825, written around the same time that he returned to work on *Faust: Part Two,* and in the wake of catastrophic floods on the North Sea coast, he observes that "where man has taken possession of the earth and is obliged to keep it, he must be forever vigilant and ready to resist."[32]

Following the Boxing Day tsunami of 2004, to name but the most deadly of recent disasters, we, too, can be in no doubt about the potential violence of the elements. Today, however, in the postmodern era of global warming, it is increasingly against conditions of our own making that we need to guard ourselves: the unintended consequence of our endeavor to take possession of the earth. In these circumstances, the redemption of Prometheus should perhaps be premised on a more thoroughgoing transmutation of the ceaseless striving of autoconstruction into a gentler ethos of ecopoetics that is more respectful of the potentially resistant agency of the other-than-human. To thus redeem Prometheus as ecopoet would be to figure a rather different model of the postmodern from that which prevails in poststructuralist cultural and critical theory. While acknowledging the ways in which our identities, perceptions, and artifacts are shaped to a greater or lesser extent by language, culture, and social relations, a theologically inclined ecopoetics would reframe human creative and emancipatory endeavor as a mode of participation in the more-than-human song of an ever-changing earth, comprising a forever partial and inadequate response to the call of the other, human and otherwise, in whom, and through whom, we might be brought face-to-face with a trace of the divine.[33]

∿ Toward a Deleuze-Guattarian Micropneumatology of Spirit-Dust

LUKE HIGGINS

The task of perception entails pulverizing the World, but also one of spiritualizing its dust.
—GILLES DELEUZE, *The Fold: Leibniz and the Baroque*

Christian ecotheologians of the last four decades have been pointed in their critiques of the metaphysical hierarchies lying at the foundations of Christian thought. Traditional cosmological mappings place God at the top of a metaphysical ladder that descends to the human male as God's primary "image," then down to the human female, animals, plants, and, finally, inanimate matter. This theological model of creation has been rightly criticized for its justification of human (not to mention male) domination of the earth. Less attention has been paid to the way this hierarchical model also functions to bind humans and nonhumans into "essentialized" roles or identity constructs that limit the modes of relationship possible between them.

The work of ecofeminist theologian Sallie McFague exemplifies a rather typical, critical approach toward this "Great Chain of Being." She attempts to balance divine transcendence with a reinvigorated notion of divine immanence, conceiving creation, metaphorically, as the "body of God."[1] God is no longer a cosmic ruler, fundamentally separate from creation, but a "Spirit-being" infused within the full plurality of creaturely reality. Divine transcendence, however, is by no means abandoned in her model: "In this body model, God would not be transcendent over the universe in the sense of external to or apart from, but would be the source, power, and goal—the spirit—that enlivens (and loves) the entire process and its material forms. The transcendence of God, then, is the preeminent or primary spirit of the universe."[2] For McFague, locating the divine on this "macrosphere" of reality unites divine immanence and transcendence in such a way that they are able to mutually enhance one another.

While there are compelling aspects to McFague's alternative theological model of creation, I want to ask whether her renewed emphasis on the immanence of God goes far enough in dismantling the environmentally destructive logic of the Great Chain of Being. To begin with, God is still conceived as an entity occupying the highest level of reality—a preeminent "macro-Spirit" from which all creatures derive their life, their goal, their very reason for being. Multiplicity and diversity are affirmed but only insofar as they can be traced back or added up to a higher, "macrotranscendent" unity. If the divine continues to function as the guiding logic, or telos, for all that lies beneath or within it, does not the *singularity* of the *transcendent* divine consistently end up trumping the *multiplicity* of an *immanent* divine? In short, I wonder if this model's renewed emphasis on immanence really moves us forward if the immanence is always *immanence to a transcendence*. My contention is that this logic of "macrotranscendence" can only hinder ecotheology's effective engagement in the political, ethical and spiritual work of ecological transformation.

This argument will be supported in part by the insights of philosopher of science, Bruno Latour. Latour provocatively argues in his *Politics of Nature* that political ecology should have nothing whatsoever to do with "nature."[3] For Latour, the concept of nature carries with it a whole slew of illegimate, socially damaging assumptions held over from the West's Enlightenment heritage—the chief of which is the idea that the earth's constitutive entities can be objectively understood and ordered in advance from some unitary, "neutral" position of authority. If we are to develop a truly democratic political process capable of transforming human modes of relationship to nonhumans, invoking totalizing concepts for the Whole—of which "nature" is a quintessential example—can only set us back. What is demanded is a collective process of decision-making able to put up for grabs again *who* gets included in creation and *what* we can all become *together*.[4] Latour's critique of the concept of "nature" highlights problems inherent to a theo-logic of macrotranscendence. The latter's tendency to view creation's dynamic multiplicity only under a unitary, divine "umbrella" risks essentializing the identities, roles, and relationships among the earth's diverse creatures. A god that functions as the guarantor of a particular (albeit idealized) order of creation will tend to inhibit rather than inspire our collective capacity to both creatively imagine and politically realize new, more life-giving forms of relational becoming.

Perhaps a more radical critique of the Great Chain of Being is called for—one that does not merely tinker with its structure but turns it on its head. Instead of seeking the divine on the level of what we have called the macrotranscendent, what if we turn our gaze toward the smallest, most dynamic microelements of reality imaginable? What if we envision the divine not as an infinite singularity but an infinitesimal multiplicity? The divine, then, would no longer function to validate from on high a single structure or pattern of ecological relationships. Doing away with this single, divinely sanctioned "mold" for ecological interaction would not mean denying our ability to distinguish between healthier and unhealthier—or in theological terms between more sinful and less sinful—ecological relationships. It *would* mean, however, that solutions to the various problems of how to live together as diverse creatures on this biosphere could not be mapped out in advance or approached under the auspices of a single, all-encompassing agenda.

Indeed, this essay proceeds on the assumption that healthy ecological relationships *cannot* be held to a single, static pattern of interrelation but instead must be flexibly negotiated and renegotiated in an ongoing process. This is by no means to devalue ecological stability, but rather to recognize that, in complex systems, stability precisely depends on a certain flexibility, or "improvisation," in that system's relational balancing. This essay will suggest that ecotheology is best served not when it theologically endorses some given relational structure of the Whole but when it views the divine as the *inspiritor* of relational *becoming*. Gilles Deleuze and Felix Guattari, particularly in *A Thousand Plateaus*, provide important conceptual tools for helping us to rethink Spirit on a *micrological* level as the *Spirit-dust* that animates the creative, interrelational becoming of the cosmos.[5]

Such an attempt to think beyond categories of divine macrotranscendence need not abandon the theological imperative to think divinity as the deepest source of our ecological interconnection. What I suggest is that we relocate this divine "source" from an *overarching macrosphere* to an *underlying microsphere*. I believe Deleuze-Guattari's intriguing notion of the "plane of immanence" might suggest to us a new way of conceptualizing an ultimate, or cosmic, connectedness that does not submit the universe—better, the *pluriverse*—to any predetermined, stultifying pattern of order.⁶ A "micropneumatology of Spirit-dust" would imagine Spirit not moving downward into the plurality of creation from a higher, transcendent realm of unity but, rather, upward/outward from an underlying, constitutive sphere of dynamic microparticles. Our orientation to the divine would thereby shift from a vertical to a horizontal axis (though these spatial metaphors can only apply in a limited sense), drawing us to focus on what we can become here and now, through our concretely emerging relationships. My hope is that a pneumatology conceived along these lines might free us toward new, co-creative modes of becoming better able to nourish planetary creatures of all kinds.

RHIZOMATIC MULTIPLICITY: BECOMING-ANIMAL AND BECOMING-MOLECULAR

Always look for the molecular or even submolecular particle with which we are allied.
—GILLES DELEUZE AND FELIX GUATTARI, *A Thousand Plateaus: Capitalism and Schizophrenia*

In their introduction to *A Thousand Plateaus*, Deleuze-Guattari distinguish between two strategies for "organizing" multiplicity: the "rhizomatic" and the "arboreal." Whereas the latter unfolds along clear, branching genealogical lines of filiation, the former moves—like grass or waterlilies—in a transverse fashion across heterogeneous lines or patterns of structure. The arboreal derives the many from the One: all branches can be traced back to a single root or trunk. The multiplicity of a rhizome, on the other hand, is a substantive unto itself, ceasing "to have any relation to the One as subject or object, natural or spiritual reality, image and world."⁷ A rhizome's pattern of coherence cannot be traced back to any line of genealogical descent or derived from any transcendent unity.

It is formed along a creative "line of flight" that springs up between aggregates that had previously pursued heterogeneous paths of development. Rhizomes cannot be understood as sets of structural correspondences between existent entities, but are rather "blocks of becoming" that require entities to venture the very terms of their existence.

Deleuze-Guattari's model of the rhizome frees up a way of conceiving ecological connectedness that does not reduce the many to the One. Relational becoming, for them, does not predictably unfold along pathways precharted by some transcendent logic. Rather, it spontaneously originates a pattern of continuity or consistency purely immanent to the specific relational field at hand. If the theo-logic of macrotranscendence negotiates ecological multiplicity in arboreal terms, the rhizome might suggest an alternative strategy. But before describing this alternative in any depth I need to explicate Deleuze-Guattari's notions of "becoming-animal" and "becoming-molecular"—two interrelated concepts that are the primary inspiration for my micropneumatology of Spirit-dust. A brief sketch of Deleuze-Guattari's unique cosmological mapping in *A Thousand Plateaus* will clarify how these concepts fit into their larger philosophical framework.

Deleuze-Guattari find at the base of all reality not a solid foundation, atomistic or otherwise, but a kind of seething magma of pure singularities coming and going out of existence at "infinite speeds." Correlating to what they call "chaos," it is a level of reality occupied by "unformed, unstable matters, by flows in all directions, by free intensities or nomadic singularities, by mad or transitory particles."[8] This chaos or virtuality is *laid out*, or *gathered up*, within a single, universal plane that retains the infinite speeds of virtuality while "giving the virtual a consistency specific to it."[9] This "plane of consistency" (or, in *What is Philosophy?*, "plane of immanence"—terms I will use interchangeably) is best understood not as a Platonic collection of ideal forms so much as an infinitely saturated intersection of all concrete forms.

"Molecular" particles form out of, and pick up their movement trajectories along, this plane of consistency. These molecular particles then get *stratified* or bound into the next sphere of reality—that of *molar aggregates*—the first sphere it would be accurate to describe as *organized*. Molar aggregates subject the free movements of molecular particles to a transcendent "plane of organization," congealing their possibilities into what are called "strata": "Strata are Layers, Belts. They consist of giving form to matters, of imprisoning intensities or locking singularities into systems of resonance and redundancy."[10] Deleuze-Guattari often describe this stratification process in negative terms—the free, creative motions of molecular particles become oppressed by a falsely unifying logic of organization. At other points in their work, however, they make clear that it is not the mere existence of strata that is problematic but only the illusion that *particular* stratified, molar structures are inevitable. In other words, the deleterious effects of strata begin when they deny their own dependence on a highly indeterminate flow of molecular particles that could always have given (and could always still give) rise to something entirely different.

Deleuze-Guattari's intertwined notions of "becoming-animal" and "becoming-molecular" describe processes in which our access to this dynamic, micrological sphere of reality ("grounded" in the plane of immanence) gives rise to new, creative forms of becoming. I am particularly interested in the way these concepts might challenge ecological philosophy and theology to rethink the modes of relationship that are possible between humans and nonhumans. Becoming-animal can be defined in Deleuze-Guattarian terms as the forming of a rhizome between humans and animals. As I have made clear, the rhizome's coherence—that which holds it together, or "surveys" its multiplicity—cannot be derived from, or traced back to, a higher unity encompassing its separate components but is better described as a "line of flight" *between* them. That is to say, the "line of flight" creating the rhizome originates in a logic not transcendent but purely immanent to the aggregates at play.

Deleuze-Guattari are emphatic in their insistence that becoming-animal is something qualitatively different from merely imitating an animal. For Deleuze-Guattari, imitation-mimesis is too static and reductive an understanding of relational becoming. The logic of mimesis assumes that we can differentiate between those actions that belong to us essentially and those that are simply "put on" as an imitation of something else. Rhizomatic relationships go beyond imitation in that they always consist of a *becoming*—a line of flight emerges between two entities in order that *both* might become something fundamentally different in the process.[11] Becoming-animal would precisely *not* entail imitating the *molar* subjecthood of a given animal but, rather, accessing that animal on a *molecular* level. It is for this reason that becoming-animal can only take place through the process of becoming-molecular.

Becoming-animal involves descending, or "tuning" into, the dynamic micromultiplicities that compose our realities.[12] In communing with our own constitutive molecular movements we become able to free a certain line of flight connecting *these* molecular movements to those of an animal. In other words, this process entails going *beneath* our static, human molar forms and making something else out of our constitutive elements. Instead of being bound to predetermined, transcendent principles of organization (in this case, those that dictate to us what it means to be human) our molecular fields interweave with other, nonhuman molecular fields, creating new alliances, new blocks of becoming. Deleuze-Guattari discuss "becoming-dog" in the following quotation, but what they describe could be applied to a variety of human/nonhuman relationships—perhaps, most interestingly for our purposes, *un*domesticated animals.

> Do not imitate a dog, but make your organism enter into composition with *something else* in such a way that the particles emitted from the aggregate thus composed will be canine as a function of the relation of movement and rest, or of molecular proximity, into which they enter. Clearly, this something else can be quite varied and be more or less directly related to the animal in question. . . . You become animal only if, by whatever means or elements, you emit corpuscles that enter the relation of movement and rest of the animal particles, or what amounts to the same thing, that enter the zone of proximity of the animal molecule. You become animal only molecularly.[13]

In this example, becoming-animal opens up a unique sphere of intimacy between the human and nonhuman. It is less a matter of doing what a dog does than discovering one's "inner dogness"—and in the process putting up for grabs what "humanness" and "dogness" mean to begin with. This is brought out well by Deleuze-Guattari's description of Alexis the Trotter's act of *becoming-horse*: "Sources tell us that he was never as much of a horse as when he played the harmonica: precisely because he no longer needed a regulating or secondary imitation."[14] "Becoming-animal," then, is to enter a plane of continuity with the nonhuman world such that nonhuman rhythms, patterns, even particular wisdoms, enter dynamically into our own becoming.

Becoming is always a becoming-molecular in the sense that we must always begin from where we are micrologically—the rhythms and speeds of our molecular flow. From here we can open our awareness to those beings with whom we are in intimate ecological relationships and find ways to shift our life-patterns in relation to them. Although Deleuze-Guattari tend to discuss becoming-animal and becoming-molecular in the context of more subjective, aesthetic types of experiences, the concept's political significance should not be overlooked. Becoming-animal necessarily entails a renegotiation of those boundaries that separate and respectively define human and nonhuman entities. It subverts any anthropocentric perspective that views humanity as the highest rung of an evolutionary ladder (in scientific terms) or as the exclusive image of the divine (in theological terms)—nonhumans relegated to mere fodder for our self-propagation. It aims at opening human experience to a larger, more shared continuum of life, one that makes possible new forms of alliance between humans and nonhumans.

Earlier I discussed Latour's compelling argument that political progress on environmental issues can only take place when we are able to venture again (and set into a collective decision-making process) both *who we are* and *what we can become* together. The process of becoming-animal functions to do just this: it forces us to question the "nature" of our humanity and opens up new, strategic alliances with nonhuman entities. In short, becoming-animal/becoming-molecular can function as both spiritual and political strategy, helping us to re-envision and remap our relationship to the nonhuman.

THE PLANE OF IMMANENCE AS DIVINE MATRIX FOR ECOLOGICAL CONNECTEDNESS

We will say that the plane of immanence is, at the same time, that which must be thought and that which cannot be thought. It is the nonthought within thought. It is the base of all planes, immanent to every thinkable plane that does not succeed in thinking it. It is the most intimate within thought and yet the

absolute outside—an outside more distant than any external world because it is
an inside deeper than any internal world: it is immanence . . .
—GILLES DELEUZE AND FELIX GUATTARI, *A Thousand Plateaus:*
Capitalism and Schizophrenia

Rather than seeking the divine matrix of ecological interconnection on a higher plane of organization or transcendence, we might follow Deleuze-Guattari's gaze downwards toward the immanent microplane that constantly gives rise to new relational possibilities. If the ecotheological model that I presented in the first section of this essay (represented by McFague) sought the divine principle of interconnection in a macrotranscendent Spirit of the universe, I will suggest instead that Spirit-dust moves or dances on the plane of immanence or plane of consistency. The process of becoming-molecular I have been describing derives its creative energy and inspiration from the plane of consistency. It functions as the very basis for all abstract, creative leaps between and among multiplicities, but not by reducing the dimensions of the world's plurality. Rather, it cuts across all of them, making possible the creation of new multiplicities with new combinations of dimensions: "The plane of consistency is the intersection of all concrete forms. Therefore all becomings are written like sorcerer's drawings on this plane of consistency, which is the ultimate Door providing a way out for them."[15] The plane of consistency can thus be understood as a kind of underlying matrix that connects all things. But it functions not as a macroplan or an all-encompassing vision so much as a "skeleton key" capable of opening a creative doorway between any combination of multiplicities. It provides the ground for a certain *kind* of unity within the universe but not in the form of a central or guiding telos:

> There is therefore a unity to the plane of nature. . . . This plane has
> nothing to do with a form or a figure, nor with a design or a func-
> tion. Its unity has nothing to do with a ground buried deep within
> things, nor with an end or a project in the mind of God. Instead, it
> is a plane upon which everything is laid out, and which is like the
> intersection of all forms, the machine of all functions; its dimensions,
> however, increase with those of the multiplicities of individualities
> it cuts across.[16]

Deleuze-Guattari invoke an almost apophatic awe in the face of this shared matrix: "that which must be thought and that which cannot be thought." The plane of consistency is closer, more immanent to us than we are to ourselves. At the same time it draws us out of any static "version" of ourselves that may take hold, constantly launching us into new modes of becoming.

Deleuze-Guattari's concept of the plane of consistency allows us to conceptualize a divine matrix of ecological connectedness that does not function by the logic of macrotranscendence. The plane of consistency is able to describe both our connectedness and our constantly unfolding differences in a way that does not derive the latter from a foreclosed vision of the former. It gathers up even as it spins out all of reality's multiple multiplicities. While it is a kind of ultimate intersection of reality, it is found not at the apex of a *universe* but at the base of a *pluriverse*.[17] Most importantly, it functions to turn us toward actual possibilities for new becomings and alliances with the nonhuman world.

SPIRIT-DUST: THE "GROUNDS" FOR A MICROPNEUMATOLOGY OF BECOMING

Then the Lord God formed man from the dust of the ground, and breathed into his nostrils the breath of life; and the man became a living being.
—GENESIS 2:7

A micropneumatology of Spirit-dust would cease to understand the divine as a unitary macroentity providing the *universe* (rather than a *pluriverse*) with a singular, transcendent goal or telos. Spirit-dust first of all conveys the inherently irreducible multiplicity of the divine presence—a presence uniquely correlated to the irreducible multiplicity of creation itself. The theological tradition of the Holy Spirit is in a sense already equipped for this task—of all the members of the Trinity it most powerfully symbolizes God's intimate, dynamic relationship with every piece of creation. It also lends itself most easily to being reconceived in pluralized, decentralized terms.

The use of the word "dust" conveys not only the Spirit's fundamental multiplicity, but also the sense in which all of creation's myriad multiplicities are formed in and through this divine "matter." The book of Genesis describes the creation of humanity from "the dust of the ground." Womanist ecotheologian Karen Baker-Fletcher uses the biblical metaphor of dust to describe our "bodily and elemental connection to the earth": "Dust includes within it water, sun, and air, which enhance the vitality of its bodiliness and ability to increase life abundantly. So dustiness refers to human connectedness with the rest of creation."[18] The concept of "Spirit-dust" reminds us that the level on which we access the Spirit *is* the level on which we are interwoven with all of creation. We are connected to and responsible for one another not because we are all children of a higher God or common descendents on a single genealogical tree, but because we are all composed of the same creative Spirit-dust.

The traditional liturgical formulation "from dust to dust" might take on a whole new meaning here, expressing our indebtedness not to inert matter, but animate, inspirited microparticles—what Deleuze-Guattari call the molecular. If Spirit is life, then what is truly living about us is not something we can identify with any particular molar structure—human or otherwise. Spirit-dust is the life-giving force underlying and giving rise to *all* of reality. Spirit does not *come down* to form our various molar identities—rather it bubbles up to reinspire our molecular becomings.

Spirit-dust thus draws us into the transgressing of those boundaries that isolate us from our environment and rigidify our patterns of ecological relationship. Although ecotheologian Mark Wallace takes a different approach to pneumatology than the one presented here, many of his descriptions of the Spirit's action in our current world are pertinent: "The Spirit is an agent of 'creative formlessness' who dangerously foments 'boundary transgression' and the dissolution of 'order' into 'formlessness.' Thus, in an ecological age the Spirit is working to subvert our privileged boundaries between human and nonhuman species."[19] Paying attention to the movements of Spirit-dust would entail discovering new convergences and intersections connecting our molecular flows with those of nonhuman entities. Spirit-dust draws us back to the plane on which our ecological continuities can be productively engaged. It is simultaneously the deepest, most intimate thing within us and that which points us furthest outside of ourselves.

My hope is that this micropneumatology of Spirit-dust has begun to chart a course for ecotheology that no longer relies upon an ecologically debilitating logic of macrotranscendence. The concept of Spirit-dust may itself function as a kind of rhizome between theology, philosophy and ecology—one that helps us realize, with Latour, that any truly life-giving process for composing and recomposing a collective world must be willing to let go of a prematurely unified conception of the Whole. Spirit-dust calls us to attend to the ever-opening possibilities for creating new alliances, new rhizomes, new blocks of becoming between human and nonhuman ecologies. Its inspiriting, animating force is what sweeps us into our pluriverse's dance of becoming.

❧ Specters of Derrida: On the Way to Econstruction

DAVID WOOD

The future can only be anticipated in the form of an absolute danger.
—JACQUES DERRIDA, *Of Grammatology*

Never have violence, inequality, exclusion, famine, and thus economic oppression affected so many human beings in the history of the earth and of humanity.
—JACQUES DERRIDA, *Specters of Marx*

And provisionally, but with regret, we must leave aside here the nevertheless indissociable question of what is becoming of so-called "animal" life, the life and existence of "animals" in this history. This question has always been a serious one, but it will become massively unavoidable.
—JACQUES DERRIDA, *Specters of Marx*

ANIMAL RIGHTS AND ENVIRONMENTALISM

I do not claim that Derrida explicitly saw environmentalism as the next step for deconstruction. It could be argued, indeed, that the direction he took cuts against environmental concerns.[1] In an extension of our broad responsibility for the human other, he has on a number of occasions attempted to articulate a face-to-face relation to his cat.[2] There is, however, a well-documented tension between those who take up questions of individual animals' rights and/or their well-being, and those who pursue environmental issues. The animal rights advocate will rescue the bison trapped on the ice; the environmentalist will think of the bear and her cubs who depend for their survival on such unfortunate accidents. Derrida does indeed problematize the ethical focus on the privileged individual (his cat), but he does so by asking how we can justify ignoring all the other (individual) cats. He does not talk about mice or birds, or the snakes so unwanted in Egyptian homes that the Egyptians domesticated cats.

In taking this path, Derrida follows Levinas in seeing the movement from the ethical to the political in terms of the importance of the *third*, who is always implicit in the otherwise privileged face-to-face relation, muddying the waters. Still intact is the implication of a potential personal relationship with a discrete individual—that the other typically has a face, and that it is hard to know how to deal with the *many* others whose faces are indistinguishable from those we happen to meet. Arguably there is a residual humanism in this approach, as is also the case with Levinas. The first stage of otherness, at least, opens us up to creatures a bit like us, or those with whom we share our lives. Moreover, the ethical focus on the unidirectionality of obligation, on the gift that seeks no return, that (in Levinas, for example) is even compromised by that possibility, suggests an awkwardness in thinking deep interdependency with other life forms, which is surely our condition. But we may properly ask whether Derrida does not offer us elsewhere the resources to take our thinking further. In her essay "The Preoriginal Gift—and Our Response to It," elsewhere in the present volume, Anne Primavesi argues that Derrida worries about the logic of the gift "not as a rejection of the possibility of gift but as a conscious, formalized subversion of the prevailing capitalist logic of exchange relations." My sense is that Derrida goes further, and casts doubt on what we might call the "purity" of the gift in whatever setting it occurs, that it is never entirely free from the flows of interaction and exchange. Primavesi's account of the preoriginary gift offers us a powerful way of understanding our very existence as life-forms in a community of living beings as a gift from the past, an endowment, or what Heidegger might call an "original indebtedness." It is perhaps in this light that Heidegger will come to translate *es gibt* (there is) in a more literal way as "it gives." The recognition of our original dependency on the whole life process, both in an evolutionary sense, and in terms of our current sustenance, is an important counterweight to the humanism implicit both in a rigid economy of exchange, and in understanding the space of the ethical in terms of a pure gift.

This question of humanism arguably taints even the very word "environment," in its suggestion that we concern ourselves with what surrounds *us*.[3] Strategically, and historically, there is some justification for this focus. In many ways it is our blindness to, and lack of interest in, the impact of our ways on "what surrounds us" that is the source of the problem. But the problem is no longer that we are polluting "our surroundings" but that we are transforming the earth in ways that are deleterious for it, and not just for us. Derrida does not, to my knowledge, write much, if at all, about the environment. In *Specters of Marx* he writes of both *earth* and *world*. But in each case, the primary focus is on human beings. Thus, "[N]ever have violence, inequality . . . affected so many human beings in the history of the earth and of humanity. . . . [N]ever have so many men, women and children been subjugated, starved or exterminated on the earth."[4] It is at this point that he pauses to set aside "the indissociable question of . . . so-called 'animal' life," as in the epigraph above, in a reprise of Levinas's move that treats the face of the animal as derivative from that of the human, casting serious doubt on issuing any credentials to the snake.

The chapter in which the passage above occurs (chapter 3, "Wears and tears: [tableau of an ageless world]"), begins *"The time is out of joint. The world is going badly."* Here the "world" is that of the New World Order proclaimed by liberal triumphalism. "World" here means the human project (of freedom and Enlightenment). Derrida need not be blamed for framing things in this way. He is responding precisely to an exceedingly human way of understanding planetary history. And it is in this context that he will enumerate the ten plagues of this New World Order.[5] The implication of this account is that these plagues threaten both the attainment of such an order and the credibility of the very idea of such an order. While the seven biblical plagues of *Revelation* are eerily close to some of the ecological threats that face us, Derrida's plagues are all plagues that beset human institutions, especially those that threaten the possibility of realizing such fundamental values as peace and justice.[6] We do not need to apologize, however, for suggesting that the destruction of the earth should be listed as the eleventh plague.[7] Not only does it have an importance at least as great as the other ten, its interruption of the human institutional space that the others occupy is something of a second-order plague. Derrida speaks of messianicity without messianism as the unpredictable arrival of something or someone, unpredictable both in its timing and in its outcome. The implication is that this arrival will be our salvation, or at least be positive. But if we set aside that presumption for the moment, the destruction of the planet, the destabilization of its sustaining life-processes is an intrusion on our human project as powerful as the appearance of the refugee on our doorstep. Environmental destruction gives us a wake-up call of epic proportion, and is surely a candidate for the status of *arrivant*. For it arrives *as* something that has been excluded, much as Freud describes the return of the repressed. In the "Exergue" of *Of Grammatology*, Derrida writes, in a passage that I have chosen as the first epigraph to this chapter, that the future can only be anticipated as an absolute danger. With that seemingly apocalyptic remark, he was drawing attention to the shift from what we might call a "phenomenological" model of meaning (which would privilege the voice) to one centered on *writing*, in which meaning (as well as history and the future) would escape the control of guiding human intentionality. The emergence of ineliminable paradox, and of structures of aporia, reflects the absence of any overall synthesizing power.

The prospect of the collapse of that assurance is akin to what Nietzsche called the "death of God," and, from the point of view of the project of meaning, it can only appear terrifying, out of control, "absolute danger." Writing infects meaning from an exteriority that turns out to be immanent to meaning, and not exterior at all. Something analogous, at a second level, might be thought to be happening with the current environmental threats; they seem to be coming from the outside. And yet they reflect the structure of what the CIA calls "blowback," the facial dampness that follows spitting into the wind. What seems to come from outside are the cumulative consequences of our own actions—as ye sow, so shall ye reap; what goes around, comes around. Here we meet a parallel with part of Derrida's analysis of 9/11, in which he questions the idea that these attacks came from "without."[8] This sense that the "outside" may reflect a certain blind construction of inside/outside that needs to be re-evaluated parallels the sense that the environment may not be, as we suppose it to be, what surrounds us; it may equally be what pervades us, or what has a precarious life of its own.

EXTERNALITIES

In various early essays,[9] Derrida wrote of the need for a double strategy: immanent critique and the step beyond, working within the closure of metaphysics, and attempting a creative leap outside its borders. And in many other places, he will destabilize the assured boundary between inside and outside, not least in various performative hesitations over the idea of a preface ("Hors Livre"). We might reconstruct what is going on here in the following terms: when we think naturally, literally, we suppose that the inside is inside and the outside is outside. But as soon as we realize that inside/outside operate within a signifying space, the clarity of that distinction starts to break down, and so, too, do the moves (metaphysical, ethical, political) that presuppose the stable operation of this distinction. Interestingly enough, the very distinction between the way inside/outside operates at the level of nature and signification itself breaks down. We only need to ask ourselves about how this distinction applies to a living being to find ourselves at sea. Not only is it dynamic (respiration, excretion, digestion, and so forth), but it is also multidimensional. Think of the ways in which we appropriate intimate, proximate, and dwelling spaces as part of our "interiority." This makes clear that something like the structure of the trace is already operative "in nature." It does not wait for language to come on the scene.

And yet it is equally clear that the demand for operative demarcations between inside and outside continues unimpeded, tied up, as it is, with questions about responsibility, respect, and integrity, but also with property, exchange, and profit. And it is with these latter considerations, economic in a literal sense, that a certain deconstructive lens can shed light on our environmental predicament. It may be that what we are tempted to think of as a metaphysical illusion—the clear demarcation of self and other, inside and outside—is, as Kant suggests elsewhere, not simply a mistake, but a misunderstanding about the *scope* of a distinction that, up to a point, or under certain conditions, we cannot avoid making. And although one might think that our entering into relations of exchange both with others and with the world would compromise any strong sense of inside/outside, it could be argued that it is precisely by such a distinction that such exchanges are regulated, subject to some sort of law. And thus *enabled*, in an economy, in which time, effort, and energy are productively directed and calculated. It is such an economic space that, as I understand it, Heidegger describes in *Being and Time* as the space of the "in order to," a causal-practical nexus.[10] And it is because all living beings, including humans, are engaged in other modes and forms of exchange, as well as in relations hard to reduce to exchange, that the inside/outside relation turns out to be multiply laminated and undecidable. Being born into a legacy of genetic inheritance, and a "life-world," as we have seen, means that inside/outside is originally compromised by before/after. The "beginning" is, as Derrida would say, an "always already," which fundamentally subverts any oversimplification of the economic space in which we dwell.

It is in such a way, I believe, that we can come to understand a fundamental driving force behind our environmental crisis. For what is true of personal exchange-relations between humans is carried over into those legal persons we call "businesses." And whether or not we are dealing with extractive industries, where the primary "exchange" is with nature, or with processing or service industries, the basic law of success is to maximize profit and minimize cost. And the secret of minimizing costs is to externalize them as much as possible. "Externalize" here means passing on the costs to someone else, or better, to *something* that will not notice this happening. This is the mechanism that links profit to pollution, that accounts for the pumping of toxic waste into rivers, lead into the atmosphere, garbage into landfills, and on and on. This is not at all mysterious. An immense number of people who change their own motor oil pour the old oil down storm drains, where it enters the waste-water system—with dire consequences. This externalizes the cost of the transaction that would otherwise involve a trip to the recycling depot. The natural world, especially air and water, is the prime candidate for being the recipient of these externalities. And practices that were once harmless turn dangerous, even lethal. When no one lives downstream, pissing into the river has a negligible effect. But strings of townships engaged in such practices on a macroscale on the same river will destroy all the life in it. What has happened? What has happened is that there is no downstream any more, no outside, no elsewhere. And yet individuals and businesses continue to survive and flourish to the extent that they continue to maximize the externalization of their costs. Without tough and well-enforced environmental regulations, it is cheaper to pollute and risk fines.

On the model I am sketching here, the business-technological-industrial world is devoted to the creation of what I would call *toxic identities* that flourish to the extent that they can excrete their waste products into a relatively cost-free outside. Another word for that outside is "the environment." We may think that ethical and political implications flow from this account, and they do. But the analysis of profit-making entities as cost-externalizing devices is meant to be metaphysically neutral! It may be the case that for this or that business, recycling will pay off—for example, recovering precious metals from one's waste. But that is a matter of chance. The iron law of competition requires that costs be maximally externalized. Businesses that do not do this will go to the wall.

On this analysis, we can see how the inside/outside distinction is *not* a natural property, but a highly constructed one, essential to the persistence of these artificial entities, businesses, and even to the exchange-relations we enter into as individuals. Inside/outside is in this sense, as Derrida will say, undecidable. But the effort to decide it, to determine it in this or that case, is central both to the current ways in which we conduct our economic life, and also to the environmental destruction that comes in its wake.

UNSETTLING THE PRESENT

We soon come to see that this reconfiguration of outside as inside is as much a temporal phenomenon as a spatial one. Here, Derrida's sense of the past haunting us, especially that past that we thought we could bury (as in the thinking of Marx), is apposite. We frequently hear the claim that if we stopped all carbon dioxide emissions today, the atmosphere would continue to warm into the foreseeable future from the effects of past carbon dioxide emissions. And in environmental thinking, there is no shortage of concrete ways of understanding the idea that our time is out of joint. In this is a truism that the present is overlaid by past and future. This is enough to wean us off an assurance that we can seal off this or some other present as a secure basis of meaning. But not only, as we have suggested, are accounts with the past far from settled, we are equally unable to accurately predict vital aspects of the future. And the various scenarios that are being projected set the stage for wars—some ideological, others physical.[11] Never has the future been so indeterminate. And this epistemological shortfall has a dramatic consequence.

There are many destructive environmental trends about which we will not be able to be *completely certain* until it is too late to do anything. It is remotely possible that current global warming, for example, is just part of a natural cycle. Those who fear otherwise argue, on the basis of the precautionary principle, that we need to act as if it were true, for to wait to find out could be fatal—a bit like Pascal's Wager in reverse. It would be foolish, for example, to set the bar of proof so high that one would not stop smoking until one had proof positive of its carcinogenic potential.

WHAT CAN WE *LIVE WITH?*

But there is yet another issue that needs to be raised here. The question of certainty (or uncertainty) about some predicted dire consequence of an environmental character is often inseparable from whether the consequence in question would indeed be dire. Some have suggested that global warming would be good for (American) business interests. And others have wondered whether dramatic species-loss really matters. Part of what is at stake here is a certain resistance to recognizing what we could call our "sustaining dependency" on the natural world, as well as our genuine inability to calculate the shape and extent of that dependency. We breathe in and out, and in so doing, affirm, at every moment, our need for air. When we imagine that we could *live with* this or that dramatic change—we could adapt, it would bring back the pioneer spirit, our natural human (or American) creativity to have to do so—we not only regress to a blatantly anthropocentric view of the world, signing up for the ecological equivalent of a gated community in which we are fine with our isolation, we also work with a model of our own independence that is almost certainly flawed.

Here, it might be thought, is the problem with at least a version of religious belief. When Wittgenstein or Schleiermacher emphasizes our fundamental *dependency*, and traces religious belief to this recognition, it might seem unimportant what name we give to this dependency. ("That on which we depend we may call God," says Wittgenstein.)[12] Although for those who would be happy to see the back of the autonomous subject by whatever means the religious move might be as good as any, it is a short-sighted position, and at best a place-holder. If we begin (as at one point Irigaray does) with breathing, and work outward from there, we come to see that dependency cannot and must not be capitalized into a single relation to another being, but is essentially multifaceted, and multivalent, and, in so many ways (and senses), material. Man does not live by bread alone. But a gesture of welcome, the touch of a hand, the scent of a flower, the song of a bird, even the envelope with the pay-check, are no less material. Derrida's faith without faith, or religion with-out religion,[13] is one way of capturing this sense of a necessary dependency, that we are always leaning into the future in hope, and that the dehiscent present will always gesture forward. No doubt it is tempt-ing to give a name to *what we relate to* in this way. First the *to come*, then God, and now, it seems, the *earth* is another candidate for that on which we depend, that to which we relate, even as it exceeds our knowledge and control. For those of a religious persuasion, the battle over God becomes a battle within the space of God. There have, through the ages, been many theological attempts to conceptualize a pluralistic reciprocity, challenging the idea that our personal relation to God exhausts the shape of our dependency. The danger of funneling the God-relation into a pri-vate link with the Creator is that it can all too easily lead to a passive complacency about the impact of our practices on the created world that, in a real sense, sustains us. We can be grateful to those who now advo-cate "creation care," for example, and who convert their faith into an active practical environmental responsibility.[14] The need to materialize and pluralize our constitutive relationality can be grafted onto Derrida's original critique of the philosophy of presence. And the need to avoid the false solution that would *substitute* a relation to a personal God, or savior,[15] for an adequate recognition of, and response to, our complex material dependency is an extension of that critique.[16]

Derrida works with, and takes for granted, what we could call a "dif-ferential" model of identity. This was announced, most effectively, in his early *Différance* essay. This allows him to step back from every consti-tuted ideality, every identity-preserving boundary, and draw attention to the instability or permeability of such boundaries, the contexts on which they depend, the economy they reflect. (While he does not emphasize, as does Foucault,[17] the historical contingency of origination, I see no reason why he should not do so.) This focus on the primacy of difference and differentiation opens up a deep connection with environmentalism that is worth pursuing.

THE RESERVOIR OF DIFFERENCE

One of the most shocking environmental statistics is the rate of species-loss. This rate is estimated—and there can only be estimates—at between 50–150 a day,[18] and it is accelerating. Now in the course of natural evolutionary change, there will always be creatures and species evolving and dying out. That is not the shock. What is troubling is that the current rate is said to be between 1,000 and 10,000 times as fast as what would be occurring in the absence of humans. We need to go back 65 million years to when the dinosaurs disappeared to find pale comparisons. By the next century, half of our planet's species will be gone. We are now witnessing the fastest rate of extinction in the history of the earth. Why is this happening? Habitat destruction, overgrazing, logging, water pollution, atmospheric changes, and our general blindness to the consequences of human activity are chiefly responsible. But is there really anything wrong with this? If we visit the paint store, and bring back a large number of swatches, don't we expect to discard most of them when we figure out which we need? Could we not think of the superabundance of species in the same way? Aren't many of them just redundant? Evolution began as a biological process, but with humanity it has itself evolved into a new project, the project of freedom. Can there really be any comparison between the value of one creative, self-aware human being and all the beetles on the forest floor?

Luc Ferry's charge of "ecofascism" against the *soixante-huitards* (the '68ers) is perhaps driven by the sense that the Enlightenment project of freedom is under threat.[19] But the threat does not come from a group of French philosophers quite as much as from the history that has been made in its name. The value of freedom has become inseparable from the logic of sacrifice by which those who claim to be promoting this value can justify bringing death and destruction to soldiers and innocent civilians alike.[20] Equally, the project of freedom can clearly be used to justify the mass destruction of other living species. If a new order of value begins with the human, then the subhuman becomes at best a resource for the realization of the human project. We might perhaps err on the side of caution in allowing other species to die out, just in case they harbored useful compounds from which we might make life-saving drugs (Amazonian plants), or were useful in keeping other noxious creatures in check (a few wolves), or brightened up our parks (birds with pleasant songs), or delighted our sense of the diversity of life. But in principle, we should not mourn the loss of the surplus of species that have aimlessly adapted themselves to niches that are disappearing. We need to stop being sentimental. Things change! Life is flux!

DEPENDENCE AND INTERDEPENDENCE

Although there are other, more hopeful Christian traditions,[21] such sentiments as those I have just alluded to are only reinforced by the usual reading of the well-known passage in the first chapter of the book of Genesis: "And God blessed them [man and woman], and God said unto them, Be fruitful, and multiply, and replenish the earth, and subdue it: and have dominion over the fish of the sea, and over the fowl of the air, and over every living thing that moveth upon the earth."[22] Here, too, without invoking freedom, the suggestion seems to be that animal life has been created for the benefit of humans, and that its value is derived from that role. This reading has recently been blown out of the water by Catherine Keller's magisterial deconstruction of the broader orthodoxies of power and domination that this *creatio ex nihilo* interpretation of Genesis so effectively sustains. As she points out, "The current hoard of CEOs and fundamentalists who cite the verse to justify ecological exploitation invariably omit the next verse. 'See I have given you every plant yielding seed that is upon the face of the earth, and every tree with seed in its fruit; you shall have them for food' (1:29) It is a vegetarian domain Elohim has offered."[23] Keller's work unfolds a processual ecological metaphysics (in the best sense) in which the human domination of nature is not a manifest destiny legitimated by scripture, but a hermeneutic betrayal of the creative complexity of the original texts.

It sometimes seems that there is a willful misunderstanding of attempts to displace the privilege of the human. What has been called "antihumanism" is *not* antihuman, but rather an attempt to interrogate the privilege of the human as the unthematized point of departure for reflection. It is not to deny that (as far as we know) it is only humans that *can* reflect. It is indeed to affirm that it may be one of our distinctive powers to step back from a narrow understanding of our privileged position. The argument is that too great a self-assurance about our uniqueness, originality, and independence of agency may blind us to the truth of our condition.

Drawing on both Heidegger and Derrida we could describe such a condition as one of "constructive dependency." In each case, a certain decentering of origin creates a space in which a fundamental relational dependency becomes visible (and a textual space in which this thought can be articulated). For Heidegger, this first appears in his account of *Dasein* as "being-in-the-world" (not as consciousness set against the world),[24] and later as a being whose agency arises through a capacity to respond to what language opens up: "Man speaks in that he responds to language."[25] For Derrida, "a writer writes in a language and in a logic whose proper system, laws, and life his discourse by definition cannot dominate absolutely."[26] If we were to translate this model into a natural setting, we would have to say that our agency (and identity, and even dominion) rests on a capacity to relate to, and respond to, and negotiate productively with, the natural world around us. But what kind of claim is this? Constitutive dependency can be thought of in rudimentary material terms: air, nutrition, sensory stimulation, without which we would not make it through to the next day. But it would be hard not to agree with Thomas Berry that

> we begin to understand our human identity with all the other modes of existence that constitute with us the single universe community. . . . Every being is intimately present to and immediately influencing every other being.
>
> Without the soaring birds, the great forests, the sounds and colorations of the insects, the free-flowing streams, the flowering fields, the sight of the clouds by day and the stars at night, we become impoverished in all that makes us human.[27]

This may sound like an enlightened anthropocentrism, but it is just as much a recognition of our fundamental coexistence with the rest of life. The pleasure we take in the birds and insects and stars is in part a pleasure at their independence from us! And if we came to see our capacity to recognize other creatures as having an intrinsic value, or a value not dependent simply on their use-value for us, would not this human achievement at least *compete* with the value of that freedom by which we think we can license ourselves to sacrifice subordinate forms of life?[28]

As humans, we do not typically set out to kill off entire species (with the exception of some diseases, like smallpox and malaria, and many species of wolf). Rather we neglect to protect their conditions of life (habitat). Or we hunt or trap or fish them to extinction. We do not need to engage in moralism to find this problematic, especially given the scale of extinction to which we have already alluded. But how should we best frame the problem? So far we have proposed two competing frames, the first unashamedly anthropocentric, in which the only limit we might impose on extinction would rest on a given species' amenity value to us, plus a bit of a margin for uncertainty. The second would see us humans as interdependently coexisting with other species, with our value as humans resting on our capacity to celebrate that coexistence, and our desire to preserve it. I now want to propose something of a basis for this second frame, one that taps into Derrida's articulation of the privilege of difference, and differentiation, over any essentialist account of identity.

WHO "WE"?

When we look at the stars, we are rightly awed by what we are seeing. When we realize that the light has taken so long to get here that what we are seeing may no longer exist, the astonishment only increases. And when we further reflect that for all their size and magnificence, they cannot see us, and that they are most likely huge clumps of swirling dead matter, we may well turn our astonishment back on ourselves as living, sentient, human beings. We creatures may be made of stardust, but any living being is of a complexity that inorganic matter cannot approach. If anthropocentrism is a kind of myopia, I want to argue that biocentrism is a kind of full-spectrum seeing, a capacity to respond to other life-forms that is only possible for us because we are ourselves life-forms. What other life-forms offer us is the opportunity for the affirmation of difference, other ways of organizing reproductive complexity.

It is odd. Creationists who believe that God created "every living thing that moveth upon the earth" ought to be appalled at our destruction of God's work. But we do not need to be creationists, or even essentialists, to lament the destruction of species. Darwin did not think a "species" was definable: "It is really laughable to see what different ideas are prominent in various naturalists' minds, when they speak of 'species.'"[29]

Nowadays, biologists understand "species" in various ways, such as inter-breeding lineages, ecological lineages, and phylogenetic units. In each case, however, what we know we are dealing with are population units with unique genetic characteristics that have evolved over millenia, consisting of individuals differentiated internally, from each other, and from other species. To celebrate the existence of species is to celebrate deep difference. The loss of a species is the loss of an irreplaceable developmental history, not to mention a piece of an interconnecting ecological puzzle. But it is also the loss of a future possibility of transformation, adaptation, differentiation. We may be tempted to think of biodiversity as something of decorative significance. Of course we do delight in butterflies and swallows. But it is much more than that—it is the pool of the possibilities of life differentiation and transformation. We have no evidence that this occurs anywhere else but here on earth. And it is an adventure of which we are, perhaps obscurely, only a part. John Donne once wrote, "Any man's death diminishes me," and the same is true for the deaths of other species. It is not just that their extinction diminishes us. It diminishes the web of life to which we each belong.[30]

Obviously part of the point of using this language is to induce a sense of connectedness between human beings and other beings that may not be at the forefront of our everyday experience. Indeed, we try not to share our domestic and intimate lives with noxious bacteria and viruses, with ticks, leeches, headlice, bedbugs, hornets, mosquitos, rodents, spiders, or snakes. Hospitality has its limits. But the argument never was that we might not legitimately want to keep clean and avoid disease. All creatures protect their bodily integrity in that way. The argument is that what we take to be "other," what we categorize as utterly alien, as dispensable for our existence, as outside our space of concern, may reflect a myopic prejudice. "We" are human. But equally, "we" are white, "we" are (often) men, "we" are "semiaffluent, English-speaking academics," and so forth. Who are "we" really? "We" mammals? And yet when focused on a frog breathing we notice the rise and fall of our own chest, what is it we are registering? Who are "we"? If *you*, my friend, see nothing, are *we* a "we"? Is there not something vital we do not have in common? We intellectuals are quite rightly afraid of anthropomorphizing, but perhaps we lean over backward so far that we eliminate a quite legitimate biomorphizing, legitimate because being alive gives us imaginative and projective access to all kinds of living beings. Even when they are very unlike us in body or habit, we can deploy schematic analogies to forge bridges.

When Derrida speaks of animals, he tends to use scare quotes, not least because the very word (which he plays with as *animot*) encourages a homogenizing differentiation from the human, as if being different from us made them similar to each other. Are not humans essentially different from animals? Well, which humans, and which animals? And why do we suppose that there are differences and then *essential* differences?[31] What Derrida is encouraging is a radical de-essentializing, and a persistent suspicion, of the language of anthropocentric convenience. But we cannot pretend that there is not a play of both analogy and difference in the extension of our response-ability to other living creatures. This was something of the point of his reading of Levinas in "Violence and Metaphysics."[32] What I have done is to propose a biocentric pathway for such analogizing, even as we must give our morphological imagination free rein. I do not need to be able to imagine being a sixty-foot squid to respect its right to exist. But if I have tried to chart new analogical pathways to other species, do we not need to hold onto a certain stable sense of our own species? Those opposed to Darwin demonstrate a high level of anxiety at this point. And it is worth sharpening what Derrida would call the "undecidable" dimension of this issue rather than aiming for a premature resolution.

The idea that the human species has a lineage that connects it in evolutionary time with the higher apes is a threat to those with a certain understanding of essentialistic identity, one that cannot allow that something quite new could develop in time from something different. This however, is a genuine cognitive mistake. We know a butterfly develops from a chrysalis. We know there is a qualitative difference between a bike kit and the fully assembled bike. In neither case is the product compromised by its origin; rather, it is made possible by it. Resistance to evolution is sufficiently explained by the fear that we may not have complete control over those aspects of our animal ancestry that might still lie within us. The need to stabilize the boundaries of our own species is also political in that it serves to unify the various human races under the common legal framework of human rights (which is a good thing). But would this strategic political consideration arise if we accorded apes and monkeys an appropriate respect?[33] Can we imagine a world in which respect for all other beings in their deep singularity was sufficiently firmly ingrained that our own species-identity no longer operated as a license to kill, mistreat, and neglect others? In *The Other Heading* Derrida takes up the question of European privilege, the kind of privilege that would give Europe a pre-eminence in human history.[34] He concludes that this could only be justified if Europe were to offer unconditional hospitality to the "other," were to welcome the rest of the world. This argument could be reworked as a way of reformulating the distinctive privilege of the human—that we alone are perhaps capable *both* of extraordinary destructiveness and blindness to the fate of other beings, *and* of the far-reaching responsibility, compassion, and hospitality toward other living beings.[35] Can we consistently claim the privileges of the human when we deal with other creatures in such a *beastly* way?[36]

JUSTICE BEYOND REPRESENTATION

In *Force of Law*, Derrida writes that "there is no justice except to the degree that some event is possible which, as event, exceeds calculation, rules, programs, anticipations and so forth. Justice as the experience of absolute alterity is unpresentable."[37] Elsewhere he will speak of going through the undecidable as the condition for responsibility. And again, he will explain this in terms of recognizing that there is no formula, no algorithm that can decide for us. The attempt to mark a space—perhaps the space of the ethical—beyond calculation, beyond representation, is a persistent theme of Derrida's later writing. There are times when one might be forgiven for understanding these remarks in an existential vein. But there is nothing quite like the environmental landscape for shaking one from this interpretation.

In *The Greening of Ethics*, Richard Sylvan writes that, "increasingly, environmentalists are appalled by the pronouncements from orthodox economists."[38] What appalls them is the application of cost-benefit analysis to environmental decision making. And it appalls them because it presupposes that environmental impacts can be quantified to the point of being successfully costed. Why is this a problem? It is a problem for at least two reasons. First, the value of an environmental amenity is assessed entirely in human terms—the value for us—as if there were no other source or center of value. And second, this assessment is typically undertaken by measuring people's short-term, surface preferences, rather than their long-term interests. Thus, given a typical "discount rate" by which our preferences next year are weighed at 7 percent less than this year, after cumulative discounting, our preferences ten years hence cease to count. And this itself measures not our informed judgments, but our surface preferences—what we would vote for here and now. So there are perhaps three levels of the failure of representation here: first, that it is wholly anthropocentric; second, that it makes no effort even to be grounded in truth rather than preference; and third, that it is myopically short-term.[39]

Going through the undecidable here is a multistage process. Even economists disagree enormously about what figures to use when trying to evaluate environmental impacts. And I have read reports in which their difficulty in deciding a figure led to them assign a zero value to a certain amenity in their equations, as if zero represented matters more adequately. But after that, environmentalists face a huge dilemma—whether to accept the demand to calculate, to play the economic game, or to hold out for a place at the table for nonquantifiable, or incommensurable, values. And of course it would be another mistake to "represent" this situation as one in which delicate nature is being violently trampled on by brutal humanity. There may well be many conflicting calls on our conscience, each of which resists quantification.

Undecidability never was offered as a helpful decision procedure, and we should not be surprised when it draws us into fractally proliferating spaces. What environmental dilemmas make clear, however, is that these are not the personal dilemmas of an existential conscience. These are, dare one say it, the problems posed by the real when it bites back against the poverty of our attempts adequately to represent it.

THE PARLIAMENT OF THE LIVING AS THE DEMOCRACY-TO-COME?

There is nothing outside of the text (il n'y a pas de hors-texte).
—DERRIDA, *Of Grammatology*

The absolute present, nature, that which words like "real mother" name, have always already escaped, have never existed; that what opens meaning and language is writing as the disappearance of natural presence.
—JACQUES DERRIDA, *Of Grammatology*

Nostalgia runs all through this society—fortunately, for it may be our only hope of salvation.
—DONALD WORSTER, *The Wealth of Nature*

Our image of the French philosopher as an intellectual, an activist, and a habitué of the Parisian café, a Socratic figure as much at home protesting and engaged in public discussion, may have a lot to do with the charisma first of Sartre and then Foucault. But the commitment to the life of the city, and a certain reserve with regard to nature, is much more widespread and of longer standing. The Sartre of *La Nausée*, for whom the writhing shape of the root of the tree in the park is a challenge to our very conceptual grip on the world, for whom the slimy is a source of disgust, is inheriting at some distance a Cartesian sensibility whose attitude to nature is taken to its limit in the geometrical gardens of Versailles. The fact that France's foremost twentieth-century anthropologist, Claude Lévi-Strauss, for whom the opposition between the "raw" and the "cooked" was to become emblematic, spent only months doing the fieldwork that sustained a lifetime of writing should not surprise us. We may, then, not find it surprising that Derrida can write, as in the heading epigraph above, that "the absolute present, nature, that which words like 'real mother' name, have always already escaped, have never existed; that what opens meaning and language is writing as the disappearance of natural presence."[40] Derrida is here endorsing the logic implicit in Rousseau's writing. If we take it as evidence that deconstruction inherits a kind of suspicion with regard to nature, we may conclude that the prospects for "econstruction"—a deconstructive environmentalism—are not bright. When Derrida wrote, only a few sentences earlier (and, again, as in the epigraph above), that "there is nothing outside of the text," is it not clear that environmentalism has little to gain from what many of Derrida's critics have argued is a self-regarding textual idealism?[41]

For some while now I have engaged in productive dialogue with deconstruction while keeping my environmental concerns on the back-burner. Even as I have no illusions about the human capacity to hold in mind, and even pursue, quite incompatible lines of thought, the question I want to pose here is precisely whether green deconstruction need be a philosophical monster, an oxymoron. Or whether, conversely, this is a natural alliance. Might not a living and developing deconstruction find itself quite at home thinking through the quandaries of environmental concern? Or might not environmentalism provoke a certain materialistic mutation within deconstruction? As will become apparent, what I am calling "materialism" is nonreductive and relational in character and convergent with, rather than opposed to, what has come to be called a "theology of becoming."[42]

Let me begin by amplifying the two questions in a way that will help move us forward. First guiding question: *Might not a living and developing deconstruction find itself quite at home thinking through the quandaries of environmental concern?*

We may think of deconstruction as a kind of antinaturalism, one that takes seriously complex non-natural structures and relations—from logical aporiae to undecidability and infinite responsibility. We may think that these are a far cry from the accelerating rate of species-loss, global warming, deforestation, the depletion of natural resources, the hole in the ozone layer, the sucking dry of deep aquifers, the dangers of catastrophic changes in weather patterns, and so forth. These all seem like very real, concrete phenomena, dangers we do not need philosophy, let alone econstruction, in order to know that we need to address them. And each one of these "dangers" is what Beck and Latour call a "thing,"[43] that is, not simply a sharply defined object of knowledge, but an issue, a concern, a site of complex relationality. As much as these are what we call "natural" phenomena—who could doubt the naturalness of a tornado that uproots trees, sucks windows out of skyscrapers, throws cars across town?—what is at issue for us is to understand their cause, the probability of them occurring, the true danger they do pose, how far we are really responsible, and how they might be averted. In other words, while it is a crucial matter of intellectual integrity to acknowledge that we are talking about real phenomena whose reality can be measured in the unmistakable currency of death and destruction, these phenomena are *issues* for us because of further questions about the adequacy of our knowledge and control, the models we deploy to engage with these phenomena, the social practices that enable and disable appropriate responses to these problems, and the deep difficulties we face in trying to think through the contradictions they throw up. The irrepressible reality of these phenomena does not in the end tell us what to do, or how to think about them. To the extent that deconstruction trades in this complexity, it might be thought to be just what environmentalism has needed.

Second guiding question: *Might not environmentalism provoke a certain materialistic mutation within deconstruction?*

Where it has not been a formal charge it has often remained a suspicion: that deconstruction reflects a kind of academic detachment from the real. Reading "there is nothing outside of the text" a certain way reinforces this suspicion. We might suppose that this would present an obstacle to the fully blown materialism to which any environmentalism is surely committed.

In the afterword, "An Ethics of Discussion," to *Limited Inc.* and in many other places, Derrida is at pains to insist that "text" does not mean words on a page, and that the term "con-text" might have been less misleading.[44] And he will go on to add that while all meaning is contextual, no context is ever fully saturated or complete. Central to deconstruction, too, is a refusal of any absolute distinction between meaning and force, expression and indication, intentional and causal. What we think of as a privileged realm of human meaning-giving activity is always exposed, internally and externally, to what it seeks to exclude, which we could call "the real," "the outside," even "nature." We *could* call that exposure the "materiality of meaning." This word "material" is worth pausing over, because it is the source of much misunderstanding. On one reading, to insist on the material is a kind of reductionism, one that begins with complex meaningful, even spiritual, phenomena, and reduces them to cruder material forces. "Freud reduces everything to sex" would be an example. But it is a quite different reading that would insist on the multiplicity, complexity, and multidimensionality of material forces. This approach, for example, would agree with Marx where he understood "relations of production" as material forces, even if he could not subscribe to a single, root-cause foundationalism, a "last analysis." Material forces here would cover a wide range of phenomena, from the effect of automotive lead poisoning on child brain development to the role of economic interests in motivating discretionary war. And the point of going *material* is not reductive at all, not an attempt to reduce apparent complexity to some single, underlying material cause, but rather to disclose and explore the wealth of the real.

Deconstruction had its roots in Derrida's taking up of an early materialist critical reassessment of Husserl by Tran Duc Thao. But it is true that deconstruction focuses less on the *matter* of materiality than on the significance of the irreducibility of the material, the limits and paradoxes of ideality. We might say that deconstruction is a strategic rather than a substantive materialism. It acknowledges and welcomes the interruption wrought by an excluded materiality. In this way, we may suppose, it is well positioned to help us think of environmentalism not as a positive science, but as a challenge to any science blind to its necessary ideality.

The two quotes from Derrida's *On Grammatology* that appear as epigraphs to this section lead into a discussion of the logic of supplementarity, in which something supposedly complete (such as nature) needs a supplement (writing) that will expose the original completeness as a lack.[45] This reminds us, if we needed reminding, that environmentalism cannot just be the champion of a lost purity of nature. It has to reckon with the inseparability of our thinking about nature from the theme of loss, alongside the *real* loss reflected in the very many species of creature that are becoming extinct every *day*. Bill McKibben's *The End of Nature*, for example, argues that man-made atmospheric changes alone mean that nothing on earth is any longer in its natural state, unaffected by the human presence.[46] But it would be foolish to conclude that because there is no pure "nature" any more (if there ever was) we do not need to make distinctions between different kinds or levels of loss. Donald Worster opens *The Wealth of Nature* with an unashamedly nostalgic evocation of the natural abundance that was once America.[47] We will consider later how to evaluate this attitude.

As a final note, consider, too, how we might rework Derrida's idea of a "democracy-to-come" within an environmental context.[48] What does Derrida mean by this phrase?

[Not] simply a future democracy correcting or improving the actual conditions of the so-called democracies, it means first of all that this democracy we dream of is linked in its concept to a promise. The idea of a promise is inscribed in the idea of a democracy: equality, freedom, freedom of speech, freedom of the press—all these things are inscribed as promises within democracy. . . . We don't have to wait for future democracy to happen, to appear, we have to do right here and now what has to be done for it. That's an injunction, an immediate injunction, no delay. . . . [W]e do not confine democracy to the political in the classical sense, or to the nation-state, or to citizenship.[49]

Derrida connects the ideal of equality within democracy to that of the fraternal bond, and looks to find ways of overcoming that limitation. It is in this context perhaps instructive that in his famous poem "The Canticle of Brother Sun," St. Francis of Assisi (1182–1226) addresses himself to Brother Sun, Sister Moon and Stars, Brother Wind, Sister Water, Brother Fire, and Sister Mother Earth, drawing Sun and Moon and the four elements into an expanded fraternal nexus, and praising God *through* each. The question I would pose here is whether a "democracy-to-come," an expanded fraternity that broke free of the political, of the nation-state, that perhaps points toward what in *Specters of Marx* Derrida calls a "New International," might come to embrace the nonhuman inhabitants of the planet.[50] They are clearly interested parties, stakeholders, as we have come to say, in the fate of the earth, even if they have no voice. As Christopher Stone argues,[51] this is no more an impediment to their being represented in a court of law than it is for other fictional legal persons, such as corporations, or other voiceless persons, such as infants and the mentally impaired. Of course, nothing would prevent various parties from claiming to represent the beetles or the elephants. But genuinely trying to represent the interests of a species, or a region, et cetera, sets a standard that a court can take seriously.[52] This is not to say that there will not be countless impossible cases, aporias, dilemmas, conflicts of interest. But that is true of the earth itself—nature is as much a battle zone as a cooperative community.

The claim is not that a parliament of the living would bring an end to violence.[53] It could only hope to manage an economy of violence. But it might prevent or ameliorate what Derrida calls "the worst violence." He writes that "one must combat light with a certain other light, in order to avoid the worst violence, the violence of the night which precedes or represses discourse."[54] It is perhaps ironic that we are marshalling Derrida on the side of the voice when he has striven mightily against phonocentrism, the philosophical privilege of the voice. But the paradox is only apparent. For what becomes abundantly clear in the case of animal rights is that having a voice is *not* some original natural phenomena, but essentially bound up with representation, just as it is for those humans of whom we say that "they need to be given a political voice." Is it intelligible to think of living beings more generally as deserving of the ethical or political concern to "give voice to the voiceless"?[55] Aren't they voiceless in quite a different sense from humans who are being ignored, or from infants or impaired humans, who one day will, or once did, or might in principle have had a voice? It is hard to see why. Many creatures clearly do have voices—we simply cannot understand their calls and cries. And everything that lives has interests that can be met or frustrated. It is hard to doubt that being poisoned, or losing the habitat necessary for survival, is against the interest of whatever creature suffers this fate. Derrida's sense of a "democracy-to-come," one that would break free from the constraints of the nation-state, one that would pursue justice beyond the rules laid down in advance, one that would take seriously the need to represent the interests of all earth's stakeholders, could surely embrace this difficult but necessary ideal of a parliament of the living. It may be that its decisions would only be enforced once it becomes clear that, to paraphrase Benjamin Franklin, if we do not hang together, we will surely hang separately. But even the thought of such a body begins to animate virtual voices in the human head, voices of creatures whose spectral presences haunt us in so many ways—species that have died out, flocks and herds we breed to eat, animal companions we live with, even the sports teams we name after animal totems. How long can we refuse to acknowledge all these ghosts, just as we balk at acknowledging the source and character of our own animating energies?

CONCLUSION

We began with two guiding questions: first, might not an econstruction—a living, developing, and materially informed deconstruction—find itself quite at home thinking through the quandaries of environmental concern?; and second, might not environmentalism provoke a certain materialistic mutation within deconstruction?

I have tried to argue this both ways—that environmentalism finds itself in an often problematic and aporetic space of posthumanistic displacement with which deconstruction is particularly well equipped to offer guidance. But I have also contended equally that environmental concerns can embolden deconstruction to embrace at least what I have called a "strategic materialism," or the essential interruptibility of any and every idealization. Another way of understanding a deconstructive embrace of materialism would be this: we tend to think of matter and spirit, or matter and mind, as somehow opposed, and hence as unable to be thought of in the same space. But we can, I believe, get beyond this reductive understanding of opposition without falling into the arms of dialectical synthesis. The model of the Möbius strip allows for the idea that radical opposition can be combined with deep ontological continuity: at every point two sides but one surface. This gives us a beautiful way of representing our relation to nature—opposed, in some sense, and yet at the same time continuous. Deconstruction allows us to think in this kind of space.

And deconstruction's critique of presence leads effortlesly to the strange temporalities of environmentalism: we need to act long before we can be sure we need to, and it may only be our grandchildren who will benefit. The dangers we face are from the accumulated impacts of past practices. And we are dealing with a singular sequence, the history of the earth, that takes to a whole new level the familiar idea that human history is ideographic—concrete, not rule-governed, and not to be repeated.

I have also suggested that the problematic antihumanism of the Generation of '68, far from being a dangerous ecofascism, is precisely adapted to our current situation, one in which the whole privilege of the human, as a well-meaning but often toxic terrestrial, is quite properly being put into question. I have argued for the renewed privilege of the human, by analogy with the privilege that Derrida conditionally accords Europe, if the new human can be understood as embodying a proper respect for otherness, and difference. And I have suggested that, problematic as it may be, we might be able to extend Derrida's "democracy-to-come" to the (imaginary) parliament of the living. Derrida agreed that environmental destruction needed to be on any short-list of the plagues of the New World Order. I hope I have begun to show how deconstruction-as-econstruction helps us address some of the complexities it raises.

❧ Ecodoctrines: Spirit, Creation, Atonement, Eschaton

Sacred-Land Theology: Green Spirit, Deconstruction, and the Question of Idolatry in Contemporary Earthen Christianity

MARK I. WALLACE

I enter a swamp as a sacred place—a sanctum sanctorum.
—HENRY DAVID THOREAU, "Walking"

This pneumatological materiality, far from effecting a spiritual disembodiment, a flight from the earth, suggests in its very birdiness a dynamism of embodiment: lines of flight within the world.
—CATHERINE KELLER, Face of the Deep

Christianity often acts like a "discarnate" religion—that is, a religion that sees no relationship between the spiritual and the physical orders of being. Historically, it has devalued the flesh and the world as inferior to the concerns of the soul. In the history of the church, the earth was considered fallen and depraved because of Adam's original sin in the Garden of Eden; many early theologians rejected marriage as giving in to sexual pleasure; and greatly revered saints and martyrs starved their bodies and beat themselves with sticks and whips in order to drive away earthly temptations. Pseudo-Titus, for example, an extracanonical exhortation to asceticism from late antiquity, urges Christians to cleanse themselves of worldly pollution by overcoming fleshly temptations: "Blessed are those who have not polluted their flesh by craving for this world, but are dead to the world that they may live for God!"[1] Christianity has been conflicted about, and at times at war with, the genuine human need to reconcile the passions of worldly, physical existence with aspirations for spiritual transformation.

In fact, however, Christianity is *not* a discarnate religion. On the contrary, beginning with its earliest history, Christianity offers its practitioners a profound vision of God's fleshly identity through its ancient teaching that God at one time embodied Godself in Jesus—God became incarnate. Long ago God poured out Godself into the mortal body of one human individual, Jesus. But that is not all. Christians also believe that since the dawn of creation, throughout world history and into the present, God *in and through the Spirit* has been persistently infusing the natural world with divine presence. The Spirit is the medium, the agent, or, in terms more felicitous for a recovery of the Bible's earth-centeredness, the *life-form* through which God's power and love fill the world and all of its inhabitants. Through green Christian optics, we see that the gift of the Spirit to the world since time immemorial—a gift that is alongside and inclusive of Jesus' death and resurrection—signals the beginning and continuation of God's incarnational presence. As once God became earthly at the beginning of creation, and as once God became human in the body of Jesus, so now God continually enfleshes Godself through the Spirit in the embodied reality of life on earth. In this sense, God is carnal, God is earthen, God is flesh.

In this essay I take up the question of Christianity's earthen identity by way of a biblically inflected, nature-based retrieval of the Holy Spirit as the green face of God in the world.[2] Taking my cue from the Bible's definition of the Spirit according to the four cardinal elements, I begin with an analysis of how the Spirit reveals herself in the scriptural literatures as a physical, earthly being who indwells the earth—even as the earth enfleshes the Spirit.[3] To make this point I develop a case-study of the Crum Creek (a local watershed near my home and workplace) as a Spirit-filled (albeit degraded) *sacred place* because it continues to function as a vital if threatened habitat for a wide variety of plant and animal species. But if it is the case that the earth embodies the Spirit's power and love for all things, then whenever this fragile, green planet—God's earthen body, as it were—undergoes deep environmental injury and waste, it follows that God in Godself also experiences pain and deprivation.[4] Since God and the earth, Spirit and nature, share a common reality, the loss and degradation of the earth means loss and degradation for God as well.

This model of sacred-land theology raises two troubling criticisms that I will seek to address here. On the one hand, some environmental deconstructionists question appeals to nature per se in formulations such as mine for, as they see it, betraying a crude essentialism that fails to account for the founding interpretive assumptions that shape one's experience of the natural world. On the other, some Christian ecotheologians question appeals to earth community as sacred, holy ground—the site of God's earthen presence—as surreptitiously idolatrous. My response to these criticisms will be to develop a rhetorically rich, rather than essentialist, celebration of Christianity's quasi-animist understanding of God's Spirit as both *beyond* all things and radically enfleshed *within* all things. I will conclude that it is crucial for the vitality of the planet, and the health of our own species, to reimagine theologically the mutual interrelationship of Spirit and earth. A deep-green recovery of Christianity's central teaching about the unity of God and nature is essential to awakening both a sense of kinship between our kind and otherkind and the concomitant desire for the well-being of the land that is our common home and destiny. Unless we can experience again a spiritually charged sense of kinship with the more-than-human world, I fear that the prospects of saving our planet, and thereby saving ourselves, are dim and fleeting at best.[5]

LANDSCAPES OF THE SPIRIT: THE CONTEST BETWEEN SPIRIT AND FLESH

While I maintain that Christianity's primordial identity is fundamentally nature-centered and body-loving, it is no secret that this thesis has historically been at odds with a residual Platonist tendency within Christian theology to devalue, even demonize, the realities of body and world. Many of the church's most influential early thinkers were enamored with Plato's controlling philosophical metaphors of the body as the "prison house" or the "tomb" of the soul. The fulfillment of human existence, according to Plato, is to release oneself—one's soul—from bondage to involuntary, bodily appetites in order to cultivate a life in harmony with one's spiritual, intellectual nature.[6] Origen, the third-century Christian Platonist, literally interpreted Jesus' blessing regarding those who "have made themselves eunuchs for the sake of the kingdom of heaven" and at age twenty had himself castrated.[7] Consistent with the theology of Pseudo-Titus, Origen became a virgin for Christ who was no longer dominated by his sexual and physical drives—he became a perfect vessel for the display of the power of the Holy Spirit over bodily temptations.[8]

This long tradition of hierarchical and antagonistic division between spirit and matter continues into our own time—an era, often in the name of religion, marked by deep anxiety about and hostility toward human sexuality, the body, and the natural world. And yet, the biblical descriptions of the Holy Spirit do not square with this oppositional understanding of spirit and flesh. Granted, the term "Spirit" does conjure the image of a ghostly, shadowy nonentity in both the popular and high thinking of the Christian West. In her earlier work, for example, Sallie McFague argued that the model of God as Spirit is not retrievable in an ecological age. She criticized traditional descriptions of the Spirit as ethereal and vacant, and concluded that Spirit-language is an inadequate resource for the task of earth-healing because such language is "amorphous, vague, and colorless."[9] Later, however, McFague performed the very retrieval of pneumatology she had earlier claimed to be impossible: a revisioning of God as Spirit in order to thematize the immanent and dynamic presence of the divine life within all creation.[10] "[T]he spirit of God [is] the divine wind that 'swept over the face of the waters' prior to creation, the life-giving breath given to all creatures, and the dynamic movement that creates, recreates, and transcreates throughout the universe. Spirit, as wind, breath, life is the most basic and most inclusive way to express centered embodiment."[11]

McFague's recovery of scriptural Spirit-breath language underscores how the biblical texts stand as a stunning countertestimony to the conventional mind-set that opposes Spirit and flesh. Indeed, the Bible is awash with rich imagery of the Spirit borrowed directly from the natural world. The four traditional elements of natural, embodied life—*earth*, *air*, *water*, and *fire*—are constitutive of the Spirit's biblical reality as an enfleshed being who ministers to the whole creation God has made for the refreshment and joy of all beings. In the Bible, the Spirit is not a wraith-like entity separated from matter, but a living being, like all other created things, made up of the four cardinal substances that compose the physical universe.[12]

Earth, Air, Water, Fire

Numerous biblical passages attest to the foundational role of the four elements regarding the earthen identity of the Spirit.

(1) As *earth* the Spirit is both the *divine dove*, with an olive branch in its mouth, that brings peace and renewal to a broken and divided world (Gen. 8:11; Matt. 3:16; John 1:32), and a *fruit-bearer*, such as a tree or vine, that yields the virtues of love, joy, and peace in the life of the disciple (Gal. 5:15–26). Pictured as a bird on the wing or a flowering tree, the Spirit is a living being who shares a common physical reality with all other beings. Far from being the "immaterial substance" defined by the canonical theological lexicon, the Spirit is imagined in the Bible as a material, earthen life-form who mediates God's power to other earth creatures through her physical presence.

(2) As *air* the Spirit is both the *vivifying breath* that animates all living things (Gen. 1:2; Ps. 104:29–30) and the *prophetic wind* that brings salvation and new life to those it indwells (Judg. 6:34; John 3:6–8; Acts 2:1–4). The nouns for Spirit in the biblical texts—*ruach* in Hebrew and *pneuma* in Greek—mean "breath" or "air" or "wind." Literally, the Spirit is pneumatic, a powerful, air-driven reality analogous to a pneumatic drill or pump. The Spirit is God's all-encompassing, aerial presence in the life-giving atmosphere that envelopes and sustains the whole earth; as such, the Spirit escapes the horizon of human activity and cannot be contained by human constraints. The Spirit is divine wind, the breath of God, that blows where she wills (John 3:8)—driven by her own elemental power and independent from human attempts to control her—refreshing and renewing all broken members of the created order.

(3) As the *living water* the Spirit quickens and refreshes all who drink from her eternal springs (John 3:1–15, 4:14, 7:37–38). As physical and spiritual sustenance, the Spirit is the liquid God who imbues all life-sustaining bodily fluids—blood, mucus, milk, sweat, urine—with flowing divine presence and power. Moreover, the Water God flows and circulates within the soaking rains, thermal springs, ancient headwaters, swampy wetlands, and teeming oceans that constitute the hydrospheric earth we all inhabit. The Spirit as water makes possible the wonderful juiciness and succulence of life as we experience God's presence on a liquid planet sustained by nurturing flow patterns.

(4) Finally, as *fire* the Spirit is the *bright flame* that alternately judges evildoers and ignites the prophetic mission of the early church (Matt. 3:11–12; Acts 2:1–4). Fire is an expression of God's austere power; on one level, it is biblically viewed as the element God uses to castigate human error. But it is also the symbol of God's unifying presence in the fledgling Christian community where the divine *pneuma*—the rushing, whooshing wind of God—is said to have filled the early church as its members became filled with the Spirit, symbolized by "tongues of fire [that were] distributed and resting on each one" of the early church members (Acts 2:3). As well, like the other natural elements, fire is necessary for the maintenance of planetary life: as solar power, it provides warmth and makes food-preparation possible; as wildfire in forested and rural areas, fire revivifies long-dormant seed cultures necessary for biodiverse ecosystems. The burning God makes alive the elements of the lifeweb essential for the sustenance of our gifted ecosystem.

God as Spirit is biblically defined according to the tropes of earth, wind, water, and fire. In these scriptural texts the Spirit is figured as a potency in nature who engenders life and healing throughout the biotic order. The earth's bodies of water, communities of plants and animals, and eruptions of fire and wind are not only *symbols* of the Spirit—as important as this nature symbolism is—but share in the Spirit's very *nature* as the Spirit is continually enfleshed and embodied through natural landscapes and biological populations. Neither ghostly nor bodiless, the Spirit reveals herself in the biblical literatures as an earthly life-form who labors to create and sustain humankind and otherkind in solidarity with one another.

Running rivers, prairie fires, coral reefs, schools of blue whales, equatorial forests—the Spirit both shares the same nature of other life-forms and is the animating force that enlivens all members of the life-web. As the breath of life who moves over the face of the deep in Genesis, the circling dove in the Gospels who seals Jesus' baptism, and the Pentecostal tongues of fire in Acts, the Spirit does not exist apart from natural phenomena as a separate, heavenly reality externally related to the created order. Rather, *all* of nature in its fullness and variety is the realization of the Spirit's work in the world. The Spirit is an earthen reality—God's power in land, water, and sky that makes all things live and grow toward their natural ends. God is living in the ground, swimming through the oceans, circulating in the atmosphere; God is always afoot and underfoot as the quickening life-force who yearns to bring all denizens of this sacred earth into fruition and well-being.

A SACRED PLACE: SOJOURNING IN THE CRUM CREEK

I turn now to an analysis of the Crum Creek watershed, at the edge of the Swarthmore College campus near my home and the place where I work, as a case-study to illustrate my overall thesis concerning green pneumatology. Crum Creek winds through a thirty-eight-square-mile area of land that sits on the western edge of suburban Philadelphia. This area is a network of streams, wetlands, and aquifers that supplies two hundred thousand households and businesses with drinking water as well as being a discharge site for wastewater effluent and a natural floodway for storm water events. The watershed is a scenic retreat for persons in the Philadelphia area who need a place of refuge from the strains and stresses of urban life. And it is an important habitat for many native plants and animals.

A variety of species of wildlife relies on the Crum Creek watershed for food and habitat in which to raise their young. Scarlet tanagers migrate from Colombia and Bolivia to lay their eggs in the old-growth forests surrounding the creek area. Spotted and red-backed salamanders are two of the twelve or so species of amphibians that live within and along the banks of the creek and its tributaries. Monarch butterflies migrate from Mexico to the open meadows of the watershed area, where they roost to feed on milkweed plants and lay their eggs. Ancient southern red oaks survive in a section of the Crum Woods near the Swarthmore campus in an aboriginal forest relatively undisturbed by white settlement. American eels migrate downstream through the creek every fall to lay their eggs in the Sargasso Sea near Bermuda; in turn, their offspring then swim upstream to mature in the same creek area where their parents began their own journeys out to sea. And showy, large-flowered trillium wild-flowers fade from white to pink each year in the deep, rich woods of the watershed.[13]

The Crum Creek near the Swarthmore campus is my favorite site for passive recreation and easy walking meditation. Living in a world awash in parking lots and strip malls, I find it healing and restorative to be able to take refuge in the dark quiet of the woods. Henry David Thoreau writes about the art of getting lost, the vertiginous pleasure of abandoning oneself to a natural place without the artificial supports of urban maps and street signs. "Not until we are lost do we begin to understand ourselves," says Thoreau.[14] Today many of us travel with cell phones and global positioning devices so that no one need go missing and become confused about where they are. But in taming wild places and making them the quantifiable objects of our measurement and control, we have done harm to our basic humanity, our basic animal nature. We are animal beings at our core. Our need for sleep, hunger for food, drive for companionship, and desire for sex are telling signs of our carnal natures. To be sure, we are animals that are self-aware and self-conscious, animals whose conscience can burn with shame and guilt, animals who create art, engage in science, and produce grand mythologies that map the cosmos and set forth the roles each of us should play. But we are animals all the same.

To be divorced from our fleshly, bodily natures—not to see and hear the mad rush of a swollen river in the early spring or the smell of moist leaf litter in the autumn in the woods around us—is to be cut off from the vital tapsprings that make us who we are. We live and work in fixed-glass, temperature-controlled buildings sealed off from the natural world; we transport ourselves in fossil-fuel machines that require ever-widening incursions into undisturbed habitats; we eat processed food that has been genetically manipulated, irradiated, and then sealed in airtight packaging in order to preserve its interminable shelf life. We have replaced lives lived in sustainable harmony with the rhythms and vitalities of the natural order with soul-deadening, consumption-intensive lifestyles that leave us emotionally depleted and spiritually empty. We need untamed places to return us to our animal identities, and I am deeply grateful for the role the Crum Woods plays in my own return to the wildness within me.

The Crum Creek is a celebration of the natural amity that characterizes the human and the more-than-human spheres of existence. It is a place of scenic beauty, sensual delight, and spiritual sustenance. Like the ancient groundwater aquifers in the woods that are recharged by winter snows and spring rains, the depths of my own inner life are recharged by regular sojourns along the forested banks of the streams and tributaries that make up the watershed.

But in spite of its natural beauty and seeming health, all is not well with the Crum Creek. There are many threats to the biodiversity and well-being of the creek area. Overall development pressures pose the largest perils to the integrity of the watershed. In the upper portion of the creek area, housing construction, shopping centers, office parks, and parking lots have fragmented natural habitats and increased the amount of paved areas, leading to storm water runoff problems. In the lower portion of the creek near Swarthmore College, continued institutional development by the college along the edges of the watershed has created the same sorts of problems. Ironically, while Swarthmore College has been a relatively benign caretaker of the woods near its campus for many generations, in recent years the college's growth pattern has made it a threat to the preservation of species and habitat in the lower Crum Creek. This troubling growth pattern entails cutting down edges of the forest preserve to open up space for college facilities. Since the 1960s new townhouses for faculty, expanded student dormitories, additions to existing academic buildings, new access roads, and construction of surface parking lots have shrunk the perimeter of the forest. These past and possible future uses of forest near the college campus raise troubling questions about the long-term health of the Crum Creek watershed.

The Crum Creek as the Wounded Sacred

Degraded but still robust, wounded but still alive—the Crum Creek watershed is an impaired wildlife area that continues to supply water, food, and other basic elements to the many communities, human and nonhuman, that flourish alongside and within its banks and streams. Though the Crum Creek suffers regular abuse, to me it is a sacred place, a place where I am nourished and affirmed in my religious quest, a place where I find God.

But does it really make sense to say that the Crum Creek is a *sacred place*? Today our common discourse has expanded to make almost anything we do and believe in sacred. Special periods spent with family is "sacred" time. The important responsibilities assumed by law enforcement officers or child-care workers is a "sacred" trust. And almost anyplace one might venture—from a graveyard to a churchyard, from a memorable site in one's childhood to a battlefield or even a football stadium—can be a candidate for a sacred place. But if anything or any place can be sacred, then what is not sacred? If the term is so elastic as to include virtually any activity or place we might imagine, then does the term any longer carry any significance?

I grant that to honor the Crum watershed as a sacred place appears, at first glance, to continue to expand the use of this term to include locales that might not obviously appear to be sacred sites. The Crum Creek is not a built religious structure like a church or a temple. It is not a time-honored legacy site such as a war memorial or historic battleground. It is not even a widely recognized natural place of extraordinary beauty and grandeur, such as the Grand Canyon or Yellowstone National Park. Nevertheless, the Crum watershed is a living system that supports an astonishing wealth of native wildlife, and insofar as it continues to function as a vital habitat for a variety of species and their young, it is a "sacred" place.

Health and vitality are the highest ideals that make life on earth possible and worth living. The preservation of species-richness, which directly supports the stability and productivity of diverse biological communities, is the supreme value that nurtures human and nonhuman flourishing on our fragile planet. A place where God especially dwells, a place that is "sacred," is a place where ecosystem diversity is protected so that the miracle of self-regulating species-development is allowed to thrive. God as Spirit inhabits the biotic support systems on which all life depends, invigorating these systems with divine energy and compassion. The Crum Creek is not a pristine watershed; it will not win any virgin forest or clean water awards. But it is a site for the landed sacred, a place where God is alive and present because it is a small, and increasingly rare, patch of earth and river in harmony with itself that supports the well-being of its living inhabitants.[15]

Wherever there are places left on earth where natural ecosystems are in balance with their surroundings, there is God's presence. God is the giver of life, the sustainer of all that is good, the benevolent power in the universe who ensures the health and vitality of all living things. The Crum Creek watershed—battered and degraded though it may be—continues to function as a balanced and self-sustaining network of life-giving habitat for plant, animal, and human well-being. The life-giving role the Crum Creek performs is divine in the truest sense of the word because it describes precisely the role God performs in and through the earth: to give life, to make all beings come into fruition, to sustain the zest and vigor of creation. In this sense, the Crum Creek and God are one because they are both sources of life and health for earthen beings. To say, then, that the Crum watershed is a sacred place does not debase the meaning of the word "sacred" by designating just any such place as sacred or religious based on personal whim or fancy. On the contrary, to celebrate the Crum Woods as a sacred place is to drop to one's knees on the ground, and extend one's arms to the sky, in order to honor this place of God's indwelling as one of the remaining life-giving habitats on our planet that make our existence, indeed the existence of all of us, possible.

The Crum Creek is sacred, indeed, but the Crum Creek survives today as the *wounded sacred*. Envisioning Spirit and the Crum habitat as one opposes the classical theological idea of God as unchangeable and apathetic in the face of the suffering and turmoil within creation that God birthed into existence. God's Spirit is not a distant abstraction but a living being who subsists in and through the natural world. Because God as Earth Spirit lives in the ground and circulates in water and wind, God suffers deeply the loss and abuse of our biological heritage through our continued assaults on our planet home. God as Spirit is pained by ongoing ecosqualor; God as Spirit undergoes deprivation and trauma through the stripping away of earth's bounty. As the earth heats up and melting polar ice fields flood shore communities and indigenous habitats, God suffers; as global economic imbalance imperils family stability and intensifies the quest for arable land in native forests, God suffers; as coral reefs bleach into decay and whole ecosystems of fish and marine life die off, God suffers; and as stream quality and wildlife habitats endure further degradation in the Crum watershed, God suffers. When we plunder and lay waste to the earth, the Spirit suffers as God's presence on a planet that is enduring the loss of natural resources and cascading species-extinction. The Spirit is the injured sacred, the enfleshed reality of the divine life who grieves over what may become a lost planet, at least for human habitation and that of countless other species. As the Spirit is the suffering God, so also is the body, so to speak, of the Spirit's worldly presence, the earth itself, the wounded sacred. Together in a common passion and common destiny, the Spirit of God and an earth scarred by human greed body forth the wounded sacred in our time.

In the green Spirit perspective suggested here, God's vulnerability as a fleshly being and damage to the Crum watershed are one and the same reality. Even today, the Crum Woods are one of many surviving networks of life-giving habitat that manifest God's bounty and compassion in the earth. But the Crum Creek also displays the Spirit of God's deep and abiding suffering in our present time as well. As toxins from ruptured sewer lines and storm water leech into the creek, as the edges of the forest are cut down to make way for more suburban sprawl and commercial and institutional growth, God's Spirit experiences the loss and depredation of this delicate watershed in the depths of Godself. God is harmed by what we do. God is injured by the ways in which we despoil the natural systems that have supported life in many bioregions, including the Crum Woods, for tens of thousands of years. Spirit in love with the land—God in friendship with this small strip of Pennsylvania greenway—are co-determined, fellow sufferers in a unified effort to bring sustainable well-being to earth community. The Crum Creek is a small but important member of the Spirit's earthen body; as is all of creation, this forest fragment is part of the body of God's material presence. When the Crum Creek suffers, God suffers as well, reminding all of us to travel lightly on the earth as we participate in the evolution of particular ecosystems, including the evolution of this particular watershed.

THE CHALLENGE OF DECONSTRUCTION: IS NATURE REAL?

If nature is the primary focus of an earthen theological perspective, then what do we mean by the term "nature" when we valorize it in this way? Postmodern deconstruction questions assumptions about the seeming self-evidential character of landed reality and thereby challenges green spirituality to re-examine its basic identity and suppositions. The postmodern project seeks to show how knowledge about the world is generated through language and culture. If the task of an earlier modernism was to uncover the nature of reality as stable and ordered, the task of postmodernism is to destabilize or deconstruct notions of so-called reality by laying bare the ways human understanding of the world is always already a product of culturally embedded interpretive activity.[16] Postmodernism's challenge to religious ecology, then, is to question whether green theology's "turn to nature" betrays a crude understanding of the

natural world as a self-evident set of facts, when the meaning and significance of the purported facts that make up reality are actually imposed
upon the world based on prized cultural assumptions. There is no empirically obvious "raw" nature that tells us what reality is really like; rather,
we import into the natural world socially mediated presumptions about
our proper role in nature in relation to the wider world around us.[17]
Weighing in on the postmodern constructivist side of the discussion is
William Cronon, a history professor at the University of Wisconsin, and
a group of interdisciplinary scholars whose 1994 seminar with Cronon at
the University of California at Irvine explored the theme "Reinventing
Nature."[18] Cronon and his colleagues argue that "nature" is a term
loaded with cultural baggage. Nature is not a fixed reality but a value-
laden concept whose meanings undergo dramatic shifts over time. Ironically, there is little if anything that is natural about nature because both
the term "nature" and the reality to which it corresponds have been
ineluctably shaped by human desires and imaginings. Cronon writes:

> Popular concern about the environment often implicitly appeals to
> a kind of naive realism for its intellectual foundation, more or less
> assuming that we can pretty easily recognize nature when we see it
> and thereby make uncomplicated choices between natural things,
> which are good, and unnatural things, which are bad. Much of the
> moral authority that has made environmentalism so compelling as
> a popular movement flows from its appeal to nature as a stable
> external source of nonhuman values against which human actions
> can be judged without much ambiguity. If it now turns out that the
> nature to which we appeal as the source of our own values has in
> fact been contaminated or even invented by those values, this would
> seem to have serious implications for the moral and political author-
> ity people ascribe to their own environmental concerns.[19]

Cronon's point is that ecoactivists' unexamined assumptions about the
nature of nature—which they assume to be a self-revealing order of being
imbued with inherent worth—mask the highly imaginative constructions
of nature crafted by human cultures. The ecoactivists' essentializing orientation toward nature is generally taken for granted. Of course, this
orientation is pragmatically useful as it serves as the basis for the moral

exhortations by environmentalists to take up the plight of nature and fight against the destruction of our planet home. Thus the understandable fear for some environmentalists is that by moving away from regarding nature as a self-evidential good to redefining nature as a cultural construction, the grounds for action in defense of the earth will be effectively undermined.

Not only do Cronon's critics regard his constructivism, therefore, as undercutting green activism, but they also regard his position as a stalking horse for pro-development forces opposed to systemic environmental protection. By examining nature within the confines of cultural relativism and projectionism, Cronon provides intellectual backing for conservative policymakers and developers who maintain that nature is not an absolute good to be defended at all costs but, instead, a resource to be used to serve the ideal of human flourishing. Nature is morally ambiguous, since "nature" is not "naturally" good or bad. Cronon's critics argue that if you say nature is not an intrinsic, self-evident good that cries out for our protection, you are then giving tacit permission to antienvironmental forces to exploit the natural world in whatever way they deem necessary in order to serve their commercial interests.

The argument of Cronon's critics is that (a) once nature is deconstructed as a project of discourse with (b) no inherent capacity for providing criteria to adjudicate which projections are better than others, then it follows that (c) nature can be used and abused to serve selfish human ends since its true essence is always filtered through the lens of imaginal activity. If nature is not a fixed, objective fact that tells us what to do, then it becomes a candidate for exploitation and depredation. This is the main point environmentalists make in opposition to Cronon. In a trenchant criticism of Cronon and his ilk, deep ecologist George Sessions writes that "for most postmodernists, there is no standpoint beyond human cultures . . . there is no objective truth—all theories and statements (even by scientists) reflect only the interests of power elites; and that since Nature is a human construction, humans can 'reinvent Nature' . . . in any way that suits our immediate interests and desires."[20] Earth First! and Wilderness Society cofounder Dave Foreman writes similarly that "the irony of Cronon is that he is the kind of intellectual the anti-wilderness populists decry in their red-faced anti-intellectualism, yet he gives these people arguments to use against wilderness (and they *are* using Cronon's arguments)."[21]

But it is important to note Cronon's claim that his attention to the value-laden character of nature discourse is intended to strengthen, not undermine, environmental thinking and action. Cronon believes that the commitment to earth healing will become more honest and effective when environmentalists learn to become more nuanced and self-reflective in their use of the term "nature" in their moral language and ethical engagements. Though unadulterated nature is beyond our reach, Cronon's hope is that by examining the rhetorical constructions of nature within human cultures, environmentalists will become more sensitive to the various uses *and* abuses to which the idea of nature has been put. With this new sensitivity in place, the fight for nature can proceed with renewed vigor as it attends to the terminological and political shades of meaning employed on behalf of—and against—the integrity of the natural world.

In my mind, while Cronon and the Irvine group are at pains to argue that the concept "nature" is a human idea, they are not thoroughgoing idealists. Their opposition to naïve realism—the commonsense notion that the external world is fully knowable apart from cultural mediation—does not entail subscription to the extreme antirealist or idealist position that the material world either does not exist or is fundamentally unknowable. Cronon and his supporters do not argue that physical objects and places like rocks and trees and wetlands have no reality apart from our perception of these objects and places; rather, their point is that the cultural meanings we attach to these things are largely a human affair and cannot be divorced from any supposed inherent qualities these objects and places have apart from our interpretation of them. In spite of the dismissals by his critics, Cronon for his part labors to make clear his desire to hold in tension the *constructivist* thesis that the concept of nature is a cultural invention with the *realist* notion that the actual material world exists independently of being perceived. In this sense, Cronon is neither, philosophically speaking, an extreme idealist nor, politically speaking, opposed to preserving the integrity of the natural world. He says:

Asserting that "nature" is an idea is far from saying that it is only an idea, that there is no concrete referent out there in the world for the many human meanings we attach to the word "nature." . . . Yosemite is a real place in nature—but its venerated status as a sacred landscape and national symbol is very much a human invention. The objects one can buy in stores like The Nature Company certainly exist in nature—but that does not begin to explain how they came to inhabit some of the most upscale malls in modern America.[22]

In green theology's encounter with deconstruction, deconstruction pulls back the veil of our most prized notions—notions such as "nature" or "the world"—and show us that such notions betray our attachments to often hidden assumptions about what constitutes our picture of reality. Whenever theorists (including religious theorists) claim to delineate "how things really are," they are, in fact, representing their own presuppositions and convictions as much as they are describing the so-called real world. There is nothing wrong with this sort of value-laden mode of analysis as long as scientists, theologians, activists, and others recognize how their own personal worldviews and beliefs are implicated in the accounts they give of the realities they describe. In this vein, nature is not natural: nature is not raw and ready to be *discovered* by the individual observer because nature is always being *constructed* according to the socially mediated, partisan convictions of the observer in question.

Green Spirit theology finds this constructionist model of knowing to be a healthy tonic in its attempt to articulate an earth-friendly religious vision. Constructionism helps to keep theology honest in its reminder that all claims to reality are partial and reformable. Now fully aware that the meaning of the natural world is generated through cultural assumptions, green theology enters the public fray clear-headed about its own founding assumptions and clear-sighted in its distinctive vision of an interdependent world charged with the healing power of the Spirit in all things. Purified of its essentializing tendencies through its hygienic encounters with deconstruction, green theology takes its place in the public square aware of its need to make a strong *rhetorical* case regarding the liberatory truth of God's earthen love for all members of the biosphere and human beings' concomitant responsibility to care for creation. It is not obvious that nature is the abode of Spirit and that compassionate, sustainable relations with earth community are incumbent upon all persons, religious or not. Nature per se does not tell us anything; its meanings are created on the basis of our kinship with its regular flow patterns. But when we attune ourselves to these daily rhythms, then we can engage in the constructive task of articulating to others a vision of a renewed earth where all of God's creatures live in harmony with their surroundings.

THE CHARGE OF IDOLATRY: COBB'S CRITIQUE OF SACRED-LAND THEOLOGY

In traditional Christian thought only God is sacred. God alone is supremely absolute and sovereign over the whole created order. All other beings, while valuable as products of God's creative love and bearers of God's image, only have value and worth relative to God. The dominance of this model is entirely understandable given the important theological images in the Bible and Christian liturgy that focus on God as Lord, King, Sovereign, Ruler, Monarch, and Judge. From a monarchical vantage point the biblical message is clear: God is sovereign, just, and good, and all of God's creaturely subjects—plants and trees, human beings and other beings, ocean, land, and sky—have value and goodness only derivatively in relation to the supreme life source of God in Godself.

John B. Cobb Jr., who, along with Joseph A. Sittler, is arguably the father of Christian environmental theology, has consistently rejected this feudal view of God and the world. His 1972 book *Is it Too Late? A Theology of Ecology*, written in the wake of first-wave environmental awareness during the social justice movements of the 1960s, is a searching indictment, on the one hand, of how Christian kingly theology has paved the way for ecological destruction and, on the other, a visionary proposal for an earth-friendly theological agenda.[23] This pathbreaking book was followed, along with other works written by Cobb, by *For the Common Good: Redirecting the Economy toward Community, the Environment, and a Sustainable Future*, coauthored with Herman E. Daly, which further refines Cobb's ecological vision in dialogue with process theology, natural science, and holistic economics.[24] Cobb's process theology is an exercise in panentheism: God and the world are internally related realities brought together in a dynamic process of mutual transformation.[25] Cobb's interdependent model of the God-world relationship is the grounds of his criticism of historic Christianity's myopic focus on the salvation of human beings to the exclusion of concern for the well-being of nonhuman plant and animal communities. This anthropocentric bias has blinded Christianity to the degradation of the biosphere and the suffering of individual creatures; a new vision of Christianity in harmony with nature is the demand of our time. Cobb's move, then, to a thoroughgoing green Christianity predicated on ascribing sacred value to earth community would seem to be the natural trajectory of his thought. And in certain important respects, Cobb does share basic assumptions with this orientation. All beings, including human beings, are radically and mutually interdependent on natural systems for their well-being: for human beings to destroy wantonly plant and animal life is to threaten and diminish the life quality of all of us, human and nonhuman alike.

But in spite of these core areas of agreement, Cobb also carefully distinguishes his project from that of the religious ecology suggested here. In particular, Cobb, while investing nature with spiritual power and sacramental meaning, disagrees with the tendency in nature-based religion to honor the natural world as sacred *in itself.* While God is in the world and benevolent toward creation, God alone is sacred. It is a dangerous misnomer, even idolatrous, to confuse the Creator and the creation and to venerate the earth as sacred along with God. In a word, God alone is holy. Cobb writes:

> Nevertheless, [the sacredness of all creatures] language is, from a historic Protestant perspective, dangerously misleading. Speaking rigorously, the line between the sacred and the profane is better drawn between God and creatures. To place any creatures on the sacred side of the line is to be in danger of idolatry. For many Protestants, including process theologians, the right way to speak is incarnational, immanental, or sacramental. God is present in the world—in every creature. But no creature is divine. Every creature has intrinsic value, but to call it sacred is in danger of attributing to it absolute value. That is wrong.[26]

Cobb's case against sacred-land theology is twofold. His first objection is theological: such theology wrongly blurs the line of distinction needed to separate beings of relative value from the divine being itself, the bearer of absolute value. The specter of idolatry haunts Cobb's writings about the environment. Unless the borderland that divides Creator and creation is carefully policed, there is the danger that the value of a sacralized earth will be purchased at the price of denying the transcendence of God. Cobb is not simply speaking about the generic Protestant concern with the threat of idolatry outside of process thought; rather, he makes clear in the above quote and elsewhere that process theology shares with its Protestant conversation-partners the anxiety about idolatry in omnisacred earth theologies. Idolatry for Cobb is the confusion of realms of reality that need to be kept apart; thus his theology operates within a binary, either-or logical field: one worships either God or nature but not both. Since Christianity, in Cobb's perspective, is not an animist religion that invests the natural world with sacred, absolute value, one should worship God alone as sacred. While nature *is* charged with God's presence, according to Cobb, it does not follow that nature *itself* is a divine reality alongside or on a par with God and thereby an object worthy of our devotion and worship. To call the created order sacred, therefore, is dangerous and idolatrous: it is to run the risk of deifying and revering the earth as equal in worth and value to God. To do this is to displace God's unique role as humankind's proper object of worship and center of absolute, transcendent value.

Cobb's second objection to deep green Christianity is practical: unless one can refer to a being of absolute value, judgments of relative value are impossible to make. If all beings—everything from megafauna, such as human beings and blue whales, to microflora, such as mold spores and green algae—are sacred, if everything is equal in value and worth, then on what basis can decisions be made about what should be saved and protected and what can be used and destroyed? Without some hierarchical system that grades the relative value of different life-forms, there is no coherent foundation on which to base preservation of species, resource-allocations, food production, biomedical research, and so forth. Cobb writes that one "cannot give up the affirmation of gradations of value. All creatures have intrinsic value, but some have greater intrinsic value than others. That is to say, the inner life of some creatures is more complex, deeper, and richer than that of others."[27] Or, as he says at another point, "We believe there is more intrinsic value in a human being than in a mosquito or a virus."[28] For Cobb, God alone is sacred and the highest expression of absolute value; after God, humans, as beings of complex rationality and rich experience, are next in value in this ordering hierarchy; after humans, other communities of animals and plants are graded according to the depths of their cognitive functions and range of feelings and abilities. Without this sort of pecking order, moral decision making is impossible. For Cobb, extreme green spirituality is well-meaning but wrongheaded. By affirming the sacredness of all creation, land-based theology plunges us into a night in which all things are black and there is no way to distinguish between which use patterns are healthy and sustainable and which are not.[29]

An alternative to Cobb's axiological hierarchy is suggested by renewed appreciation of the energy-exchange and feeding patterns that character-ize diverse biological communities, a point emphasized in the previous section on the Crum Woods. In this model, supreme value is inherent in the vitality of the food-web vis-à-vis God's Spirit; it is characteristic of the natural life process of biocommunal eating and being eaten as that pro-cess is energized by divine power. Supreme value is not an attribute of one reality (God) over and against another reality (the earth) because both realities are one (dialectically understood). By the same token, dis-tinctions regarding relative value and worth are not made by first privi-leging human beings as bearers of more intrinsic value than other beings. The task of Christian earth healing would be to ensure, therefore, the health and dynamism of the life cycle rather than protect the interests of added-value beings (such as human beings) whose inner life is supposedly more richer and complex than other beings. Thus deep green spirituality is able to make highly nuanced and sophisticated *practical* judgments about use and value, but it does so in biocentric rather than anthropocen-tric terms. Such judgments are made not in relation to the putative higher value of human beings but on the basis of maintaining healthy predator-prey relationships within the food-web—that is, in reference to how energy is obtained and transmitted through a series of exploitative and mutualistic relationships among different and interconnected living things. Aldo Leopold, the early-twentieth-century Wisconsin conserva-tionist and forest advocate, alternately refers to this flow of energy ac-cording to highly organized systems of biotic relationships as the "food chain," the "energy cycle," or the "land pyramid":

Plants absorb energy from the sun. This energy flows through a circuit called the biota, which may be represented by a pyramid consisting of layers. The bottom layer is the soil. A plant layer rests on the soil, an insect layer on the plants, a bird and rodent layer on the insects, and so on up through various animal groups to the apex layer, which consists of the larger carnivores.

The lines of dependency for food and other services are called food chains. Thus soil-oak-deer-Indian is a chain that has now been largely converted to soil-corn-cow-farmer. Each species, including ourselves, is a link in many chains. The deer eats a hundred plants other than oak, and the cow a hundred plants other than corn. Both, then, are links in a hundred chains. The pyramid is a tangle of chains so complex as to seem disorderly, yet the stability of the system proves it to be a highly organized structure. Its functioning depends on the cooperation and competition of its diverse parts.[30]

Green spirituality can learn a lot from conservation biology about the crucial importance of food-webs for ensuring the future of the planet. Scientifically speaking, in the natural order, everyone is food for everyone else. Human beings, for example, both eat and then are eaten by fungi, bacteria, and often insects and other anthropods as well. Everyone is predator and prey in relation to other living things. All of us, from the smallest bug to the largest carnivore, rely on this complicated flow mechanism for our daily bread. Moreover, all beings play an equal and vital role in maintaining the integrity of the energy cycle; no one member of this integrated plant and animal community is any more important in sustaining the cycle than any other member. Theologically speaking, then, judgments about value should be based not on human needs but on keeping open the living channels of energy that make life possible. This is the point Leopold makes in his general maxim for a land ethic: "A thing is right when it tends to preserve the integrity, stability, and beauty of the biotic community. It is wrong when it tends otherwise."[31] Value accrues to the health and vitality of the food-web; it is not a property of particular organisms—in other words, more complex creatures do not get more intrinsic value than less complex creatures.

Cobb is very articulate about the role biotic interdependence plays in the life-cycle.[32] But his fear about ascribing sacredness to nature, and his human-centered value system, blunts a full turn toward the biocentric theology adumbrated here. His emphasis on subjective experience as the criterion for making comparative value judgments undercuts the power of the food-web model to make clear that *real* value inheres in the integrity and well-being of the web itself. As humans, according to this "web-first" model of reality, we should simply see ourselves as equal citizens of the biotic order—we do not possess more value than other beings. Some critics regard this subordination of human concerns to the welfare of the whole as a type of misanthropic thinking, even a kind of ecofascism in which human interests are now located in (or subordinated to) the wider orbit of ecosystemic interests. But the point is not that human happiness is unimportant in green systems thinking but rather that, without the well-being of the whole as the paramount concern, attention to human needs and interests is not possible. To put the point bluntly, if the worldwide system of energy flow patterns collapses due to ecocatastrophe of our own making, then our discussions about whether human beings have more value than other beings will seem academic at best and, at worst, contributory to the very mind-set that gave rise to the collapse in the first place.

Cobb (and others) criticize sacred-land theology as flirting with idolatry. But the witness of Scripture and Christian tradition is to the world as the abode of divinity, the home of life-giving Spirit, God's here-and-now dwelling place where the warp and woof of everyday life is sacred. All life is sacred because the earth is a natural system, alive with God's presence, that supports the well-being of diverse ecological communities. God's gift to all beings is Earth itself: this highly complex, biologically diverse system of interlocking relationships where enfleshed existence is celebrated in all its fecundity and passion. Sacredness resides in the God-given capacity of native plants and animals to stock and replenish the food-web on which we all depend. Supreme value inheres in the Spirit-infused dynamism and elasticity of the energy cycle that makes our lives and the lives of other beings endlessly rich and potent with new possibilities. The Spirit is the green face of God who animates the living food chains that make possible the flow of energy and sustenance for all of us. God is not a dispassionate and distant potentate, as in classical feudal theology, who exercises dominion over the universe from some far-removed place; rather, in and through this planet that is our common home, God is earnestly working with us to heal the earth.

And yet, as we have seen, God also suffers deeply from the agony of inhabiting a planet badly degraded and out of harmony with itself. For this reason I have said that the *green* Spirit who infuses all things with her benevolent presence is also the *wounded* Spirit who implores us, in groans too deep for words, to practice heartfelt sustainable living in harmony with the natural world around us. In a highly insightful discussion of the Spirit's relationship to nature, the apostle Paul writes that human arrogance has caused the whole creation to groan in agony as it awaits final deliverance. To update Paul's insights and correlate them with the contemporary ecocrisis, Paul's writings about the Spirit appear uncannily prescient: our hostile treatment of the earth has now plunged all of creation into deep suffering and travail. He writes: "The creation waits with eager longing. . . . [to] be set free from its bondage to decay and obtain the glorious liberty of the children of God. We know that the whole creation has been groaning in travail together until now; and not only the creation, but we ourselves, who have the first fruits of the Spirit, groan inwardly as we wait for adoption as sons, the redemption of our bodies" (Rom. 8:19–23). In the midst of the current crisis, all of creation groans under the weight of humankind's habitual ecoviolence. We feel the weight of this crisis, and we sense the Spirit alive within each of us, moaning out of pain and yearning for the renewal of a green, healthy, vibrant planet. In visceral sighs too deep for words, as Paul writes, the earthen God inwardly calls on us to care for our planetary heritage. God as Spirit agonizes over the squalor we have caused, and through her abiding earthly presence implores us to stop the violence before it is too late.

It is not blasphemous, therefore, to say that nature is sacred. It is not mistaken to find God's presence in all things. To speak in animistic terms, it is not wrong to reenvision Christianity as continuous with the worldviews of traditional peoples who bore witness to and experienced divinity everywhere—who saw and felt the Spirit alive in every rock, tree, animal, and body of water they encountered. For me it is not idolatry to enjoy the Crum Creek, degraded though it may be, as a sacred place that plays a crucial role in maintaining the health and well-being of humankind and otherkind in eastern Pennsylvania. God as Spirit is the gift of life to all creation, and where life is birthed and cared for, there God is present, and there God is to be celebrated. God is holy, and by extension all that

God has made participates in that holiness. Thus, when we labor to protect and nurture the good creation God has made, we invest all things with inherent, supreme value as a loving extension of God's bounty and compassion.

Sacred, then, is the ground we stand on; holy is the earth where we are planted. Sacred-land theology envisions God as present in all things and the source of our attempt to develop caring relationships with other life-forms. This perspective signals a biophilic revaluation *and* continuation of characteristic Christian themes. Christians speak of the embodiment of God in Jesus two thousand years ago, but now we can see the *entire life-web* as the incarnation of God's presence through the Spirit on a daily basis. Christians speak of the miracle of the Eucharist, in which bread and wine become Christ's flesh and blood, but now we can regard the *whole earth* as a living sacrament full of the divine life through the agency of the Spirit who animates and unifies all things. Christians speak of the power of the written word of God, in which God's voice can be heard by the discerning reader, but now we can view *all of nature* as the book of God through which one can see God's face and listen to God's speech in the laughter of a bubbling stream, the rush of an icy wind on a winter's day, the scream of a red-tailed hawk as it seizes its prey, and the silent movement of a monarch butterfly flitting from one milkweed plant to another.

✦ Grounding the Spirit: An Ecofeminist Pneumatology

SHARON BETCHER

INTRODUCTION

Her legs tap dance under the sheets as her ten-year-old mind reels from her first contemplation of googolplex. Wanting something to keep her from flying loose in the centrifugal forces that tug at bone and body and brain, she begs: "Make me heavy, Mama!"

"Okay, sweetheart," I say. "Close your eyes." Then we begin to imagine those elemental threads that spin a hammock for mortal bodies. "Feel how the sun makes you heavy with sleep. Feel how the cool ground under the apple tree invites you to sink down into the violets and creeping Charlie. Let the earth grow roots under you and tie you to its back. Now just backfloat until the moon comes out and begins to rock you on the tides of the sea." With a little whisper of a lullaby, she falls asleep in gravity's cradle. The grace of gravity: Returned to its rhythms, we sleep like a baby.

Funny, though, when I introduce that other parental discipline to bring her in touch with this earthy wisdom, she never begs for more. Clothes droop over drawer edges. By night, books crawl out from under the sofa and rocks creep into all corners of the living room. When she tops it all off with "that attitude," I issue the parental decree: *You're grounded.* She wails, listing all those things, including infinity, so much worthier of her attention than these trivial matters. Suddenly the gift of gravity has become the curse of ground.

Truth be told, most of us are less than enamored with the contours of a mortal life. If our culture, more than any other, concerns itself with the body, the gaze with which we attend to it has hardly been loving. Rather, disgusted with our weight, we make war against our corporeal contours—as with fat and flesh and foibles, so with dirt and death and disease, those other niggling reminders of mortality. Mortal flesh: it's such a vulnerable, base, and pathetic proposition. The humus—the earthiness—of humanity can be so mortally humiliating.

When humanity falls into disgust with mortal life, this loathing has frequently been visited upon women and the earth. Because—as the story has gone—women and earth conspire in bequeathing us our history of flesh, we use women's bodies to buffer us from what we loathe about being "body"—its frustrations and disappointments, its neediness, limits, and dependencies. The messes and miseries of bodies, these we hide from our eyes by cursing the earth and by keeping women anonymous, both in name and in economic practice. When we are loathe to love the body and condemn its care with the epitaph of "dirty work," this loathing begins to circulate in each one of us, no matter our gender. Carrying the weight of disgust that humanity feels toward bodily life, we women disparage each other in the same way that we punish our own flesh. Meanwhile, men go into a charcoal-gray exile from the sensuous, imaginative, and colorful life, depriving themselves of the gifts of the life lived body to body—of intimacy and community.

Falling in love with the mortal life has not been on the top of our sociopolitical, economic, or religious agendas. Because it has not, bodies and those that remind us of the grounding and heaviness of all incarnate life—women and the rest of nature—suffer battering and poverty, as well as the more subtle digs of psychic degradation. In this essay, I will analytically engage the theological history of Spirit, that concept by which Christianity has often escaped the tug of gravity. By opening out the aversions to ground that have been carried in theologies of Spirit, I hope to release Spirit in and for "organic transcending."[1] When Spirit gets grounded, we may then circulate our life love as a groundswell, rather than as a transcendental updraft.

THINKING SPIRIT WITH THE WEIGHT OF THE EARTH

Our spirits tell on us when thoughts of "getting grounded" elicit greater fear of corporeal punishment than promise of a spiritual epiphany. Unless you consider the earth a cursed place, "grounding" someone need not be considered inherently and unreasonably punitive. Among those who are wise to the earth, "getting grounded" is actually a very ancient and widespread wisdom for "centering" and restoring spirit, for fitting a body back into its elemental and social niche. Getting grounded re-minds us to "think with the weight of the earth."[2]

Amazingly, when we get theology "grounded," what emerges is pneumatology. "Spirit" has always intriguingly carried the double resonance of a theological term borrowed from the natural sciences.[3] Spirit metaphorically draws us outdoors, beyond human referents, to view life within its ecological context. While recognizing that our speech about Spirit is metaphoric, there is always a "return to nature" in order to speak of Spirit at all. When we speak of Spirit, we speak of water, wind, and birds on the wing. Spirit in all its down-to-earthiness could offer Christians an epistemological standpoint for behaving ethically within the natural orders and complexes that we inhabit. Thankfully, we have not been able to rid Spirit of these ancient metaphoric reservoirs of the elemental.

And yet, perhaps nothing so much as Spirit has been used to avoid awareness of the material implications of Christian practice. Hiding behind claims to the immateriality of Spirit, we have been able to avoid committing ourselves to the ground, able to avoid the social and economic implications of Christian spirituality. In Western Christianity, Holy Spirit—floating supernaturally outside the gravitational field of the elemental earth—has guaranteed context-independence. Consequently, getting a theology of Spirit grounded can strike one as just so much nonsense. The relinquishment of a modernist, substantialist worldview and its replacement with the physics of an Einsteinian worldview, if not the developing sciences of chaos-complexity, can make the concept of Spirit, something like an energy field itself, seem somewhat more viable. Yet, "ground" and "Spirit" may not sit well with each other, even in post-Einsteinian minds.

When we take account of the metaphoric registration of Spirit through the four ancient cosmological elements—fire, air, water, and earth—one element has gone missing in the Western Christian iconic repertoire: earth. Within the trajectory of Christianity, in an important sense "a religion of the body,"[4] there developed "the deepest perversion of Spirit, which is anti-matter."[5] The ecological wisdom of many indigenous religions we have come to admire these days is carried in a Spirit matrix. So why and how did the Holy Spirit of Christianity become so averse to planetary life?

In presuming to keep Spirit pure and uncompromised by the recalcitrant nature of matter, we have, I would suggest, sacralized an abhorrence of humus—of earth, our own bodies, of women, and of other earth-affiliated persons. Our "perversion of Spirit" covers over our deeply felt ambivalence, even hostility, toward the contours of a mortal life. Our views of Spirit have often provided us a fantastical preserve at a psychical distance from what we cannot stomach about existence. Premised upon Platonic philosophy and formulated as a "liberative movement" against the "bondage to decay" (Rom. 8), Western Christianity has read the promise of Spirit as what breaks the chain of the planet's organic systems.

Consequently, Spirit has been "airborne," aspiring to forms of freedom that have been construed as freedom from the constraints of corporeality, freedom from death, transience, and ambiguity. Taking shape already in the Hebrew Scriptures, Spirit comes ever more through the centuries to resemble one of its four elemental coordinates: air. Through its association with prophetic inspiration and its affiliation with the prophetic writings, Spirit becomes the subservient breath of a purified logos. Such a trajectory, ever averse to getting grounded, culminates in the postmodern pre-eminence of language and text.[6] While symbolically construed as the supernaturally transcendent scope of the power of the divine, Christianity's mystification, or "spiritualization," of Spirit has been undergirded by its own abhorrence of the material, organic, biotic aspects of being. Holy Spirit has been wielded as a theological legitimation of humanity's disregard of ecosystemic life.

The name "Holy Spirit" was evoked to get life up and off the ground and so to deny our ecological dependence. Sacralizing a life-generative system that has imagined itself above organic accountability has not only led to failure in our soteriological and sacramental practice on behalf of bodies and their needs; it has resulted in an inability to account for wastes, limits, spoliation, fallowness, rest, the cyclical and seasonal passages. But such abhorrence of sentience, in the end, disarticulates Spirit's own more "humorous" qualities, its passionate love for *mortal* life.

As awareness of the ecological matrix of life now recalls us to the practice of "saving life" (soteriology) within an ecosystemic integrity, we must reinsert these categories of mortal truth—need, suffering, limit, dependence, waste, finitude—into our pneumatology, which has been to this time our escape clause from these prevailing conditions of life.[7] Rather than employing Spirit as an excuse to think against the flesh, we might think Spirit's transcendence within the thickets of organic, biotic, social, and cosmic life. When so reoriented, we may finally, with the companionship of Spirit, be able to "experience matter as privilege."[8] Only by analyzing our disgust with our own "humus" in such a way as effectively to "ground" Spirit, to think pneumatology within the contours of corporeality and ecosystems, might we come to wonder in, and gratitude for, mortal life, and to a sensibility that respects the limits of ecological systems.

REDEMPTION FROM . . . OR CREATING THE CONDITIONS OF . . . WORLD DESPAIR?

Biblical Wisdom literature, within which Spirit makes its constitutive appearance, celebrates the cosmically expansive, if immanent, intimacy of Spirit pervading, suffusing, and transpiring in and through all (Wisd. 7:22b–24). Along with that, however, sapiential thought introduced a discourse on what came in Paul to be called "the bondage to decay" or "futility" (Rom. 8:20–21). Consequently, inherent to many readings of Paul, from the early-patristic to the contemporary, there churns this contention with what scholars have simply signified as "transience." This can be heard ringing quite poignantly through even Jürgen Moltmann's creation-centered, but still eschatologically future-driven, pneumatology, *The Spirit of Life*:

> In the coming of the Spirit . . . this world is revealed as a world of death which has failed to find God and itself. Transitory time and the mortality of all the living was hitherto held to be the "natural" condition of created things, because there was no alternative; but this condition now emerges as sick. . . . Once there is reason to hope for the world's redemption, that world ceases to be seen as natural and finite. . . . Out of the general transience of things, unredeemed creation can be heard "sighing and groaning" for its liberation. (88)

On the basis of Paul's lament of the "bondage to decay" and his similarly founded conceptual binaries, perishable/imperishable, corruptible/incorruptible, mortality/immortality (1 Cor. 15:53–54), Western Christianity has encoded "transience" to name what is wrong with the earth. "Transience" became the obverse of pneumatological thought, its conceptual foil.

An overview of theological reflections on "transience" suggests that the term has lent itself as a psychic surface on which we grow barnacles of frustration in regard to human ethical endeavor within this world as well as in regard to "the nature of nature" itself. Such loathing, knotted and balled into despair, obversely lofts us into transcendental ethers. Idealist metaphysics—as the obverse of our loathings—becomes a short-circuiting of ethically discerning behavior in relation to the tragic concourse of existence.[9] With its feet planted on the eschatological horizon, Western Christianity's flighty and squeamish Spirit has been busy cleaning up the spoliative aspects of creation.

While Spirit has been greeted in Western Christian theology as cosmically immanent, Spirit also becomes an interpretive—that is, idealized—lens through which nature is viewed.[10] So while Augustine opposed the Gnostics' diminishment of the material world (hence his leave-taking of the Manicheans), he nonetheless also imagined Spirit as something like a hovercraft, floating over the sentient morass. Augustine emphatically underscores this point in book 13 of his *Confessions*, where he defines Spirit as "referring to the changeless super-eminence of divinity [moving] over all things mutable."[11] On numerous occasions he returned to the image of the Spirit in the opening verses of Genesis, obsessively contemplating the way in which Spirit "moved *over*"—that is, "*above*"—the waters (chap. 5) as over the dark abyss (chap. 6). Augustine extended this analogy to speak of the way in which the Spirit "quickens our mortal bodies" (Rom. 8:11) by hovering over our own dark and fluid inner being (chap. 14).

Augustine's comments seem in continuity with Tertullian's earlier reflections on Spirit. In his treatise *On Baptism*, Tertullian likewise imagined Spirit as an "overhang" of holiness: "It is necessary," Tertullian concluded, reflecting upon the same image of Spirit in the opening verses of Genesis, "that in every case an underlying material substance [here, specifically, water] should catch the quality of that which overhangs it

[Spirit]."[12] Spirit, it would appear, hovers just above the flux and frustration of flesh. By "overhanging" the finite realm, Spirit magnetically draws the spiritual body to its own eschatological realm, indemnifying "life" against organic process.

Similarly Luther, whose sense of the infinite indwelling the finite (hence, his panentheistic Eucharistic sensibilities) was suggestive of the sapiential theologies of earlier centuries of the Rhineland valley. While chiding Aristotelean philosophers for developing a "happy science" of essences and actions and consequently ignoring the "sad creation," Luther ended up with an all-too-happy—that is, idealist and fantastical—pneumatology.[13] In his commentary on Romans 8:18–25, Luther affirmatively observed that Paul "speaks of the 'expectation of the creation'": "Because [Paul's] soul can hear the creation waiting, he no longer directs his attention to or inquires about the creation itself, but rather to what it is awaiting."[14] While Luther unquestionably heard the groaning of creation and anticipated Spirit's broad concourse with creation, he nevertheless makes clear that Spirit's eschatological interest is in the time when creation "will not be subject to corruption."[15] Such a pneumatic economy cannot but in some way avoid sinking its sensibilities into the heterogeneous milieu of the present, the time in which our lives take place. Like Luther, Western Christian pneumatology in general, as theologian Catherine Keller has said, "forgets forward, into a supernaturally artificial environment."[16]

Pneumatological thought all the way through Paul Tillich in the twentieth century has consequently conceived Spirit's primary effects upon organic systems as passing from the divine through the human and only then into matter. Only in this indirect way does the earth "catch the quality of that which overhangs it." Quickening life over and above the dark and fluid material substrate, the flux and flow of finitude, high-minded Spirit creates an updraft of energies and ideals, premised upon the assumption that the organic and the biotic, sentient relational circulation and sensuous sympathy, are mortally polluting to spiritual mindfulness. We have thus imagined sanctificatory systems that "freed" or liberated life by fleeing the gravity of need and suffering and by abrogating the natural economy, which requires us to respect limits, finitude, dependence, and necessity. Spirit "redemptively" counters the gravitational pull of the earth, that compacted weight named "transience" Christians so wish to off-load.

This flighty and squeamish Spirit, despite its immanent intimacy, became the legitimation for humanity's disregard of ecosystemic life, for exercising creativity without accounting for ecosystemic impact. In what I take to be an anonymous charge against Hegel's "Absolute Spirit," feminist philosopher Luce Irigaray spells out the cost of this sublime conceptual "overhang": "But the absolute, unlike the most probably cosmic rhythm, kills, saps vitality, because it tears nature away from temporalization. . . . The spirit, in its perfection, does not thrust its roots deeper into the earth. It destroys its first roots. Its soil has become culture, history, which successfully forget that anything that conceives has its origins in the flesh."[17] Modern technology, bathed in this spiritual imaginary, assumed this same earth-avoidant aspiration. Indeed, Christianity, wrote twentieth-century theologian Nicolai Berdyaev, was "obliged to mechanize nature" in order to keep humanity from the "danger of communing with nature."[18]

Western theology has consequently conceived the salvation of life to be, in effect, salvation from transience. Christian pneumatologists, including Luther, have protected the qualitative difference of the divine by defining Spirit as a preserve of idealistic purity moving above this transient world. Christian theology projects an idealized eschatology from the precipice of a supernatural future and thus frees and protects itself from the sentient ambience of life. Consequently, rather than transcendentalizing the body *with respect to* our sentient condition, a system of loathings reifies and legitimates idealism through identification with Spirit. Such discourse, when weighted upon the individual human and when cathected to eschatological idealism, has left Christians simultaneously in perpetual despair of self and world.

While Spirit has been greeted as lifegiver, healer, and giver of grace and freedom, our high-minded Spirit has worked as a condensed symbol for covertly withholding from the sacred the sensuous, sympathetically related, or corporeal body. In the baptismal practice of Christianity, as shaped by Pauline discourse, one puts off the off-putting body, thereby acquiring the dissociative "ability" that novelist Christa Woolf finds endemic to the atrocities of the twentieth century, "the ability to be here and not be here at the same time."[19] Woolf wrote most explicitly of persons refusing to intervene in the Holocaust of the Jews in her German homeland. Could it be that such a disarticulation of our presence to life may also occasion our awareness of, but inability wholly to respond to, ecological crisis, caught as we are in the specular spectacular—that golden elsewhere—of consumer-based capitalism?

Jürgen Moltmann first intuited a corollary between the forgetfulness of Spirit and "creation-forgetfulness."[20] Elizabeth Johnson, working from an ecofeminist perspective, extended this observation. "The exploitation of the earth, which has reached crisis proportions in our day," Johnson writes, "is intimately linked to the marginalization of women, and . . . both of these predicaments are intrinsically related to forgetting the Creator Spirit, who pervades the world in the dance of life."[21] If we are to recover our memory of Spirit, then we need to figure out by what apparatus these destructive vectors—the disparaging of mortal life, including the denigration of women, the destruction of the earth, *and* the repression of Spirit—coincide. Passing this off as a slight case of amnesia denies accountability for the seething hostility that attends the Western historical disposition toward any insinuation of mutuality between earth, or women, and Spirit. "Forgetfulness" leaves undisclosed, and therefore unreconciled, the psychic apparatus whereby we unleashed such historic violence as the repression of prophetic and independent women and suppression of earth-affiliated religions.

In order to ground Western Christian pneumatology, and thereby situate ourselves ethically in relation to our planetary life, I propose that we engage something analogous to psychoanalysis. Our symbolic economy shows itself to have subjective, ecosocial, and corporeal effects, as farmer-philosopher Wendell Berry has observed. Describing Christianity's tendency to "separat[e] . . . the soul from the body and from the world" as "a fracture that runs through the mentality of institutional religion like a geological fault," Berry explained the ecological consequences of this religiously metabolized disposition. "This rift in the mentality of religion," he writes, "is not *like* a geological fault; it *is* a geological fault. It is a flaw in the mind that runs inevitably into the earth. Thought affects or afflicts substance."[22] Employing a psychoanalytical hermeneutic may be able to get us beyond amnesia into the mechanism of active denial, locating where and how our divine economy embeds itself in our psyches.

HUMILIATED BY HUMUS

"Could the sacred be, whatever its variants, a two-sided formation?" philosopher Julia Kristeva asks presciently, proceeding psychoanalytically to explicate the underbelly—loathing—that serves as the secret lining of the religious sublime.[23] As Kristeva has shown, psychocorporeal archaeology reveals a map of what each religion marks out, through ritual, as its borders. Psychologically below and chronologically prior to our practice of discursive reason lay psychic maps of the pure and holy that each culture draws. Loathing of the transient—of transitoriness, decay, mortality, the ambiguity of the cycle of life and death—creates the specific perimeter of Christianity's presumed domain of freedom. The affect of loathing, termed "abjection" by Kristeva, indicates that such a demarcation is psychically intact, attempting to turn us away from what is religiously polluting. Encountering those psychic borders causes us to gag, to revulse, and to return to our "spiritual dwelling." Our theology of ideal Spirit is limned, Kristeva's insights would lead one to conclude, by a loathing of our own sentient contours.

Sentient beings are feeling folds in the relational communion we name "the web of life." Process theologian Jay McDaniel describes sentience in Whiteheadian terms as "the capacity of energy events to *feel influences.*"[24] Feeling is an experience of co-inherence. To feel means equally "to extend outward toward" and "to pulse with feelings"—that is, to take in the impresses of other beings, the influence and extension, definition, and delimitation of the lives of many others. The "skin" of the body, our feeling-sensing self, binds us in the sympathetic relations with all beings that we name "consciousness." Sentience registers pain and pleasure, frustration, boredom, and satisfaction, wistfulness, wonder, and worry, anger and love, laughter and sorrow, in all their poignancy.

Our sentient liveliness emerges at the threshold of multitudinous orders and complexes. We unfold like a sprout from the soil, dependent upon the life and labor of everything from microbial mites and bacteria to photosynthetic processes and geoevolutionary tectonics, from the water and wind cycles to the food- and waste-cycles of humus, to the web of interspecies and intraplanetary habitats. The "land," as Wendell Berry notes, "passes in and out of our bodies, just as our bodies pass in and out of the land."[25] Body from body, earth from earth—as sentient beings, we live in and as an ever fluid and fluxing life-circuit, feeding upon and yielding ourselves into the earth. Within this fertile crescent, we are birthed and sustained in our vitality as an occasion of transient relations, creatively developing character of mind and the spirit of compassion insofar as we are also creatively transient.

Western Christianity, however, has engaged sentience via mortal combat. Christianity has responded to sentience, our existential status, as if it were an oppression. Spirit has consequently signified forms of freedom that breach the constraints of corporeality—hence, the conquest of death, as articulated by Luther, Hegel, and Moltmann, as also the conquest of ambiguity, as articulated by Tillich. Spirit has been for the Western Christian theological world the key to liberation from the organic realm—finitude, suffering, spatiality, and temporality, all construed here as aspects of "nonbeing." Abjection wedges a divide into this communal nexus, dividing life between the corporeal body and the soul, between ground and grace. Pushing away from what it deems disgusting—sentient co-inherence—the Western pneumatological economy creates a domain of freedom, but only by compressing its sediment into corporeality, a dead weight, a drag weight. Spirit psychically metabolizes, especially in the Pauline trajectory taken up in Protestantism, the eradication of the "futile" and "corruptible" body.

Abjection of sentience is not a repulsion of matter per se, but horror of our inherence, of our porosity, of our life in and by influx and effusion—of the fluid mechanics of the flesh, as Irigaray has put it.[26] In other words, this abjection resists the dynamic, processional quality of humus: the way living soil refuses objectification and the way land—blowing in the wind, wandering downhill, and washing up on shores—refuses property lines; the way bodies—sweating, exuding, singing, perspiring, developing, and diseasing—refuse to stay "clean and proper." Horrified by the Eucharistic liquidity of life, we have developed and carried through an articulation of Spirit an autoallergic reaction to our own humus, our mortal flesh and earthy habitat. Sublime Spirit has been opposed to futile flesh and underscored by a dissociative abhorrence of the material, organic, biotic aspects of being. Loathing insures that, while we are within the force field of the earth, we hold ourselves "apart from" the earth.

Such abjection rends the sentient matrix, interrupting the gift-giving exchange of ecosystems. When sentience is split between the idealized and the loathed, then the domain of freedom is set "over" and "above" or "over against" the transient, foreclosing any recognition of nature's ability to heal, to soothe, to save, to accompany us. That nature would now seem to be, as Moltmann puts it, "tongue-tied"; that nature would merely seem to sigh "dumbly" (like all of her supposedly "dumb" animals, that is, "dumb" cows, "dumb" sheep, "dumb" dogs, and so forth): is it due to the fact, as Moltmann has suggested, that "nature has fallen victim to transience and death"?[27] I would contend rather that if earth and her creatures appear tongue-tied, it may be because we are loath to hear them out. Living in this particular "domain of freedom," we cannot see nature as a functioning community of mutually supporting life-systems or ourselves as inherent participants in them.

Further, subject to this pneumatic economy, the human community abdicates reciprocity and mutuality within these organic life-systems. "Every time man or men seek to build an economic order at the expense of the earth," Irigaray observes, "that order becomes sterile, repressive and destructive. . . . It leads to an economic superstructure (falsely labeled infrastructure) that has no respect for the infrastructure of natural fertility." This "supernatural" structure interrupts the sustenance cycles of the earth (land, air, and water, energy), which require careful accountability and reciprocity, taking only enough for human needs, giving back to the earth what can be elementally digested—our wastes. As wastes are cut off from fertility, self is cut off from place, need from community, community from its ecological niche. The abstract idea that nature is futile in its seasons and cycles—an idea implicit to a pneumatology that takes transience as its aggravating motivation—inevitably leads us to overrun the telos of nonhuman orders, refusing them the opportunity to reproduce, evolve, and respond without human interruption. "Whether consciously or not," Irigaray concludes, "sacrificial societies perpetuate the unconsidered destruction of the products of the earth. . . . And yet ultimately these [products of the earth] and not advanced weaponry and sophisticated techniques are the only *guarantees* of exchange value."[28] They are the only possible roots for social justice as well, she later adds.[29]

GAG ORDERS

Loathings, as intimate as the psychodynamics of the personal body and as broad as the social bodies into which we are enfolded, open out a chasm of dissociation. The abject designates "what of the body falls away" in order for "the [socially conditioned] clean and proper body, the obedient, law-abiding, social body" to emerge.[30] Kristeva notes that within Western Christian monotheism what has been abjected—that is, what has been set aside in "excess" of the sociosymbolic system—are the material aspects of existence that have been maternally connoted: women (because all women have been seen at all times as maternal); the body (of which women are "projective" representatives); and, I would add, though it does not concern Kristeva, earth (that is, the communion of sentient relations). Working between Kristeva's analysis of abjection and the psychotheological history of our humiliation of the finite, earthy, and mortal, Christianity joins the corporeal body, women, and earth under one dismissive "gag order" so as to delimit the domain of Spirit's public universal freedom. In lofting the trajectory of Western Christian pneumatology, the abjection of women, the corporeal body, and earth have been employed to gain Spirit's tensive thrust.

With Christianity, Kristeva adduces, purity rituals have been ingested, interiorized, and individualized such that we become confessional beings. Christianity "invert[s] the pure/impure dichotomy into an outside/inside one," and pneumatic carnality, now identified with the "speaking being," must purge itself of its own flesh through ceaseless speech.[31] Henceforth, Kristeva asserts, "evil, thus displaced *into* the subject, will not cease tormenting him[/her] from within, no longer as a polluting or defiling substance, but as the ineradicable repulsion of his[/her] henceforth divided and contradictory being."[32] Such "intolerance" can hardly be construed as ecosocially innocent, having been bound up, as it was, with the crucifixion of the flesh, lived as the cultural loathing of, and self-loathing among, women, as conquest of disease, as battling the elements, as moral cleanliness, as geosocial displacement, and so on. Christianity has been driven by a gross intolerance of the ambiguity of sentient life, an ambiguity that it is ever "loath" to resolve; that is to say, loathing forces Christianity's hand toward action on behalf of "redemptive resolution" through the attempt to alleviate ambiguity, to become intensely persecutory in the name of Spirit.

This loathing appears to be as contemporary as it is archaic. Wolfhart Pannenberg's recent pneumatological reflections assume the old refrain: What keeps life from being fully united with Spirit are "failure and guilt, disability, disease, death." By posing such a problematic, he, too, associates "the Christian proclamation" with the "assurance of a new life that will be no more subject to death," equating Spirit with "the power of . . . finally overcoming the absurdities and adversities of the present world."[33] Michael Welker has come the furthest along the neoorthodox trajectory in attempting to heal this revulsion of the flesh: "The unclear notion of a totally numinous 'action of the Spirit,' fully removed from the flesh, has stood in the way of understanding either creation or new creation," he writes. Further, "the creative Spirit holds fleshly life together by giving a share in the medium common to all life. . . . The renewal of creation goes hand in hand with a *renewal of and change in fleshliness*." But again, that "change in fleshliness," in the end, overreaches, erases, represses, when Welker concludes that "through the Spirit and from the Spirit, earthly, frail, and perishable life . . . is removed from death and is removed from under the sway of transitoriness."[34] Despite Spirit's cosmic immanence and its hopes for the cosmos, a ridge of disavowal keeps Spirit and her spiritual brood from nesting on the earth.

Indeed, this way of splitting the labor of sentience appears to be ethically tyrannical: One gender enjoys the ungrounded mind, while the other the labor, now become suffering, of sentience.[35] Women and body, earth and earth-keeping peoples serve as buffers for this ideality of the pneumatic body, as the dumping ground for the abject excess of this pneumatic economy. "Redemption" thus conceived precludes, through the logic of sacrifice, earth's reproductive cycle, the land pyramid.[36] "This dichotomy [between spirit and flesh] has played an ideological role in sexism, racism, and domination by one class," theologian Dorothee Soelle asserts. "It was and still is," she concludes, "a tool of empire and the will to power; in this sense, it is imperialistic."[37]

SPIRIT-FORGETFULNESS

This psychoanalytic reading of Western pneumatology has suggested that a system of loathing splits sentience into the hoped for and the abhorred, into the transfigured and the transient, into the ideal and the impermanent. Positioning its hopes upon a Platonic interpretation of the Pauline notion of the "bondage to decay," Western pneumatology has been preoccupied with prying the pneumatic body or "soul" loose from corporeal limits, from its terrestrial tethers. Freedom, the key marker of Spirit's presence for those of us who have inherited the Pauline trajectory, has been construed as freedom from the constraints of corporeality, as enlightened consciousness in control of its material processes. Beginning from the premise of an idealized eschatology, we habitually refuse to resolve ourselves with finitude, with death, with suffering. We have consequently been invoking the Holy Spirit to get life up and off the ground, to produce a body that forgets its *corpus*, to seek a domain of freedom that forgets its earthly umbilicus.

Philosophical idealism—intimately related to this pneumatological posture, as evidenced by Hegel—holds sentience itself to be a humiliation. This liberative project assumes the nonacceptance of the existential situation of nature. Refusing "the nature of nature," we project against it our worst fears and divest from it our spiritual energy. Articulated through these abjections of "humus," Spirit, in generating soulfulness, occasions the transcendence *of* the sentient matrix, installing a mechanism of revulsion and separation from flesh and feeling and ontological communion. Consciousness of the finitude of life on earth has been deflected into the tyrannical grip of mind over matter, into the lofty trajectory of an imaginary that has purposefully avoided rooting itself in body and earth. And spiritual transcendence comes to preclude "the mutable body"—a status, it should be noted, that common laborers, such as farmers, migrants, machinists, and maintenance workers have shared with women. Further, by divorcing voice from flesh and consciousness from the physical senses and the drives of the somatic region, this pneumatic economy forecloses discourses on pain, suffering, need, and passion. The illusion of pure rationality has isolated pain in the flesh, and flesh consequently develops the mark of misery. An exaggerated emphasis on text has privileged a knowledge-formation of the urban and/or elite and has prevailed against the common sense of those who know the rhythm of the seasons, the signs of the heavens, the "virtue of roots and the character of animals" (Wisd. 7). Such foreclosures also cut us off from the healing capacity of our inherence in nature—how we can take comfort and pleasure in the rhythm of the seasons, in the fresh beginning to a new day, in homing.

In this vein, Berry insightfully reminds Christianity that two desouling paths have always lain open ahead of us—one trajectory suppressing sacred value by "disowning the breath of God that is our fundamental bond with one another and with other creatures," and the other, with identical results, proceeding by way of "misunderstanding and misvaluation of dust." Most of our modern troubles, Berry concludes, come because we have traveled the latter path.[38]

Paradoxically, however, Spirit and the earth degenerate together. Observing our current ecological situation, Berry asks with rhetorical bite, if "spirit is opposed to flesh" (Gal. 5:17), then given "this decline of our worldly flesh and household . . . 'our sinful earth,'" why are we not "healthier in spirit"?[39] If we cannot draw from that paradox a simple corollary between Spirit and the earth, we can make a historical observation—to live freedom as freedom from "the bondage to decay" and its "spirit of slavery" (Rom. 6–8) results not in the raising of our spirits, but in the decline of Spirit. By severing body from Spirit and by withholding our bodies from the earth, by curtailing Spirit's "allowable" affiliations with flesh, Spirit's range of motion has been severely diminished. As Spirit became abstracted from the thick relations of ecological situatedness, Spirit slipped from sight, from practice, from theology. As would appear to be suggested by the centuries of doctrinal silence (400–1980 CE), a systematic theology that levitated above organic systemic accountability could not itself think pneumatologically. Suppression of the corporeal becomes a repression of Spirit . . . becomes, to follow Moltmann and Johnson, "Spirit-forgetfulness."

SPIRIT AS PASSION FOR LIFE

To be sure, this is not all Christianity has meant by Spirit, who is love, justice, mercy, and compassion, who meets us as Shekhinah, Sophia, Christ, and Sabbath Queen. In this final section, I recuperate one of the subjugated histories of Spirit as "humility" as a first movement toward recuperating Spirit as passion for life.

Reading Christianity's sacred texts, literary critic Elaine Scarry noted (with more certainty than is yet historically warranted) that with Christian incarnational theology, the severed relation between pain and power, or ideality, appears to have been subverted, bringing with it an assurance "that sentience and authority reside at a single location and thus cannot be achieved at each other's expense." In other words, one should expect "the conferring of the authority of the spirit on the fact of sentience."[40] This suggests an appropriate horizon of concern for an ecofeminist pneumatology that desires to ground Spirit. By conferring with Hildegard von Bingen, that medieval "Sister of Wisdom," I hope suggestively to surface contours of Spirit now more sentiently disposed—a pneumatology that evokes Spirit not so much as transcendental power over ground conditions, but as passion for life.[41]

SPIRIT GETTING DOWN TO EARTH

Laid low to the earth—as woman, as one frequently ill—Hildegard, twelfth-century abbess and mystical theologian from the Rhine Valley, was cradled in the grace of gravity. There she heard what was unreasonable to those who had already crawled out of their skins by a stretch of their imaginations. Her chin pressed to the earth by confessions of submissive, feminine humility and frequent illness, she began to laugh back with good and holy humor. Humility—why, it was nothing other than the humus force of the Holy Spirit! Divine power was all muddy and moist, all earthy and *under* our noses. She laughed, with tongue-in-cheek irony, that what the "men of airs" in Paris, propounding the Infinite, found so disgraceful, humus or earthiness, was precisely where one encountered the divine Spirit—and if in humus, then in humanity, Humility's children.

Hildegard, the first female, writing theologian of the West, saw the Incarnation as the passion story of Divine Spirit from eternity and daily putting on of the "habit" of the flesh of the cosmos: "In this way God put on human form. God covered the divine Godhead with the Virgin. . . . Hence, Humility, which God formed in its height and width and depth, is also God's abode."[42] Holy Spirit, Abbess Hildegard chuckled, was in the "habit" of dressing up in humus. "Humility" named for Hildegard the in-vestment of Divine Spirit in the becoming of the cosmos, Wisdom's "green cloak."[43] Far from bowing her chin in obedient submission, "Queen Humility," as Hildegard invoked her, preferred a dress of vibrant green, the green of vines in spring. Like a child burying her nose in the perfume of her mother's flesh, Hildegard buries her nose into the earth, where she discovers the sacred showing of the salvation story.

Divine Wisdom recognized that God's love for the world, Hildegard taught, could be discerned only in humility, the heavy, fleshy, earthy habit of corporeal life in which bodies, vulnerably dependent, lean into each other. "And God looked in love, and God had a burning passion for the flesh," she wrote, paraphrasing John 3:16.[44] From before all ages, Holy Wisdom decided to dress down in the tunic of humus, to dwell in the tent of humility, to circulate love in this most vulnerable and fragile texture. Then just as every one of us has had to look to a woman to provide us that thick, weighty "habit" of a body, so did God. If God were ever to be known in the world, if God's passion were ever to arouse the world to life, then God needed woman and earth to "save" or materialize Divine Spirit. God, too, needed the heaviness of flesh and the cradle of gravity.

The divine passion for the mortal life, for the incarnation of Spirit in the flesh of the world; as a passion story, it's still completely confounding. God had a "burning passion" for what we find so difficult to love—mortal life, a muddy and mundane existence. How difficult to comprehend a belief in Divine Spirit, not in disgust of body and earth, but as desirously beholding flesh. Obviously, Spirit does not get mortally embarrassed, though we do. Holy humility proves mortally humiliating to minds that could imagine something much more sublime—say, virtual realities, where the messiness, misery, and the monotonous daily demands of bodies don't erode the paradise of our leisure. But to be sure, humility, let us admit, is a very vulnerable power, a fragile proposition on the part of Divine Spirit, trusting that we should be able to learn to lean into each other in mutuality and respectful interdependence, trusting that we should be able to break the habits of mastery and control and domination, that we might love and not loathe ground and gravity.

In an age in which we are trying to remember our ecological place, Hildegard's evocation of "Humility" can remind us of the humble contours of our earthy habitat and of the ecosystemic interests of the Spirit of life. Humility reminds us of God's inseparability from planetary life, God's presence in the depths of our corporeality—in our "humus," not just our souls. Humus has long afflicted the Western psyche with its seeming ability to resist those who wish to live "upright" lives. Viewed as dead, with no potential for a life of its own apart from the implanted seed or from the breath of Spirit (Gen. 3), humus has been—as the essayist William Bryant Logan puts it—construed as "ejecta, dejecta, rejecta and detritus."[45] Because Western Christianity has closed its imagination to metaphorically relating Spirit with earth, and since the collapse of the

"earth" metaphor of Spirit has been coordinated with planetary decimation and with the displacement of women and children, it is crucial that
we construct a pneumatology that can take into account one of the most
unique, elemental particularities of our life here on earth—humus or
dirt. Humility could thus name Spirit "getting grounded."

Hildegard, in her hymnody, imagistically evoked Wisdom-Spirit etch-
ing with her wings the "dwelling," or habitation, of life.[46] The three pairs
of wings of Wisdom-Spirit create a locality by setting in motion the vortex of love. Her encircling, encompassing wings create the contours of
cosmic circumference and compression, of height and depth and integral
intensity—the "eco," or nest, of life. Instead of fleeing the scene in transcendent abstraction, Spirit, in Hildegard's symbolic universe, is what
localizes. In such a sacred imaginary, the Spirit of Humility—a passional,
relational locality—offers itself as a protective shelter or sanctuary for the
intricate web of life. Rather surprisingly, at least for those of us raised up
to think of Spirit as without weight or dimension, Spirit provides life
with these architectural beams in order to generate a "living room."

While the term "sanctuary" in the Western Christian orientation has
connoted the walled fortress that held nature out in order to insure the
lofting and uplifting of human souls to the celestial heights above the
earth, "sanctuary" can also designate a sacred place of refuge—
specifically, ecologically speaking, a protected reserve for animals, birds,
and plant species. Humility, abrogating this dualistic practice, could name
the worldli-nest of Spirit—the "eco," or "sanctuary," of Spirit—and the
immanent and "immense power [that] holds the inhabitants of the world
in this nest."[47] Nestling, holding, and quickening lives within an ecosystemic nexus of necessarily reciprocal relations, this is where and how we
encounter Spirit. Humility evokes the energy of Spirit pervading and
suffusing life, keeping life all muddy and moist and even *mundane*,
"bound to earth" (*OED*). Spirit is what cohabitates with us so that—to
borrow a notion from the theory of cosmic self-organization proposed
by the scientist Stuart Kaufmann—we can make ourselves "at home in
the universe."[48]

If God's passion is for mortal, corporeal, material, and sentient life,
then Spirit isn't so much interested in extricating souls from the world
milieu as in rooting life in the material world. Brooding upon the relational nexus of planetary life, Spirit quickens and rekindles life through

its humble power, pulsing through life's natural, mortal, and biological concourses. It has been all too common, à la Hegel, to trace the evolution of matter into the flowering of disembodied consciousness. I am rather insisting on the praiseworthy way in which "Spirit comes to matter." Thus, Humility—as Hildegard evoked Spirit and as we may live into Spirit—may recall us to a certain respectful form of "materialism." For, as Joseph Sittler, the first North American theologian to call for an ecologically accountable Christianity, insisted, "Christianity is a material religion. . . . It affirms the reality of nature, denies that nature is evil, or only the realm of the demonic and the destructive, [and] affirms that nature can be the field of Grace."[49]

Grounding the Spirit alerts us to a more subtly creative power, a humble power, working with, and not against, the earth. Welker astutely discerned such:

> Authoritarian theologies of one-upmanship have sought to grasp and expound God and God's revelation in numerous abstract formulas: God always comes "from above," God always "proceeds," God is the "all-determining" reality. The theology of the Holy Spirit will challenge us to replace these formulas. . . . It will teach us to concentrate in a new way on seeing God's reality make its appearance in tension-filled interconnections of different realms of experience that are not necessarily compatible with each other.[50]

The power and wisdom of God, attuned to sentient vulnerability, has the character of communion and pathos and presence—of humility, not force, or might, or riches.

So getting Spirit grounded offers us a more modest, yet more daring pneumatology—a pneumatology that recognizes that life does not and cannot overcome death. Whereas the biotic was the line drawn in the sand to distinguish things divine from things of the earth—finite/infinite, perishable/imperishable, mortal/immortal, transience/eternity—pneumatology can now rethink what distinguishes the contours of Spirit and nature. Rather than viewing Spirit as in dialectical opposition to need, transience, temporality, limits, and dependence, an ecofeminist pneumatology would acknowledge such organic interdependence, biotic vulnerability, and even nature's tragic tears, without employing these as the divide between Spirit and body or between Spirit and nature. It is a pneumatology that pleads for cessation of the war against transience, limits, and corporeality. To be sure, by erasing the idealistic horizon of Christian pneumatology, we must face the fact that there is and will remain tragic and irredeemable loss. Yet far from legitimating power's infatuation with purity, Spirit groans and broods on what has been rejected, redeeming it into livelihood.

HUMILITY AS THE SPIRIT OF LIFE, OF INCARNATE LOVE BECOMING

Given the psychological effects of globalization on human persons, religious communities, and ecological niches, our own hearts (as surely as did the ancients remembered in biblical Wisdom literature) know the temptation to make an existential, rationally calculated judgment to let go of life—in more or less dramatic ways, if even by refusing to be psychically and sentiently present to the world. Since situated despair has often been that which diverted Spirit, our passion for life, into a transcendental updraft, our wisdom theologies must today find ways to address our anguish that do not yield to the abjection that has afflicted Christian pneumatology in the Pauline and Augustinian traditions and that remains a concomitant of ecological devastation, of disregard for the world and for women. It might well be, then, that Spirit as "Humility" could allow Christian theology to navigate between despair and the illusions of the specular spectacular of consumer capitalism.

A life of hope when all seems hopeless does not, after all, require the "cropped photos," the idealized images of nature, offered by this psychically split subject of classic Christian eschatology. World wonder might be set in motion by something so humble as the kiss of peace from a hummingbird, whose play in the arc of a sprinkler brings it unwittingly eye to eye with humanity. As if of such humble means, Spirit in the blink of an eye transforms the heart, breaks through the realism of current structures, breathes us into possibilities—a leveling of egoic stultifications and a flow of passion loosed in the twinkling of an eye.

Christianity as a philosophical deep materialism believes in this world. Where world-wonder cannot be deftly cut away from rupture or illusive enticement, there we find our only opportunity for becoming creatively different—as persons, as a world. There in the churning whirl, the creative vortex of always beginning again, we find ourselves gathered up into the capacious presence of the Spirit of life, of Incarnate Love Becoming still—humbly so, and on the ground.

❧ Hearing the Outcry of Mute Things: Toward a Jewish Creation Theology

LAWRENCE TROSTER

Up until recently, modern Jewish theology has not emphasized creation theology. Arthur Green has written that Jews have largely discarded creation as a theological issue because they became convinced that the origin of species or the origin of the universe is strictly the purview of the scientist rather than the theologian. As a result, any attempt at delineating a theology of creation was of little value. He says:

> But the issue of creation will not disappear so quickly. The search for meaning and the questions of origins do not readily separate from one another. When we ask ourselves what life is all about, why we live and why we die, we cannot help turning to the question of how we got here in the first place.[1]

The search for meaning may draw upon the data of the universe that science discovers but the old ideas of creation still have the power to continue to inspire and shape the way that search proceeds. Green further writes:

> When we try to understand our place in the universe, and especially the relationship of humanity as a whole to the world of nature, we find ourselves returning to the question of creation. As we seek to extend our notion of community and fellow-feeling to include all *creatures*, seeking out the One within the infinite varieties of the

many, we discover that we are still speaking the language of creation.[2]

In developing a Jewish response to the present environmental crisis, in the necessity of grounding a Jewish environmental ethic in theology, creation theology must again come to the fore.

The reason that Jewish theology has ignored creation theology, according to Green, is that the Jewish community in the twentieth century has been occupied more with the issues of revelation, religious authority, and divine providence after the Holocaust. This situation has begun to change with the rise of a religious environmental movement, and concomitantly creation theology has begun to emerge as a major topic in Jewish environmental theology. In attempting to create a new environmental theology, some Jewish thinkers have drawn on Kabbalah, the Jewish mystical tradition.[3] Since Kabbalah teaches the essential unity of all existence, it can, as a spiritual path, engender a greater concern for the environment. Arthur Green observes that "we are the One: each human mind is a microcosm, a miniature replica of the single Mind that conceives and becomes the universe. To know that oneness and recognize it *in all our fellow beings* is what life is all about."[4] Kabbalah, however, used uncritically, can be a problematic source for environmental theology insofar as it does not cherish the natural world for its own sake but rather sees it as a symbol for another, ultimate reality.[5] It is partly for this reason that Green has called for a "post-Kabbalistic" mysticism that can incorporate modern science.[6]

There is, however, another path that can be followed in formulating a new Jewish creation theology. Within the Jewish community, philosopher Hans Jonas (1903–1993) has been one of the most neglected philosophers of the twentieth century. In environmental and scientific circles, however, he is highly regarded. His writings are also esteemed among bioethicists and in discussions of the relationship between religion and science.[7] Europeans have especially shown great interest in his writing on environmental ethics.[8] Jonas is the only Jewish philosopher who has fully integrated philosophy, science, theology, and environmental ethics.

It is the purpose of this paper to utilize the work of Hans Jonas as the basis for suggesting the way in which a new Jewish creation theology can develop. A creation theology in the present time must incorporate the findings of modern cosmology and evolutionary biology as its fundamental facts. The Kabbalistic idea of *tzimtzum*, or the withdrawal of God at the time of creation, in the manner Jonas considers it, can be utilized for a creation theology that allows for God's nonsupernatural involvement with the world. Jonas's theology is also based on sound philosophical foundations, and from these foundations he also produced a model of environmental ethics. Another Kabbalistic idea, the *Sitra Achra* (Aramaic, the Other Side) can also be utilized to express the idea that God can be both the ground of law and novelty; the source of creativity but also the necessary process of creative destruction.[9]

In formulating a new Jewish creation theology, it is also important that it be can be judged as an *authentic* Jewish theology. Neil Gillman has said that "true" theology arises from those "who share a sense of tradition that has become problematic and yet holds out the promise of renewed meaning."[10] Gillman has also emphasized that the traditional liturgy is "the locus for the authoritative system of Jewish belief."[11] In the morning service the *Sh'ma* prayer is surrounded by blessings in which God is praised as Creator (both initial and ongoing), Revealer, and Redeemer.[12] These are the theological concepts through which providence or divine action is expressed. Any Jewish creation theology must therefore be seated within this larger structure. At the end of this essay I will suggest how a new creation theology connects with some modern theologies of revelation and redemption.

Any authentic Jewish environmental theology must also come to grips with and incorporate the biblical concept that humans are created in God's image. Jewish ethics is founded on this idea and it cannot be easily dismissed. While some might think that this would bring Jewish environmental theology into conflict with biocentric environmental ethics and philosophy, it need not be so, and, as will be seen, Jonas's environmental ethics is a way to incorporate both perspectives.[13]

Hans Jonas was born in 1903 in Germany[14] and left in 1934 as a result of the rise of Nazism.[15] When war broke out in 1939, Jonas joined the Jewish Brigade of the British 8th Army and was in the 1943 Italian campaign, in which he saw a great deal of action. In 1945 Jonas learned that his mother had been killed in Auschwitz in 1942. After fighting in the

Israeli War of Independence he eventually came to North America, accepting a fellowship at McGill University in Montreal, and in 1955 he was appointed to the graduate faculty of the New School for Social Research in New York City. There he remained until his death in 1993.

Jonas's own philosophy begins with his critique of modern philosophy generally,[16] which he claims lacks an ethical theory because it has "become a camp follower of Science" and has created new concepts of "nature" and "humanity." These new concepts have in turn created and supported modern technology. "All three [nature, humanity, and technology] imply the negations of fundamental tenets of the philosophical as well as religious tradition"[17] (by "religious tradition" he here meant the biblical-Jewish tradition). This is the nihilism that he feels is at the heart of modern thought.

In this modern concept "nature," the universe is no longer creation but rather a process determined by law, which is inherent in it. There is no longer a divine order. Nature is also no longer "good" in the biblical sense but indifferent to the distinction between "good" and "bad" and the world is thus purposeless, lacking values, goals, and ends.

Therefore, humanity is no longer created in the image of God. As a result of Darwinism, it is now believed that humanity arose as the result of unconcerned forces. "He is an accident, sanctified merely by success."[18] This modern concept of humanity is also influenced by historicism, which asserts that all human values are the product of each culture. Therefore, in Jonas's view, there are no universal or absolute truths, only relative and socially particular ethics. Another influence on the modern view of humanity is psychology, which claims that all the "higher" aspects of the human character are really gratifications of its base drives. "The higher in man is a disguised form of the lower," he writes.[19]

What is left of the human character is a strange paradox. Humanity creates its own values in the use of its power, which exalts it above any other creature or concern for the natural world while at the same time reducing humanity to the same level; humanity thus itself becomes an object of its own use and abuse, as power is now the only real value. There is no transcendent authority to check or control human power, which is implemented through modern technology. Nature becomes a mere object and humanity is the sole subject and sole will. "The world then, after first having become the object of man's knowledge, becomes the object of his will, and his knowledge is put at the service of his will; and the will is, of course, a will for power over things."[20] That will, once the increased power has overtaken necessity, becomes sheer desire, of which there is no limit. Jonas observes, "Some ineffable quality has gone out of the shape of things when manipulation invades the very sphere which has always stood as a paradigm for what man cannot interfere with."[21]

From this critique of modern nihilism Jonas developed an environmental ethic. This ethic is a response to the fear of the destruction of humanity and the need to create a philosophical basis for human responsibility to save themselves and the planet. His response to the environmental crisis is most fully elucidated in his book, *The Imperative of Responsibility.*[22] In this work, Jonas argued that the environmental crisis emerged from the human impact on the natural world, which is greater and more far-reaching than in any previous age. This unique and novel power comes from modern technology, which is also radically different from the technology of previous ages. Previous ethical systems are no longer adequate to deal with the moral issues now raised. "Modern technology has introduced actions of such novel scale . . . and consequences that the framework of former ethics can no longer contain them."[23]

For Jonas the lengthened reach of our deeds moves the principle of responsibility into the center of our ethical stage. His theory of responsibility, which he saw as the correlate of power, must therefore be proportionate with the range of modern power. We must also have greater foresight into the possible impact of new technology; what Jonas called "scientific futurology." Jonas also called for a "heuristics of fear" that will tell us what is at stake and what we must avoid.[24]

What is at stake for Jonas is the survival not only of other forms of life but also the very survival of humanity. In previous ages human action might lead to the elimination of a tribe or a nation, but now all of humanity is at risk. And if humanity were to be destroyed, it would not be the extinction of just another species in the evolutionary history of the universe but a cosmic disaster. Jonas believed that humanity matters to Being itself as "the maximal actualization of its potentiality for purposiveness."[25] The survival of humanity and the world is metaphysically significant and therefore a moral imperative.[26] Any technology that can put humanity at risk is immoral.

Jonas believes that in order to have an ethics adequate to save humanity it must be based on a doctrine of general being: *metaphysics must underpin ethics*. A philosophy of nature is required that will bridge the gap between the scientific "is" and morally binding "ought." It will then be possible to understand what are the legitimate objectives for human power: "the more modest and fitting goal is . . . to save the survival and humanity of man from the excesses of his own power."[27]

Jonas's metaphysical basis to his imperative of responsibility was to extend Heidegger's concept of existence to include all living organisms. Jonas did this by showing that every living creature, from the smallest microbe to humankind, shows concern for its own being by connecting to the world around it in order to stave off death and nonbeing. Thus, even in the simplest organism, there is a kind of "inward relation to their own being." Jonas saw metabolism, the exchange of matter with the environment, which all organisms must exhibit in order to survive, as the most basic expression of that organism's struggle for life. This is what Jonas called its "needful freedom." Each organism is "free," in the sense that it consistently exhibits a dynamic unity beyond the sum of its parts, and yet it is dependent on constant exchanges with its environment in order to avoid dying.[28] With each new level of complexity in evolution there is an increase of mind, which brings a new level of freedom, as well as an increased potential for pain and suffering. For example, all life requires nutrition and displays reproduction. Animal life also has the capacities for movement and desire and sensitivity to its environment that plants do not. These increased capabilities also create in animals the ability to feel pain, fear, and abandonment. With the arrival of humans in evolution, being now becomes reflective and begins to try and understand its place in the universe. This is the source of the human anxieties of existence, unhappiness, guilt, and despair.[29]

Jonas wants us to understand that by interpreting the facts of biology existentially and seeing the inherent value in the natural world beyond ourselves, we then must ethically accept our role as the guardians and stewards of the natural world. If the world is not a mindless machine of which we are but cogs, then we must respond ethically to protect the whole of life and not only ourselves. The ethics of previous ages, which were anthropocentric, are no longer good enough. The whole of the biosphere has become "a human trust" and therefore has a moral claim on us, not only for our utilitarian needs but also for its own inherent value. It is necessary to thus expand our ethical circle beyond the human good to the "good of things extra-human" and include in the human good the imperative to care for them. "No previous ethics (outside of religion) has prepared us for such a role of stewardship—and the dominant, scientific view of *Nature* has prepared us even less."[30]

For Jonas, living nature is a good-in-itself, which therefore commands our concern and even our reverence. Since all organisms are vulnerable ends-in-themselves and they all express concern for their own existence, humans have a particular responsibility as moral agents to protect them. Our difference from the rest of nature is only in kind. Humans have continuity with the rest of life in that all of the biotic community shares life's goodness, regardless of whether other forms of life serve human needs. "So our self-respect requires 'cosmic piety,' reverence for the whole of which we are part."[31] As Lawrence Vogel has noted, Jonas is not in thrall to either biotic egalitarianism or radical ecocentric holists because he believed that the human arrival in evolution "marks the transition from vital goodness to moral rightness: from desire to responsibility."[32] Jonas's focus is on the reality of human power and both the danger and the responsibility that results from such power.

Lawrence Vogel, in his introduction to Jonas's *Mortality and Morality*, discusses the latter's elucidation of the four tenets of the biblical-Jewish tradition that he feels modern nihilism has denied: that God created the universe, that the universe is good, that humanity is created in the divine image, and that God makes known to humanity what is good. Vogel contends that Jonas preserved the last three in his naturalistic metaphysics without the necessity of theology.[33] In other words, it is possible to maintain the idea of the universe as being "good," humanity as having a special place in the universe, and humanity as being inspired to create a moral order by the universe without any belief in God or that God created the universe—it is possible to answer modern nihilism without theology. It is possible to believe that there is meaning in the universe without God.

Although Jonas holds that theology is not necessary to answer modern nihilism, he observes that the first tenet—that God created the universe (which does require a theological response)—nonetheless responds to certain basic human spiritual needs.[34] We want to know that there is a loving God who created the universe and still sustains it. And we hope that goodness is not lost or forgotten in this universe, that in some way we are inscribed in what the Jewish tradition calls the "Book of Life." Only through theology can these longings be addressed. Jonas thus has sought to create a rational theology in conformity with his existentialism, his metaphysics, and with science as he conceives it. In the process of creating that theology Jonas has also to confront the fact of Auschwitz.

In 1987 Jonas published "The Concept of God after Auschwitz: A Jewish Voice," which was a revision of his earlier essay on immortality.[35] He considered these articles to be a "piece of speculative theology," by which he meant that although we cannot prove the existence of God, working on a concept of God is a form of understanding even if not a form of knowledge.[36] Jonas asserts that when we begin to consider the concept of "God," we are immediately confronted with Auschwitz. He holds that nothing in previous ideas of theodicy is of any use in dealing with the Holocaust. At Auschwitz, "Not fidelity or infidelity, belief or unbelief, not guilt or punishment, not trial, witness and messianic hope, nay, not even strength or weakness, heroism or cowardice, defiance or submission had a place there."[37] The victims of the Holocaust did not die because of their faith as traditional martyrs did; yet because of their membership in the "ancient people of the covenant," they were singled out for murder.

For a Jew, Auschwitz calls into question the central belief in a God of history who makes the world a "locus of divine creation, justice and redemption." Thus, Jonas says that unless one is willing to give up the idea of God, the "Lord of history . . . will have to go by the board."[38] Jonas then repeats a "myth" that he created for his article on immortality. In this "myth," "the ground of being, or the divine, chose to give itself over to the chance and risk and endless variety of becoming. And wholly so: entering into the adventure of space and time, the deity held nothing back of itself: no uncommitted or unimpaired part removed to direct, correct, and ultimately guarantee the devious working-out of its destiny in creation."[39]

Creation, for Jonas, begins with God's self-limitation. Jonas directly connects his myth and his idea of divinity with the Kabbalistic doctrine of *tzimtzum*, the contraction of the divine for the purpose of creation. In *tzimtzum*, the contraction is followed by the insertion of divine creative energy into the space provided—it is the beginning of the process of creation. For Jonas, however, the contraction is total: "The Infinite ceded his power to the finite and thereby wholly delivered his cause into its hands."[40] What is then left of the relation of creation to its Creator? Jonas: "Having given himself whole to the becoming world, God has no more to give: it is man's now to give to him."[41] In other words, the divine direction of creation has ended. The universe proceeds in a completely naturalistic way on its own without God's intervention.

Jonas explicitly elaborates three critical characteristics of the God of this myth. First of all, while God is still a Creator, God is a *suffering* God. Here the divine suffering is the pain that God feels alongside of the pain of God's creations and the disappointments with humanity that God experiences. As William Kaufman writes in his discussion of Jonas's concept of God, "creation is tragic."[42]

Secondly, God is a *becoming* God. "It is a God emerging in time instead of possessing a completed being that remains identical with itself throughout eternity."[43] The becoming God is affected and altered by the events occurring in the universe. If God has any relation to creation, then God must in some way become different through the process of the emerging universe.

Lastly, God is a *caring* God. "Whatever the 'primordial' condition of the Godhead, he ceased to be self-contained once he let himself in for the existence of a world by creating such a world or letting it come to be." This does not mean that God intervenes in history in the traditional sense. After Auschwitz, that must "go by the board." Instead, God has left humans with the responsibility of acting for the sake of the universe. Therefore, God has "made his care dependent on them." This implies that God is also an "endangered God, a God who runs a risk."[44]

Taken together, these characteristics imply an endangered God: "For the sake of our image of God and our whole relation to the divine, for the sake of any viable theology, we cannot uphold the time-honored (medieval) doctrine of absolute, unlimited divine power."[45] Jonas advances two arguments for the rejection of divine omnipotence. First of all, from a purely logical perspective, omnipotence is self-contradictory since absolute power implies that there can be no other object on which to exercise its power. The existence of such an other itself implies limitation. An omnipotent God, therefore, could only exist in solitude with nothing on which to act.[46]

Second, and most importantly for Jonas, there is a theological argument against divine omnipotence: absolute power can only exist in God at the expense of absolute goodness unless God is completely inscrutable. "Only a completely unintelligible God can be said to be absolutely good and absolutely powerful, yet tolerate the world as it is."[47] For Jonas a God who is good and intelligible cannot be omnipotent.

In a later essay, "Is Faith Still Possible?: Memories of Rudolf Bultmann and Reflections on the Philosophical Aspects of His Work," Jonas stepped back somewhat from the idea that God has absolutely no connection to the universe.[48] He suggests that God continues to act in relation to the universe, not through direct supernatural actions that contradict the laws of the natural world, but rather through the inspiration of certain individuals. Just as our freedom to act in the world is scientifically compatible with causality, it is possible to accept a kind of divine causality that comes into our inner self that does not conflict with the human free will.[49] It also connects Jonas's conception of creation to a concept of "revelation," though not one in which there is propositional content—rather, he postulates "revelation-as-encounter." This coincides with the views of Elliot Dorff and Norbert Samuelson, who argue that most modern Jewish theologies of revelation redefine it as an encounter with God's own self rather than specific laws and beliefs.[50]

It is possible to go even further and say our encounter with God primarily is through creation. This primary encounter results eventually in a written record of that encounter, which then in turn becomes a sacred canonized text. In the traditional Jewish morning liturgy, God is first blessed as Creator and only then as the giver of Torah. It is possible to say that it is impossible to have Torah without creation. In fact Moses Maimonides (1135–1204) saw the study of the natural world as the necessary basis for the love and awe of God.[51] Thus God can continue to influence the world through inspiration and the encounter with the divine presence in evolution does not violate the laws of nature. Evolutionary theology enriches the concept of revelation as the universe itself provides us with another source of divine guidance. By understanding the laws of nature, something of the nature of God is revealed. The imitation of God's attributes as Creator can help to lead us to a new kind of creation spirituality or environmental ethics.[52]

Jonas's theology of God is further developed in a later essay, "Matter, Mind, and Creation: Cosmological Evidence and Cosmogonic Specula-tion."[53] Here Jonas begins with the material content of the universe as described by science and how life and subjectivity emerged from matter, which then became mind. He then raises the "creative ground" of these stages, which we call "God," and whether these facts of the universe can suggest the existence of such a God. In other words, from the evidence of the universe, the material content is deduced that reveals the existence of subjectivity. This then leads to mind, which then comprehends the existence of God.[54]

Jonas uses the evidence of cosmology (that is, what the universe looks like now), moving from its outer characteristics to the inner, which means from the earlier to the later and from the most frequent to the rarest. The central question for Jonas is how can the development of the universe from its beginning to the emergence of mind be explained. What is known of the development of the universe in this direction was physically improbable in that it moved in an antientropic direction from disorder to order. Does this improbable development prove that there was a program or a design inherent in matter from the time of the Big Bang that guided its development to a higher order? If there was, then, according to Jonas, this design would be a "cosmogonic logos."[55] Jonas rejects the idea of such a logos on the ground that, if there were informa-tion in the primordial matter of the universe, it would have been the result of some kind of organization. But there was no such organization in the Big Bang, only a primordial chaos out of which order emerged. The existence of such a logos would also be too deterministic.

In Jonas's view, the laws of nature emerge in a kind of cosmic natural selection because only a self-conserving reality would survive. Order is more successful than disorder and regularity replaces arbitrary multiplic-ity. Therefore protons become the law of gravity and mechanics, hydro-gen becomes the elements, and the four basic physical forces emerge from the original "chaos." But even the durability of the natural laws is relative and not absolute. And the question remains: why did the uni-verse continue to develop order after the basic laws emerged? The Dar-winian answer would be that there is always enough disorder left over to allow for new characteristics by random events, which would be then subject to natural selection. Therefore this is the "transcending factor" that leads to higher levels of organization without a logos. The most that can be assumed is that there is a "susceptibility" to order that codes the surrounding disorder for "information."

While this would explain the rise of order from chaos, it would still not explain how inwardness-subjectivity arise from this process. Jonas points out that subjectivity, or inwardness, is ontologically a fact of being. The arrival of life can be explained from nonliving matter but not its subjectivity—life could be only a kind of chemical reaction. Yet subjectivity in life does exist and it interacts with matter and is inseparable from it.[56]

Jonas rejects both the dualistic and monistic explanation for subjectivity and mind. Dualism is logically rejected because an independent thinking substance that cannot prove its own existence is stupid, and phenomenologically it is rejected because the experiences of the body, such as pleasure, pain, perception, desire, and feeling, are part of thought. Therefore there cannot be "pure" thought. This also rules out the existence of an individual immortal soul. Jonas also rejected monism or materialism because it turns the soul and mind into instances of "epiphenomenalism," which is self-contradicting.

A different kind of monism is thus needed. Matter finds its voice in subjectivity and so it is necessary to revise what matter, beyond the external qualities that are measured and abstracted by physics, means. Therefore there is a need for a "metaphysics" of the material substance of the world, from which it follows that the matter that emerged in the universe must have had the possibility of inwardness, which did not yet exist.[57] The only thing that can be said is that inwardness was "possible" in characteristics of matter as originally formed. But even so, matter must be more than what physics describes it to be.

Jonas's line of thought gives out in the following questions: "Who (or what) 'endowed' matter in this way? What share did this 'endowment' have in the course of cosmic events?"[58] Given Jonas's theology of God and rejection of a logos, he had to believe that this arising was "an entirely neutral contingency whose occurrence involved no favoring preferment of any kind."[59] This arising of subjectivity must be inherent in matter itself and therefore there is no "plan" in the universe for the arrival of mind but rather there must be a tendency, a yearning—or what Jonas called a "cosmogonic eros"—instead of a logos. Even with this "eros," chance plays a large role. Even the existence of an earth where there are favorable conditions for life arising came about only by chance circumstances in the universe. When such favorable conditions have arisen, then the "readiness is there and subjectivity receives its opportunity." Jonas asserts that matter is "subjectivity in its latent form, even if aeons, plus exceptional luck, are required for the actualizing of this potential." Even if this "yearning," or "eros," could be causally active, we cannot know that it is so and therefore cannot use it in science.

When mind does arise, thought adds a transcendental dimension that is expressed in three freedoms of thinking that go beyond matter: the freedom of thinking for determining itself through choice of object; the freedom to transform the sensations into self-created inner images (imagination's freedom to invent); and the freedom carried by language to transcend the images of the imagination and thus allow the mind to gain the ability to grasp the idea of the infinite, the eternal, and the absolute. Only through language does mind still remain bound to the world of the senses. For Jonas, this ability of the human mind is what makes us different from the animals.[60] These characteristics of the human mind allow humans free will of conduct and moral choice: it moves humans from the "is" to the "ought." Moral freedom is both the most transcendent and the most precarious of human freedoms because it can lead to evil: "The knowledge *of* good and evil, the power of discriminating between them, is also the capacity *for* good and evil."[61]

Jonas believed that all of this was *cosmological evidence*, the characteristics of the universe as it is now seen. Is it possible from the cosmological evidence to deduce anything about the beginning of the universe (cosmogony)? Since subjectivity emerged from matter, then it cannot have been foreign to matter from its very beginning. There must have been some kind of possibility of subjectivity. Jonas admitted that this speculation is a matter of faith but he considered it to be a rational faith given the evidence. For Jonas, matter "from the very beginning is mind asleep, so we must immediately add that the really first cause, the creative cause, of mind asleep can only be mind awake." Here Jonas referred back to his earlier essay about God after Auschwitz and reasserted that such a God could only be one that "abandoned Himself and his destiny entirely to the outwardly exploding universe and thus to the pure chances of the *possibilities* contained in it under the conditions of space and time."

Why God did this is unknowable but once it occurred, it did so only immanently with no outside intervention. One of Jonas's rational proofs for such a process is in the size of the universe and the apparent rarity of life and mind. "Only a universe colossal in space and time, in accordance with the rule of mere possibilities and with no intervention of divine power, offered any chance at all for mind's coming to pass at any time or place whatsoever." And it is this rarity of life and mind that imposes upon us the duty to preserve creation: "[W]e must protect from ourselves the divine cause in the world that has become threatened by us, that we must come to the aid of the deity who has become powerless for Himself regarding us. It is the duty of power that knows, a cosmic duty, for it is a cosmic experiment, which we can wreck along with ourselves and spoil within ourselves."

As witness to God's need for our help, Jonas then quotes from the diaries of Etty Hillesum, a young woman from the Netherlands who was killed at Auschwitz (her diaries were only published forty years after her death). In them she wrote: ". . . and if God does not continue to help me, then I must help God. . . . I will help you O God . . . you cannot help us but that we must help you, and in doing we ultimately help ourselves."[62] Jonas was inspired by her words and he considered them a witness to the truth of his theology: that it is not God who can help human beings—human beings must help God.

While Jonas emphasizes the idea of divine risk and divine powerlessness, he does not fully deal with the destructive forces that are also part of the process of creation. Creation requires creative destruction: the carbon atoms that are required for the emergence of life come from the death of stars earlier in the history of the universe; 98 percent of all species that have ever existed have gone extinct in the process of evolution. The Kabbalistic concept of the *Sitra Achra*, the Other Side (of God), can be utilized to understand how this destruction can be part of an ongoing creation that is still the product of a good God. The Kabbalists believed that since all phenomena must reside in, and be a part of, God, even evil must in some way come from God. Unlike Maimonides who limited evil to human interaction, the Kabbalists saw that the destructive forces of the natural world must be a necessary part of the ongoing divine creative process. Human action can either mitigate or exacerbate this potential for destruction, which begins as a kind of creative destruction within that aspect of the divine limiting the flow of creative power. This limitation is expressed in the term "Judgment." With human evil, judgment grows and creates a kind of demonic realm within the divine. This is the *Sitra Achra*. What begins as creative destruction can become evil destruction. While we must reject the literal idea of a demonic realm within God, nonetheless humans can either use the forces of creative destruction paradoxically in a constructive way or can take those same forces, like fire, for example, and use them to destroy the world. Chaos is the ground of order but chaos unrestrained is evil.[63]

Jonas's theology also does not contain an idea of final cosmic or human redemption. He believed that the most that we can say is that God feels all that happens in the universe, is affected by all that occurs, and preserves all "perishing occasions everlastingly."[64] But while there is no loss, redemption is only within God. Jonas specifically rejected Teilhard de Chardin's idea that the universe is growing inevitably toward an Omega point.[65] For Jonas, the future of the universe is still a risk for God and the outcome uncertain.

This lack of a concept of redemption in Jonas does not necessarily negate the rest of his thought as a basis for a Jewish creation theology. But a Jewish theology of creation requires a corresponding concept of redemption: "[T]he universe originates in the nothing of creation as a process towards the something of redemption."[66] Jonas's creation theology can be supplemented by a theology of redemption similar to that of John Haught. In *God After Darwin*, Haught asked how a traditional concept of redemption could be maintained in an evolving universe, with constant death being part of the process. Haught uses process theology, which answers this problem by asserting that God is infinitely responsive to the world, "feels," and is influenced by all that happens during the process of evolution. Being taken eternally into God's own feeling of the world preserves everything that happens through evolution. Each event is redeemed from death and thus has ultimate meaning. As humans, we cannot see this clearly because we live in an unfinished universe.[67]

Hans Jonas's philosophy, ethics, and theology are a rich source for developing a new Jewish creation theology. From Jonas we can learn how to theologically use the data of how the universe evolved without resorting to a crude form of design theology. We can construct an idea of God who works through natural law. We can construct a creation theology that is neonaturalist: accepting the real value of the existence of life and mind in the universe without resort to supernaturalism. The creation theology that will emerge from such an encounter with science will be one that is deep and rich with meaning. It will connect us to the natural world and all of life. It will engender in us the grave responsibility for taking care of this incredible expression of divine creativity that has emerged after so many billions of years of risk, creative destruction, and beauty. Such a theology will hopefully lead us to action on behalf of creation to save it from the destructive forces that we have set loose. Jewish environmentalists often quote this midrash: "When God created the first human beings, God led them around the Garden of Eden and said: 'Look at my works! See how beautiful they are—how excellent! For your sake I create them all. See to it that you do not spoil and destroy My world; for if you do, there will be no one else to repair it.'"[68]

Tikkun 'olam, the perfecting or the repairing of the world, has become a major theme in modern Jewish social justice theology. It is usually expressed as an activity, which must be done by humans in partnership with God. In our ignorance, our greed, and our egotism, we have silenced many of the voices of creation. It is these voices that now cry out to us in the latest revelation found not in the text but in the earth itself.

The last public words of Hans Jonas have become a prophetic call of this revelation:

> It was once religion which told us that we are all sinners, because of original sin. It is now the ecology of our planet which pronounces us all to be sinners because of the excessive exploits of human inventiveness. It was once religion which threatened us with a last judgement at the end of days. It is now our tortured planet which predicts the arrival of such a day without any heavenly intervention. The latest revelation—from no Mount Sinai, from no Mount of the Sermon, from no Bo [tree of Buddha]—is the outcry of mute things themselves that we must heed by curbing our powers over creation, lest we perish together on a wasteland of what was creation.[69]

When, then, will we hear the outcry of mute things?

❧ Creatio ex Nihilo, Terra Nullius, and the Erasure of Presence

WHITNEY A. BAUMAN

The total structure of man's being as a creature made out of nothing roots his life beyond himself in the transcendent source of his existence, in God his Creator and preserver.
—LANGDON GILKEY, *Maker of Heaven and Earth: A Study of the Christian Doctrine of Creation*

Religion is a here-and-now conquering of nihilism and a re-creation of our world out of nothing by continually generating new metaphors and new interpretations.
—DON CUPITT, *Creation Out of Nothing*

The Indian tribes in the new world were regarded as mere temporary occupants of the soil, and the absolute rights of property and dominion were held to belong to the European nation by which any particular portion of the country was first discovered . . . as if it had been found without inhabitants.
—U.S. SUPREME COURT, *Martin v. Waddell* (1842), in Ward Churchill, *Struggle for the Land: Native North American Resistance to Genocide*

Justified by a transcendent and omnipotent Creator *ex nihilo*, imperial Christianity has been re-creating the world—as if *ex nihilo*—for the past 1500 years. The theology of an all-powerful Creator and Preserver has arguably served as the justification for a theoanthropology in which humans mimic the power of the Creator God through what Don Cupitt calls the "conquering of *nihil*" and the "re-creation of" the world.

Through this re-creation, an erasure of agency and identity takes place—as if the many spaces recreated by colonial powers had been, indeed, "found without inhabitants."[1] In this chapter, I argue that *creatio ex nihilo* can and did provide a justification for the colonial concept of individual property articulated by John Locke, along with the corollary colonial, national legal claim of *terra nullius*, or *Territorium Res Nullius*. The latter terminology suggests that there is "no prior presence" in conquered or "discovered" territories and they therefore can be owned and "made useful" through colonization. Is there an epistemological and ethical source—flowing from the erasure, through the doctrine of *ex nihilo*, of the pre-existent chaos in Genesis and that book's context within and alongside other Ancient Near Eastern creation stories—for the erasure of nature's agency in the Lockean concept of property, and the erasure of human and nonhuman agency in the concept of *terra nullius*?

FOUNDATIONS FOR A CHRISTIAN LOGIC OF DOMINATION

In *Face of the Deep: A Theology of Becoming*, Catherine Keller argues that theologies of *creatio ex nihilo* might support precisely such a colonial epistemology and colonial anthropology as I seek to excavate here.[2] Keller's work brings process metaphysics and postcolonial theory into dialogue. From process thought, she brings a critique of the classical understanding of God's aseity and corresponding omnipotence, which Charles Hartshorne referred to as "the taint of divine tyranny which disfigured classical theology."[3] Included in this critique of "divine tyranny," or the classical understanding of God's omnipotence, is a critique of *ex nihilo*. Process theologians have critiqued the notion of God's power as *power-over* and offered in place a concept of *power-with*, or power-as-lure. This critique of power flowing from the metaphysic of process has included a serious challenge to *creatio ex nihilo*.[4] From postcolonial thought, Keller brings an understanding of the power of discourse, the textual nature of all human understandings of reality, and an appreciation for the edges or interstices of chaos and order, of self and other. She brings this interstitial postcolonial and process perspective together in her understanding of the God-world relationship.[5]

Keller argues that the interpretation of creation as *ex nihilo* arises out of *tehomophobia*—her term for the fear of the deep, of the chaos called *tehom* in Genesis, the chaos that exists already when God begins to create.

This fear or anxiety of otherness, wildness, chaos, and difference is what doctrines of ultimate origins and final ends seek to alleviate.[6] In doing so, however, these ultimate origins and ends only "background" the contextual, fragmented nature of all narratives and their knowledge claims, and thus succeed in transforming this fear of difference into a universal justification for claiming ultimate truth.[7] This, in turn, leads to violence toward the other. "The abiding western dominology can with religious sanction identify anything dark, profound, or fluid with a revolting chaos, an evil to be mastered, a nothing to be ignored."[8] The denial of the eco-socio-contextual nature of any theological or epistemological claim leads to violence toward otherness and difference (coded as chaotic) in the form of "civilizing," "cultivating," "educating," "taming," or "exterminating" those differences. Thus "when religion pretends to 'systematized exhibition', it removes us both from the streets and from the deep."[9] Pretending to transcend space and time, objective, foundational knowledge claims ignore the ecological, social, and historical context of a situation or place and project the claimant's own system of thought onto it. By projecting an ultimate origin, the doctrine of *ex nihilo* serves to sever Christian beginnings from (other) Ancient Near Eastern traditions, as if "other" histories could be encompassed by the Christian One, and to project the One Christian history onto all others, thereby incorporating, ignoring, or erasing other peoples' histories. In this way, *ex nihilo* functions to deny the presence of others' truth-claims on reality in a similar way that *terra nullius* functions to deny the presence of colonized others some 1,400 years later.

Terra nullius, literally "empty land," is an eighteenth-century legal concept that justified the takeover of other peoples' lands by European colonizers. As a legal concept, *terra nullius* was used specifically to take over the lands of modern-day Australia from the Aboriginals and was only overturned as a legal justification for land-seizure in that country in 1990. It has a history, however, that is intertwined with the whole period of European colonization.[10] Val Plumwood, in *Environmental Culture: The Ecological Crisis of Reason*, argues that the claim of *terra nullius* has also meant that "barbaric" peoples need education and wild lands are free for cultivation and ownership.[11] Similar to Keller's argument that *ex nihilo* backgrounds the prior presence and legitimacy of other narratives and knowledge claims, Plumwood argues that *terra nullius* backgrounds the prior presence of relationships between peoples and land, and it denies the presence of any kind of agency to the colonized peoples and the rest of the natural world.

My argument, building upon Keller and Plumwood, is that *ex nihilo* functions epistemologically like *terra nullius*, as a justification for denying the agency and value of the "other"—both human and nonhuman—and that this leads to a process of assimilation, denial, and/or extermination of the other. It creates what Plumwood refers to as a "hegemonic centrism": that is, "a primary-secondary pattern of attribution that sets up one term (the One) as primary or as centre and defines marginal Others as secondary or derivative in relation to it."[12] Similar to the *cogito* of Enlightenment thinking, which may actually have foundational support in *ex nihilo* thinking, *ex nihilo* functions in such a way that "only what can be encompassed by unity has the status of an existent or an event; its ideal is the system from which everything and anything follows."[13] In other words, it sets up a monological epistemology.[14] This monological epistemology is in part based in the desire to realize the ideal unification of will and act found in the God who creates *ex nihilo*. This theology reduces the anxiety of living in a finite, relational world through postulating an all-powerful God who saves us from the anxieties of finitude and relation. This God, whose will and act are united, then becomes a foundation for a colonizing ethical and political approach to the world. This colonizing approach translates into an ethics and politics of civilizing other peoples and an anthropocentric view of the rest of the natural world. The world is only recognized in terms of what it means to the center, or the colonizing mind. That which cannot be assimilated, in the "best-case" scenario, is not acknowledged as real or important. In Plumwood's terms, this means a "backgrounding" of reality that does not fit into one's conception of it. In the worst-case scenario, that which cannot be assimilated is annihilated. In this way, the human-colonizer mimics the omnipotent God by defeating the chaos through assimilation, backgrounding, or annihilation.[15]

In order to illustrate how this theological logic of *ex nihilo* comes to support, mimic, and get reinforced by a national *terra nullius* approach to the "other," I first analyze how *ex nihilo* serves as a mythico-epistemological foundation for the colonizing understanding of individual property found in John Locke's work, which then becomes the foundation for the national, legal justification of *terra nullius*. In doing so, I am not suggesting linear causality between *ex nihilo* (a third-century invention), Lockean property (a seventeenth-century invention), and *terra nullius* (an eighteenth-century invention), but merely pointing out the epistemological support system these concepts provide for one another: from an origin story that posits a creator *ex nihilo*, to the way in which Lockean property mimics the logic of this One True Creator God *ex nihilo* by extending this God's powers to all individual humans equally (read, all "civilized" humans—Christians), and finally to the way in which this logic of domination is used at the national level through claims of *terra nullius*. I will conclude by offering some preliminary thoughts about Christian understandings of creation that challenge the colonial reading of *ex nihilo* and thereby weaken the arguments for a Christian justification of Lockean concepts of individual property and a national, *terra nullius* approach to the "other."

I should note here that I could have written also about how the "scientific revolution," the European logic of discovery during the colonial era, the concept of the Enlightenment, and the Cartesian *cogito* all function through an *ex nihilo* logic, but this will be the task of the longer version of this chapter (my PhD dissertation).[16] Finally, I will not focus on the rhetoric surrounding the formulation of *ex nihilo* in the second and third centuries c.e. by Irenaeus and Tertullian; there are many sources one could turn to for this discussion.[17] I will assume some familiarity with the formation of this nonbiblical doctrine, and jump right into its use by one of the founders of modern liberalism, John Locke.

LOCKEAN PROPERTY, OMNIPOTENT INDIVIDUALS, AND THE ERASURE OF NATURE'S AGENCY

John Locke was very much an exegete and theologian. Carolyn Merchant recognizes the importance of the Bible for Locke's work in *Reinventing Eden*. She notes that the justification for his *Second Treatise* on Government is found in the less frequently read *First Treatise*, which was written almost solely as a biblically based refutation of Robert Filmer's *Patriarcha*.[18] Filmer's *Patriarcha*, written around the time of the Glorious Revolution, is a treatise on government that uses scripture to justify monarchy as the divine form of government. Whereas Filmer argues that a monarchical form of government is suggested by the fact that God gives dominion to Adam over all of creation, Locke argues that dominion is given to all "mankind," including women.[19]

As Jeremy Waldron notes, "Locke wanted to attack the Filmerian view that certain particular men had the right to rule over the rest."[20] One important note here—lest I am accused of demonizing Locke—is that Locke *does* extend dominion to all of humanity, male and female.[21] Locke identifies the *imago*, and its privilege of dominion, with the intellect. He wrote: "For wherein soever else the *Image of God* consisted, the intellectual Nature was certainly a part of it, and belong'd to the whole Species, and enabled them to have *Dominion* over the inferiour creatures."[22] Though extending political dominion to all humans, male and female, his identification of the *imago* with intellect locates the *imago*, and thus dominion, with a Western intellect–reason-based anthropology and therefore aids in the civilization of non-Western "others." Evidence of this is provided by the fact that Locke struggled with who was included in the human species. Again, Waldron notes, "The species-difficulty arises [for Locke] because even if God *has* announced that all humans are created equal and commanded us to treat them as such, we still face the problem of defining the class of beings, the species-members, who are supposed to get the benefit of that commandment."[23]

Locke's extension of dominion to males and females participates in a logic resembling that of early liberal feminism; just as some early liberal feminists wanted to extend the rights of men to women, and in a way wanted the right to become "like men," so Locke merely makes all humans lords over their own domain—at least, all of those who have the common quality that Locke identifies as "intellect-reason." According to this criterion, the extent to which an individual human displays intellect-reason is the extent to which he is fully human. In other words, all "humans" are judged according to the way in which Locke defines "intellect-reason." Those who display the most reason, and in this way mimic the all-powerful Creator God in their *imago*, become little centers of control; will and act are united. Those who do not display what might be considered "reason" from a Western perspective ought to be educated, civilized, and controlled.

Locke's philosophy of equality suffers from merely extending the power of monarchy (and an omnipotent God) to all humans. This mimics Filmer's understanding of power as dominion and extends monarchical power to all people. In this sense, as far as equality is concerned, might still makes right. The problem with this type of extension of power is exemplified in the disjoint between Locke's philosophy of equality and his support of slavery. Though his philosophy hardly justifies the type of slavery that took place in the African slave trade, Locke participated in the slave trade directly (at least in an economic sense). Waldron wants to leave this paradox open-ended. He writes:

> Two Facts are clear: (1) There is nothing in Locke's theory that lends an iota of legitimacy to the [then] contemporary institution of slavery in the Americas; and (2) African slavery in the Americas was a reality and Locke himself was implicated with it, in the ways that I have described. I prefer to leave those facts where they lie, sitting uncomfortably together, than to try and resolve a contradiction between them.[24]

While I am sympathetic to Waldron's acknowledgment of the sociohistorical complexity of any given person, I want to question how it was that Locke's egalitarian thought could ever justify his own actions in support of slavery. I argue that his justification for inequality and erasure of agency may lie in his understanding of individual property, which mimics the unification of will and act in an omnipotent God, Creator *ex nihilo*.

The litmus test for the extent to which one emulates the Lockean "human" seems to lie in the extent to which she, by virtue of her "intellectual nature," improves upon raw nature, presumably one of the most *reasonable* things any human can do. Locke writes, "God gave the World to Men in Common; but since he gave it them for their benefit, and the greatest Conveniences of Life they were capable to draw from it, it cannot be supposed he meant it should always remain common and uncultivated."[25] Furthermore, he writes, "For I aske whether in the wild woods and uncultivated wast of America left to Nature, without any improvement, tillage or husbandry, a thousand acres will yield the needy and wretched inhabitants as many conveniences of life as ten acres of equally fertile land doe in Devonshire where they are well cultivated?"[26] In other words, Locke's anthropocentric and Eurocentric understanding of creation and misreading of the lack of use of the land serve to erase the agency and value of both the land in America and the Native American presence therein. Just as the exegetes trumpeting the omnipotent God ignore the existing chaos and posit *creatio ex nihilo*, so the individual human ignores prior claims in the land and cultivates it *ex nihilo*, thereby making it his own property. In this way, Locke's understanding of an all-powerful God legitimates the colonization of the "other" and sets the stage for his understanding of what sets humans apart from the rest of nature in a privileged role.

For Locke, God is all-powerful in that God's will and action are one. In his *Essay Concerning Human Understanding*, he writes, "This eternal Source then of all being must also be the Source and Origin of all Power; and so *this eternal Being must be also the most powerful.*"[27] The eternal Being must also be singular! It is also in this essay that he affirms *ex nihilo*. For Locke there can only be one source of value and power, and since matter could never produce a thinking thing, "the first eternal Being cannot be matter." How, you might ask, could nonmatter produce matter? According to Locke, if your own subjective, intellectual, conscious person came from "nothing," and you do indeed have a body, then it is possible for other material things to come from nothing. (Never mind the importance of your parents' role in your own existence!) Locke writes, "If therefore you can allow a thinking thing to be made out of nothing (as all Things that are not eternal must be), why also can you not allow it possible, for a material Being to be made out of nothing?"[28] The all-powerful, immaterial God creates all other matter, and even other "thinking things," out of nothing through sheer will.

Locke turns this theology into economic theory in his *Second Treatise*, where he defines "property" in such a way that individual humans mimic God's power. Just as the omnipotent God creates all matter out of nothing, and is thus a type of owner of creation, so, too, are humans in relationship to the rest of the natural world. He writes: "God, who hath given the World to Men in common, hath also given them reason to make use of it to the best advantage of Life, and convenience. . . . Whatsoever then he removes out of the State that Nature hath provided, and left it in, he hath mixed his *Labour* with, and joined to it something that is his own, and thereby made it his *Property*."[29] This understanding of property begs the question, what is the "state of nature" according to Locke?

Locke understands the "state of nature" as fallen due to human sin. As Carolyn Merchant notes, "humanity falls out of Eden into 'the state of nature'. . . . The state of nature retained its paradisiacal potential. But nature was worthless until they acted to transform it into a new paradise."[30] In his *Essay Concerning Human Understanding*, Locke writes, "Matter, then, by its own Strength, cannot produce in itself so much as Motion: the Motion it has, must also be from Eternity."[31] Though Locke's understanding of "nature" is more complex than this essay has room to deal with, he clearly sees "matter" in the state of nature as passive material moved by an external source. His understanding of matter combined with his view that the natural world lies in wait of improvement by human beings gives legitimacy to the colonizing and cultivation of "wild" lands. According to Locke's understanding of private property, mixing labor with passive nature helps to transform fallen nature into a new paradise.

I agree with Carolyn Merchant that Locke in this way "presents the emergence of 'civilized man' from the state of nature by domination, the appropriation of nature by ownership, and the transformation of gathered goods into items of trade and commerce as the 'natural' upward course of events."[32] Furthermore, as Horkheimer and Adorno recognized, this power over nature mimics God's power: "In their mastery of nature, the creative God and the ordering mind are alike. Man's likeness to God consists in sovereignty over existence, in the lordly gaze, in the command."[33] Like a God who creates out of nothing, so the human creates his individual property out of nothing (at least in the sense that the notion of uncultivated, raw matter erases agency and amounts to nothing); and just as God, the Creator *ex nihilo*, is the sole source of value in creation, so humans, the *imago* of the *ex nihilo* God, bring value to valueless matter.

The problem is that just as the chaos of Genesis 1:2 gets erased in theological exegesis, so the prior presence of other, non-Christian, non-European humans and nonhuman others in the land gets erased. If the "other" appears, from the perspective of the colonial gaze, as uncivilized, barbaric, or as otherwise not performing his God-given duty of transforming nature toward human ends (as defined by the European colonizer), then the colonizer has the right to educate, missionize, or civilize the "other" and/or to use his lands toward fulfilling the Christian God's will. If the "other" is thought to be too wild or uncivilized, then perhaps he does not fall under the ethical consideration demanded by the *imago*, and can then be used to help cultivate lands for the true human family. From the perspective of the colonizer, the difference of the "other" justifies denial of his humanity and thus his claims to land-use and ownership. Slavery and colonial takeover of lands are then justified. As Val Plumwood notes, "In the context of the 'new world', [Lockean Property] also provided . . . the basis for the erasure of the ownership of indigenous people and the appropriation of their lands. . . . Locke's recipe for property formation allows the colonist to appropriate that into which he has mixed his own labor, as part of the self."[34] This individual "logic of *ex nihilo*" toward land and "other" peoples is transformed into a legal national justification for seizure of lands through the concept of *terra nullius*.

TERRA NULLIUS, OMNIPOTENT NATIONS, AND ERASURE OF PRIOR HUMAN PRESENCE

"The 'state of nature', as Europeans understood it [as *terra nullius*], was a state in which humans had not yet appropriated land as property"; those who did not cultivate land needed cultivation-colonization because they were uncivilized or in the state of nature. "If [they] were still in the state of nature, then by definition they did not own their land. The land was *terra nullius*."[35] Perhaps the most influential work supporting this idea of *terra nullius* is that of the eighteenth-century Swiss philosopher Emmerich de Vattel, *The Law of Nations*.[36] Though Vattel was strongly influenced by the work of German scholar Christian von Wolff, the understanding of property he develops is clearly Lockean. He takes the concept of Lockean individual property to a national level. From the opening pages of his work, it is clear that he is very concerned with power in the form of complete sovereignty. "A nation or a state is . . . a body politic, or a society of men united together for the purpose of promoting their mutual safety and advantage by their combined strength."[37] The bulk of his work then addresses obligations that the state has to its citizens and rules for the interaction of two or more nations—including guidelines for commerce and war between nations. For the purposes of this chapter, I examine his understanding of cultivation and just appropriation of uncultivated land.

Clearly for Vattel (as for Locke and others at the time), uncultivated land is worthless land. He writes:

The sovereign [of any particular nation] ought to neglect no means of rendering the land under his jurisdiction as well cultivated as possible. He ought not to allow either communities or private persons to acquire large tracts of land, and leave them uncultivated. Those rights of *common*, which deprive the proprietor of the free liberty of disposing of his land . . . are inimical to the welfare of the state, and ought to be suppressed, or reduced to just bounds.[38]

From this quote it is safe to conclude that Vattel identifies the commons with worthless land. This thinking is right in line with the enclosure of the commons, which took place in England in the late seventeenth and early eighteenth centuries. Similar processes of enclosure took place across Europe during this same time period.[39] Vattel's denial of the value of the commons while upholding the value of common power (in the nation) presents an interesting paradox worth exploring.

If it is necessary for individual property to be defended by the nation, then one must admit that the power of the individual is not enough to protect individual property—individuals need the collective power of the state-nation. At the same time, according to Vattel's work and the trend of enclosing the commons, common property is devalued. Here individual property is valuable and common property worthless, while at the same time common power is needed and valued insofar as it secures individual property. One grounds the power needed to defend one's claims and one's being in a transcendent source: common power. Does this paradox feed the illusion of the unification of will and act at the individual level? Does it mimic the justification that Locke used, whereby the individual human is able to create property because that individual is made in the image of an all-powerful God? It is at least plausible. In this case, however, the all-powerful God in the omnipotent theology cum ethics and politics is now substituted with the sovereign nation. It is the job of the nation to use its sovereign power to ensure that nature is made to produce for individual human needs. Placing power in the sovereign nation secures the individual from the lack of power he has to consistently make nature produce or to consistently "defend" his property.

When one cannot unite will and act alone, perhaps many together can. Here the notion of individual property subsumes or backgrounds that which maintains it: collective power. The *individual* can persist in the belief that he creates and owns property *ex nihilo*. (Grab hold of those boot-straps and pull!) In Vattel's work, a further erasure of relationality and agency occurs in his justification for the cultivation of uncultivated lands. That the nation should cultivate uncultivated land is for Vattel a natural obligation.

> The whole earth is destined to feed its [human] inhabitants; but this it would be incapable of doing if it were uncultivated. Every nation is then obliged by the law of nature to cultivate the land that has fallen to its share; and it has no right to enlarge its boundaries . . . but in proportion as the land in its possession is incapable of furnishing it with necessaries.[40]

From this statement, one can see why it might be necessary to form regulations as to how and when a country is justified in enlarging its boundaries! One such justification, according to Vattel, is that a nation can expand if the territory into which it expands is *terra nullius*. "When a nation takes possession of a country to which *no prior owner can lay claim*, it is considered as acquiring the *empire* or sovereignty of it, at the same time with the *domain*."[41] Of course the problem is that a country without any history of human inhabitants rarely (if ever) exists, much less a country without nonhuman inhabitants. Vattel explores specifically the case of the "New World," where land has been claimed despite the presence of "erratic nations whose scanty population is incapable of occupying the whole," also known as the various Native American tribes. Here Vattel justifies the takeover of land by Euorpoeans with the idea that cultivation (land mixed with labor) is ownership.

> Their [Native Americans'] unsettled habitation in those immense regions connot [*sic*] be accounted a true and legal possession; and the people of Europe, too closely pent up at home, finding land of which the savages stood in no particular need, and of which they made no actual and constant use, were lawfully entitled to take possession of it, and settle it with colonies. . . . We do not, therefore,

deviate from the views of nature, in confining the Indians within narrower limits.[42]

Though the land is recognized as not completely without presence, it is treated as empty enough to allow for the justification of colonial take-over. This denial-erasure of prior presence in the Americas was greatly enabled by the genocide that took place as a result of diseases being passed from Europeans to Native Americans in initial encounters between them. About both Native Americans and Aborigines, Val Plumwood writes, "The colonized are denied as the unconsidered background to 'civilisation', the Other whose prior ownership of the land and whose dispossession and murder is never spoken [of] or admitted."[43] Furthermore, Jared Diamond notes that "throughout the Americas, diseases introduced with Europeans spread from tribe to tribe far in advance of the Europeans themselves, killing an estimated 95 percent of the pre-Columbian Native American population."[44] The illusion of *terra nullius* comes at the expense of the killing of indigenous peoples with foreign diseases from which they had no immunity.

As *ex nihilo* erases the agency of the *tehom* in Genesis, and Lockean individual property erases the agency of nonhuman nature, so the claim of *terra nullius* by nations erases the prior presence of "others" in the land. It seems to me that all three concepts suffer from the desire to unite will and act and thereby to create an impossible level of certainty and order in the face of uncertainty and chaos. In this way, they support what Catherine Keller identifies as a "theopolitics of omnipotence" that justifies imperial and colonial attitudes and arises out of the notion of an omnipotent Creator God *ex nihilo*.[45]

GLOBAL HEIRS OF *TERRA NULLIUS* AND LOCKEAN PROPERTY

Though, today, we may have seen the legal end of *terra nullius*, the concept is still very much alive and well—witness the recent discussions about the Alaska National Wildlife Refuge that suggest there is "nothing there" and thus the area is suitable for industrial development-colonization. In a White House Press Conference in March 2001, President George W. Bush stated, "It would be helpful if we opened up ANWR. . . . I think it's a mistake not to. And I would urge you all to travel up there and take a look at it, and you can make the determination as to *how beautiful* that country is."[46] Here Bush tries to negate any sort of aesthetic value in the "undeveloped" land. Again, land sitting uncultivated, unused by humans (read, industrialized humans), is seen as worthless until it is put toward human, economic ends. This comment seems to ignore the prior presence of the indigenous peoples who *do* live in that region, not to mention the presence of caribou migrating throughout it. Another, more subtle erasure is taking place here, however.[47]

Bush's comment assumes that all of the people he is addressing have the ability to "travel up there and take a look at it." This is an erasure of the presence of class differences, which, in the United States, always also means an erasure of racial differences. Of course, the majority of people in the United States have neither the time nor the money to "travel up there and take a look at it." Therefore, in this one statement, Bush erases any opposition to his point of view and thereby the presence of any *future possibilities* in the land other than oil drilling.

Just as claims of *terra nullius* are still made today at the national level, so Lockean individual property is still efficacious. Though our understanding of property has changed from (solely) a literal interpretation of labor-mixing to one that also includes capital accumulation, I think the politics and ethics of the free market are still firmly founded in the *ex nihilo*, chaos-erasing idea found in John Locke's understanding of property.

In *Globalization*, Zygmunt Bauman examines how individuals in a globalized world are falling more and more into a global hierarchy of mobility. The "global mobiles" are at the top and include those who transcend space through Internet technologies, transportation, capital, and so forth; global mobiles are not tied to any place. At the bottom of this hierarchy are the "immobile locals": those who are very much tied to a place for their economic, social, and psychological well-being. Of course the self-deception of the global mobiles is that they *do not* depend on space; they do, but merely background that dependency. In that act of backgrounding, the social, cultural, ecological, and psychological space of the immobile locals is destroyed. Perhaps the realization that the individual cannot be omnipotent-creator *ex nihilo*, and therefore cannot escape the anxieties of finitude, leads to the new faith in the global free market, which collectively acts as a creator *ex nihilo* (at least for the benefiting elites). Bauman notes, "It is this new elite's experience of non-terrestriality of power—of the eerie yet awesome combination of ethereality with omnipotence, non-physicality and reality forming might" that leads to the destruction of the majority of the immobile local peoples, and I would add animals and the rest of nature as well.[48] This type of power sounds a lot like that attributed to a Creator God *ex nihilo*. Bauman notes, however, that "their power is, fully and truly, not 'out of this world'—not of the physical world in which they build their heavily guarded homes and offices, themselves extraterritorial, free from intrusion of unwelcome neighbors, cut out from whatever may be called a *local* community, inaccessible to whoever is, unlike them, confined to it."[49] In other words, the collective market functions in an *ex nihilo* fashion, erasing contextuality and prior presences in the local places it destroys through its own remaking, reshaping, and destructive patterns.

Is this an even more complex version of the *ex nihilo* logic found in Lockean property and in the *terra nullius* approach to the "other"? In effect, agency is taken away from the individual subject and given over to the market system, which transcends subjects and places. This market then returns power *ex nihilo* to those *individuals* who buy into it: anyone can gain in the system if he pulls himself up by the boot-straps and starts working within the system. This belief, of course, denies socioeconomic, sex, race, and other historical inequalities—a denial that is exactly part and parcel to the *ex nihilo* way of thinking. The theoanthropology that unifies will and act and is based upon a theology of the Creator God *ex nihilo* takes *place* in the world in the form of violence toward those who fall outside of its system. *Ex nihilo* thinking then becomes a justification for judging all life based upon the center (read, most powerful). Might this *ex nihilo* theoanthropology be one trajectory of the Christian doctrine of creation? If so, there must be a way to think about creation from a Christian perspective without its functioning as an epistemological foundation for a logic of domination. I turn now to some concluding remarks about the doctrine of *ex nihilo* and challenges to the logic of domination therein, which, as I have argued, provides foundational support for the colonizing concepts of Lockean property and national claims of *terra nullius*.

CHALLENGING THE LOGIC OF DOMINATION: FROM *EX NIHILO* TO CONTINUOUS CREATION

There are many good arguments for maintaining a doctrine of *creatio ex nihilo*. For the most part, however, these "good reasons" seem to be tied to that classical understanding of theology that maintains God's asiety, and the colonial concept of omnipotent power found therein. For instance, *ex nihilo* is supposed to imply the equality of all peoples. As we have seen, however, who is considered a person, and how (physically, politically, and economically) powerful that person is, varies. In a sense, *ex nihilo* functions to seal off full humanity from those outside of a Christian, logocentric understanding of the world. Another argument is that *ex nihilo* is supposed to lead to the contingency of all creation upon God and that nonhuman creation is therefore not solely for instrumental human use. Contrarily, might *ex nihilo* instead function to serve as a transcendental basis for an ethic that justifies an anthropocentric and colonial attitude toward the earth and the "others" with whom we share it? God owns the world *He* created through *His* sheer will and power, and those made in the image of God are stewards or keepers of God's property by virtue of how much they mimic God's unification of will and power on Earth.[50] I do not wish to deny the usefulness of stewardship metaphors in contemporary Christian ecothought, but surely there are metaphors other than "manager" that Christians can offer.[51] Still another argument is that *ex nihilo* leads to an equal valuation, navigating between monism and dualism, of the material and spiritual worlds. Has it not instead led to the material-denying position it attempts to avoid? Is not all material made to fit into an overall (human) salvific scheme from creation to new creation, or, in some cases, from ultimate origin to ultimate end?[52]

Given the above arguments, and the epistemological support system I have articulated between *ex nihilo*, Lockean property, and *terra nullius*, it can at least be argued that *ex nihilo* has been used to support colonizing tendencies throughout the history of Euro-American Western Empire. The doctrine of *ex nihilo* arose in the context, and under the influence, of a Hellenistic paradigm, in a time of uncertainty, and from a nondominant, still-unidentified group as, perhaps, an effort to create certainty and define that group. Now that Christianity has been the religion of the empire for some 1600 years, the time for the doctrine to change has long since come.

What we need today are understandings of creation that lead us to what Susan Betcher describes as an ethics of alterity: "An ethics of alterity—as opposed to the voracious dialectic of modernity's 'Spirit'—would begin to swerve us from the heroic, redemptive path often taken in relation to an-other."[53] At the conference from which this volume emerged, an interdisciplinary group of scholars came together to discuss the intersection of poststructural, postcolonial, and environmental philosophies, theologies, and ethics. The "common ground" at the meeting was concern for the well-being of the earth and all life. In other words, the context for the discussion was the acknowledgment of current forms of ecological degradation and the social injustices that accompany, contribute to, and result from this degradation. Might we start from this common, contextual understanding to offer some grounds of hope for our future? How would a contemporary doctrine of creation recognize the value-in-the-limitation of human religious claims, the value of the diversity of all life on the planet, and the value of contextual hope for the future of life on the planet?

Fortunately for Christians, there is another source for thinking about creation that focuses more on the continual process of creation rather than the ultimate original act of creation: namely, *creatio continua*. Though perhaps less developed and only an afterthought to *ex nihilo*, I argue that this strand ought to be developed more fully in future reflections on the Christian doctrine of creation. This, however, is not the place for a rehearsal of the history of *creatio continua*. Rather, here I define it operationally as a focus on the current context of the continuing process of creation. Instead of concern about ultimate origins and ends, perhaps a Christian understanding of creation ought to be more concerned about how Christianity continually shapes the grounds that it co-creates in the continuing creation. Focusing on the present manifestation of how Christians and their theologies co-create nature-cultures, rather than on an ultimate justification *ex nihilo*, may help us stay focused on the shifting grounds of continuing creation. In this move, Christianity and Christians are not so much defined by their orthodoxy as they are by their orthopraxis. *Ex nihilo* need not be a foundation for continuous creation in the

image of likeness or sameness. This shift in reflection to continuing creation reflects an epistemological shift from global (acontextual) to planetary (ecosocial contextual) thinking. It is to some developments in this
line of thinking that I now turn.

One such move toward the type of contextual doctrine of continuing
creation I am seeking to articultate is found in Keller's relational doctrine
of *creatio ex profundis*, a creation from the *tehom* (see her contribution to
the present volume). She argues that any attempt to discover an ultimate
origin "only brings us to the boundary of our own language, to *originating conditions that have themselves originated*: an infinite regress."[54] Her
creation account begins from "grounds" rather than "foundations."

> Theology and philosophy were always fleeing the surface of the
> earth, upwards or downwards, in search of more solid foundations,
> in search of changeless grounds, or reasons, beyond the shifting
> earth. So what if in the interest of a deconstructive ecotheology
> (itself a productive aporia), we announced: *let the earth itself be the
> ground*; the ground of theology; the ground indeed even of faith in
> an incarnate Wisdom?[55]

This move from "foundations" to "grounds" is crucial—it does not participate in the *ex nihilo* logic of domination that denies ecosocial contextuality and relationality, but rather it begins from a specific context. An
understanding of theologies as constructed on (shifting) "grounds" and
the subsequent understandings of creation that arise out of them should
be focused on connectedness with the past, present, and future of "others" rather than on subsuming those "others" under one universal
foundation.

This also reflects the change from the global to the planetary perspective that Gayatri Chakravorty Spivak articulates and Keller draws upon.
As an alternative to the colonial, global mindset, Spivak offers the concept of "planetarity." She writes, "I propose the planet to overwrite the
globe. Globalization is the imposition of the same system of exchange
everywhere."[56] An ethic of planetarity, contrary to that of globalization,
operates out of a respect for differences and seeks to forge strategic alliances among differences. "Freeing itself from the orthodox forms of socialism and of religion, there is emerging a planetary spirituality of the

interstices. No locality can be located apart from its interrelations. Close and alien, intimate and systemic, they add up to the whole."[57] This planetary awareness attends to the ecosocial context of continuing creation and how humans and nonhumans come together to co-create (and co-destroy) planetary places in each moment. Rather than modeling power here after an omnipotent God who creates *ex nihilo*, this continuing co-creation model depends upon the power of (our understandings of) God creating ecosocial communities *with* human and nonhuman "others" in different times and places.

Anne Primavesi's work elsewhere in this volume also points out the need for an ecocontextual understanding of our ever-changing human reality and our human thinking (including theologies) about that reality. She writes, "Through 3.5 billion years [the evolution of living and nonliving entities before the emergence of humans] gave us the different components of our planetary physical environments that eventually generated the particular conditions from which human life emerged and became sustainable. Their continued existence remains, in effect, the necessary condition not only for our being alive but also for gift events to occur in our lives today."[58] For Primavesi, a gift-based understanding of life, of creation, means an ecologically and socially contextual understanding of creation.

Other reformulations of the doctrine exist as well, many from process thinkers and/or ecofeminist theologians, but a discussion of them is beyond the scope of this essay. It is valuable to have *many* understandings of *creatio* because we begin from many different grounds, many different terrains. The dialogue that ensues from grounded observations and conversations challenges the monological thinking of colonial global maps of reality. In their place, grounded, planetary maps offer situated, evolving truths, which create dialogue through fostering a respect for, and positive valuation of, difference. Susan Armstrong suggests that these types of "in-between epistemologies respect differences and diversities without either, as in foundationalism, assimilating them into a unitary form which ignores and represses such diversities or, as in relativism, giving up the attempt to find patterns and commonalities between elements of our experience."[59]

CONCLUSION

In closing, a concept of continuing creation—a creation open to the future—should challenge the anthropocentrism and androcentrism found in the logic of domination and suggest, rather, that we exist "not from nothing but from this ambiguous mix of precondition."[60] This suggests that we are contingent—meaning eco-socio-contextual—creatures, always existing not in some ultimate transcendent being but in histories of nature-cultures.[61] From these histories of nature-cultures our own existence emerges out of intense connectivity rather than nothing, and in, with, and for these nature-cultures that we live. Meaning and purpose, then, lie in particular planetary places, not in some far-off place or future, or in some remote point in the past. Meaning is made through dialogical interactions within and between nature-cultures. This dialogical process restores agency to nonhuman nature. Humans are "co-creators" (as many theologians have suggested). Humans, however, are also limited and cannot fully manage (or steward) the rest of the natural world; nor is the co-creation of a given reality possible without the destruction of some other reality (actual or unrealized). There will always be uncertainty about how exactly any given action will affect the rest of the natural world; but a dialogical process, rather than a monological, power-over process blinded by foundational assumptions, will provide us with a self-correcting method of negotiation.[62]

An understanding of nonhuman creation as agential, and humans as parts of (and sometimes as partners with) this continuous creative process, may help to destabilize the dominating individualistic notion of Lockean property, and challenge national claims of *terra nullius*, both of which mimic the *ex nihilo*, globalizing understanding of creation. If, as I am suggesting, questions about creation are ultimately about the world we humans are co-creating in the present—that is, ultimately about our acts in the world—then I think it can at least be argued that the logic of *ex nihilo* found in Lockean property and the concept of *terra nullius* have been the source of much destruction of creation. Challenging the symbolic support system of the logic of domination, then, might provide the turning point needed for creating a theological support system for the very life damaged and denied by the history of Lockean property and national claims of *terra nullius*.

❧ Surrogate Suffering: Paradigms of Sin, Salvation, and Sacrifice Within the Vivisection Movement

ANTONIA GORMAN

Vivisection: *The cutting of or operation on a living animal usu. for physiological or pathological investigation; broadly: animal experimentation, esp. if considered to cause distress to the subject.*
—*Webster's Ninth Collegiate Dictionary*

I have no pleasure in the blood of lambs and goats.
—ISAIAH I:II

INTRODUCTION

At the end of the nineteenth century, philanthropist Frances Power Cobbe bemoaned the nearly religious reverence bestowed upon the scientific community by the general populace.[1] Of particular concern to her was the newly emergent, yet increasingly powerful, discipline of vivisection—a discipline that promised medical salvation for humanity in exchange for the sacrifice and suffering of nonhuman animals. "To thousands of worthy people," lamented Cobbe, "it is enough to say that Science teaches this or that, or that the interests of Science require such and such a sacrifice, to cause them to bow their heads, as pious ones of old did at the message of a Prophet: 'it is Science! Let it do what seemeth it good.'"[2] In a perhaps unconscious mimesis of the sacrificial paradigm identified by Cobbe, Anna Kingsford, one of England's first licensed female physicians, announced her intention to offer up her own living body for medical experimentation on the condition that thereafter the medical community forever forswear experiments on nonhuman animals. Kingsford eventually was convinced not to follow through with her intention by her companion, Edward Maitland, who believed that the medical establishment would deride the offer as insincere and ascribe to her either "downright insanity" or "an inordinate vanity and craving for notoriety." Yet Kingsford, wrote Maitland, continued to insist that "if she could not sacrifice herself for the animals in that way, she would in some other which, if less painful, would be far more protracted."[3]

The statements of these two women point to a powerful nexus of sacred and secular salvational imagery that was coalescing around the vivisection debate in the Victorian era. While the obligations of humanity to the nonhuman world and the place of nonhumans within the salvational economy had been a consistent, albeit often a peripheral, part of the Western conversation in prior generations, in the nineteenth century the conversation took on a new tone of urgency and a new praxis of compassion that momentarily appeared poised to alter forever the relationship between human and nonhuman animals. Although different models of sin and salvation, based upon different understandings of human-animal kinship/dissociation, struggled with each other over the course of the century, as we shall see, by century's end a particular understanding and secular application of atonement Christology had gained ascendancy—an ascendancy that continues to exercise its field of force to this day. This permutation of atonement Christology (a Christology that itself was based upon the ancient Hebraic practice of sacrificing animals upon the altar of Yahweh) took as a priori the premise that the torture and crucifixion of Jesus not only was an acceptable price to pay for the salvation of humankind, it was the essential price. Through the mystification of secular and sacred actors, the logic of sacrificial atonement came to justify, indeed to dictate, the sacrifice of innocent animal victims for the secular "salvation" of the elect among humanity.

It is not my intention in this article to argue for or against atonement Christology as a religious doctrine; others before me have pointed out both the liberating and oppressive possibilities inherent within it.[4] Rather, it is my intention to lift up for recognition the presence of a particular, secularized formulation of the atonement model within vivisection in the hope that through recognition we may dislodge this model's psychic hold upon, and undergirding support for, the Western anthropocentric imagination. My own experience in the field of animal advocacy has convinced me that arguments for the subjectivity and inherent value of nonhuman animals,[5] while providing essential ground upon which to cultivate an ethos of animal care and protection, nevertheless are insufficient when faced with the power and predominance of the sacrificial paradigm that permeates not only the practice of vivisection but so many anthropocentric interactions with the nonhuman world. I suggest, in light of this, that an alternative theological understanding and praxiological application of "salvation" is needed—one that does not call for the continuing crucifixion of the nonhuman world in an always elusive quest for ever-retreating human fulfillment, but instead seeks mutual well-being and fullness of life for the entire panoply of God's creation. Fortunately, many of the ingredients needed for this task are present already within the discussions that led up to the vivisection debate. I propose an experiment in ecotheological reconstruction: by critically combining some of these ingredients with insights from feminist and relational theology and from another great biblical story of salvation—the story of Job—we may open up vistas compelling enough to refocus the Western gaze away from the logic of substitutionary sacrifice and toward a soteriological space where all God's creatures are welcome.

MODELS OF SIN AND SUFFERING: SETTING THE STAGE FOR THE VIVISECTION DEBATE

Charles Darwin

The single greatest catalyst for Victorian debates over the moral, ontological, and theological import of animals arguably was the publication of Charles Darwin's *On the Origin of the Species* in 1859. This work created seismic shifts in the Western psyche by documenting the processes of natural selection and thereby giving credence to evolutionary theories that, in one form or another, had been circulating within the scientific community for several generations.[6] Darwin's work challenged assumptions concerning the "unbridgeable gulf that divided reasoning human beings from irrational brute,"[7] a challenge made more explicit in his *The Descent of Man and Selection in Relation to Sex* (1871),[8] in which Darwin made his now famous assertion that humans are merely one twig on the primate bush of life (an image that Darwin preferred to the "tree of life" since it suggested no hierarchical arrangement).[9] The kinship of humans and nonhumans made explicit by Darwin not only raised moral concerns over the appropriate treatment of nonhuman "cousins," but also shook belief systems concerning God's ongoing presence within (or absence from) creation and elicited uncertainty over the meaning and inclusiveness of salvation, as will become clear below.

The Scientific Revolution

As with all seismic shifts, however, the conditions that gave rise to Darwinian tremors had been building for some time. The scientific revolution of the seventeenth century had created pressure on both sides of the fault line. On the one side, Enlightenment science had disenchanted the physical world in an attempt, in the words of Max Horkheimer and Theodor W. Adorno, to liberate "human beings from fear and [install] them as masters" over nonhuman nature.[10] Characterized by the writings of Francis Bacon and René Descartes, the Enlightenment contended that nature operates like a machine and hence can be understood and controlled by utilizing experiential mathematical principles.[11] Underlying this contention was the conviction that nature was "both ordered and orderly" because it had been set in motion by an orderly, rational God who had constructed it according to absolute rules and who then had withdrawn from creation, allowing it to operate according to the dictates of those rules.[12] Embedded within this contention was a faith that there was no mystery within nature that was beyond human comprehension and control and a presupposition that comprehension and control were achieved best by breaking nature down into its simplest component parts and isolating those parts for closer examination. Nature as machine was said to possess no mind or soul (interchangeable terms within mechanism) and hence no capacity to suffer. It therefore could make no claims on human compassion or moral consideration and possessed no value outside its instrumental value to humans. "Since nature," explains Warren Ashby, ". . . was seen in this way as mechanically ordered, a major category of ethics became the useful. Whatever was useful to the [human] individual . . . became the criterion of moral good."[13] Subjugating nonhumans to human control and exploiting them for human use was thus an unambiguous good from the mechanistic perspective. Mechanism's utilitarian approach, although not its claim against nonhuman subjectivity, would be skillfully woven into the atonement imagery employed by nineteenth-century vivisectionist Claude Bernard, to whom we shall soon return.

Yet while blatant utilitarianism was one perspective made possible by the scientific revolution, it was not the only one. On the other side of the fault line, scientific experiments upon animals revealed that humans and nonhumans share many physiological traits in common. This in itself was neither surprising nor morally troubling, since mechanism had consistently contended that human and nonhuman bodies are similar in structure and that it is "man's" mental-spiritual qualities, not "his" physical traits, that raise him above the animals. Neurological experiments soon revealed, however, that physical sensations are, at least in part, dependent upon mental functions. "So," explains James Turner, "if the senses operated through identical physical networks in the bodies of people and animals, what was one to conclude? If brutes could feel, they must have minds; and if they had minds . . . ?" Many people came to "believe that beasts not only [have] feeling but also a degree of reason,"[14] and thus possess the subjectivity necessary to be included within the circle of moral concern. This belief found popular expression in the romantic movement of the late eighteenth and early nineteenth centuries.[15] Although romanticism had been largely marginalized by the time of the vivisection debate, many of its assertions concerning animal subjectivity and human/nonhuman relationality remained the subtext of the Victorian animal-welfare movement.

Romanticism

While romanticism was far from a monolithic movement, it frequently exhibited what Kate Rigby has identified as an "ecocritical perspective," whereby Cartesian dualism was rejected in favor of a more integrated worldview.[16] Romantic poets such as William Wordsworth mourned over the feelings of displacement, estrangement, and despair that they saw as the inevitable result of disconnection from the nonhuman world. These feelings were heightened by the visible manifestations of mechanistic utilitarianism in the form of "modernized" agriculture that privatized the commons and drained the fens, growing industrialization with its accompanying environmental degradation and urban blight, and colonial expansion with its lack of attunement to the environmental particularities of specific places. Explicitly connecting the deadening of nature with the deadening of the human soul, Wordsworth wrote:

> For was it meant
> That we should pore, and dwindle as we pore . . .
> On solitary objects, still held
> In disconnection dead and spiritless,
> And still dwindling and dwindling still,
> Break down all grandeur . . .
>   Waging thus
> An impious warfare with the very life
> Of our own soul?[17]

In place of the damnation wrought by disconnection, atomization, and mechanization, romantics such as Friedrich Schleiermacher offered up visions of human salvation as a personal "wholeness" achieved through the self-transformative integration of the human being into the larger, agential, organic whole. "The Universe exists in uninterrupted activity and reveals itself to us in every moment," wrote Schleiermacher. "Every form that it brings forth, every being to which it gives a separate existence according to the fullness of life, every occurrence that spills forth from its rich, ever-fruitful womb, is an action of the same upon us. Thus to accept everything individual as part of the whole and everything limited as a representation of the infinite is religion."[18]

Romanticism's vision of salvation through integrated wholeness was not without its problems, however. The vision contained an implicit and often explicit hierarchical, anthropocentric bias that frequently compared the integrated planetary whole to the integrated human body and equated the human being with the body's head. Under this trope, the rest of the body could be figured as important primarily insofar as it served the needs of the head.[19] Romantic "wholeness," at its worst, could thus be as dominative an ideology as was Enlightenment mechanism. At its best, however, romanticism accepted the ambiguities, uncertainties, and mysteries within life and thereby respectfully opened itself up to the teachings of floral and faunal nature. For romantic scientists such as Johann Wolfgang von Goethe, these teachings were garnered not through the forcible extraction of universal laws from "dead" matter (through what Bacon gleefully and in apparent mockery of Spinoza had dubbed *natura torturata*),[20] but rather through respectful encounters with the agential subjectivity of nonhuman creatures. This image of nature as agential subject was bolstered by the increasing influence of the sciences of geology, paleontology, astrology, and especially the newly emergent science of biology, the discoveries of which supported and helped inspire the romantic contention that the earth—and indeed the universe as a whole—was a "dynamic, self-transforming natural order," in which process and relationality are fundamental to life itself.[21] This conclusion had implications not only for interhuman and human-animal relationships, but also for relationships between the world and God. After all, if creation provides intimations of the Creator (as even mechanism agreed), then the Divine, said Friedrich Schlegel, must be "an endless becoming" that is both imminent within and transcendent of its relational partners.[22] Yet if, as Schlegel suggests, God is panentheistically present within creation—is present throughout creation, though not reducible to creation—then encounters with finite creatures can yield intimations of the infinite, thereby allowing salvific encounters with divinity and eternity in the here and now.[23]

John Wesley and the Charity School Movement

Romantic ideas concerning human-animal relationality and the place of animals within salvation were supplemented by, and subtly changed from, the pulpit. Within England, where the antivivisection movement would be born in the next century, John Wesley was particularly attuned to these issues, although he, like the romantics, approached them predominantly from the perspective of human salvation. Wesley was concerned that the doctrines of predestination promulgated by Martin Luther and John Calvin encouraged human passivity and fatalism by rendering all human efforts irrelevant to salvation. More importantly, in Wesley's estimation, theories of predestination (he believed) conflicted with the biblical contention that God desires the salvation of all creatures and encouraged the perception that God arbitrarily decides whom to save and thus deals unjustly with creation.[24] Wesley agreed with Luther and Calvin that God's grace is freely and undeservedly given to all people independent of their works, yet departed in emphasis from these predecessors in his insistence that once grace is received, a measure of free will (lost in the Fall and transmitted through original sin) is restored. Once free will is restored, said Wesley, love of God and neighbor arises within the individual as the free expression of a newly found faith. Therefore, personal salvation, while not earned through works, can be known through works. These works, said Wesley, are characterized by scrupulous, compassionate, and never-resting service to God and neighbor.[25]

Inspired by this clerical emphasis on service and by the pairing of compassion and salvation, the Charity School movement soon arose. This movement initially focused on alleviating human suffering through institutional reforms such as prison reform and the antislavery movement,[26] but by century's end its ethos had spilled over into the arena of animal welfare. This is perhaps not surprising given Wesley's own insistence that God loves and suffuses creation ("We are to see the Creator in the glass of every creature," said Wesley; "we should use and look upon nothing as separate from God") and thus intends eternal salvation for human and nonhuman alike.[27]

COMPASSION AND REDEMPTION: INSTITUTIONAL REFORM AS ORTHOPRAXY

By the nineteenth century, the confluence of evolutionary theory, enlightenment evidence for the subjectivity of animals, romantic holism, and Wesleyan sanctification through service resulted in the growth of activities on behalf of nonhumans. Agitations for compassion toward animals first bore substantive fruit in 1822 with the passage of "Martin's Act." Named after Richard Martin, the man who had introduced the bill into Parliament, Martin's act imposed a substantial fine and three months in jail on anyone convicted of abusing horses, cattle, or sheep.[28] This law was a significant achievement, since it was the first time in history that any country had instituted national legislation against the mistreatment of animals,[29] but it initially failed to have practical effect since, prior to London's formation of the world's first police force in 1829, the pursuit and arrest of criminals was left largely to private citizens and few citizens actively pursued animal abusers.[30] This changed in 1824 when the Reverend Arthur Broome founded the Society for the Prevention of Cruelty to

Animals (SPCA). Defining its mission in secular terms as the enforcement of anticruelty legislation and the education of the masses about the importance of kindness to animals, the SPCA nevertheless left little doubt that it saw its efforts as a form of religious praxis, stating openly in its mission Declaration of 1832 that educational and enforcement activities would be "based on the Christian Faith and on Christian Principles."[31] Given the concerns over sin and salvation that permeated much of nineteenth-century charitable activity, it is perhaps predictable that the SPCA defined its praxis as much in terms of human salvation as in terms of animal well-being. Martin's Act, after all, had won passage largely through arguments that "habitual cruelty to animals predisposes us to acts of cruelty toward our own species."[32] The SPCA agreed with this argument and promised that its efforts to protect animals against suffering would promote human moral uplift and hence "promote the salvation of human souls."[33] "By the discouragement of cruelty and insensibility of heart, in the treatment of inferior creatures," wrote the SPCA, "human beings will be rendered more susceptible of kind impressions towards each other, their moral temper will be improved, and consequently, social happiness and genuine philanthropy must, infallibly, be strengthened and enlarged."[34] The aura of sacred calling promoted by the SPCA soon began to grow in the rich soil loamed by romanticism, Wesleyanism, and evolutionary theory until, by the latter decades of the nineteenth century, animal protection societies "were among the largest and most influential voluntary organizations in the Anglo-American world."[35]

Dr. Jekyll And Jack The Ripper: Crusading Against "Profane" Science

The SPCA's faith in salvation through compassionate praxis was tested, however, by the new scientific discipline of vivisection. Initially vivisection found little favor either within the general medical community or among the wider Victorian public. At the beginning of the century, reports Richard D. French, doctors not only were concerned about the moral implications of vivisection but were unconvinced of its medical necessity. Arguing that the practice was less effective than casework, efforts to improve hygiene, and studies of comparative anatomy, many doctors contended that the suffering imposed upon animals and the cruelty that the infliction of suffering might incite within the hearts of vivisectors could not be justified by the doubtful medical benefits of the practice.[36] A variation of this "inefficacy" argument still can be heard among contemporary antivivisection activists. Drs. Jean Swingle Greek and C. Ray Greek, for instance, stress the iatrogenic effects of many medications that have been developed, tested, and approved under the vivisection model[37] and point out the more promising (in their view) avenues of in vitro testing, computer modeling, and genetic research.[38] Drs. Greek and Greek, however, represent a heterodox position within today's medical community while equivalent positions were orthodox during the first half of the nineteenth century.

Influenced by romantic sensibilities, clerical calls for compassion, and medical ambivalence, the majority of England's media sources also were critical of vivisection during this period, echoing the argument put forward by the SPCA (among others) that brutality toward God's nonhuman creatures could jeopardize the human soul by inciting cruelty toward God's human creatures.[39] Vivisectors initially attempted to counter such criticisms by wrapping themselves in the mantle of impassive science, claiming, as did Lord Lister when asked by Queen Victoria to speak in Parliament against vivisection, that "an act is cruel or otherwise, not according to the pain which it involves, but according to the mind and object of the actor."[40] This claim brought cold comfort to the general public, who feared that underneath the mantle of dispassion lay an evil depravity that reveled in meting out misery. As Coral Lansbury writes, "It was not only Robert Louis Stevenson who felt that the soul of a ravening, amoral monster could be found in the breast of a kindly man of medicine like Dr. Jekyll."[41] The wider Victorian public wondered aloud if Jack the Ripper might be a vivisector who had expanded his work from animals to women and gasped in unsurprised outrage when doctors from the Westminster Hospital administered large doses of sodium nitrate to outpatients to determine if the compound was poisonous.[42] Eighteen outpatients, reports Lansbury, became ill and suffered "frightful pain," but the doctors justified their actions by stating that the few must suffer for the benefit of the many,[43] a refrain that continues to be sung by the vivisection choir today,[44] although the concept of "few" has transmogrified from a numerical designator to a signifier of diminished value (with the "few" now translating into approximately 250 million animals vivisected worldwide each year).[45]

During the Victorian era, the battle against the "depravity of science" took on the tone of a holy war as antivivisectionists fashioned themselves as soldiers of Christ engaged in a battle against the profane intellect of pure and applied science. From their perspective, science jeopardized salvation by spurning compassion and thereby confessing "the corruption of its soul."[46] The holy crusade against the "devil's work" appeared to be unstoppable when, in 1874, the SPCA, which by now had received the queen's royal imprimatur, used Martin's Act to prosecute four doctors who had attempted to induce epilepsy in a dog by injecting her with absinth at the annual meeting of the British Medical Association. The "prosecution failed on a legal technicality," reports Harriet Ritvo, "but the case received enormous publicity and generated widespread sympathy for the critics of vivisection."[47] When, however, a mere two years later, Parliament passed the Cruelty to Animals Act regulating the licensing of vivisectors, the primary beneficiaries of the legislation were, and were intended to be, doctors rather than laboratory animals. By the turn of the century, experiments on living animals had grown exponentially,[48] more than 75 percent of the press supported the practice of vivisection,[49] and the general public had lost interest in the issue.[50]

Surrogate Suffering: Medical "Salvation" through the "Sacrifice" of Animals
Why the sudden and dramatic change? Turner, French, and Lansbury, respectively, hold up a number of possibilities, including, but not limited to, issues surrounding industrialization and the loss of daily contact with the nonhuman world,[51] the growing hegemony of the medical establishment,[52] and the association of antivivisection with women (75 percent of the rank and file were female) and hence with "softheaded sentimentality."[53] While all of these factors undoubtedly contributed to antivivisection's failure, I think that they are insufficient explanations in and of themselves. Antivivisection was a moral crusade—one that attempted to alter the relationship between the human and nonhuman world and in so doing to redefine (sometimes in half-formed and only implicitly articulated phrases) not only humanity's obligations to animals, but also the nonhuman world's rightful place within salvation. Such a crusade could be defeated only by an enemy who offered an alternate and more compelling, though not for that reason more just, salvational paradigm.

Such a paradigm was most eloquently and effectively articulated by Claude Bernard, a French vivisectionist whose sacerdotal cast of mind soon won acolytes throughout Europe and America. Bernard, who once had been held up as a fiend for his practice of baking living dogs in ovens,[54] now became the high priest of the new and highly influential "religion." Positioning vivisection within the atonement tradition, Bernard spoke repeatedly of the need for sacrificial suffering and death to make life possible for the elect: "The science of life can be established only through experiment," said Bernard, "and we can save living beings from death only after sacrificing others."[55] That the beings worth "saving" were human and those appropriately "sacrificed" were nonhuman was left in no doubt by Bernard. "It is immoral," he said, ". . . to make an experiment on man when it is dangerous to him, even though the result may be useful to others." He insisted, however, that "it is essentially moral to make experiments on an animal, even though painful and dangerous to him, if they may be useful to man."[56] In this brief statement, Bernard elucidates the problem faced by contemporary animal advocates when they base their strategy solely on arguments for animal sentience (relying exclusively on a variation of Jeremy Bentham's famous aphorism, "The question is not, Can they *reason*? nor, Can they *talk*? But, Can they *suffer*?").[57] For while Bernard clearly accepts the capacity of animals to suffer pain, he nevertheless holds their suffering in low regard compared to the benefits that animal sacrifice could one day bestow upon the human race.[58] Passing over the fact that no practical clinical or therapeutic applications had come out of animal experimentation at the time of his first writings, Bernard nevertheless propounded a deep faith in the eventual capacity of vivisection to unlock the secrets of mortality and thereby to alter the processes of life itself. Animal suffering was (perhaps) regrettable, but it was nonetheless an acceptable, and indeed essential, price to pay for the possibility of eternal life—a possibility that Bernard insisted was within humanity's grasp.[59] The Research Defense Society, a professional association of vivisectors, embraced the sacrificial paradigm, making it the cornerstone of its campaign to win the hearts, minds, and souls of the English people. Opening a shop next door to the British Union for the Abolition of Vivisection, the Society posted a picture in its front window of a smiling mother with a baby upon her knee, underneath of which was captioned, "Which will you save—your child or a guinea-pig?"[60] With these words, vivisectionists ensured that temporal salvation would be the province of the (human) elect only (and, as it has turned out, only the elect within humanity) and that the more-than-human world would be sent into the deepest recesses of hell in the name of human salvation.

The "Sin" Of Surrogacy: An Additional Insight from Contemporary Feminism

While the recognition of nonhuman subjectivity was, and continues to be, a vital step in the journey toward animal protection (as such seminal animal rights activists as Peter Singer and Tom Regan so convincingly demonstrate),[61] it proved inadequate when confronted with vivisection's application of the atonement model. As Bernard's own recognition of animal suffering indicates, the subjectivity of the sacrificial victim had (and continues to have) little weight when placed on the scale opposite the elect's "greater need." Unless the dualistic paradigm of substitutionary suffering is challenged, it seems to me that the callous use of animals in laboratories and other venues of human-induced oppression has little chance of ending. Furthermore, if, as ecofeminists have long insisted, all forms of oppression are interconnected, stemming from identical paradigms of subjugation and control,[62] then vivisection is not merely an injustice inflicted upon isolated nonhuman beings (although if it were, that would be reason enough to question its legitimacy), but also a microcosm of the potential and actual effects of traditional surrogacy logic upon all vulnerable members of the human and nonhuman world. We already have seen one such effect in the example from Westminster Hospital cited above. As Victorians feared, medical science has become replete with such examples, perhaps most infamously in the experiments performed by Joseph Mengele against Nazi concentration camp prisoners, but also in examples closer to home, such as the syphilis experiments performed, without informed consent, on African American males in Tuskegee, Alabama, between 1932 and 1972. Accepting the moral legitimacy and ontological necessity of the surrogacy model thus has implications far beyond the world of animal and environmental advocacy.

Indeed, it has implications beyond the world of medical science and technology, as Rita Nakashima Brock and Rebecca Ann Parker so powerfully illustrate in their discussions of domestic violence. For Brock and

Parker, any capacity atonement Christology may have to be a source of comfort and compassion is overshadowed by its tendency to model a cruel and despotic Patriarch whose anger is appeased only through the violent death of His innocent and acquiescent Son. This model, they contend, functions to sanction parental and spousal violence and to promote passive acceptance by victimized children and wives who believe "that obedience is what God wants."[63] While insights from domestic violence are not always a perfect fit with experiences from the realm of animal abuse—it is unlikely, for instance, that animals possess the human capacity to internalize a message that Jesus' crucifixion demands an attitude of passive obedience (despite the claim by some Victorian vivisectionists that dogs willingly sacrificed themselves upon the medical table because of their love of and devotion to humanity)[64]—it nevertheless is clear that similarities can be found between the two fields. As we have seen, both vivisectors and domestic abusers have used the atonement model to justify their violence. In addition, domestic abusers often incorporate animal abuse into their reign of terror as a way to signal their dominance over others.[65] Thus while Brock and Parker may be writing in a different context than my own, I nevertheless feel great connection with their insistence that violence, ultimately, cannot lead to salvation. "What if the consequence of sacrifice is simply pain, the diminishment of life, fragmentation of the soul, abasement, shame?" they query. "What if the severing of life is merely destructive of life and is not the path of love, courage, trust, and faith? What if the performance of sacrifice is a ritual in which some . . . bear loss and others are protected from accountability or moral expectations?"[66]

THE SALVATION OF JOB IN A RELATIONAL WORLD: RETHINKING SURROGATE SUFFERING

While a deconstruction of secularized atonement is a necessary first step in loosening its grip upon the throat of marginalized members of creation, any challenge to the sacrificial paradigm eventually requires the construction of alternate images from which to model "salvational" praxis. Before I begin to attempt a constructive move, however, I would like to echo the words of Marion Grau, who warns that "ambivalent texts," such as those concerning atonement, "remain complicated and complex and can never become 'safe.'" As a result, such "texts" require "many images and many hermeneutical locations to read them from."[67]

The remainder of this essay, therefore, offers one possible and tentative reading of praxiological salvation, but makes no claims that this reading will answer all needs from all hermeneutical locations. Rather, it is my hope that while what follows necessarily will be partial and incomplete, it nevertheless will offer one vantage point from which to see other roads toward a more just and ecologically sustainable world. With that caveat, I would like to begin by suggesting a return to some of the insights encountered earlier in this essay. The challenge to Cartesian dualism and its mechanization and atomization of the nonhuman world mounted by the scientific discoveries of the Enlightenment and Darwinian theories of evolution is, I think, the necessary place to start. As I have already intimated, however, arguments against mechanism and for nonhuman subjectivity now have become commonplace within the environmental movement in general and the animal protection movement in particular. This is not to say that these arguments are not vital—indeed, I believe they are essential, for without them the only basis for ecological and/or animal protection is human self-interest. The nonhuman world does have instrumental value for humanity, of course, and utilitarian arguments do have immense emotional power. Indeed, if the history presented in this essay is any indication, self-interest is perhaps humanity's greatest single motivator for compassionate engagement with others. Yet arguments for compassionate praxis based on enlightened self-interest are swayed too easily by other, less compassionate utilitarian arguments, as we have seen in the case of vivisection. My reluctance to rehash arguments for animal subjectivity, therefore, is not based on any disregard for the importance of such arguments or a belief that their further articulation and development is unnecessary, but rather on a realization that that ground has been well traveled by others.[68] I begin, therefore, by accepting rather than rearguing the proposition that was inspired by Enlightenment scientific discoveries and Darwinian evolution that nonhuman animals possess a subjectivity that is relevantly similar, although not identical, to human subjectivity. They thus are appropriate subjects of moral concern.

The second insight from history that I think bears revisiting is the romantic notion of salvation as "wholeness." While the pairing of salvation and wholeness has a long and hallowed tradition in the West (Plotinus, for example, believed that salvation would be the experience of the simultaneity of past, present, and future in a unified temporal "whole,"[69] while Augustine claimed that life in Jesus would allow us to "be in every respect perfect, without any infirmity of sin whatever . . . until wholeness and salvation be perfected in us"),[70] we owe the romantics a special debt of gratitude for lifting up human salvation as dependent upon integration into the more-than-human planetary whole. Despite the value of romanticism's image of salvation as enacted and encountered through inter- and intrahuman temporal relationships, however, its vision of "wholeness" must not be appropriated uncritically. The totalizing tendencies of the wholeness metaphor have too many parallels with atonement logic, implicitly and often explicitly accepting the moral legitimacy of sacrificing the vulnerable and innocent for the good of the elect—as we saw, for example, in romanticism's willingness to sacrifice the planetary "body" for the perceived benefit of the human "head." Arguably, it is precisely these parallels that allowed antivivisection to succumb to vivisection's utilitarian, utopian promises. Fortunately, the relational theology of Alfred North Whitehead offers us a way to retain romanticism's embrace of creation as a partner in salvation while ameliorating the potential suffocating effects of that embrace.

According to Whitehead, the Western world (including its romantic manifestations) all too often commits the "fallacy of simple location"— the idea that matter is characterized by simple location in space and time. Animate and inanimate creatures, according to this fallacy, are said to "be *here* in space and *here* in time . . . in a perfectly definite sense which does not require . . . any reference to other regions of space-time."[71] This theory, contends Whitehead, ignores the fact that creatures take in reality (frequently noncognitively) through a process in which an awareness that takes place *here* in this place is also an awareness of something or someone in another place (I am here on my porch, for example, hearing the chirps of baby sparrows who are nesting over there).[72] Second, simple location says that lapses of time are accidental to the character of a creature.[73] Yet, says Whitehead, if something or someone existed throughout a stretch of time with no inherent reference to any other times, then it would follow "that nature within any period does not refer to nature at any other period." If this were true, argues Whitehead, there would be no reason for memory to have evolved within nature.[74] From a Whiteheadian perspective, then, "wholeness" may be too static and insufficiently relational a concept for useful appropriation and perhaps should be replaced by concepts such as "mutual dependence" or "interrelationality." Reality, clarifies Whitehead, is made up of vibratory ebbs and flows of energy that ingress into one another in such a way that no one being can be said to exist independently of others. All creatures, from this perspective, are so intimately constituted by their mutual ingressions that every creature is implicated in the existence of every other and "every location involves an aspect of itself in every other location."[75] We can see examples of this, for instance: in Einstein's theory of relativity where even time and space are mutually implicated; in chaos and complexity theories, which reveal that even small changes in initial conditions iterate and reiterate into wild and unpredictable consequences; in the theory of evolution, which contends that we all have a common origin and have exploded into diversity through the influence—the literal "flowing in"—of each upon the other, all upon the whole, and the whole back into the many and diverse one; and within the biblical assertion that all creatures live and move and have their being in the common vivifying breath of God.[76] Given these insights, we must affirm, as historian of science Donna J. Haraway phrases it, that "the one fundamental thing about the world [is] relationality."[77]

Yet if relationality (or, in Whiteheadian terms, "interrelationality") is fundamental, then "salvation" never can be static, for its promise of physiospiritual healing and interdependent "wholeness" will shift with the shifting relations of time and space, available only in given moments and particular locations as the plenitude of the fully embodied past and the enticing lure of the conceptual future become fully integrated into the rich complexity of the ever re-creating "now." Yet each particular and always partial concrescence of the salvational promise is not, for all its uniqueness, an isolated occurrence. Rather, it becomes its own small change that iterates and reiterates in often unpredictable and chaotic ways, so that the singular Jewish figure of Jesus in the singular historical period of first-century Palestine can, indeed, transcend his temporal-physical locatedness and thereby transform the world.

From this position of connection, multiplicity, and complexity, the suffering of the "one" necessarily becomes the suffering of the many, just as the suffering of the many inevitably effects the well-being of the one. Surrogate suffering, therefore, is untenable, for it assumes the ontologically impossible possibility of isolating suffering from the individual and community in which it arose. Stated differently, in a relational world, sin cannot be placed upon the back of a scapegoat to be carried off into the wilderness without consequences being incurred not only by the scapegoat, but also by the wilderness, the sinner, and the community at large.

With these insights from Darwinian evolution, romanticism, and Whiteheadian metaphysics, it is possible to read the story of Jesus through the lens of that other biblical story of suffering and redemption, that of Job. Via the "comforting whirlwind," to borrow Bill McKibben's joyously eco-evocative phrase,[78] God brings Job redemption, not through a promise to end suffering in general or even Job's suffering in particular (tribulation, after all, is an inevitable part of creaturely existence), or through a validation of God's justice (Job's pain, the story makes clear, is not "deserved" in any meaningful sense), but rather through a repositioning of humanity within the larger creation. Job's self-pitying ululations are quieted by the divine who coaxes and cajoles, and possibly even bullies, this self-absorbed representative of humanity to move beyond his isolation and narcissistic self-obsession and behold the miraculous plenitude and profoundly lovable subjectivity of the nonhuman world. Catherine Keller has argued persuasively that "YHWH's speech to Job may be read as an exegetical iteration of the creation narrative canonized in Genesis 1,"[79] as an explanation that "human becoming looks cramped and cancerous—unless we collude more wisely with the elements, the plants, the beasts and each other."[80]

I find this a deeply helpful insight, yet ask myself, if the story of Job may be seen as a proto-Midrash on the creation narrative, as Keller suggests, might it not also be seen as an anticipation of—a sort-of pre-Midrash on—the re-creation promised by Jesus? Surely birth and rebirth have shared resonances. If this is so, then might we not understand Job's own rebirth—his healing, his salvation—as occurring only after he viscerally experiences and thereby newly understands his place within the created world? By seeing, by feeling, by understanding the integrity, wild complexity, and lovability of the nonhuman world, Job enters into a right relation with creation and thereby is able to re-establish—or more accurately, establish anew—right relation with his fellow humans. Is not this, after all, the promise held out in Jesus' Kingdom of God—a promise not of an anthropocentric, anthropomorphic basileia, but of a gleefully undomesticated, interrelationally creative universe where all God's creatures are fed and healed and comforted? If this is so, then we need add only the Wesleyan insistence that salvation is accompanied by active and compassionate engagement with the world in order to ground our hope for a more just future in which human and nonhuman alike are "saved" together, or not at all.

❧ The Hope of the Earth: A Process Ecoeschatology for South Korea

SEUNG GAP LEE

What I have hoped is to draw such a picture of sounds of wind and water, spring flowers blossoming in sunshine, and salamander laying eggs at the edge of the clean water.

—BUDDHIST NUN JIYUL, in a letter to the government of South Korea

On June 2, 2006, the Supreme Court of South Korea gave final clearance for the South Korean government to continue tunneling through Mount Cheonseong for a new high-speed rail line; it did so without asking for or obtaining an Environmental Impact Assessment. Although environmentalists claim that adjacent ancient marshlands and the more than thirty protected species that inhabit Mount Cheonseong will be adversely affected by the construction project, the Court sided with the government's decade-old official Environmental Impact Assessment, despite a unanimous February 2005 resolution from a parliamentary committee on construction and transportation advising a new assessment.

Following in the wake of the Saemangeum decision earlier in 2006, the case dealing with Mount Cheonseong—or "Salamander case," as it has come to be known—pits the union of economic progress with land development against preservation of natural habitat.[1] As Koreans have been increasingly faced with the effects of deforestation, shrinking wetlands and the decimation of farms, and a decreasing water supply in conjunction with increasing levels of pollution, some of them have engaged in activist campaigns in order to stop the rapid disappearance of natural habitats. Notable among these activists is Jiyul, a Buddhist nun who has led many civil protests. On account of her belief that nonhuman animals and habitats are viable subjects in lawsuits, Jiyul's work has helped lead to the popularization of the "Salamander case." A type of Korean salamander was indeed a plaintiff in the case to prevent tunnel construction through Mount Cheonseong. The prominence of the Salamander case has forced Koreans, Korean Christians included, to reckon with the condition of their environment. In contrast to Buddhists like Jiyul, Korean Christians have not, as a group, been at the forefront of environmental preservation. To the contrary, they have generally supported economic progress without regard for environmental impact. This essay is an outgrowth of my deep concern about the suffering ecology of the Korean peninsula, which is currently facing its most serious crisis in history. Literally all South Korean environments have been abused and exploited, both by the Korean people themselves through national land development and by transnational corporations in conjunction with economic globalization. While rapid economic growth in South Korea has brought political stability and fiscal opportunity, these accomplishments have had significant human and environmental side effects. As seen in the completed project at Saemangeum, changes in the wetland ecosystem are progressing more radically and rapidly than even environmentalists predicted. Such dramatic change raises the specter of an uninhabitable Korean peninsula, an apocalyptic image outlined by each waft of factory pollution, each explosive charge detonating through the Korean earth to make way for a new rail line. And South Korean Christians have been slow to respond.

ECOLOGICAL CONSCIOUSNESS AND PRACTICE IN KOREAN CHRISTIANITY

In this essay I intend to articulate why South Korean Christians have not, as a lot, joined Jiyul in her protest—why they supported the Saemangeum project and other economic expansion efforts that had horrific environmental impacts. I argue that, despite various ecological programs and activities, tremendous obstacles remain to attracting the wide and strong participation of church members in environmental issues. Conflicts are continually arising due to certain distorted views of congregation members, caught up, I believe, in conventional theologies and faith styles, especially regarding creation and the future of the world. Accordingly, for an ecological ethic to function effectively in the South Korean churches, an "ecological spirituality" is required, through which it would be possible to agree that the ecological crisis is not just a problem of scientific, technological, political, or strategic consideration, but is, in fact, one of radical theological consideration.[2] In this vein, it is imperative that we critically reflect upon at least three problems in the faith and praxis of the South Korean churches, including mainline and conservative, or fundamentalist, churches.[3]

Above all, Christianity in South Korea is exalted by the myth of growth in size. It praises productivity, effectiveness, stability, and sensible enjoyment, while downplaying asceticism and concern for the weak. As South Korean society has passed through stages of industrialization, urbanization, information-related transformation, and now globalization, Christian churches have subjected themselves to the logic of global capitalism. Today, evaluations of the *quality* of ministers and integrity of their proclamations are offered mainly in terms of *quantity*—of membership growth and the size of church buildings. Though "the road of the cross" is preached, and the "evangelical faith" is emphasized, in reality, the faith of church communities is directed toward individual blessedness, material wealth, and the supposed separation of church and world.

The case of Yoeuido Full Gospel Church, the biggest Protestant church in South Korea, will serve to illustrate my point.[4] Reverend Yongkee Cho's "Five-Fold Gospel" is one of personal salvation, with associated belief in the Holy Spirit, divine healing, divine blessing, and the Second Coming of Christ. The "Three-Fold Blessing" is Cho's way of articulating the spiritual blessing, the daily blessing, and the blessing of health. In the historical transition from overwhelming poverty to capitalist "success," Cho's teaching about the "blessing" has motivated Korean Christians to pursue material wealth. The hope for the "blessing" of material wealth, accelerated by a certain faith motive, has led to the rapid growth of the Yoeuido Full Gospel Church, which has set the standard for Korean Christian faith and church growth. In the process, the Christian prophetic ethic of communal responsibility to establish social justice has been all but jettisoned. Blind acceptance of the ideology of economic and land development, including the unmerited profits gained by land speculation, has been justified in the name of the gospel of blessing.

On the other hand, the gospel of the Second Coming of Christ, refer-ring to the belief in the apocalyptic prophesies of the end of world, has influenced church members to accept dualistic binaries of church and world, humanity and nature, et cetera. In Cho's version of the Second Coming, the "rapture" is not of individuals, but of the church. With the rapture, the Holy Spirit will leave the earth; the world will subsequently stand under judgment and will be destroyed. Therefore, the salvation is not salvation *of* the world and history, but *from* the world and history. This pneumatology has no relevance to the Spirit of the whole creation, let alone the continuation and preservation of the life of ecosystems. The tendency of church teachings and theologies to apply a logic of economic development to the growth of faith communities has encouraged certain distorted theologies, especially with regard to God's relation to the cre-ation and the future of the world.

Korean churches tend to regard nature as God's creation in terms of the *anthropocentric dominion* over the earth. This sense of human domi-nance has been inherited from the theology and practice of the early missionaries in the country. The biblical injunction to "dominion over the fishes of the sea, and over the birds of the air, and over every living thing that moves upon the earth" (Genesis 1:28) has led to the idea of *unilateral* human lordship over creation in the theology and praxis of most Christian churches. In such theology, to rule over and to use nature is seen as the will of God, therefore as the criterion of virtue. Moreover, many people believe that it is God's will for human beings to promote human happiness as much as possible by utilizing all of the given natural resources surrounding them. Thus, nature, especially as land, is consid-ered as the object of human conquest and domination.

Under this theology of dominance, people have little difficulty believ-ing that other animals and creatures exist only as means to human ends. They can hardly consider the possibility that the misuse and abuse of other animals and creatures could be judged as destructive and immoral from the perspective of their religious faith. Their Christianity does not teach them the hard truth that people have been oppressors of nature. Rather, religious teachings within the churches unfortunately miss the simple lesson that human beings, from the perspective of the *Imago Dei*, should be committed to the stewardship of God's creation. In this regard,

Christianity is responsible for the popular ethos of anthropocentric dominion in the country, and the consequent use of all creation for entertainment, pleasure, and economic development that has permeated a modernizing South Korea.

The problem of dominance is especially due to the cultural influence felt in South Korea from the potent mix of Western civilization and religion. For example, the conventional Christian theological emphasis upon the transcendence of God has strongly supported the tendency to separate nature from human beings and make it the object of human domination. The theology and practice of the mainline South Korean churches and believers are therefore not free from the criticism of Lynn White Jr., historian of technology and medieval culture. In his classic critique, White explains that the current situation of ecological crisis reflects the tendency to combine anthropocentrism with human dominion.[5] In so many respects, the South Korean churches have been following the example of Western Christianity as the most anthropocentric and earth-dominating religious culture the world has seen. This tendency to divide human beings and the world of nature is radically different from traditional Asian worldviews, in which humans are part of nature, and humans, animals, and natural things together form an interconnected web of life.

A further problem within the South Korean churches is their lack of ecological consciousness concerning the use of natural resources, as shown in their blind submission to the path of *egoistic consumerism*. Under the circumstances of limited natural resources and land, people have been pushed into ever mounting levels of competition, especially in the process of economic development. The process of industrialization itself, made possible by the development of modern science and technology, has achieved progress at least in part through the conquest of natural forces. Furthermore, the effects of globalization fostered by the capitalist system have resulted in the waste of natural resources without consideration of the reproductive possibilities of nature throughout the peninsula. To our shame, people within the churches are operating under the same mechanistic view of nature and the concomitant mentality of capitalistic consumerism.

The trend of capitalistic consumerism within Christian communities has been enhanced by the egoistic individualism that has been concretized in the emphasis upon materialistic blessedness and the quantitative growth of local churches. South Korean Christianity's egotistical individualism sheds light on why South Korean Christians have been clinging to the notion of individual salvation while ignoring societal issues, such as economic injustice, environmental destruction, and so forth. For example, in South Korea, more and more local churches, and even individuals, are constructing various types of retreat places in mountains and country areas. In the process, they are developing many mountains and fields of the country without regard for environmental impact. Christianity in South Korea is thus privatizing nature in the name of salvation and blessedness, and thus speeding up environmental destruction in rural areas. Christian faith communities have been remiss in their ethical teachings about their duties to their neighbors and to future generations.

A third problem of the South Korean churches is their negligence in approaching the issue of the ecological crisis and the future of the world. This is most specifically due to their *deterministic escapism* regarding the future of creation, a product of traditional beliefs such as "apocalyptic anticipation" and "pentecostal hope."[6] According to conventional theology, the future of creation is predestined by God, and will finally be redeemed with the Second Coming of Jesus Christ. The problem with pentecostal hope, especially the belief in the imminent return of Jesus Christ, appears in the tendency to accept and use, without reservation, the abundant gifts of God's creation. Additionally, even among the mainline churches, hope for the future of the world depends upon the final judgment and intervention of God. The inevitable result is that Christian believers do not concern themselves with being agents of social change because of their apocalyptic anticipation. In this mode of thinking, the kingdom of God is portrayed as an idealized future of ahistorical, atemporal bliss.

Following the trends described above, the South Korean churches generally can, and do, detach themselves from the multiplicity of urgent issues of secular society.[7] This pentecostal zeal in the faith and practice, giving exclusive emphasis to the transcendental dimensions of the world of God, supports the proverbial "eat, drink, for tomorrow we die" mentality. Ironically enough, this is so even among the most sincere believers who are willing to recognize the reality of nature, and to be responsible in caring for the threatened earth. In other words, eschatology taught in the concepts of traditional theology more often than not leads faithful

Christians in the South Korean churches to escapism, that is, to a tendency to ignore historical responsibility for the future of this present world.

A PROCESS ECOESCHATOLOGY FOR SOUTH KOREA

Here, in order to challenge Christian believers in South Korea to transform their ecological theology and practice, and to responsibly engage themselves in various efforts to protect the country from environmental destruction, I propose a process eschatological ecoethics. This ecoethics includes specific requirements for a Christian ecological spirituality in the South Korean churches, requirements that I derive from a process-relational eschatological perspective.

Human Responsibility in the Openness and Uncertainty of Future Possibilities

Process theology, in its eschatological perspective, teaches that the future is radically open and uncertain. God is perfectly responsive to all happenings, and in the process of God's urging and luring, that is, in the workings of the primordial nature and persuasive power of God, we human beings are free to respond—or not—to God's "aiming." Accordingly, what happens within the relationship of God and humankind ultimately shapes the future. Thus, there is no assurance about the final result of human decisions. The future is therefore genuinely and radically open, especially in the sense that future possibilities will become realized depending upon God's creative "aiming" and humankind's cooperative response.

Because of this vision of the future, process theologians cannot be confident that the needed changes will occur soon enough to avoid catastrophe for humankind. Open possibilities entail that the options of the world are open to both good and evil. Consequently, a process worldview does not support the idea of an already-planned end to the world, as a part of God's creation, as has traditionally been taught in many branches of theology. Rather, in the process vision, the future is open, uncertain, and thus most risky. This eschatology emphasizes human responsibility and human effort in the establishment of meaningful relations between God, humanity, and nonhuman creatures in this present world.

Yet, as mentioned above, one can find a contrasting deterministic escapism to be typical of the eschatological faith in South Korean churches, which promulgates the complacent belief in final exemption from the tragic results of natural exploitation. This blind focus on an otherworldly spirituality has discouraged South Korean churches from responding positively to national, sociopolitical issues, such as the ecological crisis threatening the future of the people as well as the country as a whole. For instance, environmentalists in South Korea insist that, given the government's policy in favor of nuclear power, the present authorities have dared to put the burden of radioactive nuclear waste into our descendants' hands. The South Korean Christian denominations and churches do not, however, acknowledge their responsibility for the suffering that these future generations are destined to face. Rather, even while faced with ecological concerns, it seems that they are still content with popular apocalypticism and believe that "salvation is ultimately God's work and that final redemption will only come with 'glorious appearing of our Lord and Savior, Jesus Christ.'"[8]

In their theology and practice, the South Korean churches, overwhelmed by dualistic, transcendental, and deterministic worldviews, tend to define "salvation" or "redemption" as issues ultimately beyond this world, and thus certainly beyond South Korea itself. Yet they need to recognize that the world exists for the generations of the future as well, and thus the present generation must bear a serious responsibility for the condition of creation. As Clarence L. Bence concludes, "We cannot share the underlying pessimism of many . . . writers, who [have] adopted the 'lifeboat theology' of Dwight Moody, seeking only to snatch souls from the sea of life while consigning the social and political structures to destruction."[9]

Accordingly, an eschatological religiosity for the South Korean churches that fosters a more relevant ecological ethics should enable a sense of openness to future possibilities instead of any determinism about the ultimate course of creation. The notion that God is in control, or the belief that "the new"—whatever form it takes—that is to come will be the fulfillment promised by God rather than destruction entails a deterministic view of the future, and as a result, a blind optimism about the relation of God and evil. From the perspective of process eschatology, in contrast, nothing is lost; there is no escape from the effects of human decisions, and there is no guarantee of good things in the future. This process-relational theistic belief calls upon humankind to participate in ethically responsible actions for the better future of this world, God's creation. Overall, the process eschatological vision of an evolving creation with openness to future possibilities challenges conscientious Christians in South Korea to be more ethically responsive, both to their critical present situations in terms of ecological stewardship and to the well-being of future generations, including both people and nonhuman ecosystems.

The Cooperation of God and Humanity in the Process of Continuing Creation
The process-relational eschatological vision also contributes to the repositioning of human beings in relation to the rest of nature. Above all, the teaching of Whiteheadian ecotheology seeks to overcome any dualistic way of thinking about the relation between human beings and the non-human creatures of this world. Process thought sees human beings as part of a larger community that includes all creatures. John Cobb Jr. maintains that "for process theology, as an ecological theology, human beings are part of nature. . . . Humanity is seen within an interconnected nature."[10] In other words, process theology teaches that human beings, as part of nature, are interacting with the rest of creation. In this way, a process-relational eschatological ecoethics—or ecoeschatological ethics—is necessarily contrary to, and undermines anthropocentric systems of, ethics in antiecological Christian traditions.

Another key component of process eschatology regarding ecological consciousness, therefore, is the notion of the interconnectedness and interdependence of God and all creatures, including humans and all living things. The Whiteheadian tradition speaks of "the internal relatedness of living things to their environment." In Whitehead's view, all entities are related, and entities in proximity influence and impact each other so that, in the process of organic life, every actual entity has its own relevance in the whole process of becoming. Accordingly, it can be said that since the mutual participation of all entities, including God, changes the nature and direction of each entity, the direction of the creation's future is not yet determined, as noted above. In other words, the meaning of each individual life and event, that is, each actual "entity" and "occasion" in

Whiteheadian terminology, becomes significant in the sense that the future direction of creation remains open to each entity's self-determination.

Considering the insight above, a reformed ecological eschatology for the Korean context requires a holistic ecological worldview, according to which nature is rediscovered as a dynamic relational system inclusive of all human and nonhuman beings. Looking back at my own Asian background, this holistic worldview is in line with one of the principles of the Taoist tradition.[11] According to the teachings of this tradition, all things are linked to the Tao.[12] As Jung Young Lee remarks in his attempt to understand the Christian concept of God from the Eastern perspective of the Taoist tradition, "It is this vital and dynamic God who is finally responsible for the living and changing universe."[13] Accordingly, there is no way to separate the world and God, and, thus, all things are interrelated. That is, the fundamental unity of all, *in* and *through* the Tao, is central to Taoism, which emphasizes a harmonious relationship of the world with the Tao. The approach of Taoism to nature is ecologically strong precisely because of its teaching that nature is not an object to be conquered or dominated; rather, nature is our valued friend to which we, as parts of nature, should be attuned. On this topic, Sang-Won Doh writes:

> To see nature as an external object out there is to create an artificial barrier which obstructs our true vision and undermines our human capacity to experience nature from within. . . . This kind of ecological self-awareness which is testified to by contemporary science and which has Far Eastern cosmological roots is needed to overcome our ecological crisis which has been caused by Judeo-Christian anthropocentric and individualistic (or ego-selfish) traditions.[14]

A holistic ecological worldview challenges Christian believers to recognize that we are connected to the rest of creation both materially and spiritually, in the sense that, if we work *through* and *with* other beings in creation, if we respect and care for them, we will benefit both materially and spiritually.

In this regard, consider, for example, the forests, one of the world's most precious natural resources, which have nurtured human beings by providing them with a basis not only for building civilizations but also for cultivating spirituality. The Indian Chipko movement, comprised mostly of women, working for the protection of trees is exemplary of the sort of reciprocal ecological commitment I have been discussing.[15] This movement had its origin in a conflict about forest resources, concerning an unlimited deforestation of the Himalayas and other areas in India, with disastrous consequences for the environment in the form of landslides, floods, erosion, and an acute shortage of drinking water.

An essential awakening to the interactive character of our relationship with the world, offered by the process eschatological vision, begins with the rediscovery of a participatory rather than an exploitative relationship between human and nonhuman nature. Living things should be liberated from the oppressive control and dominion associated with the mechanistic subject-object dualism by which the ecological consciousness and practice of modern Christianity has been influenced since the Enlightenment.

The concepts of the interconnectedness and interdependence of God and all creatures, including human beings, has many implications for an ecological theology for Christianity in South Korea. First of all, these notions, when taken to imply mutual participation in the continuing processes of creation, teach that God and human beings are "co-creators," stressing that God is engaged in a continuous creative process, and the nature of creativity is expressed *in* and *through* every actual occasion, as well as by God. This also implies that God is not unilaterally responsible for any happenings in the time to come. God is not a parent who can be expected to "make it all better," and God is not the primary actor or "puppet-master," who is to presuppose all future directions and final results. Rather, human beings, taking their own personal responsibility, are invited to play a vital role in creating a new and better future. They are challenged to turn in actual practice to recognition of sociopolitical, ecological issues, not simply through the quiet study of ancient texts or the search for spiritual perfection through the practice of established rituals, but rather by adopting a more vigorous participation in God's continuous creation. Isabel Carter Heyward argues that God needs human beings to do the work of redemption in this torn-apart world.[16] She argues, in a way familiar to process-relational theologians, that God is limited in power and God needs people to exercise what powers they have for the good of creation.

Therefore, the fullness of God's relationship to the world should not be limited to the part of God, "one-sidedly." Rather, seen from the process perspective, there is eschatological certainty that this universe, the house of human beings and all of nature, will be protected by both God and human responsibility. The vision of the Kingdom of God, an eschatological reality, is to be realized through human participation in God's continual "aiming" and persuasive works of creation. In other words, the kingdom of God results from the relational cooperation of God with human beings and all creatures through the transforming of present realities; the Kingdom of God is to be realized through the relational actualization of divine activity and human freedom.

Accordingly, we are not allowed to leave the future to God alone. Those of us in the South Korean context can find great insight in Catherine Keller's words: "To leave the future to God means in fact to leave it to the overpowering systematic forces, to what the apostle Paul called 'the powers and principalities,' gods such as 'free market principles.'"[17] For the South Korean churches and their members, particularly in regard to ecological concerns, leaving their destiny to God alone means, in the opinion of many, leaving it to imperial powers such as the United States. The reality is that, with respect to the issue of the ecological threat of nuclear plants and armaments, the South Korean churches and believers exhibit a disturbing tendency to leave the final choice to foreign powers and the highest South Korean authorities.

Of course, the evangelical faith of the majority of South Korean believers rules out the possibility that the self-decision of human beings could be an ultimate factor in the kingdom to be realized in the future. They *do not want* to believe that God can unilaterally ensure nothing. At present, it *is not possible* for them to accept the idea that no one can guarantee whether God's desire for human beings will ever, in fact, occur. They *cannot accept* the conclusion that the very survival of the human race depends primarily on whether we ourselves make the right choices.

South Korean Christians may agree with Jürgen Moltmann's concerns with process theology. Regarding the process-relational perspective on the cooperative work of God and human beings in the continuing creation of the world, Moltmann warns against a possibility of "divinization of the world or nature" in process theology, in which the Christian hope can be *reduced* to something that can be realized by human activity, or to something that is limited within this present history. We are, however, made in God's image, and that image includes creativity. We rightly understand the true meaning of "dominion" (Gen. 1:28) as creatively caring for and preserving our earth. As noted previously, human activity is best understood as "co-creating" with God. Process eschatological teaching on ecojustice holds that, in our freedom, we have the power of self-determination, and we may choose whether or not to follow the initial aim, whether to follow one path or another. In our ongoing responsiveness to God and to other realities, we may choose good or evil, ultimately affecting the destiny of creation. Thus, in the South Korean churches, the process message that true relationships among God, nature, and humanity can be built upon degrees of freedom and decision pushes church believers to be more responsible for the preservation of creation as God's *creatio continua*. More specifically, we Koreans on the peninsula must preserve the future of our country for our survival and posterity. The South Korean churches and believers, in their awakening eschatological spirituality, in becoming more responsible for the destiny of the world, must take responsibility for the ecological crisis of their country, rather than yielding to the intervention of the "absolute power of God" or the political power of neighboring—or, in the case of the United States, distant—countries.

Furthermore, another key message from process eschatological ecospirituality for our ethical awakening is reverence for the diversity of ecosystems in nature. That is, the idea of cosmic interdependence places emphasis on considering and loving others in the world. Thus, an adequate environmental ethics should value individual entities in themselves and in their connectedness as part of a purposeful natural order. Ecological ethics should be based on the idea of the sanctity of all human and nonhuman lives. Since all living beings are mutually indebted to one another, it is a human being's duty to pursue mutual co-existence and mutual benefit in our relationships with other sentient beings.

Unfortunately, people in South Korea are well accustomed to the system of conflict and competition in their small society. There is a shameful belief that in order to live for oneself, one has to shun another being, and accept that action without reservation. The belief means that one's own survival is possible only at the expense of others. This situation of all-encompassing competition leads to devastation and destruction in nature, especially in the modern political and economic climate of the country, where the ideals of growth and development have been highly emphasized, as in many other developing Asian countries. People must know, however, that, in order to survive the present dilemma of ecological destruction, their lives must stop the vicious circle of mutual conflict and competition permeating society, in both secular and nonsecular settings. Therefore, we who are concerned about these issues need to extend an ecological spirituality, and beyond the religious communities themselves, which should lead to the awakening of the rich connectedness of all beings in nature. Bokin Kim demonstrates that Buddhists in Korea are progressive in this regard. He describes connectedness as a cyclical process of mutual benefit:

> For example, a human corpse decays underground, making the soil rich. Then the grasses that grow thick around the spot will be made into compost. The compost will grow rich crops, and people eat those crops, which supply blood and flesh as the sources of their energy for actions and life. Every autumn trees shed their foliage, which is gradually absorbed by the soil and is degraded by nutrient-extracting microbes. In the spring these nutrients are returned to the tree and form fresh foliage.[18]

Recognizing interdependence, South Korean churches need to recognize the crisis that our interconnected ecosystem confronts, and not only at the valuable wetlands of Saemangeum. The South Korean churches, as eschatological communities reflecting the Kingdom of God in history, must find their responsible role within the human community, and, at the same time, their role in relation to the community of the whole of nature, especially through resisting the distorted, development-centered, egoistic tendency to place humankind over and against all other creatures in a manipulative, dominating, exploitative way.

Finally and in concluding, it should be reiterated that a sociopolitically responsible theology in South Korea is one of care for the threatened earth. In the process-relational view of God, God preserves all, and in God all things move toward the vital harmony that is the divine purpose. That is, God affects everything and is affected by everything that occurs and by what happens to everything. In short, God is the experience of all the experience of all creatures. Thus, what matters to the individual also matters to God. Therefore, when we cause some suffering to any animal or any part of nature, we, in fact, cause suffering to God. As Gregory Moses writes, "Thus, what hurts the environment, in almost a literal sense, hurts God. In such circumstances, loving the Lord my God with all my heart and soul and strength and mind, and loving my neighbor and everything else as myself are inevitably part of the same story."[19]

David R. Griffin, citing Matthew 25,[20] emphasizes the point: "If God shares the feelings of *all* his creatures, then whatever we do 'unto the least of *these*' we do unto God. This idea, added to the idea of the intrinsic value of all individuals, provides an additional basis for Christians to develop an ecological ethics."[21] In other words, God is always growing, because God experiences all that every entity experiences, and prehends all that happens in the world. As the universe grows, so God grows. Furthermore, according to the process-relational concept of the inner life of God, because of the experience and prehension of God, the universe is God's body. Sallie MaFague maintains that this model of the world as God's body, in contrast to the monarchical model encouraging attitudes of militarism, dualism, and escapism, encourages holistic attitudes of responsibility for, and care of, the vulnerable and oppressed. McFague explains: "The model of the world as God's body encourages holistic attitudes of responsibility for and care of the vulnerable and oppressed; it is nonhierarchical and acts through persuasion and attraction; it has a great deal to say about the body and nature."[22] Thus, when we embrace this eschatological ecospirituality, a new shape emerges for humanity. As McFague maintains, we are "decentered as the point and goal for evolutionary history and recentered . . . as the stewards of life's continuity on earth."[23] The new place and vocation that have been given to us are shaped by a new calling, "the calling to solidarity with all other creatures of the earth, especially the vulnerable and needy ones."[24]

This process eschatological idea of the ecological life of God teaches us how we must deal with the marginalized in society as well as in nature itself. In South Korea, for example, an adequate ecological theology that is intended to work on behalf of ecojustice needs to see clearly both the plight of people and of nature. In South Korea, nature should be defined as the *minjung* of the *minjung*, whose liberation is the goal of Minjung Theology.[25] For, as Sang Sung Lee concludes, "without the liberation of the ecosystem, there can be no liberation of human beings because every oppressed and oppressing group is unshakably related to each other."[26] In South Korea (and elsewhere), where people have been oppressors of nature, all ecologically threatened and dying neighbors within creation are merely the objects of marginalization and exploitation. In the process-relational perspective, God values all, even "the little ones"—the nonhuman as well as human—because they, in and of themselves, are also intrinsically valuable. In the hoped-for ecological practice of faith among the South Korean churches, it should be remembered that God considers the marginalized intrinsically valuable in and of themselves; we, too, are challenged to experience them as valuable *in* and *of* themselves.

Marjorie H. Suchocki also helps us realize that the South Korean churches need to hear anew even such traditional texts as that of Paul, declaring that "all creation groans in travail as it awaits its redemption."[27] In this age when all creation longs for justice, for care according to the image of God, and for the full realization of the Kingdom of God, the South Korean churches must hear the cries of the fragile, just as Jiyul persists in listening to the silent sufferings of the salamanders. As Suchocki notes, the conscientious believers in the country need to confess: "We finally hear in earnest our responsibility to nature. We realize that we, too, are nature, and that our caring cannot be restricted with impunity to sisters and brothers in the human community but must extend toward the earth and sky."[28]

WORK FOR THE EARTH: CONTINUITY OF WORLDLY HISTORY AND THE KINGDOM OF GOD

A relevant Christian hope should carry a formulation that can draw forth Christian concern for the present world, including the poor, the oppressed, and the exploited earth. As Rosemary Radford Ruether asserts, we need a "historical eschatology" that takes seriously the realities of our *present* temporal existence and which holds up a "just and livable" society as normative in *this* age, rather than subjecting ourselves "to the tyranny of impossible expectations of final perfection."[29] In such a "historical eschatology," the notion of God derived from Greek metaphysics—that of an atemporal, unchanging God incapable of suffering—on which traditional theologies have depended, must be rejected in favor of a God who is more intimately related to process, temporality, and history. In other words, Christian faith or Christianity's set of doctrines—and above all, its doctrine of eschatology—should be compatible with an affirmation of this present, historical world. An adequate Christian theology must rethink its eschatology and doctrine of God, and demonstrate the compatibility between eschatological belief in God and affirmation of the world. In this regard, process thinkers rightly "find many apocalyptic formulations disturbing at this point," because they believe that "apocalyptic interpretations picture a future so disconnected from the present as to weaken concern to realize what possibilities the present holds."[30]

In most Christian theologies of the future of the world, however, an exclusive emphasis on the "not yet" of God's future tends to degenerate into utopian thinking. People easily come to expect the intervention of that which we may call "the act of new creation by God." That is, as discussed above, apocalyptic faith has often stimulated quietism regarding ethical, sociopolitical issues, although it also has often been considered as an impetus to passionate activity. This tendency applies easily to the South Korean churches in general; they believe that an apocalyptic way of thinking about the future is the most genuine Christian way. On this point, Cobb raises a critical question: "Is apocalypticism normative Christianity?" He answers definitively:

> If it disconnects what God does in the world from our responsibility to be God's agents, it is not Christian. Passive waiting for God to solve our problems is not Christian. The invalidation of the meaning of genuinely historical events as a locus of divine actions is not Christian. Abandoning our normal responsibilities in the world to await a supernatural incursion is not Christian. Supposing that we can calculate the day and the hour when an apocalyptic event will occur is not Christian.[31]

Thus we need to turn to the process belief in the resulting reality of God's consequent nature, which affirms that the Kingdom of God lies in a continuation of the temporal world. That is, the process eschatological perspective teaches that the new does not displace the past, nor does it take place in discontinuity with it, but rather creatively transforms it. The "eschatological continuity" between the "already" and "not yet" Kingdom of God might imply that the apocalyptic act at the end of this age will not be one of total annihilation of the world but rather one of total transformation of the world. Therefore, this process vision of the *eschaton*, or the Kingdom of God, includes acknowledgment of worldly reality, resulting in an ethic that is adequate with respect to valuations and moral directives that affirm creational and worldly life. Process eschatology teaches us that we must not diminish the theological emphasis on this worldly history in our attempts to challenge our ecological religiosity. Furthermore, in the process notion of eschatological continuity based on God's consequent nature, it is guaranteed that noble human efforts will not be wasted. Finally, there is hope for the future, which is open for us both individually and collectively, because as individuals and groups change, God incorporates those changes into the shaping of the future. In particular, the process vision for the future asks human beings to be more ecologically responsible for God's *creatio continua*.

Catherine Keller thus asserts, "The ultimacy, the *eschaton*, of eschatology designates the radically inclusive spatio-temporal edge of our existence and therefore the shared future of our earth-creaturely existence."[32] In another article, considering the link between ecology and eschatology, Keller argues that eschatology, meaning "discourse about the ultimate, or the end," therefore is "discourse about the collective encounter at the edge of space and time, where and when the life of creation has its chance at renewal—that is, it is about the present."[33] Indeed, our eschatological missions are not toward "a life *after* life but life itself."[34] As Keller concludes, "home, taking on the edginess of eschatology, is not the 'end' but the earth as it is threatened."[35] For Keller, therefore, "our responsibility for the new creation . . . is, to participate in our finite, interconnected creatureliness with metanoic consciousness—that is, facing up to the *man*-made apocalypse, resisting the North American array of post-utopian cynicisms, pessimistic determinisms, reactionary Christian messianisms, and business-as-usual realisms."[36]

Theologically, faithful Christians in Korea can no longer separate their own destiny from that of this present world, that is, from the whole of nature. An ecologically adequate Christian eschatology would oppose any utopian "end" of history. An ecologically relevant eschatology for the South Korean Christian faith must detach itself from an exclusive emphasis on the discontinuity between this life and the next. Also, the so-called pious Christians inside the churches must be warned against the tendency toward a private salvation without serious consideration of the future of this present world. An emphasis on the ecological future of this temporal world is rightly a matter of faith—of evangelical passion—in the sense that we move toward the ultimate as humankind *within* history. Considering the future of South Korea, the ethical issues surrounding the nuclear power industry raise the need for an ecoethic aimed at bridging the generations.[37] The South Korean churches must approach the issues of nuclear industry and weapons as questions crucially related to future generations.

In conclusion, in a process-eschatological spirituality, Korean Christianity is challenged to turn its emphasis to the earth as its true home. As Keller writes, "the recycling of eschatology becomes precisely a means of the *metanoia* of theology itself—returning the earth."[38] In the vision of a new ecological eschatology, the ground of the new community that people can hope for and work for is *here: their home, our earth, our land,* suffering as it is from development and exploitation. Even in our eschatological faith, the earth is still our home. Keller explains, "We are here to claim, to defend, and to renew our earth home, the inhabited whole. This is the task of the green ecumenacy."[39] Keller gives a warning against the "unearthly eschatology" draining energy away from our earth-home and encouraging us to live in an orientation toward a many-mansioned heavenly home.[40] As Keller warns, interest in another life must not steal its energy from this life.[41] I believe that this warning is applicable for the South Korean churches and believers who are responsible for the ecological crisis, because in their practice of eschatological faith as an escapist utopianism "distraction from the earth complies with the destruction of the earth."[42]

The believers in the South Korean churches must depart from their hope for returning to a home that is supposedly prepared by God, not here and in this world, but in another world beyond. Rather, the earth, facing the threat of total destruction by a multidimensional ecological crisis, is our home. Accordingly, the eschatological ecoethical spirituality that the South Korean churches should restore includes a determination for turning back to home, to the earth. The ecological task the South Korean churches are facing is, as Keller aptly terms it, our *"home*work." Now, Keller concludes, "The question is whether Christian hope—in collaboration with other, earth-friendly traditions—can also energize the needed *home*work."[43]

When challenged by a process-relational eschatology and its ecoethical implications, with its world-affirming emphases upon the uncertainty of future possibilities, the cooperation of God and humanity in the continuation of creation, and the preservation of this present world in eschatological realities, our new ecospirituality and theological ecoethics might express the relevance of Christian faith to conscientious people in the present world, who need an ultimate ground for their "ineradicable confidence" in the final worth of human and nonhuman lives. Similarly, ethical perspectives coming from process eschatology challenge the South Korean churches and believers to seriously consider such ecologically relevant teachings as open possibilities, human contribution, and the continuity of history. These teachings are relevant to the future of people and of creation in the country. For South Koreans, both believers and nonbelievers alike, a decision must be made regarding a relevant ecological ethics, "not just for the sake of an abstract future," but for "a new community" already beginning to form in the practice of ecojustice.[44]

❧ Ecospaces: Desecration, Sacrality, Place

⋄ Restoring Earth, Restored to Earth: Toward an Ethic for Reinhabiting Place

DANIEL T. SPENCER

INTRODUCTION

The sun strikes the rust-orange tailings of the now-abandoned copper mine as the group of worshippers breaks into song at the Easter sunrise service, celebrating the renewal of their faith and of their commitment to restoring the toxic landfill beneath their feet. Hundreds of miles to the east, flames crackle across the open field, recently planted in corn, now returning to tallgrass prairie, as college students and community members fan out to shepherd the fire through the dried grasses and recently re-established native plants. Amidst the Rocky Mountains several states to the west, local citizens pause from uprooting exotic and noxious weeds from the open space of their beloved and protected mountain slopes to take in the spring green in the valley below. Nearby, schoolchildren gather in a barn-turned-laboratory to learn to identify native flora and fauna before heading out into the floodplain fields along the river to help with the annual inventory of the local wildlife refuge in an effort to return lands and waters to a more native state. In the city downstream, residents of an abandoned warehouse, now become sustainable housing, tend a rooftop vegetable garden and rip up old concrete slabs and noxious weeds that surround the building, replacing them with ecologically adapted native plants.

In locales as diverse as the alpine valleys of the North Cascades in Washington State, the Minnesota tallgrass prairie, and the hillsides of western Montana, the meadowlands and wetlands of New Jersey, and the everglades of Florida, equally diverse gatherings of local community members are restoring damaged and degraded ecosystems at the local and regional level while reinhabiting these places with an expanded sense of home. In doing this, they reintegrate the human community into the broader natural world within a vision of social *and* ecological sustainability. Making up the grassroots component of the burgeoning field of ecological restoration, these communities and thousands more like them illustrate with particular vigor and promise the vision of *sustainable community* that ethicist Larry Rasmussen makes the cornerstone of an "earth ethic" in his groundbreaking work, *Earth Community, Earth Ethics*.[1]

Ecological restoration has grown rapidly in the past twenty years as a science, a philosophy, and an ethic. As a philosophically grounded ethic, restoration sees nature and humanity as fundamentally united and seeks for ecologically sustainable ways that human communities can participate actively in nature. Some environmentalists, however, are cautious, even skeptical, about having restoration become the basis of an environmental ethic, seeing it as simply the latest justification for ongoing human intervention in natural systems, rather than learning to adapt our communities to the constraints of ecological systems. Below I explore some of the dimensions of sustainable community that these local efforts at ecological restoration exemplify, and argue that they illustrate a particularly promising component of an ethic of sustainability rooted in place. Such practices in turn can ground a broader restoration ethic of reinhabiting our locales in ecologically sustainable ways that are themselves restorative: in restoring the earth, *we* are in turn restored *to* the earth. These practices and the ethic they ground can help us to rethink and reshape our views of humanity and the humanity-nature relationship that is implicated in so many contemporary social and ecological issues.

In my book, *Gay and Gaia: Ethics, Ecology, and the Erotic*, I sought to develop the basis of an ecojustice ethic of right relationship by reclaiming and revisioning both the erotic and the ecological at all levels of our lives— from our deepest, most intimate relationships with self and other to our location in an evolutionary, expanding cosmos.[2] Given the contemporary ecological crisis that contains and frames all other issues of human concern, ecological ethics must become the grounding for *all* ethics; what is needed is a fundamental shift from an anthropocentric, human-centered worldview to an ecocentric, all-of-life-centered worldview.

I use the concept of *ecological location* as both an *analytical* and a *descriptive* term to help us understand the complex ways each of us is located *both* socially *and* ecologically within societies, cultures, and ecosystems. As a *normative* term, ecological location suggests ways that we *should* be located that respect the ecological integrity of the biotic communities (and their members) that also contain us.[3] My concern has been to rethink within an ecocentric framework the important liberationist attention to social location and socially constructed categories of nature and human nature, connecting them now with ecological theory and ethics.

Central to this concern is recognizing the participation and agency of nonhuman members of biotic communities and to find ways to integrate their agency and integrity into ethical method. "Ecological location" expands "social location" to include both where human beings are located within human society and within the broader biotic community. Importantly, it includes other members of the biotic community—and the biotic community itself—as locatable, active agents that interact with and shape the other members of the ecological community, including human beings. Just as "social location" is an anthropocentric term that helps us to pay attention to how human identities are multiply formed with respect to various lines of human difference, "ecological location" is an ecocentric term that recognizes that human epistemologies—how we see and interpret the world—are also shaped by our relationship with the land and other creatures in our broader biotic environment.

While this definition emphasizes the descriptive and analytical dimensions of ecological location, I have continued to explore the normative dimensions of what ecological location might say about how humans—as individuals and in communities—*ought* to relate to the broader biotic communities and ecological processes in which we live, move, and have our being. Clearly one cannot do this without paying close attention to how we understand humanity and the humanity-nature relationship (or, more accurately, *relationships*, since this will necessarily take on the particular dimensions of particular places, cultures, and ecological realities that may or may not hold true for other places and cultures).

Most recently I have become involved in the practice and movement known as *ecological restoration,* which I see as providing one promising framework for an ecologically sustainable and ecosocially just ethics that integrates aspirations for human liberation and justice with respect for the ecological integrity of the places and biotic communities we inhabit. Although my thoughts are still preliminary, ecological restoration has much to contribute—as both a practice and ethical framework—to an ethic of sustainability and justice that honors the wisdom of the earth and the integrity the earth manifests. By working to restore the ecological integrity of the earth, in the process we restore *ourselves* to the earth's integrity. I begin with some preliminary remarks about epistemology, and then turn to an examination of ecological restoration itself—what it is and some of the philosophical issues and debates it has fostered. I then turn to ecotheology, and offer some comments about the possibilities of "restoration" as a Christian theological metaphor that can be reworked within an ecological and evolutionary framework to ground a restoration ethic.

PRELIMINARY REMARKS ABOUT EPISTEMOLOGY

Because ecological ethics in general, and ecological restoration in particular, draw heavily on the natural sciences in shaping an ethic, it is important to say a word about epistemology. The works of feminist historians of science Donna Haraway and Sandra Harding are useful in developing a critical approach to science that engages four primary stances: a critical scientific realism, identity politics and insights, social constructionism and deconstructionism, and standpoint privileging of perspectives. These help to generate a social-ecological hermeneutics of suspicion that takes seriously both scientific and social claims but deals with them critically, and they provide the necessary methodological grounding for a liberationist approach to ecological ethics.

A brief further word about each may be helpful. To be "liberationist," an ecological ethics must take seriously the knowledge claims of the natural sciences while recognizing that science is always a human practice reflecting critical cultural values and assumptions. In my work on social and ecological issues I am committed to liberationist politics and theories, such as feminism, antiracism, postcolonialism, and "lgbt" liberation,[4] with their reliance on identity politics and a hermeneutic of suspicion to highlight issues of justice. Liberationist identity politics have demonstrated how a strong sense of group identity both empowers liberation struggles while revealing aspects of social reality too often obscured by the dominant social practices and discourses. Developing a hermeneu-tics of suspicion out of these perspectives is critical for interrogating "ob-jective" narratives of science and nature that too often hide the values, assumptions, and power relationships that produced them and may inad-vertently be reproduced by them. The danger of an uncritical use of identity politics is their tendency to reproduce rigid boundaries and fixed social identities that result in new patterns of exclusion and invisibility.

Hence an important epistemological complement stems from social constructionist perspectives on science that critique positivist and realist epistemologies that reduce what can be known to empirically testable phenomena and conflate the truth of the scientific account with the real-ity it tries to describe. The problem with a sole reliance on constructivism is the danger of ending up in a kind of postmodern relativism, where every position is seen merely as a narrative of power, devoid now of grounds for making evaluative judgments between competing claims. Scientific narratives are too influential for an ecological ethics to merely dismiss them; rather, a critical appropriation is what is needed.

To arrive at such a critical appropriation, the insights of standpoint theory are helpful. Donna Haraway's influential essay "Situated Knowl-edges: The Science Question in Feminism and the Privilege of Partial Perspective" shows how this epistemological privilege comes from criti-cal reflection on social structures "from the underside" to generate criti-cal knowledge necessary for social liberation.[5] This in turn raises the difficult theoretical issues of how the agency of nonhuman actors occurs and is articulated within standpoint perspectives.[6] Yet shifting our grounding from fixed identities and "natures" to locatable affinities allows us to take advantage of the many strong points of standpoint theory without becoming trapped in rigid, fictional identities that do not accurately reflect the fluid experiences and shifting contexts that inform them.

With these positions as grounding, we can and must develop a critical appreciation for science and scientific narratives. While science is always a human practice and discourse that necessarily reflects the social and ecological locations that produce it, it is not reducible to these—it is not simply about power and control, as some constructionists have argued, but about producing credible accounts of the natural world. Therein lies its power: "Scientists are adept at providing good grounds for belief in their accounts for action on their basis. Just how science 'gets at' the world remains far from resolved. What does seem resolved, however, is that science grows from and enables concrete ways of life, including particular constructions of love, knowledge, and power."[7] Ecological restoration is grounded deeply, but not uncritically, in science. Restoration of perceived ecological integrity to the places and biotic communities we inhabit is always a human judgment and action, fraught with as many social and ecological contradictions and paradoxes as other social practices. A critical epistemological grounding is imperative to avoid merely reproducing dominant social narratives and destructive practices in acts of restoration. To this we return below.

IMPLICATIONS OF ECOLOGICAL LOCATION FOR ETHICS: A RESTORATION STANCE ROOTED IN ECOLOGICAL RESTORATION

The biggest story of the twentieth century, as Rasmussen notes in *Earth Community, Earth Ethics*, is the fundamental change in the relationship of the human world to the rest of the earth: never before have humans been powerful enough to alter or destroy planetary life-systems. Rasmussen then poses a challenge: "If the great new fact of our time is that cumulative human activity has the power to affect all life in fundamental and unprecedented ways, then *what ought to be* is precisely what needs to be taken into account. . . . How *ought* we to live, and what ought we to *do* in view of a fundamentally changed human relationship to earth, a relationship we only partially comprehend?"[8] This question lies at the heart of philosophical and ethical reflection underlying the practice of ecological restoration.

Rasmussen argues that an earth ethic must favor "a downward distribution of economic and social power and heightened status for all forms of life, human and other."[9] It is just these features that make community ecological restoration such a promising element of an earth ethic. While often working in tandem with more highly technological forms of restoration, focal restoration practices are intensely local, drawing on local economic and social power to ground the practices. And they are premised precisely on a deep respect for all forms of life, restoring the ecological integrity of ecosystems through respect for both the indigenous flora and fauna and the ecological processes of these lands. By grounding the practice of ecological restoration in ecological location, the social dimensions of human community are reintegrated into an ecological framework in accord with ethical norms of ecological sustainability, social justice, responsible human agency, and respect for nonhuman nature.

So what is ecological restoration, and why do I find it so promising for framing a just and sustainable ethic for our postmodern times? In 2002, the Society for Ecological Restoration (SER) defined ecological restoration as "the process of assisting the recovery of an ecosystem that has been degraded, damaged, or destroyed."[10] Implicit in this definition is that human agency has been the primary, if not exclusive, source of the degradation of the ecosystem to be restored—hence the question of moral responsibility for restoration has emerged as a key component of ecological restoration.

As a philosophically grounded ethic, restoration sees nature and humanity as fundamentally united and seeks for ecologically sustainable ways that human communities can participate actively in nature. This is reflected in an earlier SER definition of ecological restoration as "the process of assisting the recovery and management of ecological integrity. Ecological integrity includes a critical range of variability in bio-diversity, ecological processes and structures, regional and historical context, and sustainable cultural practices."[11] The last two elements—historical context and sustainable cultural practices—acknowledge both that humans have long been an inherent part of many, if not most, ecosystems, and that any viable notion of restoration must include the social dimension of sustainable human cultural practices. Advocates of restoration thus see it combining "the best elements of two forms of environmentalism: the conservationist's willingness to participate in the ecology of natural habitats and the preservationist's insistence on the inherent value of these habitats and ecosystems, independent of their value to humans."[12]

Ecological integrity becomes a key concept and value in this practice, integrating scientific knowledge with human judgments. Because ecosystems are dynamic, changing over time, the variables listed above that determine the degree of ecological integrity will also vary—but *within* a critical range that allows ecosystems to evolve sustainably. Ecological degradation occurs when human (and other) impacts exceed this critical range. Deciding *where* within this critical range to restore an ecosystem is a human judgment, relying on several other values and criteria the restoration community must articulate.

Hence advocates understand restoration as an ethic that links engagement *in* nature with respect *for* nature, one that overcomes the humanity-nature dualism. They argue for human participation within natural systems while respecting the ecological constraints of those systems. All of these features exemplify what Rasmussen proposes for sustainable community as the heart of an earth ethic: expanding the moral framework of community from human community to the entire earth, and locating humans responsibly within both the greater earth community and our local ecosystems and bioregions.

Some environmentalists and environmental philosophers are cautious and others deeply skeptical about having restoration become the basis of an environmental ethic, seeing it as simply the latest justification for ongoing human intervention in natural systems. If the combination of modern science and technology have been *responsible* for the current ecological crisis, they ask, *why* should we trust this combination to resolve the problem, no matter how noble the intentions of restorationists? They argue instead for an ethic of ecological humility, focusing on constraining further human alteration of natural areas, and preserving and protecting remaining wilderness.

The Australian environmental philosopher Robert Eliot began the debate with his 1982 essay, "Faking Nature." Here Eliot articulated his concern that restoration is simply the latest way to justify human intervention in and alteration of nature—that it does *not* "restore" nature but rather produces a forgery, a "fake" nature. He argued against what he termed the "Restoration Thesis": "the destruction of what has value in nature is compensated for by the later re-creation of something of equal value," but the compensation is unsatisfactory, resulting in what Eliot calls "restoration as forgery": the attempt to replace something of value with a fake. Eliot went on to argue that restoration is actually a Trojan horse for industry because it serves to justify and rationalize the exploitation of nature: we can extract value from nature and then restore and replace the original. Hence, the argument goes, we can get *more* value out of nature by developing it rather than preserving it: both the *economic value* from developing it, and we retain its original *natural value* by restoring it.[13]

A decade after Eliot, Eric Katz in his essay "The Big Lie" argued that restoration is misguided and a dangerous distraction for environmentalism. Restoration is simply further human hubris: a justification of continued domination of nature. Restoration is a human artifact, not natural; it exists by human intention and design and reflects human power *over* nature, an imposition of anthropocentric values. The result, according to Katz, is the increased *humanization* of the natural world: the limitless expansion of human power to mold and manipulate nature. Restoration becomes the pre-eminent device for managing nature, rather than having humans adapt themselves to nature's rhythms.[14]

Philosophers of restoration Andrew Light and Eric Higgs have addressed Eliot's and Katz's arguments in several writings and conferences.[15] Environmental activist Adam Rissien ably summarizes the debate as follows:

[T]he main arguments against restoration are that it will provide industry an apologia for damaging ecosystems; it will take away from preservation efforts; it is the practice of human hubris; and it ultimately robs nature's inherent value. Restorationists respond by explaining restoration is not an excuse for exploitation; it does not seek to replace originality that ought to be preserved wherever possible; it can provide a way to restore human relationships with nature; and it is not a practice of human domination that robs nature's inherent value.[16]

Eric Higgs, environmental philosopher and former president of the Society for Ecological Restoration, has given perhaps the most detailed and sustained philosophical examination of issues surrounding ecological restoration in his 2003 text, *Nature By Design: People, Natural Process, and Ecological Restoration*. Here, as in an earlier influential article, Higgs asks, "What makes for *good* ecological restoration?" He identifies four key components: "ecological integrity," "historical fidelity," "focal practices," and "intentionality," or "wild design." As discussed above, "ecological integrity" refers to the wholeness of ecosystems—all the natural components and processes "that allow an ecosystem to adjust to environmental change."[17] Because nature is dynamic, practitioners of restoration seek not so much to return an ecosystem to a snapshot condition in one particular time, but rather to what they term its "historical range of variability" deduced from "reference conditions" drawn both from historical records and remnant ecosystems.[18] Key to this notion is the understanding that ecosystems have their own dynamic integrity that often has included human participation, but that this integrity can be, and too often has been, degraded or destroyed by intensive or industrial-scale human activities. Commitment to "historical fidelity" means intentionally valuing the predisturbance conditions of the ecosystem, and seeking to return it to this range of conditions—which may include or exclude ongoing human participation and presence. "Focal practices" emphasize local community participation in restoration activities to reintegrate human individuals and communities into ecosystems and place. Finally, these efforts all involve human "intentionality"—that is, design—but paradoxically in order to restore *wildness*: nature's integrity to be able to function sustainably *free* of large-scale human manipulation.

An important contribution to this debate is Albert Borgmann's philosophy of technology. Borgmann argues that technology is no longer a neutral means that humans employ to seek particular ends, but rather has become the overall cultural framework through which we "take up the world."[19] That is, we live in a world of human artifice where contemporary life is constituted technologically—technology itself has become the pervasive pattern and means that shapes how we experience and relate to the world. One result of technology as framework is the ongoing commodification of the world—both natural and human. We live in a context of what Borgmann terms "hyper-reality," where the loss of direct engagement with the natural world leads to a loss of moral commitment to it.[20] There is a profound irony here, Borgmann notes. The allure of technology is its double promise of liberation and enrichment: it liberates us from the drudgery of menial labor that has constituted human experience from Adam and Eve's expulsion from the Garden in order to give us time and leisure for more enriching activities. Yet as Eric Higgs observes, "Technology fails to deliver on the promise of liberation and happiness primarily because it *distracts* us from things that matter . . . things of *enduring significance* that have 'commanding presence, continuity with the world, and centering power.'"[21] Rather than engaging the world directly through meaningful activity such as work, within the contemporary context of the technological commodification of everything, the primary experience of the world becomes consumption. As the natural world becomes a commodity for consumption, moral regard for the natural world dissipates.

Borgmann's antidote for this comes through the recovery of what he calls "focal things" and "focal practices"—things and activities such as shared meals or the arts that are imbued with meaning and located within a network of social and natural relationships requiring our engagement and focus. He argues that such things exist within a dense context of relationships; the communion between self, thing, and world generates meaning in our lives. Commodification separates things from their context, depriving us of focus and meaning, ultimately leaving us bored and distracted—and oblivious to the underlying social and economic practices that constitute our society. Increasingly separated from nature, we also see less and less the systemic social and economic practices responsible for degrading and destroying nature. We can resist this through intentionally recovering focal practices that center our lives on meaningful things and activities—the very things that Larry Rasmussen has argued create and maintain sustainable community.[22]

Eric Higgs applies Borgmann's framework to ecological restoration to distinguish between two trends in the field. First, *technological* restoration works largely within the culture of technology and the commercialization and professionalization of restoration practices to "manufacture" restored "natures"—a process that can lead to the further commodification of nature. Second, what Higgs calls "focal," or "participatory," restoration is more local and grassroots: it integrates local populations and their cultures into a participatory process of engaged relationships between people and ecosystems. Higgs is deeply concerned that the pervasiveness of the technological paradigm will lead to the further commodification of nature *and* ecological restoration as a practice. Rather than restoration becoming a means and a practice for communities to develop appropriate "ecological humility" to reintegrate ourselves into ecologically sustainable and restored landscapes, restoration becomes the latest technological "fix," whereupon we use science and technology to re-engineer nature in our image. Artificial wildernesses and "improved" natures of the sort found at, for instance, Disney World's "Wilderness Lodge" in Orlando, Florida, employ the tools of ecological science and technology to create new configurations of "nature" that increasingly shape our very notions of what counts *as* nature—and alleviate us from having to rethink our place within natural ecosystems. The rapid growth and professionalization of ecological restoration as a field creates its own momentum in this direction and threatens to disempower community participation by leaving restoration in the hands of licensed experts and professionals. Focal-participatory restoration presents an antidote to this by anchoring restoration in local communities and their reinhabitation of place.

Higgs is aware that we are not faced with an either/or dichotomy between technological and focal restoration. "A merger of the two, technological and focal, is necessary," he writes. "They should not be mutually exclusive options. Imbuing restoration with scientific rigor and clarity is essential. We need more and better science to understand the processes of weedy invasions, seed survival, successional pathways, long-term resilience, and so on. Conversely, technological restoration needs broad engagement to ensure the success of ambitious projects."[23]

Higgs's focal restoration exemplifies my notion of "ecological location," which straddles both sides of the traditional "is/ought" dichotomy. Ecological location serves as an analytical tool for how humans relate to nonhuman nature by integrating social categories into an ecological framework. Ecological location also can provide ethical guidelines based on ecological sustainability and social justice for *how* this integration *should* take place. Ecological restoration exemplifies one promising path toward a more ecologically sustainable and socially just reintegration of the social within the ecological. It can help to reintegrate humans into the ecological fabric in a constructive ethic of hope and ecological agency. It does this by having human individuals and communities refocus on *place*, on the *particularities* of *local* places, both ecological and social. In that sense, participatory restoration practices serve as one important response to the massive, anonymous, and consumptive momentum of economic and cultural globalization—they recognize the particularity of both human and nonhuman agency as they interact ecologically and socially in particular times and places. Some today argue that such local practices—long the norm for human societies—will once again emerge as normative as globally we pass "Peak Oil" and the age of inexpensive fossil fuels,[24] making a globalized economy prohibitively expensive and ecologically ruinous.[25]

TOWARD A RESTORATION ETHIC: THEOLOGICAL AND PHILOSOPHICAL IMPLICATIONS

In this final section I sketch some possibilities for restoration not merely as a set of practices that can restore ecological integrity to particular places, but also as a framework itself for a sustainable and ecosocially just ethic—that is, a restoration ethic. Such a proposal necessarily raises important theological and philosophical issues about the set of divine-human-nature identities and relationships that this volume addresses.

More than any other advocate of ecological restoration, long-time practitioner William Jordan has sought to work out the parameters of a restoration-based environmental paradigm. In his *The Sunflower Forest: Ecological Restoration and the New Communion with Nature*, Jordan argues for a new environmental paradigm based in the practice and insights of ecological restoration. Jordan argues "for the value of restoration not only as an effective act—a way to recreate and maintain landscapes—but also as an expressive and symbolic act—a context for the creation of meaning."[26] For Jordan restoration is a praxis that both generates and grounds ethics, a process where the actual values of the ethic are created and refined:

> Such acts . . . then become contexts for the creation of values such as community, beauty, and meaning, which in turn provide the foundation on which a system of ethics may be built. The point is that without this first, emotionally demanding process of value creation, talk of ethics will always be just that—talk, and talk that has little hold on either the imagination or the conscience because it takes values for granted and fails to provide ways of coming to terms with the most problematic aspects of the experience underlying the process of value creation.[27]

A strength of Jordan's work is its grounding in anthropological insights: he develops, for example, the need for ritual and liturgy to accompany and ground community restoration efforts. In so doing, he points to both the ecological *and* social dimensions of restoration: "Environmentalists will continue to explore the value of restoration in its human dimensions—the dimensions of process, experience, and performance. Restoration will emerge as a paradigm for learning about the natural landscape and for our relationship with it."[28]

Jordan's insights cohere with a liberationist theological paradigm, which also must be rooted in liberative praxis. Theologically, restoration has long been a dominant metaphor in Christianity, with its emphasis on God's saving grace through Christ, restoring wholeness and integrity to humanity's relationships with God and neighbor. But too often this understanding of restoration has come at the expense of the earth: either the earth's integrity is absent or ignored and thus serves merely as a passive stage on which the post-Eden drama of human salvation-restoration is played out, or it is sacrificed in apocalyptic, end-of-the-earth scenarios where the destruction of the planet is a necessary prelude to Christ's return and saving action. A theologically grounded restoration ethic retains the restoration of integrity in both human-divine and human social relationships, but expands these to include the earth and reframes them ecologically. Restoration now is grounded in ecological integrity: restoring and living *within* the earth's ecological communities and processes.

Joseph Sittler, so often a generation or more ahead of his contemporaries in understanding the theological implications of ecological challenges, began to sketch an *ecological* restoration theological framework in his 1954 essay, "A Theology for Earth."[29] There he argues, "The central assertion of the Bible is that [the Creator-Word] . . . drives, loves and suffers his world toward restoration . . . 'The Word became flesh and dwelt among us.' To what end? That the whole cosmos in its brokenness—man broken from man, man in solitude and loneliness broken from Holy Communion with his soul's fountain and social communion with his brother— might be restored to wholeness, joy, and lost love."[30] Sittler images Christ as "the restoring God-man," who integrates God, humanity, and nature in the central restoring symbol of bread and wine.

For this restoration ethic to be grounded theologically in Christianity, however, several central biblical-theological themes must be revised, reworked, or simply rejected in light of newly emergent ecological understandings of the earth. Much of this work is already underway in ecofeminist theology, including the work of Rosemary Radford Ruether,

Anne Primavesi, and Catherine Keller (see their contributions elsewhere in the present volume).[31] *Restoration of the earth* cannot be to a mythical, unecological, and static Garden of Eden, but rather to the ecological integrity of complex and evolving biotic communities, bioregions, and processes to which humans must adapt themselves. Similarly, *restoration of human communities* is not to some earlier biblical or mythic state that reproduces earlier social contradictions and oppressive social structures along the lines of gender, sexuality, race, class, and physical ability, but rather to liberationist and feminist paradigms of mutuality, justice, and equality, of which we see prophetic glimpses in the biblical tradition. And *restoration of the humanity-nature-divinity relational matrix* must draw on insights from the world's many religious and spiritual traditions as well as contemporary experience in order to restore an ecologically informed and grounding understanding and experience of the sacred.[32] Hence a restoration ethic is also an ecojustice ethic—integrating both ecological and social justice concerns within a single framework.

This reworking—or restoration?—of classical theological themes and doctrines in Christianity will not be without significant tensions and difficulties, as an ecological reframing of any religion calls into question many of the foundational historical values and assumptions that either gave birth to the religion or were instrumental in its development. Yet this is the task that Mary Evelyn Tucker and John Grim (see their contributions elsewhere in this volume) are calling religions to take up as they challenge them to enter into their ecological phase:

> The emergence, then, of the world's religions into their ecological phase and their planetary expression implies not simply reformation but transformation. For as they identify their resources for deeper ecological awakening—scriptural, symbolic, ritual, and ethical—they will be transforming the deep wellsprings of their tradition into a fuller expression. As they adapt their traditional resources and adopt new resources, they are creating viable modes of religious life beneficial not simply for humans but for the whole Earth community. This involves initiating and implementing new forms of the great wisdom traditions in a postmodern context and may involve opposition to certain aspects of modernity (such as relentless consumption) and change in other aspects of modernity

(such as emphasizing individual rights, especially property rights, over communal responsibilities).[33]

Necessary to this ecological reframing and reworking is the notion of "restoration" itself. Once religions and the human story are resituated within the history of the cosmos itself, restoration ceases to be either anthropocentric or ahistorical, but rather ecological, evolutionary, and cosmological.[34] "Restoration" implies restoring something that has been lost. In this ecotheological context, what has been lost is the very integrity of the earth itself, and the integrity of our human place in that earth community and story. Hence at its deepest level, "restoration" means restoring the earth's ecotheological integrity and restoring ourselves to the earth's ecological and evolutionary story and integrity. More concretely this means restoring human ways of living to the mutuality and interdependence with other life-forms and life-processes out of which we evolved and on which we continue to depend for our well-being. We need not fear a loss of human distinctiveness and creativity in the process; they are as much a part of the earth story as natural selection itself. To the extent that every creature has evolved in distinct and creative ways to fill unique ecological niches, all creatures are distinctive. What needs to be restored, however, is human distinctiveness and creativity *within* the limits of ecological and planetary sustainability—and that task itself will call upon the very limits of our creativity and imagination.

Additional key areas of theological reworking will involve what Christianity has to say about the future, or eschatology (what Catherine Keller terms "discourse about the ultimate or the end"), and soteriology, or thought about salvation.[35] A restoration ethic understands "salvation" as a restoration of integrity and wholeness—and very much a *this*-worldly process of the sacred infusing the material. Hence, as Keller argues, "*a responsible Christian eschatology would be an ecologically sound eschatology, one that motivates work to save our planet.*"[36] God's saving grace is understood as what allows the human community to be reintegrated into the goodness of God's entire creation, that which God has pronounced *good*. This is very much an act of restoration—of restoring our place within the earth community, of restoring the ecological and divine integrity of that community itself. Insights from process theology can help move

Christian thinking about God from a supernatural interventionist "sky-god," who saves humankind from participating in nature[37] in an apocalyptic end-time, to a panentheistic understanding of the sacred that participates in the unfolding of creation through ecological and cosmological processes.

CONCLUSION

The points outlined above are preliminary, designed to foster conversation and open up new possibilities, rather than to present a wholly formed "restoration ethic." My intuition, however, is that a restoration ethic can make a critical contribution to the many social and ecological challenges that we face in these times. It is an *ethics of praxis*: it begins by healing the humanity-nature rift by immersing ourselves in creatively reinhabiting our places through acts of restoration. In the process we reaffirm that human beings are very much a part of the natural world, though a part with distinctive abilities and perils. It is an ethic of *hope and agency*: we belong to this planet and our agency is critical both to our well-being and the well-being of the planet, as is the active agency of the other members of the biotic communities we inhabit. It is an ethic of *place*: participatory restoration practices always take place in concrete ways in specific places, and require an increasing knowledge of, intimacy with, and commitment to those places, their biotic communities, and ecological processes. That is, it is an ethic of *home* and *homecoming* that celebrates the diversity of our homes—and our communities within them—and the practices needed for them to survive and thrive. It is an ethic of *responsibility and response-ability*: we have both the moral obligation to restore what we have damaged, *and* the moral capability to respond as we deepen our understanding of commitment to the particular places we inhabit. And this sense of responsibility opens up the possible development of a genuine *partnership ethic* to replace our traditional anthropocentric stewardship ethics: in recognizing the particularity of places and the agency of the other members of the biotic communities that inhabit them, we open ourselves to partnering with nonhuman nature in a genuinely just and sustainable ethic.[38]

Finally, a restoration ethic is an *ethics of wonder*: in immersing ourselves in our ecological homes, in gaining the intimate knowledge and familiarity of these homes that ecological restoration requires, we are opened anew to the wonder of this planet we call home. Mary Evelyn Tucker calls this transformation the "restoration of wonder": "a comprehensive re-visioning of what it is to be human on a finite planet amidst infinite immensities. We have the possibility to envision ourselves now not only as political, economic, or social beings, but also as planetary beings embedded in and dependent on nature's seasons, cycles, and resources. . . . Central to this great transformation of the religions into their ecological phase is the reawakening in the human of a sense of awe and wonder regarding the beauty, complexity, and mystery of life itself."[39] In this sense, a restoration ethics is fully an "ethic of Ecosophia"—rooted in and drawing on the wisdom of the earth, while committed to restoring and sustaining the very ecological integrity that allows this wisdom to flourish.

❧ Caribou and Carbon Colonialism: Toward a Theology of Arctic Place

MARION GRAU

It is our belief that the future of the Gwich'in and the future of the caribou are the same. We cannot stand by and let them sell our children's heritage to the oil companies.
—From "Protect the Sacred Place Where Life Begins," a brochure published by the Gwich'in Steering Committee

TIME AND PLACE

There's a prophecy, it's called voice from the north, there's gonna come a time when a voice from the north is gonna rise. When that voice from the north rises, it signifies a time for human kind to change their ways.
—FAITH GEMMILL, Gwich'in Steering Committee

The time has come, the *kairos* is here. The Arctic places of the world, the regions of glaciers, permafrost, snow, wind, and ice are changing. The people of those lands have said so for decades, before Western scientists were able and willing to acknowledge the reality of global warming. A large factor in climate change is a widespread, systemic dependency on carbon-based fuels to power infrastructures, industries, and homes across the world. Fossil fuels have been the main power source to fuel commerce at home and abroad. This energy does not come without severe costs for the places it is extracted from, transported to, refined at, and, finally, consumed. What are the implications for people of faith across the world? What visions of a different life do we need to empower action? How might a theology of Arctic place inform and affect theologies of place and action in contexts across the world?

In this essay, I offer one particular case study of a place up north. It is the hotly embattled Arctic National Wildlife Refuge (ANWR), a remote, roadless region in the bush of northern Alaska. ANWR is located several hundred miles north of the city of Fairbanks. Some have contended that "the 1.5 million–acre coastal plain of the Arctic Refuge is the most fought-over chunk of wilderness in the world."[1] One main source of the opposition to drilling in ANWR comes from indigenous Athabascans, the Gwich'in nation in particular, who fear that fossil fuel extraction will significantly impact the life and migratory patterns of the Porcupine caribou herd that is the foundation of their life and culture. They have been in contact with Anglican and Episcopal missionaries and priests since the 1860s and continue to affirm that affiliation as part of their tribal identity today.[2] In 1988, the Gwich'in Steering Committee was formed to address the increasing threat to the coastal plain in ANWR, the calving grounds of the caribou.

In recent years, and especially under the leadership of the Right Rev. Mark MacDonald, Bishop of the Episcopal Diocese of Alaska, the Episcopal Church has supported its Gwich'in members in their fight for survival and subsistence at the local, regional, and subsequently national levels. Indigenous peoples have responded to biblical messages as mediated by missionaries in myriad ways. The case of the Gwich'in is particularly fascinating insofar as they merge indigenous knowledge with Christian faith, and challenge all Christians to become more aware of the colonial heritage of Christianity and its complicity with cultures of exploitation.

A theology of place takes form as a theopolitics striving beyond imperial conditions.[3] It knows its biblical traditions of prophecy and protest—it listens to voices that represent the needs of the land and people, and witnesses to the unsustainability of ideologies, technologies, and lifestyles, as well as encourages needed changes of heart and hand. During travels within the Diocese of Alaska, and to native communities such as Arctic Village, I have been struck by how land, faith, and economic, social, and ecological concerns are interwoven for Alaska Natives and other indigenous peoples, and how distinct, though not unrelated, their concerns are from those of their siblings in faith in other rural, but also in suburban and urban, contexts. The challenges faced by the Athabascan Episcopalians

I encountered in Anchorage, Fairbanks, Minto, and Arctic Village manifest as "transdisciplinary," that is, there is not "one" problem that could be easily isolated and then delegated to either scientists, social workers, environmentalists, or ministers and theologians. Rather, even upon casual observation, a nexus of issues becomes visible that necessitates a wide-ranging vision: the colonial heritage of missionary theologies and church bodies; the task of contemporary theological work in facing this heritage; the issues of economy and ecology as they intersect with culturecide and ecocide; the survival of indigenous Episcopal Christians in a capitalist global culture that continues to mark native lands as "corporate sacrifice zones,"[4] from which the energy source for continued globalization will be extracted; the intransigence of fossil fuel dependency across cultures and societies; the lack of commitment to develop the viable alternatives; and the need to find visionary paths for people to meaningfully combine a variety of ways of being within a place.

When the problems we are confronted with are thus highly compounded, our constructive proposals for theology, for faithful and determined action, ought to be transdisciplinary as well. Such a nexus of issues might be helpfully conceived as the "rhizome" described by Deleuze and Guattari. Hence, the approach I am employing here is best considered a map fostering "connections between fields," removing the blockages and entering into the rootweb through "multiple entryways."[5] A theopolitics then must look closely at the nexus of issues at hand, as best they can be discerned, indicate how theological questions and concerns are interwoven with the very fabric of this weave of life, and develop a vision for faith, thought, and practice.[6] Such a theopolitics offers a rendering of what is spiritually, economically, and ecologically at stake in this particular place, outlines some key components of a theology of Arctic place, and shows how the red-hot, ongoing controversy about ANWR exposes the spiritual crisis of sustainable human life on this planet.

A PLACE UP NORTH

The Arctic Refuge "war" takes on real and symbolic value of historic and global significance and needs to be understood in the context of global (spatial) and historical (temporal) scales for its value as a precursor of events to come.
—DAVID M. STANDLEA, *Oil, Globalization, and the War for the Arctic Refuge*

Though David Standlea, in *Oil, Globalization, and the War for the Arctic Refuge*, may have overstated the global visibility and importance of the resource war around the Alaskan Arctic, it is certainly a "uniquely American environmental battle" and has, for many Americans of the United States, perhaps become the most visible and persistently recurring case of ideological division around an environmental issue in many decades, perhaps in the history of the country.

The issues concerning Arctic place span across great distance: oil, a local resource, is desired and embattled in contexts far beyond Alaska. Fossil fuel use has caused climate change that is experienced disproportionately in Arctic regions, where its effects have been reported by indigenous Alaskans for decades. Each place on this planet is vitally connected to the Arctic, which generates and sustains present climatic patterns and ocean currents. Continued dependency on oil not only endangers the local patterns of survival and subsistence of the native Gwich'in; in the larger scheme of things, each ecosystem that is connected to the Arctic is at stake as well. Hence these two related questions: What is at stake for the Gwich'in Athabascans in particular? What is at stake for those of us living in other places of the world?

The battle surrounding ANWR is becoming an endless struggle that reveals a number of issues crucial not only in the immediate context of Alaska. It takes place within a nation-state with aspirations to imperialist unilateralism based on a capitalist system, whose superior economic and military power is fueled by the extraction of carbon-based fuels from increasingly endangered locations. Alaska's particular location and history makes it an unusually clear example of a colony governed so as to facilitate a "land-grab," then internalized as a state to facilitate resource-extraction, indigenous genocide, and, more recently, environmental cultural imperialism.[7] The outcome of this particular controversy is likely to have significance in how the United States will further develop policy and act on issues of the environment generally, the rights of indigenous peoples, and energy strategies. It will have an effect on how Christians in the United States will perceive their relationship to creation, to notions of stewardship, how they might develop their often too-otherworldly theologies into more incarnate theologies of place. As Anne Daniell writes, "If one of the reasons for engaging in theology is to better discern, appreciate, and learn ways of caring for that which is sacred in our lives, then places, as relational nexuses of which humans are a constitutive part, are an essential subject for theology."[8]

It is for this reason that observing how some of the controversies around this particular place are framed and which important themes and issues need to be included in mapping the struggle will likely help us read, write, listen, pray, teach, and act in the future. The fight of the Gwich'in to prevent oil drilling in a fragile environment shares several similarities with other struggles around land, ecosystems, and indigenous rights elsewhere in the world.[9] The Gwich'in are fighting for the land as for their lives, their culture and its survival. The sacred, sacramental body of the caribou, for the Gwich'in a major source of their life-substance and energy, is endangered. Evon Peter, former chief of the Arctic Village Gwich'in states: "The animals, the rivers—we're essentially a voice for things that cannot talk. We don't see ourselves as separate from those things. If the rivers and animals are poisoned, the poisons will work their way into us, too."[10]

Some have argued that the Christian identity of the Nets'aii Gwich'in of Arctic Village "represents communal *resistance* to colonial domination, as well as the ability to blend past and present traditions."[11] When asked what could be done to support the Gwich'in in their struggle against drilling for oil and the protection of the Porcupine caribou herd that is their life-blood, the answer the Gwich'in gave was perhaps surprising: Help us to restore our church.[12] Steven Dinero, a professor of human geography who assisted the Gwich'in in rebuilding the historical Episcopal chapel in Arctic Village, states that "the degree to which Native Americans have embraced Christianity should not be constructed as acceptance of assimilation or willingness to participate in their own cultural destruction."[13] Beyond the distinctions of academia and faith commitments, people have come to understand that for the Gwich'in, the restoration of a church building—seemingly a submission to colonial influence—that symbolizes part of their history, identity, and work of their ancestors embodies a part of their ability to maintain a core strength in their resistance to the stampede of the foot soldiers of profit.

THE LAND AND THE PEOPLE OF THE LAND: "THE SACRED PLACE WHERE ALL LIFE BEGINS" (*GWATS 'AN GWANDAII GOODLIT*)

The caribou are God's way of giving us life. They are too sacred for self-interest. Politics is not in our (traditional) system. When the caribou migrate they use the energy of this planet. Now the energy is all screwed up. It is a matter of spirituality and the global environment. You can't rely on money out there; these are the cycles of nature. Without the cycle of life, without ANWR, the cycle of life is over, all for money.

—Gwich'in native, as quoted in David M. Standlea, *Oil, Globalization, and the War for the Arctic Refuge*

Alaska natives are some of the earliest human inhabitants of the Americas. Some have distinct cultural ties with indigenous Siberians, having come across via Siberia and the Bering Strait from Eurasia.[14] Many of them have subsisted on coastal whaling and fishing, while others have survived on caribou, and still others on the fish of the inland rivers and streams. Athabascans, one of five distinct Alaska native tribal groups, call themselves "Dene," the people, and are kin to Athabascans in Canada and to the Navajo in the Southwest United States, with whom they share linguistic and cultural features.[15] The modern history of Alaska is marked by periods of resource-extraction and environmental exploitation generally: first colonized in modernity by Russians, who traveled there for beaver skins, it was then sold to the United States, at which time mining for gold in the Yukon became a major attraction. The most recent and longest-lasting type of resource-extraction is that for fossil fuels at Prudhoe Bay, and now, potentially, ANWR, against which expansion both the Gwich'in and environmentalists have fought for a decade and a half.

The ancient calving grounds of the Porcupine caribou herd of northeastern Alaska and northwestern Canada are known by several names, depending on who has done the naming and what context and place they speak from. "Ivvavik" is an Inuvialut Gwich'in term for a "place for giving birth," and one might say that the story of this place is a real-time and real-space parable of the earth as an "ecosystem that does not know political boundaries,"[16] a nexus that reveals truths about our pasts, presents, and futures that are worth considering.

The direct impact of oil drilling, oil production, and oil transportation (which are often conveniently left out of government calculations of environmental impacts) are only a small fraction of what threatens the ecological value of ANWR: the longer-range impact of global warming is also disproportionately affecting the Arctic, and Alaska, thus rendering this space-time an important one for people all over the world to watch

and learn from. Where global warming is concerned, indigenous peoples—or better, "people of the land," and land at that they want to continue to belong to—have been and remain the proverbial "canary in the
coal mine." While Western economic myths of "independence" may
have left their believers ever more enslaved to destructive processes and
mind-sets, the people of the land are, of course, not immune to the
dangers imported by industrialized capitalist systems. This is the case for
peoples of the Arctic around the world as the Arctic "has been transformed into the planet's chemical trash can, the final destination for toxic
waste that originates thousands of miles away. Atmospheric and oceanic
currents conspire to send industrial chemicals, pesticides, and power-
plant emissions on a journey to the Far North."[17]

Years ago, during a legislative hearing, Alaska Senator Frank Murkow-
ski infamously lifted up a white placard, arguing that the ANWR was
nothing but this—white, empty, nothing, useless—most of the year: a
tabula rasa, a *terra nullius*,[18] unused, and in need of "use." Maps reveal
the politics and ideology of the mapmaker. Other maps diverge: birthing
place of the caribou, nesting place of countless bird species that travel
thousands of miles to reach it every year: this place is far from useless in
the mind-map of the people of that land. What happens if we lay these
maps on top of each other rather than conveniently erase the details of
that which is inconvenient to us?

The debates on whether the ANWR—or any place, for that matter—is
"pristine" shows a problematic use of that term, as well as a tendency to
either/or logic that doesn't correspond to the complexities a given location's reality. The logic goes thus: if a place is not "pristine," then why
resist fully exploiting it? If the native peoples are not fully "authentic"
and "pristine" in their own lifestyles, why ought one stop disrupting
ecosystems and cultures? The linguistic power-politics of "purity" often
emerge around notions of what is "pristine" and "authentic." Pro-drilling
discourse tactics can dispute the worthiness of protection by resorting to
devaluing land as "empty," and people as "inauthentic," because they
have adopted some modern ways of life, or have become dependent on
the Alaska Permanent Fund, a dividend paid to Alaska residents on account of oil money sloshing through the Alaska government. "Pristine,"
no, but these silent pollutants highlight even more how much is at stake!
Debates about the "purity" of culture, nature, and so forth deftly cover

over key issues at stake for the Gwich'in, as for many in more densely populated places, among them: How ought we to live within this place and time? How ought we to "rightly" use resources that might be made available to us—that is, neither destroy nor completely deny ourselves their values? And what do human and nonhuman members of a landscape, plants and animals, need to survive and thrive? Trying to find answers to these questions is where the debate becomes complicated and embattled. And it reveals many favorite myths, half-truths, lies, and assumptions, as well as great wisdom and reliable truths. To sift through them may not promise easy or final answers, but the debate itself reveals much that shows where the questions themselves may not yet be probing deep enough.

THE ECONOMICS OF COLONIZATION

The Eskimos undeniably got a good deal in the Native Claims Settlement Act, but it was good only insofar as they agreed to change their way: to cherish money, and to adopt the concept (for centuries unknown to them) of private property. . . . The forest Eskimos' relationship with whites has made them dependent on goods that need to be paid for: nylon, netting, boat materials, rifles, ammunition, motors, gasoline.
—JOHN MCPHEE, *Coming Into the Country*

The particular form of "land-grab" that took place in Alaska was distinct from the expropriation processes witnessed in the "lower 48" of the United States. In its own time, the Alaska Native Claims Settlement Act (ANCSA) was hailed as the most liberal settlement ever achieved with Native Americans, granting large amounts of money and land to a new entity: native for-profit corporations.[19] Under the pressure of corporate entities desiring access to native lands, this treacherous settlement converted "native claims" to almost all land in Alaska into shares distributed among tribes for a limited time (that is, shares of ownership in the corporations that cannot be passed down to descendents). Thus, many tribes have been reorganized into for-profit corporations. Effectively, then, many tribal entities are now structured as businesses, and in many cases tribal decision making is as profit-driven as that of any other business.

The Gwich'in and various environmental groups have been fighting against oil development for over fifteen years, and in 2005 one of the main walls of defense caved in: the U.S. House of Representatives attached an amendment permitting oil drilling in ANWR to the bill authorizing the federal budget; at the time of writing, however, the amendment has not been approved by the Senate, its defeats often coming at the last moment, and by a hair's breadth. Most recently, the amendment was taken out of the budget bill just before Christmas 2005.[20]

Another, contrasting example of how Alaska natives have interacted with the oil industry is the coastal Inupiat village of Kaktovik. Here, the tribal leaders and native corporations have been pro-oil and have, at least in the short term, benefited from it. Even the Inupiat in Kaktovik, however, frown upon off-shore drilling, which would impact their subsistence whaling and the native village of Kaktovik (as opposed to the for-profit village corporation) is now openly opposing development in ANWR. One of the superficial dynamic processes that would seem to be pertinent here is the classic conflict between short-term economic benefits and long-term cultural survival. But the issue is more complex.[21] Since no return to a precontact, "pristine" society is possible or even desired at this point, survival, for land and people, will need to be creatively reinvented from hybridized forms of knowledge, theology, and technology.

Is drilling in the Arctic economically and ecologically viable? Oil-extraction will be increasingly difficult and expensive, and the new technologies developed to accommodate an ever more desperate scarcity of oil promise to be even more environmentally harmful than their predecessors. "In other words," Sonia Shah writes, "the industry could perform both as a lumbering giant and a nimble parasite. No ecological niche, no matter how tiny, fragile, or beloved, could consider itself safe from the drills."[22] Shah argues poignantly that the increasing precariousness of extracting oil without inflicting exponential damage on the environment will both be more expensive, more dangerous, and more subject to "cutbacks," complacency, and shortcuts because saving money will be crucial in order for future oil-extraction to be even close to financially feasible within a profit-driven economy.[23]

The oil industry knows that the "end of oil" is coming,[24] and may not necessarily be inclined to push oil-extraction at all costs. Corporations are required to pursue profit. If they can no longer profit from oil, and other forms of energy come more strongly to the fore, the push to drill may cool off. But fuel cells and other viable alternative-energy types are still a long way from being feasible and sustainable, and thus frantic measures to extract the remaining oil with ever more environmentally invasive procedures are still possible, even likely.[25] In March 2006, Senate Energy Chairman Pete Domenici (R-NM) and the Captain Ahab of Alaska oil drilling Senator Ted Stevens (R-AK) pressured yet again to draft a plan to drill in the refuge.[26] At the time of writing, only one oil company, Exxon-Mobil, continues to actively pursue drilling in ANWR.[27] That is, it is conceivable that even if the U.S. Congress voted to open the region for drilling, no company would decide to take the risk or make the necessary investment of time and money. It should be noted that despite industry protestations that "least-impact" oilfield technology would be employed in the extraction process, spills and accidents occur constantly, not just when they make the news. The continued impact of the Exxon Valdez oil spill in Prince William Sound is cautionary in this regard. The corporation's response to the disaster—comprised mostly of propaganda, "green-washing," and a plethora of unfulfilled promises—does not induce great trust for further oil-exploration in the region.[28]

The Gwich'in did not participate in ANCSA, and are organized right now mostly in seventeen villages strewn across what is often called by Euro-Americans "the bush." They strongly oppose drilling in ANWR and want to see the Arctic Slope, where the Porcupine caribou herd's calving grounds are located, protected. As previously noted, the Gwich'in rely on caribou for their survival, though others argue with disturbing seriousness that they should just switch to industrialized foods and move to Anchorage. Caribou are nomadic, and over time can change patterns of migration to some degree, and their herd sizes rise and fall at various times for various reasons. What the Gwich'in and scientists are most worried about is a permanent, irreversible decline, not trying to stem the more-or-less common fluctuations within populations.[29]

More recently, there is a new movement among native Alaskans to contest the validity of the ANCSA as a policy that has served to separate the people of the land from the land by promising short-term financial gain: "The Alaska Native Oil and Gas Working Group rejects the Alaska Native Claims Settlement Act as an illegitimate infringement on our right to sovereignty and self-determination. The [ANCSA] has allowed the 'takings' of our aboriginal lands to be exploited by natural resource extraction corporations with no commitment to building and maintaining sustainable and healthy communities."[30]

As has been the case elsewhere, the relationship between indigenous Alaskans and environmentalists has frequently been uneasy. Where ANWR is concerned, alliances have often been shifting and unstable. In the past, many environmentalists, following the model of John Muir, have focused on preserving nature without any people—"nature as wilderness." Perhaps it is one of the lamentable consequences of industrial societies that they can no longer imagine a healthy, working integration of land and people. This view of nature is often combined with a paternalism that assumes the superiority of modern science over many generations of knowledge of the land that has allowed many indigenous societies to survive and prosper against great odds—it is both extremely unwise and lacking in respect. This thinking separates ecosystem health from human rights, considers humans as being outside of any given ecosystem. Nonindigenous Christians would be well advised to take a lesson from the problematic relationship between natives and environmentalists, whose green cultural imperialism often is based on an unshakeable belief in the beneficence of modern science. Paired with assumptions of the superiority of specific forms of Christian faith (usually one's own) over another's, patronizing relations often prevent mutual learning and giving.

Neither is it helpful to romanticize all indigenous peoples as essentially more "close to the earth" than we are. Throughout history, there have been many human societies that have lived above the carrying capacity of the ecosystems they lived in, and these cultures, such as those of the Easter Islands, the Anasazi, and Norse Greenlanders, all died out.[31] When ancestral wisdom of how to live within the limits of the land disappears or is displaced, ethnic identity comes to matter less, and thus one's relationship to the environment changes, becoming instead a religious or cultural value.

Today, native Alaskans experience increasing difficulties in their attempts to survive at even a subsistence level. Many Athabascans in Alaska mix subsistence with some cash-credit income. They might hunt, fish, trap during part of the year, and work as fire fighters, in the oil fields, or in the tourism industry for the rest. This split has had consequences on the technological know-how of indigenous Alaskans: internal-combustion engines are slowly replacing slower dog-sleds, to take one example. In addition, the commercialization of dog-mushing has driven up prices for sled dogs so that they have become unaffordable in comparison with a snow machine. Yet high prices for reimported gas—Alaska has oil but not refineries, and there are few fuel service stations in the bush—in combination with the expense of consumer goods that have to be flown in on small propeller airplanes are increasingly making it both economically viable and necessary for Alaska natives to use their traditional technologies and subsistence practices when appropriate. It remains to be seen whether such creative adaptation of elements of native culture holds the promise for sustainable, hybrid subsistence-cash economies.

The lures and comforts of consumer society and its attendant addictions have long since made their inroads into native communities. Fighting for survival and for subsistence also means fighting off high unemployment, drug problems, diabetes, obesity, and suicides, the latter three all epidemic in indigenous Alaskan society, as also among Native Americans in the lower forty-eight states. So, too, dependency on oil has become a problem: the amount of engine exhaust I inhaled during a visit to the Gwich'in Steering Committee in Arctic Village, where we were carted around in two-stroke ATVs, was not only rough on all our lungs, but a pungent reality check into how dependent we all are on oil, and by implication, how difficult it will be to redirect our lives and economies toward renewable sources of energy, and thus to reconnect more deeply to the land.

The impact of expropriation and cultural genocide has marked the bodies of the Gwich'in, as those of other indigenous peoples worldwide: alcoholism, drug abuse, diabetes, suicide, and domestic violence are decimating this native community. Without a sense of identity, ancestral wisdom, purpose, and the grounding stability of access to, and use of, land, any people is lost. What ensues is widely known, but not always recognized as a consequence of culture-cide and genocide. With the destruction of land-people-culture connections, survival with some degree of physical and mental integrity becomes a great, often insurmountable, challenge. Disrupted or destroyed connections to ancestry and land manifest numerous destructive effects: culture and language loss, the loss of youth through migration to urban contexts, widespread depression and anxiety.[32] Hence, the loss of the caribou, which provides the foundation of the existence of the Gwich'in culturally, spiritually, and existentially, is experienced as the latest incident in a long history of cultural genocide.

"They won't kill me," says Gwich'in Norma Kassi, "but if we lose the caribou, it will be the death of our culture."[33]

The woundedness of a place—of a land and its inhabitants, human and nonhuman—necessitates that constructive theological proposals include prophetic and pastoral dimensions in order to be therapeutically effective. What would allow us to develop a "sustainable spirituality" for the Arctic, one that viewed it as an essential part of creation?[34] If we want to look for healing, we have to consider land and people, body and soul together. We have to appreciate the historic destruction, the continuing distortions of power as well as the gifts we can offer one another: First Peoples *and* more recent arrivals to the Americas. Given these concerns, what theological tools can help faith communities live amidst the twin realities of shrinking oil supplies and global warming?

TOWARD A THEOLOGY OF ARCTIC PLACE

Almighty and everlasting God, in giving us dominion over things on earth, you made us fellow workers in your creation. Give us wisdom and reverence so to use the resources of nature, that no one may suffer from our abuse of them, and that generations yet to come may continue to praise you for your bounty; through Jesus Christ our Lord. Amen.
—BOOK OF COMMON PRAYER, ACCORDING TO THE USE OF THE EPISCOPAL CHURCH, 1979

The recently deceased Native American theologian Vine Deloria has argued that "Christianity itself may find the strength to survive if it honestly faces the necessity to surrender its narrow interpretation of history and embark on a determined search for the true meaning of human life on this planet."[35] He touches upon one of the most challenging tasks for theology and for many churches today: their ambivalent missionary heritage with regard to indigenous peoples and to the environmental consequences of modern colonialism. In light of his observation, it is crucial to ponder the forms Christian commitment should take in present and future, but with a historical consciousness that learns from the past and is capable of imagining a different future for Christian faith communities.

The story of Athabascan Episcopalians represents a fortuitous case of a gentler inculturation, largely due to the particular charisma of the missionaries involved. One resident of Arctic Village reports, "We were the last ones to be contacted by the so-called Columbus discovery. . . . The Russians came from [the] north, and the French from the south," but the remote Gwich'in semi-nomadic villages were encountered by Europeans only quite recently.[36] The Rev. Robert MacDonald, of Ojibwe background, began his missionary work in Alaska in 1862 and stayed there for forty years. He studied the Gwich'in language and created an orthography using the Bible and Book of Common Prayer. Due to his efforts, records of the language in its mid-nineteenth-century form have survived into the present. He translated many hymns into Kudth, a root dialect that allowed different Gwich'in bands to communicate with one another.[37] Using the vernacular to frame the message of the Gospel greatly aided its adaptation by the Gwich'in. Gwich'in Sarah James, for example, "still believes in the old ways, she says, carrying those fish in through the back door . . . and giving nearly ceaseless thanks to the land—but 'now I believe in the Episcopal Church. Now I believe in Jesus. It teaches us good things we already practiced.'"[38] The Rev. Albert Tritt, the first Nets'ai Gwich'in Episcopal Priest and a native of Arctic Village, was particularly effective in his missionary activities, due in part to his knowledge of Athabascan culture and language. The chapel he built between 1910 and 1917 plays a central role in the life of the village to this day.[39]

A more recent Bishop of the Episcopal Diocese of Alaska, the Right Rev. Mark MacDonald (no relation to the Rev. Robert MacDonald mentioned previously), has become caught up in the controversy over drilling in ANWR between Euro-American Episcopalians and Episcopalian tribal groups. He has continued to remind the Episcopal Church in Alaska and the national body of the church to continue in their support of the Gwich'in in their struggles. In 2005, the Episcopal Church co-sponsored and co-wrote a document on the issues at stake in the decision to drill or not to drill in ANWR.[40] The church's standpoint meaningfully connects human rights issues with environmental issues, a helpful advancement over some environmental organizations' tendency to strive to preserve places without people.[41]

Whether or not there will be drilling in ANWR, the ecosystems of the Arctic have already been compromised significantly. There is no return back to Eden, but there are possibilities for preserving broken and compromised landscapes and to work toward healing them and us together. The Arctic is a "wounded sacred" that we must come to recognize in its sustaining and healing capacity along with its compromised and endangered integrity as "degraded but still glorious, wounded but still alive."[42] Moving away from the romanticizing rhetoric of "purity" in much of the anti-drilling literature, we need to understand that the "wounded sacred" is as much worth our love, care, and action as some fantasy land "untouched by human hand." Recognizing a place and a people (including ourselves!) as wounded and sacred does not give us permission to continue in bringing about its degradation. In fact, the "wounded sacred" calls us even more to change our ways, so the entire ecosystem, humans included, can survive and flourish. Also part of the "wounded sacred" are the indigenous peoples holding out against drilling. Author Rick Bass recalls a sermon by Trimble Gilbert, native Episcopal priest of Arctic Village:

> "We are the last people," he tells his congregation—young, old, and in-between. "I hope you understand that. All the people with money are against us but we don't want to lose our culture. Because we don't have money, they think we are a poor people." He speaks of the meek inheriting the Kingdom of God. "When the time comes, maybe we're going to be the ones who go first," he says, into that Kingdom. "For sure, we try to save the earth."[43]

The true wealth here is not that of money, but land, which will guarantee future survival for many generations to come. Maintaining the integrity of the land, not temporally limited financial interests in what is extracted from it, is key for the survival of the Gwich'in.

The wealth of the coastal plain, as Alaska's Serengeti, is not in the least diminished by the fact that it is populated by caribou only for a short time during the summer months, during the calving season, as some pro-drilling Alaska politicians have argued. The worth of the land cannot be measured by how "efficiently" it is "used" by animals, humans, or others. Fallow time and winter time are the Sabbaths of nature and the ancestors of all our cultures honored those periods. If we fail to respect the land today, we do so at the peril that the land will neither carry us and our descendants, nor its other inhabitants into the future. If this huge landscape is there for animals "only," then so be it. Barry Lopez suggests that "the advantages of these dismal regions . . . are several. The number of predators is low, wolves having dropped away from the herds at more suitable locations for denning to the south. Food plants are plentiful. And these grounds either offer better protection from spring snowstorms or experience fewer storms overall than adjacent regions."[44] ANWR would thus provide safety and sustenance during a time of extreme vulnerability to the survival of the herd. Global warming has already negatively impacted the chances for survival for the herd, however:

> In recent years, the herd had been delayed on its northern migration as deeper snows and increasing freeze-thaw cycles make their food less accessible, increase feeding and travel time, and generally reducing the health of the herd. At the same time, river ice is thawing earlier in the spring. Now, when the herd reaches the river, the river is no longer frozen. Some cows have already calved on the south side and have to cross the rushing water with their newborn calves. Thousands of calves have been washed down the river and died, leaving their mothers to proceed without them to the calving grounds.[45]

As noted earlier, for the Gwich'in the caribou functions as a form of sacrament, a reminder of the immanent transcendence of the life-force that sustains them, even of consubstantiality with the life-giving animal: "In Gwich'in creation stories, the caribou has a piece of Man's heart in its heart, and Man has a piece of [the] caribou's heart in his, so that each will always know what the other is doing. It's that time of year where we have to go out and gather all the meat that's going to last us throughout the wintertime, until the springtime. It's a really important time of year. It's an exciting time of the year."[46]

In response to reading Rick Bass's account of hunting caribou out of Arctic Village,[47] seminarians compared the significance of caribou to the Gwich'in to the sacrament of the Eucharist. They argued that the sacrament is a deep life-force around which a community gathers, depending on it with a deep need and hunger that nothing else can satisfy. The caribou would then be bread and blood, a material place in which God is revealed. If a sacrament is what gives life to us—and there is sacramental import to the elements of nature—then to take away the caribou, they argued, meant to take away the existence of the Gwich'in, to deprive them of what allowed them to be in touch with spirit and matter, and so survive.[48] Indeed, "life-blood" is another term that could be, and has been, used to describe the meaning of caribou to the Gwich'in.[49]

Ironically, for contemporary consumer societies oil has a sacramental quality that competes with the sacredness of other life-grounding substances. In this industrial society, oil fuels and greases the wheels of the system. Here, then, are two materials with sacramental qualities in contest with each other. How we use the resources that are sacramental and will be sacramental to us in our lives will have great impact on future life and its survival on this planet. Substances we depend upon can quickly devolve into objects of dependency—the result is substance abuse. This takes different forms in different contexts. In urban contexts, oil-dependency contributes to apathy and inaction, creating large numbers of moral bystanders, silent beneficiaries of the addiction, if not outright dealers and pimps.

Gwich'in Episcopalians have produced vibrant hybrid identities that include a strong consciousness of being Episcopalian Christians. Archaic and polarizing notions of "uncivilized heathens" or "noble savages" or "saints" are unhelpful, because they ignore the complexities of reality and of human beings. Left-leaning industrialized peoples concerned with the survival of the Gwich'in must first rid themselves of the well-meaning, if often guilt- and shame-ridden, romanticism that obscures the reality, indeed the humanity, of those they wish to help:

> These two ideas, Progress and the Noble Savage, are both creations of the Western imagination. Proponents of both ideas conveniently assume that Native cultures are static and unchanging, that the Native peoples themselves are locked into the past. Such an assumption can become self-fulfilling: if Native peoples are not allowed the means to deal with present-day problems on their own terms, their cultures may in fact tend to become degraded and static.[50]

The Gwich'in, in their ongoing struggle for survival, have developed forms of activism reminiscent of the "trickster" figures common to many native cultures, using the means at hand, and the seductive technologies of colonial cultures.

Gerald Vizenor declares that if we look at Native survivance (survival and thriving), we find far more than stories of "real destitution" and oppression. He claims that non-Natives rarely understand the "native survivance stories" that "are renunciations of dominance, tragedy, and victimry." He extols the use of such stories, claiming, "Many natives were the past masters of trickster stories and, in that aesthetic sense, always prepared to outwit missionaries, social scientists, and manifest manners."[51]

A closer look at these issues reveals in its embodiment what often seems oracular and abstract postcolonial jargon: hybridity, ambivalence, and mimicry manifest in double-edged, often crazy-making incarnations that all facilitate survival. In this struggle, cultural changes and strategic accommodations abound. Thus, survival also entails the discernment of what kinds of changes will truly foster cultural and environmental continuance, and part of resistance to cultural imperialism is—ironically, perhaps, to many observers—cultural flexibility. The flexibility and dynamic nature of the Gwich'in, for example, can be illustrated by pointing to their ability to integrate their own version of the Christian faith into their identity.

The Gwich'in's use of cyberactivism in their efforts evokes Donna Haraway's "cyborg" manifesto, where "a cyborg world might be about lived social and bodily realities in which people are not afraid of their joint kinship with animals and machines, not afraid of permanently partial identities and contradictory standpoints."[52] Tricksters, those ambivalent agents in so much Native American narrative, might be helpful descriptors, if not models, for such hybrid agency, rather than manifestations of the Devil, as they were purported to be by some missionaries.[53] The recovery of trickster stories and trickster agency may be especially crucial in Western theological recognition of life's ambivalence:

It might be argued that the passing of such a seemingly confused figure marks an advance in the spiritual consciousness of the race, a finer tuning of moral judgment; but the opposite could be argued as well—that the erasure of trickster figures, or the unthinking confusion of them with the Devil, only serves to push the ambiguities of life into the background. We may well hope our actions carry no moral ambiguity, but pretending that is the case when it isn't does not lead to greater clarity about right and wrong; it more likely leads to unconscious cruelty masked by inflated righteousness.[54]

Taking the ambiguity of our actions as a given, how may we discern the best remedial approach to oil-dependency, global warming, the spiritual and cultural survival of all people, and especially those of indigenous peoples? Trickster agency does not always lead to the desired result. Willie Hensley's move to push for native claims to land, which later resulted in ANCSA and the corporatization of almost all Native Alaska,[55] can remind one of a trickster who tricked himself into the hands of his enemies. The Gwich'in, who refused to settle for money in return for land claims, on the contrary bring to mind the trickster who refused to be pulled into this economic scheme. Their ambivalently hybrid economic and theological cultures; their savvy use of the Internet and other communication technology; the nationwide tours through classrooms and lecture halls; the use of grant monies for micro-business and ecological defense—all of these evidence a cultural intelligence and adaptability that is easy to admire. Further, the visibility they were able to gain as part of the Anglican-Episcopal community indicates still another way in which the potential instruments of subjugation are being used for cultural survival. Jesus, a trickster in his own right, told parables of the Kingdom of God, of who is rich in it, and who is poor. For Jesus, as also for the Gwich'in, it is not the people with the money who possess true wealth.

Some native tribes use the proclivity for gambling as a way to gain access to political and economic power, while a number of Plains Indian tribes are working to democratize power production by using wind and solar power that is abundantly available for energy independence.[56] Trickster agency is alive and well in many indigenous and multiethnic cultures.

CONCLUSION

At the time of writing, the political tug-of-war surrounding drilling in ANWR sees the area still closed for oil-drilling. On December 22, 2005, the U.S. Senate again blocked an attempt by Alaska Senator Ted Stevens to attach drilling in ANWR onto a military spending appropriation bill, hoping it would not be detected; it was, and was subsequently removed.[57]

But the fate of the refuge, and of the Gwich'in, remains very much in the balance, and seemingly forever teetering on the brink of destruction.

Life—as many humans know it—begins on this planet, and derives from water, earth, air and fire. Life is elementally created, brought forth from the Spirit, the breath of life. For those of us who recognize in this breath of life a Creator God, perhaps even a Trinitarian one, our loyalties to the land that has fed us have not always been in tune with our continued interrelationality and, yes, dependency on the earth's nurturing material presence. Many humans live in passive semi-detachment from the material reality of their bodysouls, their vital connection to the elements. Many native peoples, however, are struggling to maintain these ties, despite the powerful technologies that would seem to promise domination over, and security from—rather than coexistence with—humans' vital connections to earth, water, and air.

> Even as distinctive places are dying (or perhaps because of this fact) people seem to be craving connection to place. . . . We desire to travel to places that boast "a sense of place," to places that promise a kindling of the spirit or calming of the nerves. We pine for places that remain unsullied (at least on the surface) by pollution: the scarcely populated island with its pristine beaches; a cabin deep in a national forest; urban areas where residents recycle and take care of their parks.
>
> We desire the experience of places, hoping to be immersed in their *genius loci*, even if the ways by which we travel to them, and the practices we undertake in them, are injurious to their theological and cultural integrity.[58]

We engage in ersatz cultural practices in a futile effort to overcome our own lostness and fragility of identity, our eroded sense of place. The cultural, economic, and ideological battles around the Arctic are related battles in a larger war to preserve an ecosystem, and one that includes people, a fact seemingly lost on many in the industrialized world who might otherwise be seen to be in sympathy with the wider aims of the affected indigenous population. The struggle of the Gwich'in calls us to remember our own ancestors and their far stronger roots to land than are ours today, and it is but one example of the resurgence of indigenous knowledge that is occurring across the globe.

So: can we go beyond our old simplistic views of indigenous peoples and recover, in all its complexity, our ancestors' deep groundedness in the earth, its places and rhythms? Can we do so without reviving our tendencies to territorialist, nationalist, and xenophobic constructs? The struggle of the Gwich'in is our own. If we recognize one, we must recognize the other. We can do this, not by repeating the habits of cultural imperialism, but rather by honoring the right of the Gwich'in—and not only theirs—to self-determination, to live on with the hybrid wisdom born of the past and the present. They are our sisters and brothers in faith, and they are asking only that our understanding of faith open itself to their own, deep vision of life in the world.

❧ Divining New Orleans: Invoking Wisdom for the Redemption of Place

ANNE DANIELL

This essay may be read more as an invocation, even as a prayer, than an argument. It is meant both to invoke and evoke, to summon the spirit of hope and to instigate incarnations of wisdom in the "City that Care Forgot."[1] It ponders the fate of New Orleans after hurricanes Katrina and Rita in the late summer of 2005, and calls for compassionate wisdom—indeed, for the realm of God—to take root in this place. Composed during the first several months following the deluge and destruction of approximately 80 percent of the city, this essay is reflective of the news and emotions that surfaced during that time. Having previously lived in New Orleans, I had (and still do have) strong emotional, cultural, and spiritual ties to that place. I thus wrote this essay with great yearning for New Orleans's redemption, filled with a faith that divine wisdom would again take root there.

As is true for other situations where extreme events have changed the life of whole communities, there have been varying interpretations of and evolving responses to the inundation of New Orleans. This essay reveals just one person's thoughts, feelings, and longings in relation to this natural and human-caused disaster, experienced during a particular period of time. Yet as I write this preface nearly a year after the storm (and a month after moving back to New Orleans), those of us who call this place home remain overwhelmed with the storm and its consequences. We continue to be astonished and dismayed by the amount of devastation and debris still present throughout the city. In New Orleans, much gutting of homes and businesses has yet to be done; many decisions concerning whether to rebuild have yet to be made; crucial municipal services still are not universally restored; and violent drug-related crime has been escalating in abandoned neighborhoods. At the same time, we are delightfully surprised every day by signs of renewal: people are constantly working on their homes; neighborhood groups meet at churches to work tirelessly on renewal plans for their communities; church and other volunteer groups from across the nation and around the world continue to volunteer their time to gut and clean houses and hearts are lifted every time another "We're Back" or "Proud to Call New Orleans Home" sign appears outside a business or residence.

The renewing of New Orleans will be a long process (some say it will take at least a decade), and the tough issues associated with this process are not going to be easily resolved. So I believe that the following ruminations, though they speak to the events of a particular time and place, will carry a relevant charge for some time to come and will also resonate with people of other places, who face similar questions of how to live wisely in the places they depend upon. I thus hope that while being an invocation of wisdom and compassion for a very specific place, this essay at the same time transcends that specificity by inspiring prayers of healing and practices of transformation for other places. As the Jesus of Matthew (23:37) and Luke (13:34; 19:41) longs for the healing of his beloved Jerusalem, so may we feel stirred to incarnate healing, wisdom and compassion in the places we love.

WISDOM'S WITNESS

In the ominous predawn hours of August 29, 2005, as the winds and waters of Hurricane Katrina surged upon the Gulf Coast of Louisiana and Mississippi, I was having a fitful sleep in Baton Rouge, some seventy miles northwest of the soon-to-be inundated Crescent City. After listening warily for hours to limbs cracking, winds howling, and rain pummeling, an eerie calm settled over my neighborhood. In those first surreal moments of people venturing forth to discover what damage had been done, unbelievable rumors began to circulate that the city of New Orleans—a place nearly completely surrounded by water (such that early on in the colonial period it was dubbed the "Isle of Orleans"), with over 50 percent of its land below sea level—had made it through the storm wounded, but relatively dry. Somehow, people whispered in astonishment, the predicted deluge had not come.

Yet as neighbors pushed aside debris, this illusion also lifted. A report surfaced, passing swiftly from one person to the next, that one, maybe more, New Orleans levees had been breached. Water was pouring into the city. Before long, nearly 80 percent of New Orleans would be inundated, with floodwaters continuing to rise. Only the swathe of high ground known as the "natural levee," created over the centuries by sediment deposits from the Mississippi River, plus a few other small areas of naturally raised terrain, would remain flood-free. The deluge of this place had begun, and I wondered, what would be left once the flooding was done? Would wisdom still lurk there, "in the markets . . . crying aloud in the street" (Prov. 1:20–21)? Would there be anyone left to heed her call? O New Orleans, New Orleans . . . "How often would I have gathered your children together as a hen gathers her brood under her wings, and you would not!"[2] Who would remain and who would return to incarnate care in the "City that Care Forgot"?

Half a month later, with an exhausting rescue effort just barely completed and the draining of the city underway, Louisiana Governor Kathleen Blanco made use of a state legislative address to beseech the world: "Hear this, and hear it well . . . We *will* rebuild!"[3] Yet the exasperation in her voice belied the conviction of her words. After the near total evacuation of the city, and the sheer ruin of massive neighborhoods such as Lakeview, Gentilly, and the Lower Ninth Ward (described post-Katrina as the equivalent of "a war zone"), it was not at all certain when, or even if, a "new" New Orleans would be born. So I wondered, once the waters receded, who would return? Who would move on? What cultural intangibles, what spirits, were now and forever gone? Could there even be any wisdom in rebuilding in such a vulnerable, flood-prone place? The draining of New Orleans had begun. And I wondered, what would be left when all the draining was done?

On September 15, a day after Governor Blanco's entreaty to the world, New Orleans' mayor, Ray Nagin, announced his plan for repopulating the city. Faced with the reality of a ghost town, Nagin's desire to do *something* proactive surely can be understood: only by encouraging people's return could their hopes for rebirth begin to grow. Yet state and

national officials immediately deemed Nagin's repopulation plan premature. Just two weeks after the city's inundation, New Orleans hardly
could be considered safe. Although the doomsday prediction of a "toxic
soup" indelibly poisoning all had not come to pass, still pollutants of all
sorts—raw sewage, household cleaners, insecticides and herbicides,
leaked oil and gas, and much else—had commingled with the floodwaters and contaminated the soil. Official test results varied widely, with
the Environmental Protection Agency (EPA) and the Louisiana Department of Environmental Quality (DEQ) finding very low contamination
levels, while private and nonprofit groups reported more hazardous conditions.[4] Then, too, there was the city's horrid stench, so often commented upon by reporters and believed to be related to the widespread
respiratory problems afflicting journalists embedded in post-Katrina New
Orleans. In addition to uncertainties about the city's potentially hazardous environment, very few essential services had been restored: traffic
signals were not working, most hospitals were not functioning, emergency services were sporadic, and phone service was nonexistent. And
this is to say nothing of the absence of electricity and sanitation services.
If there had been an outcry concerning the miserable conditions endured
by stranded New Orleanians awaiting rescue for many days following
Hurricane Katrina, how much stronger would the uproar be if these
same people now were invited back into harm's way?

Even more worrisome was the fact that the Gulf Coast's "high hurricane season" was not yet over. Beset with broken levees and wounded
wetlands, the city would be likely to flood again should another hurricane pass nearby. And this is precisely what happened. Mayor Nagin was
forced to rescind his repopulation plan before it had gotten underway, as
Hurricane Rita's dangerously close path became evident.[5] In addition to
the serious damage done to Louisiana's southwestern coast and to the
eastern edge of Texas, some of New Orleans's worst-hit neighborhoods
were inundated again by the impact of Hurricane Rita.

The aftermath of hurricanes Katrina and Rita revealed to the world
the disastrous intersecting realities of racism, classism, poverty, and environmental degradation in New Orleans. It became clear to all that it
would take more than good intentions to "bring New Orleans back."[6] It
indeed would take a wise and compassionate midwife, even Wisdom

herself, to bring forth incarnations of the *common good* in the Crescent City.

WISDOM IN THE WETLANDS

When healthy, Louisiana's coastal wetlands function as a sponge-like buffer, soaking up storm surge and ameliorating the brunt of turbulent winds and waters as tropical storms move inland. Yet over the last century, Louisiana's "first line" of hurricane protection has been disappearing at a staggering rate. Coastal scientists report that the state loses between twenty-five and thirty-five square miles of wetland—an area roughly the size of Manhattan—every year,[7] and that more than a million wetland acres have disappeared since the 1930s.[8] New Orleans is part of an estuary ecosystem, a place where a major freshwater source, such as a river, meets and interacts with a body of salt water, such as an ocean bay. While soil subsidence and land erosion are common to any estuary system, in the case of coastal Louisiana human folly has greatly exacerbated these processes. In an estuary system that is building land, erosion and subsidence are counteracted by the river's action of depositing silt and sand during intermittent floods. This creates a give-and-take, land-loss–land-gain relationship. Yet because the Mississippi River has been constrained by artificial levees from north of Baton Rouge almost to the river's mouth, this mighty river has been deprived of its land-replenishing faculties. Crucial habitat has been destroyed for those creatures that live within the various estuary echelons, including for those human communities dependent upon the estuary's bounty.

No longer the recipient of new sediment from the river, the area of land first built up as part of the Saint Bernard delta, between 4,000 and 2,000 years ago, and more recently (beginning about 1,000 years ago) as part of the Balize, or "birdfoot," delta—is now sinking swiftly into the sea.[9] If the Mississippi were still allowed to replenish the land, the greater New Orleans area likely would not have suffered from hurricanes Katrina and Rita to the extent that it did: there would have been a stronger natural buffer to protect it. Yet beginning in the late nineteenth century (in response to major river floods), the U.S. Army Corps of Engineers began the massive project of constructing artificial levees along the Mississippi River. As early as the 1970s it was becoming evident that Louisiana was swiftly disappearing into the sea, and that it was faulty human

intervention—namely, the Corps of Engineers' "levees-only" policy—
that was keeping the river from fulfilling its natural inclination to rebuild
the land.

In addition to the deleterious effects of endless artificial levee construction, the oil and gas industries have further damaged the wetlands by
cutting thousands of miles of canals and channels through them in order
to facilitate shipping and daily industrial operations. These waterways
have torn apart the root system of estuary flora, the very "glue" holding
the soils together. Saltwater from the Gulf of Mexico thus more easily
penetrates the area, resulting in the deterioration of low-salt and freshwater habitat, as well as further erosion. Industry, it may be noted, has also
played a role in levee building. Throughout the twentieth century, many
businesses built manufacturing plants along the Mississippi River between Baton Rouge and New Orleans, forming what has come to be
called Louisiana's "chemical corridor," or less flatteringly, "cancer alley."
Abundant freshwater for cleaning and cooling purposes, as well as easy
access to a major navigational path, made the Mississippi River seem like
an ideal location. Rivers, however, occasionally jump course. At some
point in time, a major flood enables a river to create a new principal path
to the sea. A new delta then is created, leaving the previous delta to
further erode and subside, becoming an ocean bay. The industries forming Louisiana's chemical corridor, as well as the city of New Orleans
itself, have found they have much to lose should the Mississippi River
overtop its levees, creating a new course to the sea.

Although public awareness of Louisiana's deteriorating coastal wetland
ecosystem had been growing during the final decades of the twentieth
century, and while a few small-scale restoration projects had been implemented with initially positive results (that is, land-gain), the national attention given to this situation was far too little, much too late. Numerous
studies have been funded and published over the past fifty years, all
showing that some of the river's natural processes of flooding and sedi-
ment-deposition should be restored or at least mimicked. In effect, their
lesson has been that people have much to learn from the intrinsic wisdom of the wetlands. To heed this place's wisdom would mean to discern
its patterns, learning to understand the connections between different
natural systems, such as how the river builds and replenishes the land,

and how replenished land, in turn, benefits human communities. *Incarnating* the wisdom of this place then would mean to develop ways of working with rather than against those patterns. Engineering projects that attempt to implement the wisdom of the wetlands can be—indeed, have been—designed, fashioning a system that allows for both controlled flood protection and sediment deposition. It remains to be seen whether Louisiana and the nation will work together to incarnate this wisdom.

To discern and incorporate the wisdom of a place does not mean abdication of all human innovation and technology. Human beings are "nature," too, after all, and our desire for safety from flooding, for example, would seem to be a natural inclination. What's more, one of the truths to be discerned from the Pontchartrain Basin estuary and the Mississippi River delta is the necessity of transformation for any healthy estuary. Over the past 5,000 plus years, the Mississippi River has created all of southeastern Louisiana, producing new deltas roughly every thousand years. (Rapid change in geological time!) If this need for transformation is a crucial part of the wisdom taught by the estuary, New Orleanians shouldn't feel they have to shun all innovative engineering, waiting for the next flood to destroy them. Yet if there is truth in the biblical notion that humans are made in the image of God, perhaps part of that divine image is the human capacity to discern and incorporate the wisdom of places; i.e., to learn how we may bolster our own well-being by mimicking and sustaining, rather than using up and completely controlling, the larger ecosystems of which we are a part.

To embody wisdom is to be attuned to the inherent patterns of a place, including its risks, and to live in the wisest ways possible in light of those patterns and risks. In the case of New Orleans, people should have been better informed about the deleterious effects that a major hurricane could have upon their city, especially upon their differently elevated neighborhoods. At a minimum, there should have been city regulations ensuring that houses built in low-lying areas were elevated to withstand hurricane-level flooding. Although it is becoming more and more evident that shoddy building practices by the Army Corps of Engineers is largely to blame for the most severe levee failures related to Katrina (and yes, the Corps should be held accountable for the folly it has wielded), still people who live in low-lying and coastal areas should know that no amount and strength of levees ultimately will hold back every storm and all flood waters. It should fall on educators, ministers, and especially elected city officials to continually remind people of this fact. It is a further travesty that the city and country did not have a better emergency shelter and evacuation plan in place. To have lined up only a few days' worth of emergency rations, to have delegated insufficient shelter space, and to have had no bus and rail evacuation plan ready to implement is beyond irresponsible. Safety and shelter in the face of possible disasters *is* a social justice issue, especially when nearly half of a city's residents are impoverished and lack independent means for evacuating.

Yet no one can claim that city and state officials had not been amply warned of the possibility of such a devastating flood. During the past decade, numerous front-page news stories describing the ecological destruction of Louisiana's coastal wetlands and the probable disaster scenarios that could unfold in relation to such environemental destruction had been published in local papers. University scientists and nonprofit groups such as the Coalition to Restore Coastal Louisiana had repeatedly informed people, including local and national public officials, of the disaster that might come to pass if New Orleans were directly hit by a category three or stronger hurricane. Coastal advocates had repeatedly explained the benefits of a healthy wetlands ecosystem, not only for wildlife but also *for the people*, and had lobbied tirelessly for coastal restoration. Apparently, however, such warnings were often met with closed ears and hardened hearts.

Yet now, more than ever, the need for a vital wetlands ecosystem and for rebuilding in an ecologically sustainable manner ought to be viewed as paramount, and possibly life-saving in relation to future hurricanes. It should be evident that the welfare of the people, especially of the sick, elderly, and poor, is bound up with the health of the land. As Katrina revealed, natural systems do not choose to devastate the rich over the poor, nor vice versa: houses in both Lakeview (typically white and wealthier) and the Lower Ninth Ward (typically black and poorer) were razed by the impact of the deluge. Still, those people most likely to be stranded as the floodwaters rose were the poor, the sick, and the elderly. They will be the ones most likely to be stranded again. (An oft-heard answer as to why so many people did not leave the city before Katrina hit is that they had no means of transportation, and no extra money on hand to pay for gasoline or a hotel.) Wise ecological living thus should be viewed as a way to insure the welfare of *all* the people, rich and poor, young and old, vulnerable and strong.

SOPHIA IN THE SECOND LINE

On the same day that Mayor Nagin prematurely launched his repopulation plan, President Bush addressed the nation from a hauntingly empty Jackson Square, with the majestic St. Louis Cathedral, a prominent French Quarter landmark, as his backdrop. Bush gallantly pledged wholehearted federal support for rebuilding New Orleans and the Louisiana-Mississippi Gulf Coast, promising to rebuild these places such that they would be "even better than before."[10] Bush assured Gulf Coast residents and the entire nation that the funds for repairing infrastructure, building affordable housing, and providing loans to small businesses would be procured. What's more, Bush further promised that displaced New Orleanians and Gulf Coast residents would be given first opportunity to take positions associated with cleaning up and rebuilding devastated areas.[11]

In retrospect, Bush's words seem to have been more self-serving than an honest promise of economic help for the Gulf Coast. That evening, however, his speech exuded hope. Remember that in mid-September 2005, it was not at all clear how poor evacuees would be able to return to their homes. How would people bused to Houston and San Antonio be able to re-establish themselves in New Orleans, especially if their jobs, houses, possessions, and neighborhoods, for all intents and purposes, no longer existed? Bush's promise of affordable housing and jobs for the displaced thus brought a glimmer of hope to those who could not imagine a way to return.

It therefore is all the more disgraceful that, even a year after the President's speech, little if any affordable housing has been readied and few evacuees have been helped by the government to acquire jobs relating to the city's cleaning and rebuilding efforts. The city's demographics have changed, from being about seventy percent African-American pre-Katrina to a majority-white population two months after people were given the green light to return.[12] As is so often the case, it is the poor who have suffered the greatest economic, cultural, and personal losses in the aftermath of such a great disaster. Not surprisingly, those with no means to leave when called to evacuate the city now are the ones with the least ability to return. Yet the cultural and spiritual ties of the poor to New Orleans equal, and in some cases surpass, those of the rich.

Bush, or rather his speechwriters, knowingly infused great expectation into the end of his speech by employing the image of a venerable creole tradition, the "Second Line." Its roots in the "jazz funeral" traditions of New Orleans' oppressed, poor, and African-American community, Second Line ritual nevertheless is a symbol of hope and celebration for *all* New Orleanians. Although Bush's demeanor gave no indication that he himself had ever been "caught up" in second-line spirit, still his speechwriters had chosen the image well. For those who "know what it means to miss New Orleans," just recalling the Second Line ignites an energy of rejuvenation. The Second Line might well be considered a kind of sacrament: familiar and beloved musical strains (its roots in the music and cultural performance traditions of the Mardi Gras Indians and New Orleans Social Aid and Pleasure Clubs) incite the embodied signs of umbrellas bobbing and handkerchiefs swirling, enlivening people with a shared spirit that transcends the separateness of individuals and social groups. Through the Second Line, healing energy is poured out upon the city.

According to a popular theory, the term "second line" originally referred to the brass band and crowd of people that followed the hearse during a funeral procession. As the hearse and the family of the deceased depart for the cemetery—an event known as "cutting the body loose"—this "second line" of musicians and community members continues on. The band then breaks into what has come to be called "Second Line music," which evolved from a coalescence of European marching tunes and West African syncopation and call-and-response patterns. Second Line music was, and still is, a kind of postcolonial performance: over time it has served and influenced different social groups in a diversity of ways, while also becoming the music and ritual that encapsulates the spirit of this place, bringing New Orleanians together as a people *of* a place.

During the colonial period, African-based music and ritual were vehicles for preserving and transforming West African religio-cultural practices, even as these practices were being influenced by Latin Catholic and Native American folkways.[13] Congo Square, located on the edge of the original city (today's French Quarter) and within the boundaries of to-day's Louis Armstrong Park, is the spatial node where these African-based, creole (meaning born of this place) traditions first evolved. On Sundays in Congo Square, those people descended from Africa, both slave and free, would gather to exchange information, stories, goods, and foodstuffs, as well as to share religious and cultural traditions. New Orleans jazz music and the city's own version of Voodoo religious practice are two of the traditions that felt their first glimmerings in Congo Square. When Congo Square was shut down by the city at the end of the nine-teenth century, those cultural performance societies known as "Social Aid and Pleasure Clubs" and "Mardi Gras Indians" emerged as the vehi-cles through which these African-based, creole traditions continued to evolve. Over time, and in large part through the greater community's participation in New Orleans's annual Carnival celebration, Second Line music has become "Mardi Gras music" and in turn the music that ignites a sense of place. It invokes celebration and healing in rich and poor, black and white, old and young alike. It is "a joyful noise" for the people of this place.

In a prayerful post-Katrina article for *Sojourner* magazine, essayist Danny Duncan Collum describes New Orleans as a sacred place, or, as he puts it, "America's Holy City." He adjures that anyone serious about saving the soul of America "had better get involved with saving the city of New Orleans . . . since much of the soul that America has left resided, until recently, in that poor, beautiful, and suffering city."[14] Collum points to New Orleans's history of poverty and suffering, and especially to the rich cultural traditions born of this place's suffering communities, as the taproot of this place's holiness. While I personally believe there is a vari-ety of other significant influences—Euro-Catholic (and other) religio-cul-tural practices; the geographical shape and topography of the city; and the dual magnetism of river and sea—that contribute to New Orleans's distinctive place-based spirituality,[15] I agree with Collum that the charac-ter of this place has been significantly shaped by the rhythmic rituals that first flourished in Congo Square. Obviously, the "richness" of those who gathered in Congo Square during the colonial and early postcolonial peri-ods had little to do with worldly possessions (the vast majority were slaves and descendents of slaves) but rather with that energy that can be embodied but never captured and contained, that is, with spirit.

In New Orleans, *spirit of place* is manifest in the way Second Line music stirs the body; in the way the humidity wraps everyone within a common medium such that people "navigate their streets like fish";[16] and in the way the countering (though not contradicting) pulls of river and sea create a swirling, rather than a grid-like, directional sensibility. The Second Line reminds us that, not withstanding its very real failures and inequities, pre-Katrina New Orleans was saturated with spirit, at times even exuding the "holiness" that Collum names. Epitomizing the type of ritual that generates joy from sorrow and redemption from suffering, the Second Line is a sign for the much needed hope and faith for the for the rebuilding and healing of this place.

WISDOM'S PRAXIS: CELEBRATION AND COMPASSION

The spirit of the Second Line surpasses economic, racial, and cultural differences. This does not mean that such differences, sanctioned for centuries by inequitable economic and social practices, thus suddenly disappear from the community. Obviously, this has not been the case for New Orleans. Though infused with spirit, New Orleans has remained one of America's most racially divided and economically impoverished cities. Still, wisdom might be gleaned from attention to the way Second Line music brings diverse peoples into a shared energy of redemptive celebration. At this time of recognizing New Orleans's devastation and yet dreaming of its renewal, perhaps such wisdom could be brought to bear on the plans and practices of rebuilding this place.

The significance of the Second Line reveals that there is a quality to this place more interesting and more enriching than economic growth or social progressiveness (though New Orleans surely could use a big dose of both!) by themselves can produce. This quality, this "place-vibe" or spirit, has been conceived through an incredibly rich and painful cultural history, always evolving in relationship to this place's distinctive deltaic topography and estuarine ecology. The crucial *communal* aspect of the Second Line—the way it incorporates everyone into its enlivening energies, from family members and friends to passers-by—reminds us that it will take a community deeply rooted in, and committed to, this place to incarnate the kind of spirit capable of healing New Orleans. The rebirth and redemption of this place in the ongoing aftermath of hurricanes Katrina and Rita will take what I call "communal *caritas*," an embodied praxis of wisdom and compassion that incarnates the Second Line spirit of this community, making a way out of a seeming no-way. The communal *caritas* of this place must be "creole"—inspired by wisdom intrinsic to this place—and thus attuned to the interconnected rhythms that make up this place.

CARITAS AND THE CITY THAT CARE FORGOT

Recalling the power of the Second Line reminds us that what New Orleanians yearn for in their quest to return to and rebuild New Orleans is not just the restoration of their physical homes, getting their old jobs back, or reclaiming their possessions. For so many people, such possibilities are forever gone. What people desire in their return to New Orleans (though, of course, they also need affordable housing, rewarding jobs, functioning city services, good schools, clean air and water, among much else) is this intangible, yet still very much *felt*, "place-vibe" or spirit. Such a spirit will re-emerge through the relationships, rituals, rhythms, institutions, geographical patterns, and ecological processes that coalesce to make New Orleans a distinctive place. Yet what we all must *do* with this "place-vibe" is to nurture and amplify it in such a way that it can become a sustaining, healing, and wisdom-encouraging energy, enlivening and inspiring all who are involved in the rebuilding of this place.[17]

Wisdom is inseparable from compassion, the quality of caring and extending that care to a more wide-reaching situation than that encompassed by one's own immediate concerns.[18] Such wisdom is related to the theological figure, *Hokmah-Sophia*,[19] who is described variously in the Hebrew Scriptures and New Testament as God's chief artisan, spritely and compassionate companion, and beloved child.[20] As *Caritas-Sophia*—the ever incarnating Spirit of God Herself—wisdom is continually at work and at play throughout the universe, quickening, healing, and caring for the interrelationships that make up different places, each in their distinctiveness.

Now more than ever, incarnations of *caritas*, or compassionate wisdom, are needed in this place. Care must be embodied and nourished through concrete practices. Yet, like spirit and wisdom, care is not a thing. Care is not an item one can purchase or throw away; it is more a divine quality, a gift we encounter and experience, and which we, in

turn, may incarnate in our expressions of care for others. Needing restored ecological integrity and economic and racial justice, New Orleans
would be well served by a theology that emphasizes *caritas*. Truly wise
life-ways could be encouraged, and certain wise and compassionate practices, such as zoning requirements, building standards, and environmental rules, could be enforced. Community-wide educational programs that
teach people about the eco-geology of this place, including the risks,
benefits, and responsibilities associated with living here, could be offered.
Such an educational effort would be both wise and caring in that it would
make people aware of possible hazards, while at the same time helping
them to practice wise, sustainable life-ways that work with, rather than
against, the ecological patterns of the region. Another concrete practice
of *caritas* would be to insure that all evacuees be given the opportunity
to return, should that be their desire, by creating a sufficient amount of
mixed-income, hurricane-safe housing in the more highly elevated areas
of the city, offering residents below the poverty line the first opportunity
to move there. (This would not, of course, solve the problem of how to
create a sufficient amount of gainful employment opportunities in the
city.) Countless incarnations of caritas, indeed a concerted effort of en-
dowing this place with practices of social and environmental wisdom and
compassion, are needed to transform New Orleans, the "City that Care
Forgot," into the Crescent City that remembered to care. The wisdom of
the wetlands and the redeeming power of the Second Line may together
provide a way for imagining caritas as "creole," that is, as organic to this
place and thus as appropriate religious imagery for inspiring the wise
rebuilding of this place. Thus would caritas be viewed not as a religio-
ethical ideal imposed from the outside, but as a divine energy already
inherent to this place, such as in the ecological wisdom of Louisiana's
coastal wetlands and the redemptive energy of New Orleans' Second-
Line music and ritual. Let us pray, then, that as New Orleanians return
and rebuild, we may be lured into incarnating caritas, building upon and
multiplying the patterns of wisdom, celebration, and compassion already
embodied in this place.

∿ Constructing Nature at a Chapel in the Woods

RICHARD R. BOHANNON II

INTRODUCTION

The worlds in which we live are ripe with images of nature: as our mother, as a wilderness to be tamed, as a divine gift to be stewarded, or as a pure and idyllic garden. Nonhumans appear to us as "nature" largely (or perhaps entirely) through such systems of human, symbolic constructions, often replicating or justifying power relations between humans. Nevertheless, while nature comes to us mediated through human symbols, it is not reducible to our social constructions.[1] Nonhumans—both living and inanimate—continue to exist apart from human representations of them.

Perhaps even more significantly, however, we humans also "see" and tangibly experience nonhuman nature as something constructed into built environments made useful for ourselves. Trees are transformed into lumber, mountains into asphalt, oil into plastic. The social construction of "nature," in other words, often happens through the physical manipulation of it. Human constructions of the nonhuman into "nature" thus do not happen merely through symbols and representations; the nonhuman world is continually and *literally* constructed into new objects and terrain for human use.

It is these kinds of construction, physical and yet also signifying, that shape human action, continually narrating how we smell, touch, feel, and see the nonhumans around us, whether they be trees, asphalt, or plastics. Particularly in the built environment, we doubly "construct" nature, creating stories and symbolic constructions about nonhumans (and thus humans) through our physical manipulation of them, our material constructions. On a very basic level, for instance, the transformation of nonhumans into building materials also, and more abstractly, transforms them into economic resources; a tree in a forest is part of "nature," and a tree in a hardware store is a piece of lumber with a fluctuating price based on supply and demand. In many instances our constructed environments are undoubtedly necessary, as they provide needed shelter and protection. Especially in contemporary industrialized societies, however, our buildings all too often serve to distort or to blind us from our relationships with the nonhuman world, creating a dichotomy between inside, humanized spaces, and outside, naturalized spaces. With *religious* architecture, furthermore, the stakes are raised. Not only do we play out our understandings of how the divine relates to the human—encoded, of course, in the human power-relations that regulate style, materials, and physical possibilities—but we also sacralize what it means to be human and nonhuman.[2]

While no building can represent all religious architecture—just as no congregation represents all of Christianity or all religions—the bulk of this essay will nonetheless focus on a small but significant building, Thorncrown Chapel, and use the work of Bruno Latour and Pierre Bourdieu to question how the "human" and the "natural" become constructed through the built environment. While Thorncrown is not entirely alone in departing from normative Christian architecture, in its relative uniqueness the chapel provides a helpful and more transparent example of how "humans" and "nature" are constructed in religious architecture. This is due in part to the chapel's influence and fame in American religious architecture in the twenty-five years since its completion, but, more significantly, because the building's fame derives precisely from its relationship to its immediate environment. As a building that, in the architect's words, "aligns itself with the attributes of nature,"[3] the chapel provides a challenge both to the lines often drawn between humans and nonhumans, as well as to what it means to "construct" those boundaries.

THORNCROWN: ACTORS AND ASSOCIATIONS

Thorncrown Chapel is a relatively small structure set in the woods near the tourist town of Eureka Springs, in the Ozark Mountains of Arkansas, with seating for one hundred people amid glass walls and wooden beams. Intentionally hidden behind a thin veneer of trees and a curving path, it is marked by three features that dominate the simple, rectangular building: a complex, geometric series of wooden trusses; a gabled metal roof, reminiscent of both Japanese and local architecture; and daylight, eagerly welcomed by a large, central skylight and walls of translucent glass. Run by evangelicals but not affiliated with any denomination, it does not house a regular congregation, though Sunday morning worship is held in the summers. Other than being a place for quiet contemplation, its most consistent, year-round religious function is for "destination" weddings. Thorncrown is also arguably the most significant structure built by the architect E. Fay Jones, and both he and the chapel received several prestigious awards from the architectural community after the chapel's completion in 1980. The chapel has become an icon of what might be considered a new style of religious architecture, with Jones and his firm subsequently designing several chapels and churches that strive to be

integrated into their natural surroundings through similar design choices.[4]

The general massing of the Thorncrown Chapel is taken from traditional Gothic architecture—Jones specifically cites Sainte Chappelle in Paris (completed in 1248) as his inspiration. Instead of stained glass, however, light is filtered through the forest at this chapel in the Ozarks; instead of intricate stone work and massive columns, a complex rhythm of sleek wooden beams holds up a minimal roof. However different the materials, the thrust of the chapel remains the same as its medieval forerunners: an upward movement toward heaven. The building itself is an arrow. The interior space likewise brings out a vertical movement, as the tree-filtered light and repetition of lattice-like beams together draw the eye up, and as a series of central columns, elevated off the ground, give the sense that the chapel might just float away into the heavens.

The verticality of the building and its modernized Gothic language are not to be overlooked; it reminds us that Thorncrown is not just about our human response to the nonhuman, but rather is much more about the divine-human relationship. For the evangelical founders of Thorncrown, it is not a place simply to contemplate the forest, but to contemplate *creation*, to behold the beauty of God and his son, Jesus, through the wooded sanctuary. The chapel's Web site, for instance, begins with "You walk down a wooded trail, move around a curve and there stands a majestic glass chapel—so beautiful—so close to God."[5] By thus connecting the forest with encountering the divine, they are not just constructing "nature," but a religiously normative understanding of how humans should relate to the nonhuman world. Thorncrown is demanding respect for the trees. God is to be found in the forest, it tells us, and God's people can encounter Godself in a uniquely intimate way amidst the light-filtering branches and leaves.

The layout of Thorncrown—its entrance, its series of recognizable church pews (split by a central aisle and facing a raised platform with a podium), and its sleek metal cross rising out of the ground immediately beyond—also quickly makes the visitor aware that this is a Christian space, that the chapel does not deviate too far from the norms of evangelical church architecture despite the lack of "walling out" the world. In other words, in a social context full of lingering suspicion about environmentalism among many theologically conservative Christians, a rather traditional arrangement of the interior, liturgical space, reminds visitors that it is the Creator, and not creation, being worshipped at this chapel.[6]

This evangelical unease with linking humans or God too closely with nature becomes more apparent when the chapel is contrasted to an additional building on the same property. Thorncrown Chapel's popularity, coupled with its small size, quickly gave cause for a second, much larger sanctuary—the Thorncrown Worship Center—to be erected nearby for bigger gatherings. This second building, while also designed by Jones, feels much further removed from its physical environment; while the lattice-work of trusses and the skylight repeat those of the smaller chapel, the side walls are windowless. The congregants, facing forward on a floor sloping toward the front, look out through a large window of clear glass, beyond which is not the forest at eye level (as with the original chapel) but a panoramic view of the neighboring valley. The chapel's simple, gabled roof is also exchanged for a prominent steeple at the newer Worship Center. This shift in design provides at least two functions: it affirms more strongly that Thorncrown is Christian, and it draws out a clearer distinction between "humans" and "nature." The nonhuman world becomes something to be viewed at a distance, with the congregant safely distinct from the natural world below.

A wide array of human social forces—such as white, Southern evangelicalism; U.S. Christianity more generally; the tourist industry, with its attendant middle-class economics; and modern architecture in the United States—is at play in the construction and maintenance of the chapel. I would argue, however, that these various human factors are not the only significant ones shaping the experience of Thorncrown. Something as seemingly static as a church pew, for instance, is an actor insofar as it influences those sitting in it to place their bodies in certain positions, and for those walking by it to take its presence into account when passing. Moreover, this pew is in relationship to those who built it, to human social systems that previously developed the notion of the pew, to the physical environment of the trees or metal ores from which the pew is derived, to the people interacting with the pew and nonconsciously deciding that it should in fact be treated as one treats an object designated to be "pew," and so on. These nonconscious assumptions about pews are perhaps the most effective: these normal pieces of church furniture are not intentionally interacted with, and yet they highly structure the space, how humans use their bodies, and the social relations between congregants and officiates. The pews order how the visitor is expected to act in and perceive the space; like a Gothic cathedral, the intent is to quiet the visitor into humility and awe toward something greater—whether it be the Christian God, the light passing through the windows, the forest, or simply the impressive architecture.

These pews are thus members of what Bruno Latour often calls the "collective"—that is, all that is perceived to exist, all things, objects, persons, humans and nonhumans—and, as such, are social actors.[7] This is the growing conviction within the developing body of social theory often labeled under the awkward moniker Actor-Network Theory (ANT). Jonathan Murdoch gives the following concise description of ANT:

> We can consider Actor-Network Theory to be both co-constructionist and ecological: it is co-constructionist in the sense that it emphasizes relations and the way that discrete entities or beings emerge as these relations are consolidated; it is ecological in the sense that it seeks to overcome any underlying distinctions between natural and social entities, thereby extending agency to nonhumans as well as humans.[8]

Agency or action should not be confused here with *intent*, however. While there is clearly no intent on the part of the dead trees-turned-pews, they nonetheless continue to act within the networks that comprise the space, ordering it and how the humans within it relate to each other and to the variety of nonhumans around them. With its two rows of rigid benches all facing forward, it is, for instance, a place where you watch and observe services or weddings, but not where you converse with others.[9] Likewise, the wooden trusses, the forest outside, the parking lot, the tile stone floors, the glass, and the heating and cooling systems are all relevant actors.

This emphasis on "actors," however, is perhaps misleading; the importance of including nonhumans becomes more evident, and seems less arbitrary, when the focus is shifted from human and nonhuman social beings as actors, and toward networks and the associations or relationships between them. "Basic social skills," Latour writes, are only a small "subset of the associations making up society."[10] Or, as an earlier reviewer of Latour remarked, "Sociology, to regain its lost status, must define itself as the science of associations," and should give up its Durkheimian quest for social facts.[11] The need for understanding the social through a focus on associations and networks becomes additionally apparent when we raise a simple question: what is a nonhuman? Is Thorncrown itself, or the forest surrounding it, a nonhuman; is a single pew, a single truss, or single nail, a nonhuman? These questions dissipate when the associations, instead of individual actors, are the focal center. The network of objects and history that combine into Thorncrown Chapel has its own coherence, and likewise the overlapping network of trees, cloth, human hands, and history that form a church pew gives it coherence. ANT thus carries a great deal of resonance with basic ecological insights in its emphasis on the interconnectedness of all beings.

Within this schema, we live in constant connection with a wide array of objects; whether or not they might helpfully be considered an "actor" depends largely on one criterion: "Does it make a difference in the course of some other agent's action or not?"[12] Latour makes a further, ironically anthropocentric, claim for the necessity of including humans in social analysis: it is only humans that solidify their social "asymmetries" through the presence of nonhumans.[13]

Nowhere is this more apparent than in the built environment, where towns, cities, and homes have long been physically constructed so as to delegate people into certain races, classes, and genders, thus reinforcing their differences. In discussing gender and the construction of homes in Kabyle society in Algeria, for instance, Pierre Bourdieu makes a parallel observation:

> The mental structures which construct the world of objects are constructed in the practice of a world of objects constructed according to the same structures. The mind born of the world of objects does not rise as a subjectivity confronting an objectivity: the objective universe is made up of objects which are the product of objectifying operations structured according to the very structures which the mind applies to it. The mind is a metaphor of the world of objects which is itself but an endless circle of mutually reflecting metaphors.[14]

In other words, there exists a circular relationship between human social power-dynamics and the physical artifacts we create, allowing for the continual reproduction of unequal power-relations. Bourdieu's and Latour's methodologies are strikingly different, and perhaps even antagonistic, but as the quotation above reflects, Bourdieu's general theoretical system and his understanding of a social field as a "network of relations" lends itself some complementary overlap with ANT.[15]

In this framework, the social world (and here, like most social scientists, Bourdieu is talking specifically and exclusively of a *human* social world) is composed of a variety of fields positioned in relationship to each other in terms of their respective amount of power or capital within a culture's larger sphere of power. Across the span of his career, Bourdieu looks at a variety of fields, such as the French academy, religion, or the art of photography, and at how they strategically position themselves within the larger social network to gain economic, political, and symbolic power. Equally important, however, is how these fields create forms of symbolic capital within themselves, agreed upon nonconsciously by individuals, who then attempt to maximize their amount of capital. One's understanding of what precisely that capital is, however, depends on one's positions within a field. In his work on the education system in France, for instance, Bourdieu describes very different examples of "habitus,"[16] and thus different perceptions of symbolic capital and of social possibilities, between the economic and social elite, on the one hand, and rural peasants, on the other. The power dynamics within fields thus lead them toward continually reproducing themselves—and, indeed, throughout his work Bourdieu emphasizes self-reproduction of social systems, conceiving of social structures as both structured (something preexisting that we find ourselves born into) and structuring (something we continually reconstruct in our daily actions).

Most importantly for understanding Thorncrown Chapel and the dynamics of how "human" and "nature" are defined and constructed, the borders of the fields are constantly in flux, and "border patrol" can become a primary function of actors within a field. Bourdieu writes, "The question of the limits of the field is a very difficult one, if only because it is *always at stake in the field itself* and therefore admits no *a priori* answer."[17] At Thorncrown, then, what we have is a re-enactment of, and partial challenge to, power struggles between humans and nonhuman nature, a dance around the boundaries between them.

HUMAN/NATURAL FIELDS OF POWER

Stretching the notion of "fields," I want to ask: If all things, all entities, can be understood to be actors, as Latour and ANT would claim, then what might the ramifications be for a field theory such as Bourdieu's? Can one talk of "human"-ity as a field? Can the nonhuman serve as a field itself? (Or as a variety of fields?) Field theory does indeed seem to break down when talking about interaction between trees, rocks, forests, and most of the nonhuman world, and can reach a level of unhelpful absurdity. Unlike ANT, field theory does necessitate some level of interest and intentionality, even if only on a nonconscious level. Nevertheless, I would contend that in presuming two concepts—that nonhumans are social actors interacting with humans, and that nonhumans can be humanized, or civilized (for instance, trees can be made into church pews)—field theory can be elaborated and allow for humanity itself to be seen as a "field" in a symbolic and literal power play with nonhuman nature.

By using the term "humanity" here, however, I do not intend to signify the entirety of our biological species, but rather humanity and the field of humanity as encompassing particular, culturally specific social constructions of "human"-ness. In Western cultures (though this surely is often true, in different ways and with different dynamics, of other societies), understandings of humanity as defined in contradistinction to nature have frequently been partial, narrow, and exclusive views. When we in Western cultures define the "human," we most often define a normative humanity that leaves out substantial portions of the species *homo sapiens*; those included under, or excluded from, the sign "human" are ordered by their proximity to, or elevation above, the natural. This has been the foundational insight of much of ecofeminism, which has demonstrated the symbolic and socioeconomic links between the domination of nonhuman nature and the domination of women, as well as of ecological Marxism and the environmental justice movement, which have illuminated joined patterns of ecological as well as economic and/ or racial exploitation.[18] For instance, our delineations between humans and nature, as Donna Haraway demonstrates evocatively in her studies on primatology,[19] have been necessary in order to maintain distinctions among humans. These categorizations, of course, are very much about executing an effective "border control," maintaining the perception of clarity in distinguishing between humans and nature, black and white, civilized and barbaric.[20]

The humans at Thorncrown likewise are also not simple representatives of our species—they are primarily evangelical Christians and tourists. While such wedding attendees, summer worshippers, and visitors certainly do not constitute denials of everyone else's humanity, an ideal "human" is undoubtedly implied through the class and religious assumptions present at Thorncrown. Those seeking to be married at the chapel, for instance, must submit their ceremony for approval at least four weeks before they marry, so that the chapel's ministers can be assured that the service will be "Christ-centered and scriptural."[21] In terms of architecture, the chapel's tie to elite Gothic cathedrals makes an obvious, but hardly unusual, allusion to class and European ancestry. Instead of claiming certain groups of humans as more natural, however (as Haraway describes with her work on primate studies), Thorncrown draws out how humans in the chapel should be positioned toward, and so relate to, nonhuman nature, and questions what exactly that nature is. What makes Thorncrown unique in religious architecture, then, is not so much how it creates distinctions between humans, but rather in how it brings into the foreground a variety of relationships between humans and nonhuman nature. These relationships, however, are conflicted and perhaps contradictory. Humans remain clearly dominant, and yet the distinction between humans and nature is partially dissolved.

In his monograph of E. Fay Jones's work, Robert Ivy Jr., a leading architecture critic, continually describes Thorncrown Chapel as confusing the boundaries between the human and nonhuman. "Thorncrown is elemental," he writes, "a man-made temple married to the woodland. It raises with the authority of nature in the Arkansas forest from a stone foundation to wood columns and layered branches to folded roof. . . . Thorncrown's peaked roof seems a part of the forest, its glass walls dematerialized by light and shadow. . . . It seems both man-made and natural."[22] The trees in particular work as symbols with a dual function: as representatives of "nature," nonhumans of the supposedly untamed

forest surrounding the chapel, but also as gestures toward Gothic columns or spires, as building material, and as pews.[23] This interplay of
tree and column, separated only by translucent glass, both *naturalizes the
human*, by evocatively linking the built environment with its surrounding
forest, as well as *humanizes nature* through a very literal construction with
very significant, signifying power: with precision saws and steel joins, the
trees cease to be "tree" and are instead artifactualized into "pew" and
"truss"; southern pines become 2x4s. The identity of the forest itself is
changed, shifting from one wooded hill among many, into an idyllic,
sacralized grove.[24]

Thorncrown's literal and figurative constructions of nature are ambiguous, inconsistent, contradictory. The exact boundaries between the
human and natural spaces are intentionally not clear within the architect's plans; the chapel's design calls for the roof, wooden trusses, and
stone floor to all maintain their form beyond the clear, glass walls, following a trend in contemporary architecture to blur or dissolve the lines
between interior and exterior built environments.[25] As some, following
Donna Haraway, have recently described the city as a cyborg,[26] so we
thus might also describe not only Thorncrown but all religious architecture. Haraway's interest is in the cyborg as a fictional metaphor for the
interrelatedness of humans and their machines, with human life saturated
by, and mediated through, technologies. This blurring between the
human and machine also, and importantly, blurs the delineations constructed between humanity and nature. The machines that we inhabit,
such as chapels, were formed from and by the earth itself—objects ultimately derived from rocks, water, plants, and sometimes animals—but
were also manipulated by human hands.[27] Our experiences of being
human are thoroughly enmeshed in these built environments that we
construct and inhabit. Our part-human, part-natural "cyborg" buildings
become extensions of our bodies, guiding how we breathe, feel, see,
hear, and smell the world. Put succinctly, Haraway's image of the cyborg
illuminates three "crucial boundary breakdowns": those between the
human and the animal, between organism and machine, and between
the physical and the nonphysical.[28]

Understanding humanity as a social field, juxtaposed against nature,
brings attention to the continual energy spent in reinforcing and maintaining human/nature boundaries. For Bruno Latour, while not thinking

in terms of fields, the distinction made in language concerning "nature" and "society" is a product of a "Scientific" mentality, inasmuch as it presupposes a mute, static nature that is an object for study in a one-way stream of communication.[29] "Nature" and "society" are emphatically not facts, but "specific forms of public organization," defined in direct opposition to one another.[30]

In an attempt to bypass this dichotomy, at the basis of Latour's constructive project is a "collective" of "humans" and "nonhumans," his replacement for language juxtaposing "human" and "nature." This collective is perhaps understood best as a verb, as "the work of collecting."[31] It is a dynamic gathering of humans and nonhumans, constantly in flux and being negotiated and renegotiated. Within this collective—which effectively is the conglomeration of all that is understood to exist (in distinction from simply all that exists)—lie two broad activities: bringing new members and associations into the collective, and placing those within the collective in order. Political ecologists—including environmental religious practitioners—for instance, would be a significant voice in both bringing in new members and ordering the collective. There is no whitewashing of reality here, however, or an attempt to do away with difference; the "collective" is not another attempt at a melting-pot kind of unity among all existence. Rather, Latour is seeking to provide a way for the differences between and among human and nonhuman actors to be acknowledged and appreciated, to provide an avenue for accounting for different levels of actions, from humans and primates to trees and rocks.

Thorncrown might be seen as the product of competing mentalities: it partially follows the modern "Scientific" mentality, firmly separating nature and society, while also drawing out through its construction the ambiguous, cyborg-like networks between humans and nonhumans. Similarly, Thorncrown is not a simple and egalitarian blurring of boundaries between humans and nonhumans. A valuing of the human *over* the natural is integral to how these fields get constructed, and to how physical construction and manipulation of the nonhuman is justified. Even while Jones's "organic" design at Thorncrown does much to integrate into its landscape, especially in comparison to mainstream evangelical architecture in the United States, it is the chapel itself that is the spectacle worthy of attention from tourists, wedding parties, and pilgrims.[32] Their interest lies in two constructions: the refined, civilizing humanization and physical construction of the trees into structural beams and pews, and the accompanying symbolic construction of living trees and forests, providing congregants the ability to meditate on the "outside" world (a term implying the "inside" as normal) without worrying about mosquitoes or ticks. The promotion of the chapel for weddings provides a clear example of this second construction: "Surrounded by beautiful rock formations and a canopy of trees, the chapel provides an unforgettable setting for your wedding ceremony. . . . [Y]our wedding can be outdoors and indoors in air conditioned comfort at the same time."[33] Nature is sanitized, as "beautiful" and "unforgettable," according to very human conditions. For most of its users, however, this hygienic situation allows for the rather unique opportunity to invite nature as "creation" into a church building, even if only visually.

While it makes sense to only grant humans volitional action in the construction and use of the chapel, or any other building, the framework of ANT provides an avenue for understanding the ways in which our built environments are always constructed in relationship to, and in power struggles with, nonhumans. As Latour notes how nonhuman objects give human "social asymmetries" their "steely" quality,[34] so the atypical interactions between humans and nonhumans at Thorncrown give a minor, ambiguous, but perhaps important, nudge, questioning the asymmetries in the social life not between humans, but between humans and the nonhuman world more broadly. More than nonhuman objects acting in some direct way, interaction with the trees, beams, glass and pews at the chapel provides an avenue for the human visitors to rethink their relationships with nature as creation. Thorncrown questions, albeit partially, the power-asymmetries between humans and nature, and particularly between humans and the local forest, as well as questions how humans might relate to the divine through their environments, built or otherwise. The mimicry of the forest by wooden columns carries the potential for bringing to the surface assumed and nonconscious associations that humans continually have with their built environments. While humanity is undoubtedly still dominant—the chapel is enclosed and air-conditioned—there nevertheless is an attempt made toward benevolence and appreciation.

CONCLUSION

From wooden shacks to glass high-rises, all architecture is significantly determined by the constraints of the plethora of nonhumans around us. Human action is always action in relationship with these constructed nonhumans, as we define ourselves through the distinctions that we draw between "humans" and "nature."

Thorncrown's significance lies only partially in the human and nonhuman interactions on its wooded hillside, however. In the chapel's tendency toward depicting an idyllic, sanitized, and sacralized nature, it runs the risk of devaluing everyday experiences of the nonhuman, in line with William Cronon's reservations about idealizations of wilderness.[35] Nevertheless, the intent behind Thorncrown's design was not simply to create an idealized nature, but also to be instructional. Writing late in his career about his work in general, E. Fay Jones comments:

> If what we build, our interventions in the natural situation, aligns itself with the attributes of nature, perhaps it can in some didactic or other contributing way—as model, as symbol—inspire the inhabitants to align themselves in a more beneficial and meaningful way with the natural forces, the natural conditions, the natural rhythms of life. Surely there are benefits to be derived from living close to, and in harmony with nature.[36]

Thorncrown's importance lies in challenging, albeit only partially, how one thinks and make assumptions about the world upon leaving the glass walls of Thorncrown and returning to the parking lot, to the tourist town of Eureka Springs down the street, to cars and to homes. Perhaps most relevant is how we might think about the built environment as we encounter it every day, those aspects of the nonhuman world around us that we have so thoroughly humanized that we forget their atoms once fell under the name "nature." The church pews and trusses were once parts of a forest; steel connectors and parking lots were once mountains. If the forest and the mountain are valuable—as economic resources or as sacred places—then what value must be placed on the steel, asphalt, and lumber derived from them?

In the end, Thorncrown remains a *religious* building foremost, and as such is not simply about humans and nature, but about humans and *creation*. While ultimately anthropocentric, as a chapel Thorncrown makes a distinctly didactic contribution, functioning similarly to Latour's notion of "political ecology"—the chapel gives an alternative to what it means to "build" an environment, between how the divine, creation, and humans relate to each other. In a time when large churches are more likely to mimic shopping malls than to be built in the context of their surrounding ecosystem, Thorncrown provides a modest polemic for grounding the buildings in which we live and worship, acknowledging both the human and the natural in every space.

✴ Felling Sacred Groves: Appropriation of a Christian Tradition for Antienvironmentalism

NICOLE A. ROSKOS

In the midst of an awestruck crowd, (St. Boniface) attacked with an axe one of the chief objects of popular veneration, Donar's sacred oak, which as the first blows fell upon it, the huge tree crashed, splitting into four parts, and the people who had expected a judgment to descend upon the perpetrators of such an outrage acknowledged that their gods were powerless to protect their own sanctuaries. From that time the work of evangelization advanced steadily.
—ALBAN BUTLER, *The Lives of Saints*

This story from Butler's *Lives of the Saints* initiates *The Cross and the Rainforest* by Robert Whelan. This latter, more recent book builds its polemic against contemporary environmentalism on the historical Christian precedent of cutting down trees in order to save pagan souls. It represents a corporate-funded Christianity that now implicitly champions, in its support for the globalized economy, deforestation in the name of this ancient, saintly practice of felling the sacred trees. Before examining this modern fusion of anti-environmentalism with the enduring Christian fear of paganism, let us first consider the history and context of this mandate to eradicate sacred groves.

Hebrew Scripture describes the worship of the fertility goddess Asherah in 2 Kings 17:10: "They set up for themselves pillars and Asherim [plural] on every high hill and under every great tree." Despite Judith Hadley's thorough study, *The Cult of Asherah in Ancient Israel and Judah*, which uncovers archaeological and textual evidence of the existence of the goddess Asherah, it remains unclear whether an Asherah actually was a sacred wooden pillar, a stylized tree-of-life, a tree, an image of the female goddess, or was encountered through some or all of these various forms. Archaeological evidence from el-Qom dated to the ninth century BCE confirms that Yahweh was (at least occasionally) worshipped with Asherah. Since the Hebrew Scriptures do not record any specific prophetic condemnation of Asherah worship, some scholars suggest that such worship was at one time generally accepted in Israelite society.[1]

One might speculate about a form of Asherim present in Hebrew Scripture when Abraham and Jacob built altars to the Lord under trees. At Shechem, Abraham built an altar under the oak of Moreh and received a message from God saying, "I will give this land to your offspring" (Gen. 12:6–7). Joshua made a covenant with the people and "set up the stone under the oak in the sanctuary of the Lord," also at Shechem (Josh. 24:25–26).

Hadley notes a critical turning point for the translation and interpretation of Asherah happening in later periods. From the third century BCE, the Septuagint, or first Greek version of the Hebrew Scriptures, translates Asherah as "grove," and this was taken up in the King James Version (KJV; circa 1611 CE). Hence, Hadley and others conclude that "by the time of the Mishnah (the first few centuries of the Common Era), Asherah has lost even its distinction of a humanly made cultic object, and has become any type of living tree."[2] This later translation of "grove" in the Septuagint and the KJV impacted the Christian extermination of sacred groves that were honored by Europeans.[3] The KJV was translated after much of the initial encounter between Christians and indigenous European groups. Clearly, the command to fell cultic pillars has much more limited scope than a command to fell groves of trees.

DEFORESTATION THROUGH THE EMPIRES

Although Christian history abounds with the cutting of sacred groves, this practice did not solely originate within Christianity or Judaism. The Romans cut down sacred groves as a tactic to conquer other peoples at the time of Nero. The historian Tacitus gives an account of the Roman takeover of the island of Mona, a Celtic place of pilgrimage for British Druids. Tacitus describes "Druides praying and cursing, and women . . .

dressed in . . . black, with torches in their hands and hair wildly flowing."[4] The Romans "cut down their sacred groves, and broke the altars."[5] Nor did elements of the Greco-Roman religious-philosophical worldview— the worship of Artemis (or Diana), the goddess of the forests, and the meeting of Plato's school in the sacred groves of Academus—keep the Greeks and Romans from deforesting the Mediterranean. Rather, economic and political interests reigned. Robert Pogue Harrison explains that for warfare "forests became fleets, sinking to the bottom of the wine-dark sea. . . . As for the agrarian Romans, the insatiable mouth of empire devoured the land, clearing it for agriculture and leading to irreversible erosion in regions that were once the most fertile in the world."[6] Harrison also points out that deforestation brought Ephesus to ruin, the home of the famed temple to Artemis, one of the Seven Wonders of the World. Once covered with forests of oak four thousand years ago, Ephesus was clear-cut and subject to intensive agriculture, which caused the soil to wash down into the sea. Today, its original harbors are three miles from the sea due to the accumulated silt from deforestation.[7]

When the tables of power turned and Christianity overran the Roman Empire in the fourth century CE, it continued the same practice of destroying sacred groves and deforesting the land through forced conversion and the spread of cities and agriculture. Hagiography describes legends of monks working to root out paganism through the eradication of sacred trees and groves. The French historian Montalembert describes monks entering the forests, "sometimes axe in hand, at the head of a troupe of believers scarcely converted, or of pagans surprised and indignant, to cut down the sacred trees, and thus root out the popular superstition."[8] On Monte Cassino, St. Benedict cut down a sacred grove and destroyed the temple to Apollo in order to convert the people in the surrounding area to Christianity.[9] In place of the grove, he built the first monastery. Charlemagne felled Irminsal, the Anglo-Saxon World Tree.[10] The most famous story of this sort, recounted in the epigraph above, is that of St. Boniface, who cut down the Oak of Thunor, the sacred tree of the Saxon thunder god.[11] For a somewhat more recent example, the African St. Bernard Mizeki cut down all the trees in a sacred grove in Zimbabwe in the late 1800s. His wife was weeping over the tree stumps. To console her, he carved a cross in each stump.[12]

TREE-FELLING SAINTS HONORED BY CONSERVATIVE CHRISTIANS

Imperial acts of cultural genocide and ecocide, the forced conversion of pagans and the demolishment of their sacred groves, marked paths to sainthood during the Middle Ages. These tree-felling saints are now celebrated heroes for a movement of conservative Christians.[13] Writing in the introduction to Calvin Beisner's *Where Garden Meets the Wilderness: Evangelical Entry into the Environmental Debate*,[14] Rev. John Michael Beers oddly enough likens Beisner to the courageous St. Boniface, whom he honors as one who "promoted Christianity to the indigenous pagan population by chopping down the tree, revered as sacred, that they had worshiped for generations." He then claims Beisner, like Boniface, likewise faces "today's New Age adherents in the environmental movement," who are "not unlike the pagans of a millennium ago."[15] Beers then essentializes environmentalists as those who value animal and plant life at the expense of human life. He fears that the "refusal to develop resources" would only create a huge amount of unemployment and poverty, thereby driving people to seek contraception and abortions.[16] Citing the command to cut down sacred groves (instead of sacred poles) in the King James Version of Deuteronomy 7:5 and 12:3, Beisner also suggests the legitimacy of this tactic of tree-felling as "spiritual warfare in Christian missions."[17]

This Christian movement resonates with the so-called *wise-use* movement of the political Right, which also labels environmentalists "tree-worshiping pagans" as it advocates for corporate interests.[18] Robert Whelan's *The Cross and the Rainforest: A Critique of Radical Green Spirituality* and Beisner's *Where the Garden Meets the Wilderness* are both published by the Acton Institute for the Study of Religion and Liberty, a conservative think tank and fundraiser that propagates what many would term "green-washing," or the ecological painting of economic interests under the guise of political and economic freedom.[19]

Acton uses the Cornwall Declaration on Environmental Stewardship, drawn up by the Interfaith Council for Environmental Stewardship (ICES), as its protocol for humanity's proper treatment of nature.[20] The Cornwall Declaration finds fault in the nature-based "romanticism" of environmentalists who "deify nature":

Many people believe that "nature knows best," or that the earth—untouched by human hands—is the ideal. Such romanticism leads some to deify nature or oppose human dominion over creation. Our position, informed by revelation and confirmed by reason and experience, views human stewardship that unlocks the potential in creation for all the earth's inhabitants as good. Humanity alone of all the created order is capable of developing other resources and can thus enrich creation, so it can properly be said that the human person is the most valuable resource on earth. Human life, therefore, must be cherished and allowed to flourish.[21]

The Cornwall Declaration admits to the validity of some "environmental problems" in the developing world, such as "inadequate sanitation, primitive agricultural, industrial, and commercial practices." But, it says, "global warming, overpopulation," and "rampant species loss" are "unfounded" and "speculative" concerns "of a very low and hypothetical risk," "unjustifiably costly and of dubious benefit." They fear that public policies to counter a "hypothetical threat," such as global warming, will hinder economic development and accelerate already high rates of "malnutrition, disease, and mortality."[22]

Richard Wright, a Harvard-trained biologist and author of the textbook *Environmental Science* (now in its ninth edition), comments in "Tearing Down the Green" that Acton's publications, such as *The Cross and the Rainforest* and *When the Garden Meets the Wilderness,* tap "into the (contrived science) of prominent anti-environmentalists Julian Simon, Herman Kahn, Fred Singer, and Dixie Lee Ray," and choose to disregard the overwhelming scientific evidence for ecological perils that can easily be found through more trustworthy sources such as the World Watch Institute.[23] Regarding rampant species-loss, which the Acton Institute says is "unfounded," the scientific studies collected by the Worldwatch Institute disagree, stating that "prominent scientists consider the world to be in the midst of the biggest wave of animal extinctions since the dinosaurs disappeared 65 million years ago."[24] The Acton Institute propagates its anti-environmental rhetoric with corporate funding; its board of directors represents corporations and corporate-funded conservative think tanks, among them the Wolverine Gas and Oil Corporation, the Windquest

Group, Pulsar International, and the Atlas Economic Research Foundation. Atlas, it may here be noted, is the group that provided a summer fellowship so that Robert Whelan could write *The Cross and the Rainforest*.

The Cross and the Rainforest also employs anti-pagan sentiment toward anti-environmentalism, defending St. Boniface for his tree-felling tactics, then advocating for deforestation in general. Whelan first takes issue with Thomas Berry, who believes St. Boniface's actions were quite inappropriate for our contemporary world. Berry also, to Whelan's dismay, admits "pagan" traditions have something valuable to teach us about earth consciousness: "Christians, like St. Boniface . . . cut down the oak trees deemed sacred by the pagans. . . . Today that would be absurd. The unassimilated elements of paganism have so much to offer us in establishing an intimate rapport with the natural world."[25] Whelan rebukes Berry by citing Deuteronomy 7:5, which gives "very definite instructions" to destroy trees used in the worship of other gods.[26]

Whelan then quickly jumps from the religious to the economic, from tree-worship to a discussion of the importance of deforestation in today's economy, his main concern. Does he mean to suggest that deforestation somehow prevents the rise of paganism today? And that by extension to resist deforestation is to promote tree worship? Whelan does not recognize any of the countless other reasons to stop deforestation nor that loving and protecting creation need not imply its worship. Whelan does not account for the gap between his religious and economic claims. Rather, he dismisses deforestation as a problem since the "total forested area on the globe increased from 3.52 billion hectares in 1949 to 4.05 billlion hectares in 1989."[27]

Shocked by the sheer oddity of this information, I researched it. The article cited actually discusses the rapid depletion of forests, not their increase—Whelan's statistics about forest growth are never mentioned.[28] Further research uncovered alarming rates of forest loss: over 95 percent of ancient forests have been cut.[29] According to the Worldwatch Institute, "almost half of all the forests that once covered the planet are gone."[30]

Perhaps what is most disconcerting about this misinformation, whether put forth by Acton affiliates or other organizations in the wise-use movement, is the scope of its dissemination to believing Christians. Each year, Acton organizes between three and six student conferences to promote their "Free and Virtuous Society" program. They also give away thousands of dollars in fellowships and awards, including the Novak Award, which includes $10,000, for work on integrating Christian theology with their particular conception of free-market values.[31]

This conflation between anti-pagan and anti-environmental rhetoric unfortunately, and intentionally, contributes to a discriminatory attitude toward nature in general and ignorance about forest ecology in particular. In a story about St. Jerome of Prague, who ordered the felling of a sacred grove, a woman pleaded for the grove, saying the woods were "a house of God who brought sunshine and rain."[32] St. Jerome clearly had no understanding of the indigenous wisdom being communicated about the harms of deforestation. Indeed, she was right—forests bring rain and cutting forests changes the hydrological cycle. "The high transpiration rate of forests returns water to the atmosphere to fall again as rain." Trees, through the action of their roots, also lift water that has been stored deep in the ground, contributing it to mist and streamflows.[33]

A publication of the Forest Ecology Network tells of a more recent example of Christian ecological ignorance regarding the devastating impact of yet another instance of the cutting of sacred groves. In the 1980s, when missionaries converted the Rungus of Borneo to Christianity, they were encouraged to cut their sacred groves. The Rungus feared that cutting the forest would dry up the streams. The missionaries reportedly insisted nothing of the sort would happen, and the Rungus were forced to cut them down. The Rungus were right, the stream emanating from the forest did dry up once they cut the groves. These indigenous people had an ecological sensibility lacked by the missionaries.

Acton's publication *The Cross and the Rainforest* defines forests as "just a source of wood"—as clear an example of earth-alienation as one is bound to find, and a clear sign of the author's lack of understanding of forests' vital ecological role.[34] One might conclude that an entrenched fear of paganism keeps some Christians from perceiving forests in any way that goes beyond the latter's use-value, or from appreciating their ecological value to life-systems. In so doing, they overlook abundant biblical evidence that portrays trees as much more than "just a source of wood." Are they alarmed by Abraham's setting up an altar under the oak of Moreh (Gen. 12:6–8), and by the stone Jacob set up under the oak at Bethel to mark his covenant (Gen. 35: 1–15)?[35] It may also be worthwhile to point out that despite recent efforts among some Christians to denigrate the ancient association of the sacred with trees, most modern Christians themselves retain the residues of this association in their religious practice, whether in the form of Christmas trees, Mistletoe, or worship in Gothic-style cathedrals.

EMBRACING "TREES" IN THE CHRISTIAN TRADITION

Gothic architecture, with its distinctive ribbed vaulting, is thought to have been modeled on a forest. Consider Chartres Cathedral, started in 1194. Medieval tradition tells that it was built upon the site of a Druidic sacred grove, thus explaining why its ceiling was named "the forest."[36] Thus to worship in a Gothic cathedral is, in a sense, to worship in a grove, albeit a highly stylized, which is to say humanized, one. This gives a new significance to the term "cathedral forest," "an analogy that has its basis in an ancient correspondence between forests and the dwelling place of the divine."[37] As mentioned, pillars were used in the worship of Yahweh and Asherah; they even harbored the souls of sacred trees in the Mycenaean cultic tradition. All of which reinforces the symbolism of the columns in Greek and Gothic architectures, that "if a single column once symbolized a sacred tree, a cluster of columns may well have symbolized a sacred grove."[38]

Tree-affirming beliefs abound. Whether that of the pagan, Buddhist, Hindu, Native American, African tribal member, Christian, Jew, or Muslim, in all of these belief systems one encounters trees and forests honored as sources of divine emanation or places of revelation. The monotheist's cry of "idolatry," however, too often has been used to exterminate earth-revering traditions, forcing conversions as a tool to dominate and control peoples and lands. Now Christians again are using anti-pagan rhetoric to further desacralize nature, and thereby justify acts of deforestation in the name of a free-market economy.

What lives are actually threatened by the worship of trees? There is no greater threat to human and nonhuman existence today than the idolization of the free market. As Harvey Cox and David Loy, among others, have demonstrated, salvation is sought through the excessive consumption and material wealth of the current global economy celebrated by the Acton Institute and its ilk.[39]

But there are countering voices, willing to present the contrary evidence of a long-standing Christian appreciation of nature generally, and trees in particular. As one Christian writer observes, "For every story about a saint who cut down trees in an act of anti-pagan triumphalism, there are two stories of saints living in hollow oaks."[40] Saints Gerlach, Bavo, and Vulmar are renowned for living in hollowed-out trees. Saint Gudula loved poplars, indeed even prayed with them. Saint Nectarius "taught an entire community of nuns on the island of Egina to recognize the differing songs of trees" made by wind passing through their leaves.[41]

The Religious Campaign for Forest Conservation (RCFC) and its sister program, Opening the Book of Nature (OBN), for instance, represent a Conservative Christian contingent diligently working to protect and honor the forests, which are to be valued for their role in God's creation.[42] "Since 1998, RCFC has supported forest conservation legislation, religious statements on the value of wilderness, and worked with the World Bank on forest protection globally."[43] Fred Krueger, founder of both organizations, takes Christians out on wilderness journeys, to find spiritual insight in the forests. He explains, "When Jesus began His ministry, He didn't spend forty days in the synagogue. Instead, he went to the wilderness."[44] The evangelical Au Sable Institute is another Christian group that promotes stewardship values. In contrast to Acton, Au Sable concurs with leading scientists that a massive extinction is indeed taking place and has fought repeatedly in Congress to protect the Endangered Species Act. [45]

Historically, expanding civilizations, whether Christian or Pagan, have sought out wood for ships, palaces, fortifications, and intensive agriculture to support their cities. As seen in the epic of Gilgamesh, the evidence from ancient Greece and Rome, the Christian Empire under Charlemagne, the consumptive force of empires has historically initiated massive projects of deforestation. This was too often facilitated by desecrating sacred groves and thereby eradicating people's spiritual relationship with trees that may have prohibited such felling.

The Acton Institute and its compatriots follow in these ancient and dangerous footsteps. They promote today's unregulated market, a global empire fueled by projects of overconsumption. Today's free market encourages the free-for-all that is destroying the last of the earth's old-growth forests. And they justify it with figures like St. Boniface, their tree-felling hero and model capitalist.

Yet there are other, wiser sources that can be drawn upon, ancient sources in the Jewish and Christian traditions. In the spirit of Abraham, I challenge the Acton Institute and its like to build an altar under an oak! Do this for the love of God's remaining forests, animals, and humans. Then turn to the animist view of Isaiah as he regarded the great cedars of Lebanon. In the face of the dead king of Babylon and his fallen empire, Isaiah celebrates, for the forest is free from the devouring mouth of humanity: "The whole earth is at rest and quiet. They break forth into singing. The cypresses exult over you, and the cedars of Lebanon, saying "Since you were laid low, no one comes to cut us down" (Isa. 14:7–8).

❧ Ecohopes: Enactments, Poetics, Liturgics

✥ Ethics and Ecology: A Primary Challenge of the Dialogue of Civilizations

MARY EVELYN TUCKER

The twentieth century will be chiefly remembered by future generations not as an era of political conflicts or technical innovations but as an age in which human society dared to think of the welfare of the whole human race as a practical objective.

—ARNOLD TOYNBEE

INTRODUCTION: OUR CURRENT GLOBAL CHALLENGES

This is a powerful statement from one of the leading historians of world history. Yet we might expand Toynbee's statement to suggest that the twenty-first century will be remembered by this extension of our moral concerns not only to humans, but to other species and ecosystems as well. From social justice to ecojustice, the movement of human care and ethics is now part of ever-widening concentric circles.

Indeed, the twenty-first century may be remembered as the century in which humans laid the foundations for the well-being of the planet as a whole by embracing the entire earth-community. The future of life may depend on the largeness of our embrace, for we are now challenged as never before to build a multiform planetary civilization inclusive of both cultural and biological diversity. The call for a dialogue of civilizations is of singular importance in this regard. Without such a dialogue, not only are humans put at risk, but entire ecosystems and life-forms are being compromised.

Our particular challenge, then, in this emerging dialogue of civilizations is to identify the kind of global vision, values, and ethics that will help spark the transformation toward creating such a truly planetary civilization, which will rely on common purpose and shared vision while respecting differentiated cultural values and religious ethics. The values and ethics of all the world's religions can contribute to articulating such a sense of common destiny for the human community in the midst of a growing global ecological crisis. This was the intention of the Harvard project on World Religions and Ecology (1996–98), where we assembled scholars and environmentalists from all over the world to examine the resources of the world's religions in pursuit of a more comprehensive and inclusive environmental ethics.[1]

To create the foundations for a multiform planetary civilization we need to cherish the future of life, and thus place the welfare of the earth-community as a primary aim. It is not simply sustainable development that we are focusing on here but a sustainable future for the planet. This requires not just managerial or legislative approaches, as important as these may be, but also a sustaining vision of that future—one that evokes depths of empathy, compassion, and sacrifice and that has the welfare of later generations in mind. We humans are called, for the first time in history, to a new intergenerational consciousness and conscience—and this extends to the entire earth-community.

This is a task of considerable urgency. As the world becomes warmer, as hurricanes increase in number and intensity, as species go extinct, as air and water pollution spreads, and as resource wars heat up, there is a disturbing sense among many environmentalists and ordinary citizens that the clock is ticking toward major disasters ahead. The looming ecological crisis, with its massive scale and increasing complexity, clearly defies easy solutions. Moreover, the heightened frenzy of the global "war on terrorism" creates blindness toward the widespread terror humans have unleashed on the planet—on its land and water ecosystems and all the species they contain. Blindness is combined with enormous apathy or denial from various quarters regarding the scale of the problems we are facing. This is especially true of those living within the confines of technologically advanced consumer societies. Our task in the dialogue of civilizations is to break through these blinders to create a comprehensive vision of an achievable future grounded in shared yet differentiated ethical values. This requires awakening a sense of shared species identity that transcends yet respects our cultural and religious differences.

HUMANS AS A PLANETARY SPECIES

The critical nature of our moment is described by Mihaly Csikszentmihalyi in *The Evolving Self*. He highlights the enormous responsibility of our species at present:

> The time of innocence . . . is now past. It is no longer possible for mankind to blunder about self-indulgently. Our species has become too powerful to be led by instincts alone. Birds and lemmings cannot do much damage except to themselves, whereas we can destroy the entire matrix of life on the planet. The awesome powers we have stumbled into require a commensurate responsibility. As we become aware of the motives that shape our actions, as our place in the chain of evolution becomes clearer, we must find a meaningful and binding plan that will protect us and the rest of life from the consequences of what we have wrought.[2]

He goes on to acknowledge that the emerging consciousness of ourselves as a planetary species sharing in life's future is vital: "The only value that all human beings can readily share is the continuation of life on earth. In this one goal all individual self-interests are united. Unless such a species identity takes precedence over the more particular identities of faith, nation, family, or person, it will be difficult to agree on the course that must be taken to guarantee our future."[3] To create such a species identity is precisely the challenge of ourselves as individuals as well as an earth-community. The future of evolution is at stake if we should fail: "It is for this reason that the fate of humanity in the next millennium depends so closely on the kind of selves we will succeed in creating. Evolution is by no means guaranteed. We have a chance of being part of it only as long as we understand our place in that gigantic field of force we call 'nature.'"[4] As Csikszentmihalyi suggests, one of the crucial areas we need to explore is the depth of our evolving selves insofar as they are part of the larger matrix of life. We can have a certain measure of confidence that we will find the next season of our evolution as humans as we come to "understand our place within that vast field of force we call 'nature.'"

To find our way forward we need to rediscover the intertwined coding of ourselves as biocultural beings—that is, as beings filled with the mixed heritage of biological survival and cultural creativity. This is the imperative of our evolution as a species that will require a new ethical and cultural coding resonant with, but distinguished from, the genetic coding of evolution itself. In the light of modern science, we know ourselves to be imprinted with nature's complex coding and entwined within nature's rhythms. At the same time, our ethical and cultural coding needs to be brought into alignment with the forces and limits of nature. The heritage of the world's religions is being recovered and expanded to include an environmental ethics for a sustainable future.

COMPREHENSIVE CONTEXT OF THE UNIVERSE STORY

The comprehensive framework of evolution of the universe, of the planet Earth, and of the human provides an expansive and shared context for recognizing our common past and making possible our common future. The enlarged worldview of evolution affords a means of activating a comprehensive set of values and ethics that can point the way toward humans partnering with evolution.

It is just such a large-scale context that the preamble of the Earth Charter offers. It states: "Humanity is part of a vast evolving universe. Planet Earth, our home, is alive with a unique community of life. The forces of nature make existence a demanding and uncertain adventure, but earth has provided the conditions essential to life's evolution."[5] The preamble thus affirms that the physical, chemical, and biological conditions for life are in delicate interaction over time to bring forth and sustain life. Our response to this awesome process is responsibility for its continuity. We need to become a life-enhancing species, not a life-destroying one.

The significance of this evolutionary perspective should be underscored, as it marks a watershed in our rethinking ethics within such a vast framework. The implications of the story of evolution that we are beginning to absorb are manifold. They include a new sense of *orientation, belonging,* and *vitality.* The universe story gives us an orientation toward the vastness of time and space that evokes wonder and awe. We begin to see into the macrophase of our own being as we embrace 13.7 billion years of the universe unfolding through stars, galaxies, planets, and life-forms. We recognize that the chemical components of our bodies came out of the formation of stars. We are stardust come to light in human form.

Along with such expansive orientation we are given a deepened sense of belonging to the universe and to the planet. We are grounded and connected to the planet as we share in our dependence on the elements of air, water, and soil for our survival. The universe story thus decenters humans amidst the vastness of the universe and recenters us as part of, not apart from, the great community of life. In particular, it highlights our role as a species among other species, all radically dependent on the earth for our sustenance and well-being. We are recognizing anew that we belong to the earth-community.

This perspective gives us a reinvigorated vitality for caring for and participating in earth processes. Our partnering with evolution becomes an expression of our comprehensive compassion for all life, human and nonhuman. To encourage the future flourishing of life is the destiny of humans as they participate in what the Chinese Confucians have called "the transforming and nourishing powers of heaven and earth." A re-awakened zest for life is what will carry us forward as we align ourselves with these cosmological powers. With such alignment and energy for the continuity of life we are able to create new and sustainable forms of human-earth relations. These are already being expressed in diverse fields of education, religion, government, economics, agriculture, medicine, law, technology, design, and architecture.

With sustainable technologies and design, with ecological economics and politics, with environmental education and ethics we are learning how to assist evolution and to participate in the myriad processes of universal powers. If human decisions have swamped natural selection because of our planetary presence as a species, we can learn how to become aligned again with evolutionary flourishing. In what we protect, in what we build, in what we eat, in what we cherish we will find the animating principles of universal evolution that also ground culture and guide humans in our creation of communities. We will become partners once again with evolutionary processes. That is the hope for a genuine dialogue of civilizations.

REVISIONING AND EXPANDING ENLIGHTENMENT VALUES

Within the framework of the universe story we are beginning to ac-knowledge that our common ground is the common ground of the earth itself. Survival of species and the planet depends on this. Adaptation for survival is necessary for all species and thus is especially crucial now for humans. This adaptation will be less biological than cultural. It involves a shift in vision and values from a Western Enlightenment mentality that emphasizes radical individualism to an earth-community vision of environmental ethics for a shared future.

This is a central challenge for the dialogue of civilizations and will require an expansion of ethics. The Enlightenment values of "life, liberty and the pursuit of happiness" need to be reframed, not just to suit the human person and individual property rights but to include the larger earth-community. Moving from anthropocentric to biocentric values is essential for enhancing human-earth relations.

Thus in designating "life" as an important value we now use it to include all life—other species and ecosystems, as well as people at a distance and future generations. Up to now we have developed ethics regarding life in the human community in order to address the problems of homicide and suicide, even genocide, but not biocide, ecocide, or geo-cide. This shifts us from viewing nature simply as a resource for human use to nature as source of life. In short, we are moving from viewing earth as commodity to earth as community.

"Liberty" can also be seen as not simply a matter of individual rights, but as including human responsibilities to the larger whole. We are called from personal freedom to communitarian care. From celebrating radical individualism we move toward kinship with all life ranging from local to global.

With regard to the "pursuit of happiness" we need to move away from individual acquisition and consumption as the highest good. Rather, we need to understand that, as the Earth Charter says, "when basic needs have been met, human development is primarily about being more, not having more." This perspective calls us from private property as an exclusive right to embracing the public trust of land and water and air for future generations as a sacred trust.

INTEGRATING ETHICS AND PRACTICES: THE EARTH CHARTER

Such a framework, extended beyond Western Enlightenment values, provides a context for humans to see interlinked problems along with interconnected solutions. This is what the Earth Charter aims to do as it delineates a simple but viable blueprint for a sustainable future. The Earth Charter arose within the urgent, and sometimes conflicting, agendas of the United Nations with regard to protecting the environment and assisting sustainable development. It resulted in both a document and a movement that can be seen as a vital contribution to the international dialogue of civilizations.

The Earth Charter came into being over a decade and was officially drafted by an international drafting committee from 1997–2000. Since the United Nations Conference on Environment and Development (also known as the Earth Summit) was held in Rio de Janeiro, Brazil, in 1992, the UN has identified the socio-ecological crisis as a critical global challenge. A key document of this conference, "Agenda 21," highlighted sustainable development as a central goal of the earth-community.[6] Since that time the United Nations has held seven other major international conferences to analyze our global situation and devise strategies for ensuring a sustainable future. These include conferences on women, on population, on habitat, and on social issues. This has been supplemented by the work of literally thousands of nongovernmental and environmental organizations around the world.

The Rio + 5 conference in 1997 brought together five hundred key stakeholders, ranging from leaders in business, politics, health, the environment, and education. A Benchmark Draft of the Earth Charter was issued there by Mikhail Gorbachev and Maurice Strong on behalf of an international group of two dozen Earth Charter commissioners. The Earth Charter is intended to be a blueprint for sustainable development and thus unites three areas of foremost concern: ecological integrity; social and economic justice; and democracy, nonviolence, and peace. The Earth Charter can be seen as part of a broader effort of individuals and organizations to formulate a global ethics within a context of a dialogue of civilizations that respects both diversity and commonality.

It was officially approved by the Earth Charter commissioners at the UNESCO headquarters in Paris in March 2000 and launched as an international initiative at the Peace Palace in the Hague in June 2000. An Earth Charter + 5 conference took place in the Hague in November 2005. The Earth Charter can be seen as not only a document but also a process insofar as its creation involved the most widespread consultation ever undertaken for the drafting of an international document. It continues to inspire individuals and groups as they try to envision a sustainable and sustaining future for the planet.

The Earth Charter highlights the interrelated issues of environment, justice, and peace as being at the heart of our global challenges. Against the comprehensive background of universal evolution in the preamble, the main body of the Earth Charter outlines an integrated set of ethics and practices to address these three interrelated issues. It aims to address the sometimes-competing areas of environment and development.

The Earth Charter recognizes that the future of life is impossible without ecological integrity. Life and all economic development depend on the health of the biosphere. Thus the preservation of ecosystems and biodiversity is essential, as is the careful use of nonrenewable resources and the exploration of renewable sources of energy.

To do this effectively requires social and economic equity and empowerment. The widening gaps between the rich and the poor in the developed and developing world are a cause for social unrest and can breed resentment and, ultimately, terrorism. How to close these gaps is of utmost importance. Poverty and environmental issues are closely linked in this framework. How to manage economic development for the improvement of living standards without permanently degrading the environment is also an ongoing challenge.

The third point of focus in the main body of the Earth Charter is democracy, nonviolence, and peace. It is almost impossible to achieve the goals of a healthy environment and equitable societies without democratic institutions and legal structures that encourage participation and transparency. The aspirations of millions to live in democratic societies without human rights abuses is demonstrable throughout the world. Moreover, it is becoming increasingly clear that peace among nations and across cultures will not be achievable without addressing both environmental and social issues. Thus the Earth Charter sees the importance of an empowering framework that identifies an integrated set of ethics linking a healthy environment, principles of justice, and institutions of democracy. Here are the foundations for a genuine and continuing dialogue of civilizations to ensure a sustainable future for the planet.

CONCLUSION: CHALLENGE AND PROMISE

This much-needed dialogue of civilizations is now linked to a sense of historic challenge. As the Earth Charter states: "The foundations of global security are threatened." It also observes, however, that "these trends are perilous but not inevitable" and goes on to suggest that "the choice is ours: form a global partnership to care for earth and one another or risk the destruction of ourselves and the diversity of life." This is the ultimate challenge for an effective dialogue of civilizations.[7]

The Earth Charter concludes on a cautiously optimistic note: "As never before in history, common destiny beckons us to seek a new beginning." It notes that "this requires a change of mind and heart"—of comprehensive vision and ethical values. The Earth Charter, then, exemplifies a trend toward articulating an integrative global ethics. It highlights the importance of our moment in human history. It provides an empowering context of evolution and environmental values that will steer the human community forward, toward the enhancement, not the diminishment, of life. Further reflection on the Earth Charter around the world in political gatherings, religious groups, academic settings, and environmental organizations is already beginning to foster a more unified basis for thought and action.[8]

The Earth Charter embodies the need for an expanded vision and shared values for the larger earth-community as it seeks to build common ground for a sustainable future. The comprehensive framework of the story of evolution in the preamble provides animating principles of orientation to the universe, belonging to the earth-community, and vitality in relation to life-processes. These principles forge the bonds of human-earth relations, thus sustaining the demands of relationality and restoring the wellsprings of zest. The Earth Charter becomes an empowering framework to inspire engagement in, and participation with, mutually enhancing human-earth relations. For humans to imagine and activate these relations is to bring into being the emerging contours of the future of the evolutionary process itself. This requires the participation of all cultures and religions in the dialogue of civilizations. With such a dialogue a sustainable future is possible.

❧ Religion and the Earth on the Ground: The Experience of GreenFaith in New Jersey

FLETCHER HARPER

INTRODUCTION

Since the 1992 Earth Summit in Rio de Janeiro, Brazil, the grassroots religious-environmental movement in the United States has grown noticeably. More houses of worship engage ecological issues than ever before. A small but growing number of religiously oriented organizations address the link between faith and the earth. While this rate of growth has not, in my opinion, matched the growth in the number of ecotheologians, or the prominence of ecotheology in seminaries or religion departments, it has moved from being a rarity to being part of the local religious landscape, even if a majority of congregations have still not taken on this work regularly. Thankfully, it is accurate to say that grassroots religious environmentalism has begun to earn a place on the agenda of local religious communities.

Four years ago, I became executive director of GreenFaith, a New Jersey–based interfaith environmental group (founded in 1992, at which time it was called Partners for Environmental Quality [PEQ]), a position I still hold. Since that time in 2002, and for years before, GreenFaith has engaged questions about the form that grassroots religious-environmental action should take. In this chapter I will explore GreenFaith's, and my own, experience over the past four years, looking at the most pressing questions, the most meaningful advances, and the most challenging difficulties we have faced. My goal will be to describe what has worked for GreenFaith in advancing our own version of religious environmentalism and what have been the most important moments of self-definition we have faced.

STAGES OF DEVELOPMENT

I can identify four stages of growth GreenFaith has undergone during my time with the organization, stages which I will refer to as follows:

Stage 1: Discovering Core Values and Finding a Voice
Stage 2: Hearing Responses
Stage 3: Experimenting with Actions
Stage 4: Establishing a Foundation

By listing the stages in this order, I do mean to indicate that there has been a developmental progression in our work. We needed to clarify our values and voice before experimenting with various programs. Similarly we needed to be able to compare the efficacy of these programs before we could consider how to establish foundations for our work over the long term. But it is clearly also true to state that the work of these stages has been overlapping and interconnected. For instance, once we clarified our voice and values, we were freed to experiment programmatically, but our experiences with diverse programs caused us to revisit our understanding of our identity and core values. As is often the case, growth and development do not bring consideration of certain questions to a conclusion insofar as they provide opportunities to view those questions from new vantage points.

Stage 1: Discovering Core Values and Finding a Voice
One of the first challenges GreenFaith faced in 2002 was the need to identify and communicate our core values. As I began this work I assumed that most of the congregations and people of faith with whom we would work would be engaging environmental concerns for the first time. I anticipated that their eco-awareness would mirror the generally low eco-awareness of the greater U.S. population. I assumed as well that they had never heard a sermon or taken part in religious education focused directly on the link between religion and the earth. And I further assumed that I needed to develop a clearly theological basis for our message, connecting these two worlds with clarity and depth.

As a former parish priest accustomed to leading through preaching, I worked for several months to develop a message, a sermon that introduced this topic effectively. I identified several stories that resonated well in a wide range of congregations—a story about my son and the wonder of a wilderness experience, a story about environmental racism in a post-industrial New Jersey community of color, a story about a teenage African girl transforming the perspective of an audience of thousands by reminding them of the impact of consumption choices on future generations. These stories represented themes that were genuinely religious and environmental. Together they presented perspectives that offered a holistic approach to religious environmentalism. These themes—spirit, justice, and stewardship in relation to the earth—became our core values. Through over eighty speaking and preaching engagements annually over each of the past four years, I have found that they resonate with diverse religious audiences as points of connection between religion and the environment. These themes provide a comprehensive religious vantage point from which to engage environmental concerns. GreenFaith encourages people of diverse religious backgrounds to use them to shape their understanding of the environment. We find this helps people feel comfortable engaging what can otherwise be an overwhelming set of environment-related concerns.

Stage 2: Hearing Responses

I experienced several common reactions to my early preaching and teaching. First, through discussion groups in over two dozen houses of worship and interviews with over thirty individuals, I found that all people can recall spiritual experiences that they have had in nature, with the primary themes of these experiences being awe, wonder, joy, and an expanded sense of community. I have yet to find an individual who cannot recall such an experience, usually with ease. Based on this, I believe that human experience of the sacred in nature is universal, or very nearly so.

I have also found that fewer than 5 percent of the people I have spoken with—individually and in group settings—have ever shared these stories with others in their religious communities. These stories have not been integrated into people of faith's conscious religious identities. Releasing the power contained within these memories is one of the most important contributions religious leaders can make to the environmental move-ment. "We do not protect what we do not love," as the old saying goes. Religious leaders have access to environment-based spiritual experiences that contain a love for, and appreciation of, the earth. Pastoral theolo-gians have important opportunities to shape the faith community's sensi-bilities in relationship to these experiences and to make us more aware of why these experiences remain ignored.

On a related note, I found that most people of faith do not connect their spiritual experiences in nature with the motivation to make changes in their consumption habits or their participation in civil society in rela-tion to the environment. In other words, spiritual experience in nature does not generally impact behavior toward nature. Rather, these experi-ences remain split off in people's minds or are remembered as "vacation" experiences disconnected from everyday life. This classification differs importantly from seeing these experiences as "pilgrimage" experiences or "epiphanies," which religious people acknowledge as bearing power to shape identity and behavior. Religious communities and ecotheologi-ans need to find ways to build bridges between experience, belief, and behavior so that the potential of these experiences reaches consciousness and then impacts behavior.

Thirdly, I found that the awe-filled quality people recall in their nature-based experiences stands in tension with their experience of the environ-mental movement. Generally, my impression is that many within the religious community experience the environmental movement as de-spairing and scoldingly moralistic. Several members of congregations told me that they experienced the environmental movement as "apocalyptic," a term I used until David Radcliffe of the Church of the Brethren re-minded me that apocalyptic literature, despite its dire tones, is a literature of hope that describes how good overcomes evil. The rhetoric of the environmental movement strikes most people we have worked with as lacking this hopeful dimension.

If there is a disconnection between environmental experience and per-ceptions of the environmental movement, there is an equally large gap between most religious people's definitions of social justice and their understanding of the health-related impacts of environmental degrada-tion. We found few people who understood basic concepts related to environmental justice—the fact that minority and poor communities are disproportionately impacted by poor air quality, lead poisoning, toxic chemicals, inadequate open space, and other environmental health threats. We found many people from a range of ethnic and socioeconomic backgrounds who assumed that environmental issues are a second-tier social-justice concern, a priority-ranking lower than hunger or poverty. In speaking about environmental justice to predominantly whiter, wealthier congregations, we encountered shock, sadness, anger, and rationalization when we shared stories about environmental racism and injustice. Equally disturbing was our experience in predominantly black churches, where we encountered reactions suggesting that these issues were rarely if ever engaged and that the environmental movement is perceived as uninterested in urban problems. In one black church where I preached, a member approached me after the service and said, "I think that most environmentalists care more about rare animals in Africa than they care about the children who are hungry in this community." Clearly, much work remains to build awareness of the destructive injustices that both cause and accompany environmental degradation.

Finally, despite the power and universality of the experiences described earlier, I encountered the suspicion that by mixing religion and the environment I was crossing an invisible line into nonreligious territory. By engaging the earth religiously, I was leaving sacred ground. This came to my attention in an interesting way. Prior to joining GreenFaith I worked for ten years as a parish priest and developed friendships with many clergy. After I announced publicly that I was leaving parish ministry to lead GreenFaith, I saw a colleague at a denominational event. Jokingly, he put his hand on my shoulder and said, "Well, now that you are going to work on environmental issues, you must have become a pagan."

We both laughed, and I thought no more about this comment until a few days later, when another colleague cracked a similar joke, followed by a third several days later. At that point I realized that these friends, all of whom offered their comments good-naturedly, were revealing what many people might think but not say—that anyone who becomes too close to the environment worships creation instead of the Creator. Many people of faith in the United States share a semiconscious, deeper-than-rational conviction that closeness to nature is not an attribute of a conventional person of faith. Religious leaders have the chance to serve as models that counter this negative aspect of the Western religious heritage.

So, in the wake of initial preaching and teaching on religion and the earth, I learned that "the environment" is a concept with several levels of meaning in religious communities. The earth is a source of private joy and power and a reminder of public despair and frustration. It is experienced alternately as sacred ground and profane space. Our experiences of it are strong but cordoned off, prevented from transforming the rest of our lives. The relationship between environmental concerns and the more traditional concerns for social justice is dimly understood. Indeed, grassroots religious leaders have a substantial amount of work ahead of them to correct this lack of awareness and to offer compelling models of religiously based environmental commitment.

Stage 3: Experimenting with Actions

At the same time we were developing and evaluating our values and message, we launched programs aimed at mobilizing religious action on the environment. The average house of worship's greatest negative environmental impact results from its energy use. To address this, my predecessor had secured funding for work on energy conservation and the use of renewable energy in religious institutions. GreenFaith began conducting programs and workshops on these topics, believing that it was our role to focus religious activity on high-priority environmental threats.

Through conducting energy audits and reviewing energy-conservation research, we found that the average house of worship can reduce its energy use by approximately 20 percent by implementing relatively simple conservation measures. Due in part to rising energy costs, interest in energy conservation is currently growing in the religious communities we serve.

Encouraging the use of renewable energy has been more challenging. In 2000, Green Mountain Energy Company became the first company to offer renewable energy to New Jersey consumers. GreenFaith encouraged its use. Convincing people and houses of worship to use renewable energy proved difficult. While intriguing, renewable energy carries an additional cost and, on a deeper level, asks people to make a countercultural statement through an ongoing purchase. Supporting this switching process is time-consuming, as significant explanation is required to help people understand how renewable energy is generated and distributed. Before my arrival, GreenFaith had convinced several hundred individuals and a few dozen houses of worship to buy renewable energy. Based on my experience, these seem like admirable results. Despite a growth in understanding since the 2005 hurricane season of the threats posed by climate change, convincing individuals and religious institutions to pay more for renewable energy has remained challenging.

Our experience with another form of renewable energy was different. In 2004 and 2005, we secured grant support and a partnership with an innovative solar installer called Sun Farm Network and launched a program called Lighting the Way to install solar arrays on twenty-five New Jersey religious sites. We received far more applications for the project than we had slots available, though many of the applicants had roofs that were not properly oriented or were overly shaded for solar arrays. Representatives of these sites were often deeply disappointed when they learned they could not install a solar system. Conversely, those sites that "went solar" celebrated with genuine joy. We organized solar dedication ceremonies at each site, and have subsequently found that many of these houses of worship have organized second, and even third, ceremonies to celebrate their solar array and the hopeful vision it represents.

In our energy work, we found ourselves speaking to audiences consisting of representatives of two different groups within religious communities. One group of workshop participants represented buildings and grounds or finance committees, lay persons responsible for overseeing the house of worship's buildings and budget. These people wanted to learn how to reduce their church or synagogue's operating budget and often represented conservative perspectives within their congregations. When this audience discovered that the average house of worship's greatest negative impact on the environment was through its energy use, and that energy conservation would reduce both costs and pollution, they expressed pleasant surprise. While many of these people were not environmentalists, they could identify with the traditionally conservative values of restraint, conservation, and frugality lying behind energy-conservation efforts.

A second set of participants hailed from the social action or outreach committees of houses of worship. They took part in the workshops because they believed that climate change and air pollution were matters of social concern. Their primary passions lay outside the physical boundaries of their congregations, and they often saw religious buildings as teaching tools. In relation to energy conservation, members of this group found themselves in the unfamiliar position of advocating for a practice that would decrease the congregational budget, not increase it.

We came to see this oddly mixed audience as a sign of the potentially comprehensive dimension to religious engagement of the environment. Humanity's relationship to the environment impacts all areas of life; faith communities taking on environmental work need to be prepared for this "strange-bedfellows" experience. The relative newness of religious environmentalism and the way it engages people of diverse perspectives creates, in our view, a useful disorientation that provides members of houses of worship with the chance to work practically and learn theologically together.

In addition to our energy-related work, we experimented with other programs. We conducted an interfaith environmental service in a public park and organized several "Toxic Tours," to introduce religious leaders to environmental health threats in Newark and Paterson. We developed an education program to introduce adults and teens to religious-environmental values.

During this time our board members made a decision that elevated our standing in the state environmental and religious communities and that strengthened their resolve. During our Newark environmental health and justice tour, we had learned about pollution in Newark Bay and of the glacial pace of clean-up efforts. We were invited by the Natural Resources Defense Council (NRDC) to join in a lawsuit against the U.S. Army Corps of Engineers to force a clean-up. Our board joined the suit. NRDC and GreenFaith won an important victory in forcing the Corps to redesign their dredging plan to offer greater environmental protection.

By the end of this stage of activity, GreenFaith had demonstrated the ability to define our values and articulate a message, to create appealing programs, and to partner with a globally recognized environmental group on a lawsuit. We had demonstrated a viability that now required us to build an organizational infrastructure, to create a longer-range plan, and to build a more solid financial base.

Stage 4: Establishing a Foundation

If GreenFaith entered this stage of its life-building on the successes of the previous stage, we also realized that we were in a very real way just at the beginning of our development as an organization. Much of our work in the first two stages, while foundational, had an experimental quality to it. We proceeded to launch two new programs that were more ambitious and challenging, forcing us to grow in several ways.

The first program, Sustainable Sanctuaries, was our effort to work in-depth and holistically with a select number of congregations. A total of twelve houses of worship from the Jewish, Catholic, and Protestant communities committed to complete five environmental initiatives over an eighteen-month period. They selected these initiatives from a "Menu of Options" that offered approximately a dozen programmatic opportunities divided into three categories that corresponded to our core values.

From the outset, this program forced us to increase our programming capacity. In addition to addressing energy-related issues, we developed resources to address the use of toxic cleaning and grounds-maintenance products, to recycle, and to conserve water. We refined our three-session religious-environmental educational curriculum for adults and teenagers. We developed resources enabling these congregations to advocate for reduced diesel emissions, coordinating our efforts with environmental groups around the state. The participating congregations are in the process of working through their various initiatives; the program has pushed us to a new level of programmatic capability and expanded the number of ways we can support local religious-environmental activity.

During this stage we also launched Building in Good Faith, an initiative to develop green construction, renovation, and maintenance standards for religious institutions based on the U.S. Green Building Council's LEED standards. If Sustainable Sanctuaries was a program with potential appeal and replicability outside of New Jersey, Building in Good Faith was our first program designed explicitly to be a resource for religious institutions nationwide. The program is ongoing; we have assembled a Working Group that includes religious-environmental leaders from around the country and representatives of religious institutions that have conducted green building projects. We have conducted a survey of existing religious green-building initiatives and begun writing a theology of green building.

We are now raising funds to complete written and web-based green-building guidelines, to create an educational resource to introduce the religious and environmental basis for green building to faith-based groups, and to build a "Buyer's Portal" that would enable religious institutions to purchase environmentally preferable building supplies at a discount.

Both of these programs have forced us to change our understanding of how we will work with religious institutions. We have been accustomed to responding quickly and personally to inquiries for support from houses of worship. We have learned over the course of administering these two programs that this is not always possible. We have been contacted by more houses of worship for information about green building than we can respond to, and we have had to be satisfied with pointing people to existing resources. In the past, we believed that personal contact—often in person—with an individual house of worship was critical to our success. We have begun to realize that this is not always possible, that our job would, in an increasing number of instances, shift toward that of resource-provider, not personal colleague. We are exploring how to maintain a balance between personal contact and response to increasing demand for our services.

We have also realized that we have not developed ways of measuring our impact. We can measure our activity levels by counting our speaking and teaching engagements, workshops, or solar panels installed. But we have not yet developed the sophistication to define more clearly the impact we have wanted to have, the measurable difference in the lives of people and institutions that we wish to create. We have begun working with an expert in the design and evaluation of environmental education programs to address this issue.

The final aspect of this stage of development has been an increased commitment to fundraising. We have realized that strong fundraising skills will provide a foundation for our future. We have worked with a New Jersey fundraising consultancy program called Partnership in Philanthropy, and this work now consistently occupies 40–50 percent of the executive director's time and a growing percentage of the board's time.

GRASSROOTS RELIGIOUS ENVIRONMENTALISM AND FUNDING

An overview of GreenFaith's recent work would not be complete without a more in-depth examination of the sources of funding that have made this work possible, and the motivations of our various funders.

GreenFaith's work has been supported primarily by funding from two sources. The first is private foundations with a long-term commitment to environmental issues. These foundations have funded GreenFaith, as they and their foundation peers have supported other grassroots U.S. religious-environmental groups, because they see an opportunity to expand the community of support for environmental protection. Their motivation is explicitly environmental, and only secondarily, if at all, religious.

GreenFaith's second primary funder has been the New Jersey Board of Public Utilities (NJBPU), an innovative and aggressive agency that supports the expansion of renewable energy and energy-conservation efforts. NJBPU's Clean Energy program (as have some of its corresponding agencies in other states) has funded outreach and education on energy issues in the religious community. As with our foundation supporters, the motivation behind this funding is related to energy and the environment, not religion.

Notably absent from GreenFaith's major funders are religious institutions. While a growing number of individual houses of worship have supported GreenFaith's work, these institutions and their denominational judicatories, including various religious funds established for purposes of outreach or facility maintenance, have provided under 10 percent of GreenFaith's total funding. This relatively low level of support, which to my knowledge is typical of religious-environmental organizations, raises interesting issues for the religious-environmental movement.

It indicates first that the religious community has not yet entered the stage of owning this movement financially in the way, for instance, that it has clearly owned its involvement in issue areas such as poverty, hunger, and low-income housing through the raising and investment of substantial religious funds. Given this, one of GreenFaith's priorities is developing mutually satisfying funding relationships with religious institutions on both a congregational and denomination-judicatory level.

It suggests secondly that the religious community has yet to identify environmental work as a core part of its identity, an activity to which it should commit a portion of its limited resources. Clearly, religious institutions will need to invest funds in environmental work for ecological activity to become widely and deeply integrated into U.S. religious life; foundations and government agencies will understandably and rightly question the return on their investments if they do not see a corresponding financial commitment from their religious partners. Given this, a second strategic priority for GreenFaith is working with religious institutions to integrate care for the earth as a core aspect of their mission, so that funding and volunteer support for ecoreligious work can be clearly justified.

CONCLUSION

In concluding, I would like to offer three observations about the challenges and opportunities that face GreenFaith, and the needs we see in the grassroots religious-environmental field.

First, we increasingly see the need to develop training for religious leaders to help them integrate environmental work into their vocational identity. The topic of religious environmentalism is popular enough at this point, we believe, to support expanded educational opportunities for religious leaders, including one-day educational programs and lengthier offerings, such as the one at Drew Theological School that was the impetus for the present volume. Leadership education should include information on biblical, theological, homiletic, and liturgical approaches to the environment, training on green-building maintenance, and on engaging issues of environmental justice. Religious leaders should also be trained to learn about their local and regional ecosystems and to connect with the environmental groups in their area. GreenFaith has received a substantial leadership grant to develop this kind of training, and it will be one of our highest priorities.

Second, we see a need for the development of standards to ensure the substance and effectiveness of the growing religious-environmental movement. Currently there are no standards or commonly understood best practices that govern this movement. This allows congregations to claim "greenness" if they have undertaken even minor efforts in this area. This lack of objective standards could make the religious-environmental movement vulnerable to accusations of "green-washing," the practice of claiming environmental benefits where few such benefits actually exist, though this is certainly not the intention of any of the congregations we have worked with. For the long-term credibility and strength of this movement, we believe it is critical that a larger conversation about standards and best practices in religious-environmentalism be undertaken.

Finally, we see a need for religious leaders and theological educators to work collaboratively to identify increasingly powerful actions—verbal and otherwise—that use religion's symbolic power to break down the artificial and dangerous walls that our culture has erected that separate humanity from the earth. During the civil rights movement, religious leaders of all races and traditions marched and preached and protested and organized to affirm the dignity of people of color and to protest racism. The marches and nonviolent acts of resistance shifted the world's collective conscience. Now, religious leaders of all kinds have a similar opportunity to lead the United States toward an environmentally sustainable and just future. In our view, there are few higher callings for religious leaders at this time in human history.

❧ Cries of Creation, Ground for Hope: Faith, Justice, and the Earth Interfaith Worship Service

JANE ELLEN NICKELL AND LAWRENCE TROSTER

BACKGROUND

This service took place during the Ground for Hope conference at Drew University in Madison, New Jersey, on September 30, 2005, and was designed by a Methodist minister and a Conservative rabbi so as to reflect the various faith groups who were represented at that conference and the message of interfaith cooperation and work that was central to the keynote addresses by Mary Evelyn Tucker and Jay McDaniel that followed immediately after the service. We asked students from various racial, ethnic, and national backgrounds to participate, so that the human voices represented spanned the globe and included all the continents (except Antarctica). We hope it can provide resources and inspiration for other interfaith services celebrating the earth.

Those planning such services should consider the values and texts that are common to the religious groups involved and look for images that cut across lines of culture and faith. It is important to avoid any rituals that have a specific meaning in one tradition that might overshadow their more general meaning. For example, we chose not to invite people to smudge their faces with dirt because of that act's association with the Christian service of Ash Wednesday. Including elements that appeal to all five senses can help transcend the limitations and biases inherent in language, while also celebrating the goodness of the material world and our human capacity to enjoy that goodness in diverse ways. What follows is the slightly modified text of the worship bulletin.

Cover of Bulletin

Cries of Creation

First Inside Page

ABOUT THE LAND

About ten thousand years ago, the last ice age pushed snow, ice, boulders and soil from Canada down across New York State and northern New Jersey. It scraped out what would become, upon its retreat, large swamps, such as the Great Swamp National Wildlife Refuge, in several directions from Madison and finally came to rest, leaving its residue to form the ridge south of our campus and depositing three large blocks of ice on our campus.

As these blocks melted over the next centuries, they left two large amphitheater-shaped depressions, called "kettle holes," deep in the heart of the Drew Forest. Each depression is about one hundred yards across and perhaps thirty yards deep. The smaller depression or bowl that we call Tipple Pond, south of the Drew Learning Center, was also formed the same way, and it periodically fills with water when we have a big rain.

Someone once asked a geologist at Drew, "How far is it to Canada?" He answered, "The closest Canadian soil is about eighteen inches straight below your feet."

—DAVID GRAYBEAL, *Professor Emeritus, Drew University, Madison, New Jersey*

✤ Cries of Creation

Participants are invited to stand as they are able where indicated by an asterisk below.

I. CRIES OF GRATITUDE: GOD'S CREATION

Call To Worship
The call is sounded by a Japanese Buddhist bell, on loan from the
Tibetan Buddhist Learning Center, Washington, New Jersey.

Prelude
"Ocean"
excerpt from *Deep Presence: Meditations on a Wild Coast,*
produced by Dan Kowalski and Kurt Hoelting

Presenting and Blessing the Elements of Creation
Text from 1 Genesis
As each element of creation is presented, participants join in
God's affirmation of its goodness.

Reading begins over film segment.

R1: In the beginning when God created the heavens and the earth, the earth was a formless void and darkness covered the face of the deep, while a wind from God swept over the face of the waters.

Film segment ends—brief pause

Day One:

R2: Then God said, "Let there be light"; and there was light. And God saw that the light was good; and God separated the light from the darkness. God called the light Day, and the darkness Night. And there was evening and there was morning, the first day.

During reading, two presenters light candles on the table.
Music begins with Native American flute and Australian aboriginal didgeridoo duet and continues under all readings. After candles are lit, R2 leads the blessing:

R2: Let us praise God who has created light:

P: **Praised are You, Eternal One our God, who rules the universe, creating light and fashioning darkness, making harmony and creating everything.**

Day Two:

R3: And God said, "Let there be a dome in the midst of the waters, and let it separate the waters from the waters." So God made the dome and separated the waters that were under the dome from the waters that were above the dome. And it was so. God called the dome Sky. And there was evening and there was morning, the second day.

And God said, "Let the waters under the sky be gathered together into one place, and let the dry land appear." And it was so. God called the dry land Earth, and the waters that were gathered together God called Seas.

And God saw that it was good.

During reading, one presenter comes down center aisle with blue basin, places it on center table. Two presenters carry in pitchers of water, pour into basin, place pitchers on table.

Three presenters carry in bowls of dirt, place one on each table near stones. When all is in place, R3 leads the blessing:

R3: Join in our blessing for the waters and the dry land:

P: Let us praise the Source of Life, Spirit of the world, through whose word all things exist.

Day Three:

R4: Then God said, "Let the earth put forth vegetation: plants yielding seed, and fruit trees of every kind on earth that bear fruit with the seed in it." And it was so. The earth brought forth vegetation: plants yielding seed of every kind, and trees of every kind bearing fruit with the seed in it. And God saw that it was good. And there was evening and there was morning, the third day.

Presenters carry in baskets of leaves, branches, pine cones, and place on table. Other presenters carry in plants, fruit, and place on and by table. When all is in place, R4 leads the blessing:

R4: Let us praise God who created plant life:

P: Praised are You, Eternal One, Shechinah, Life of all the worlds, creating fragrant trees.

Day Four:

R2: And God said, "Let there be lights in the dome of the sky to separate the day from the night; and let them be for signs and for seasons and for days and years, and let them be lights in the dome of the sky to give light upon the earth." And it was so. God made the two great lights—the greater light to rule the day and the lesser light to rule the night—and the stars. God set them in the dome of the sky to give light upon the earth, to rule over the day and over the night, and to separate the light from the darkness. And God saw that it was good. And there was evening and there was morning, the fourth day.

During reading, Power Point™ images of sun, moon, and stars are projected.

R2: Please join in our blessing for heavenly bodies:

P: Praised are You, Eternal One our God, who rules the universe, whose power and might fill the universe.

Day Five:

R3: And God said, "Let the waters bring forth swarms of living creatures, and let birds fly above the earth across the dome of the sky."
So God created the great sea monsters and every living creature
that moves, of every kind, with which the waters swarm, and every
winged bird of every kind. And God saw that it was good. God
blessed them, saying, "Be fruitful and multiply and fill the waters in
the seas, and let birds multiply on the earth." And there was evening
and there was morning, the fifth day.

During reading, Power Point™ *images of fish and birds are projected.*

R3: Let us thank God for creatures of sea and air:

P: **Praised are You, Eternal One, Shechinah, Life of all the worlds,
who diversifies creation.**

Day Six:

R4: And God said, "Let the earth bring forth living creatures of every
kind: cattle and creeping things and wild animals of the earth of
every kind." And it was so. God made the wild animals of the earth
of every kind, and the cattle of every kind, and everything that
creeps upon the ground of every kind. And God saw that it was
good.

During reading, Power Point™ *images of animals are projected.*

R4: Join me in blessing the creatures of the earth:

P: **Let us Praise the Source of Life, Spirit of the world, creating
many creatures and their needs; for all that You have created to
sustain every living creature, we praise You, the One whose Life
is eternal.**

R1: Then God said, "Let us make humankind in our image, according
to our likeness," . . . so God created humankind in God's image, in
the image of God they were created; male and female, God created
them. God blessed them and said to them, "Be fruitful and multiply,
and fill the earth. . . . See, I have given you every plant yielding
seed that is upon the face of all the earth, and every tree with seed

in its fruit; you shall have them for food." . . . God saw everything that God had made, and indeed, it was very good. And there was evening and there was morning, the sixth day.

During reading, Power Point™ images of people are projected.

R1: Let us join in thanking God:

P: Praised are You, Eternal One our God, who rules the universe, for creating us in the divine image.

Day Seven:

R2: Thus the heavens and the earth were finished, and all their multitude.

R3: And God blessed the seventh day and hallowed it, because on it God rested from the work of creation.

*Song of Gratitude
"God of the Sparrow, God of the Whale"
words by Jaroslav J. Vajda, © 1983
music by Carl F. Schalk, © 1983 GIA Publications, Inc.
No. 122 in The United Methodist Hymnal
(Nashville: Methodist Publishing, 1989)

Power Point™ images and hymn text displayed through first three verses of the hymn.

II. CRIES OF SORROW: OUR BROKEN COVENANT WITH GOD AND WITH GOD'S CREATION

*Call to Repentance
Sounding the shofar, or ram's horn, is an ancient act to proclaim feasts, the year of release, or as a call to repentance. It is customarily blown in synagogues to mark the end of the fast at Yom Kippur, and on Rosh Hashanah.
Rabbi sounds shofar.*

Call of the Prophet
Hosea 4:1, 3

Hear the word of the Lord, O people of Israel; for the Lord has an indictment against the inhabitants of the land. There is no faithfulness or loyalty, and no knowledge of God in the land. . . . Therefore the land mourns, and all who live in it languish; together with the wild animals and the birds of the air, even the fish of the sea are perishing.

CD of whales, wolves, and birds begins and continues underneath next reading; fade as hymn begins. Rabbi sounds shofar at end of reading.

A Modern-Day Prophet

It was once religion which told us that we are all sinners, because of original sin. It is now the ecology of our planet which pronounces us all to be sinners because of the excessive exploits of human inventiveness. It was once religion which threatened us with a last judgment at the end of days. It is now our tortured planet which predicts the arrival of such a day without any heavenly intervention. The latest revelation—from no Mount Sinai, from no Mount of the Sermon, from no Bo (tree of Buddha)—is the outcry of mute things themselves that we must heed by curbing our powers over creation, lest we perish together on a wasteland of what was creation.

—HANS JONAS, "The Outcry of Mute Things," in *Mortality and Morality: A Search for the Good After Auschwitz*, ed. Lawrence Vogel

Song of the Earth
"I Am Your Mother"
words by Shirley Erena Murray © 1996 Hope Publishing Co.
music by Per Harling, © 1996 General Board of
Global Ministries, GBG Musik
No. 2059 in *The Faith We Sing* (Nashville: Abingdon, 2000)

Confession and Repentance
During this period of silent confession, all are asked to read the community confession below. Those who wish to are invited to come to a station where there is a bowl of dirt, which you may touch as a way to reconnect with the earth, and/or take a stone as a reminder of that connection.

Community Confession by Rabbi Lawrence Troster *(read in silence)*
Lord, our Creator, we awaken each morning to the dawn chorus of Creation. Our ears hear the birds of the sky singing to the world that

they are still alive, our eyes see the flowers of the earth opening to the light of the sun. We smell the scents of the fresh morning air. How many are the things You have made O Lord, the universe is full of Your creations! And yet we ignore these sounds, sights, and smells. Instead of the birds' song we hear only the sound of cars and machinery. Instead of the sight of green, brown, and gold we see only the gray of concrete. Instead of the fragrance of flowers we smell only the sting of pollution. We experience only the fruits of our own creations. We know only of our own works, which too often have wasted Your creation and silenced many of the voices of Your choir. We think we understand the world when only a fool thinks they can fathom the depths of Your designs. May You give us the strength and the wisdom to see, smell, and listen to Your creation and be moved to protect and cherish the blessings that You have given us. May we no longer be moved by greed and destruction to waste Your world, for if we destroy it there will be no other. We now know that the destruction of Your Creation is a sin.

Chant (in unison, after the rabbi chants in Hebrew)
The tune for this chant is used for the communal confession in the service for Yom Kippur.

P: **And so for the sin / that we have sinned against You by despoiling Your Creation, / forgiving God, / forgive us, / pardon us, / grant us atonement.**

Psalm of Affirmation
This psalm is traditionally used during the Friday night Sabbath service. The Hebrew word kol *("voice") is repeated seven times, symbolizing the seven days of creation.*

Psalm 29

Ascribe to Adonai, O divi–ne beings,
ascribe to Go–d glory and stre–ngth.
Ascribe to Adonai, the glory of God's name;
Bow down to Go–d, majestic in holiness.
The voice of Adonai is over the wa-ters;

the God of glory thun–ders,
Ado–nai, o–ver, migh–ty wa–ters.

The voice of Adonai i–s power;
the voice of Ado-nai i–s majesty.
The voice of Adonai brea–ks cedars;
God shatters the cedars of Lebanon.
God makes Lebanon skip like a calf,
Skip like a calf, Si-rion,
li–ke a you–ng wild o–x.

The voice of Adonai kindles flames of fire;
The voice of Adonai– shakes the wilder–ness;
Go-d shakes the wilderness of Kadesh.
The voice of Ado–nai causes hinds to calve,
and strips forests bare;
while in God's temple all say, "Glory!"
Ado–nai, sa–t enthroned at the Flood;

Ado–nai sits enthroned as ruler forever.
May God give stre–ngth to God's peo–ple!
May Ado–nai bless God's people with peace!
Ble-̇ss God's people with peace!

III. CRIES OF JOY: GOD'S PROMISE OF RESTORATION

Words of Promise

Isaiah 55:10–13

For as the rain and the snow come down from heaven, and do not return there until they have watered the earth, making it bring forth and sprout, giving seed to the sower and bread to the eater, so shall my word be that goes out from my mouth; it shall not return to me empty, but it shall accomplish that which I purpose, and succeed in the thing for which I sent it. For you shall go out in joy, and be led back in peace; the mountains and the hills before you shall burst into song, and all the trees of the field shall clap their hands. Instead of the thorn shall come up the cypress; instead of the brier shall come up the myrtle; and it shall be to the Lord for a memorial, for an everlasting sign that shall not be cut off.

Litany of Commitment
UNEP Environmental Sabbath Program

L: We join with the earth and with each other.
W: *To bring new life to the land,*
M: *To restore the waters,*
All: To refresh the air.
L: We join with the earth and with each other.
W: *To renew the forests,*
M: *To care for the plants,*
All: To protect the creatures.
L: We join with the earth and with each other.
W: *To celebrate the seas,*
M: *To rejoice in the sunlight,*
All: To sing the song of the stars.
L: We join with the earth and with each other.
W: *To recreate the human community,*
M: *To promote justice and peace,*
All: To remember our children.
L: We join with the earth and with each other.
All: We join together as many and diverse expressions of one loving mystery: for the healing of the earth and the renewal of all life.

Closing Song
"You Shall Go Out with Joy"
words by Steffi Geiser Ruben (Isa. 55:12), music by Stuart Dauermann
© 1975 Lillenas Publishing Co., admin. by The Copyright Co.
No. 2279 in *The Faith We Sing* (Nashville: Abingdon, 2000)

Words of Blessing
At a certain point you say to the woods, to the sea, to the mountains, the world, "Now I am ready. Now I will stop and be wholly attentive." You empty yourself and wait, listening. After a time you hear it: there is

nothing there. There is nothing but those things only, those created ob-jects, discrete, growing or holding, or swaying, being rained on or rain-ing, held, flooding or ebbing, standing, or spread. You feel the world's word as a tension, a hum, a single chorused note everywhere the same. This is it: this hum is the silence. . . . The silence is all there is. It is the alpha and the omega. It is God's brooding over the face of the waters; it is the blended note of the ten thousand things, the whine of wings. You take a step in the right direction to pray to this silence, and even to address the prayer to "World." Distinctions blur. Quit your tents. Pray without ceasing.

—Annie Dillard, *Teaching a Stone to Speak*

Buddhist bell sounds to close service.

NOTES AND CREDITS

This service was planned by Rabbi Lawrence Troster, Rabbinic Fellow with GreenFaith and with the Coalition on the Environment and Jewish Life (COEJL), and by Rev. Jane Ellen Nickell, a United Methodist Minis-ter and doctoral student at Drew University, Madison, New Jersey, with assistance from Rev. Heather Murray Elkins, Richard Cox, and members of the Ground for Hope planning committee. The Native flute was played by Dean Anne Yardley and the didgeridoo was played by Dr. Lynne Westfield of Drew Theological School, Drew University.

Cover drawing by Diane Sylvain.

NOTES ON LITURGICAL ELEMENTS

Buddhist Bell
The Japanese Buddhist bell that opens our service is on loan from the Tibetan Buddhist Learning Center (TBLC), Washington, New Jersey. It not only has a beautiful sound but also has a particular significance for the interfaith gathering. It was a gift to the TBLC from Zen Master Eido Shimano Roshi of the Zen Studies Society on his 1975 interfaith pilgrim-age to Buddhist centers around the United States.

The TBLC, Labsum Shedrub Ling, was founded as the first Tibetan Buddhist dharma center in the West through the great efforts of Geshe Ngawang Wangyal, a Kalmyk-Mongolian guru who received his Buddhist training in Kalmykia (northwest of the Caspian Sea) and in Tibet. When Geshe-la founded the Center in 1958, TBLC (then known as the Lamaist Buddhist Monastery of America) served a Kalmyk-Mongolian community established in Howell Township, New Jersey, after World War II. From the beginning, Geshe-la took on Western students who had expressed an interest in learning about Tibetan Buddhism. In 1968, to accommodate this expanding interest, Geshe-la moved the Center to a hilltop property in Washington, New Jersey, where the Center remains. Before Geshe-la passed away on January 30, 1983, he appointed Joshua W. C. Cutler as director of the Center. In the following year, His Holiness the Fourteenth Dalai Lama of Tibet, always revered by Geshe-la as the spiritual head of the Center, visited the Center and advised that the English name be changed to the Tibetan Buddhist Learning Center to clearly reflect that its main activity is the teaching of Tibetan Buddhism.

The Shofar

The *shofar* is a ram's-horn trumpet and is mentioned frequently in the Bible and throughout the Talmud. It was the voice of a shofar, "exceeding loud," issuing from the thick cloud on Mount Sinai that made all the Israelites tremble in awe (Exod. 19, 20).

The *shofar* is prescribed for the announcement of the New Moon and feasts (Num. 10:10; Ps. 83:4), as also for proclaiming the year of release (Lev. 25:9). The first day of the seventh month (Tishri) is termed "a memorial of blowing" (Lev. 23:24) and as a *yom teru'ah* (day of blowing; Num. 29). This was interpreted by Jewish traditions as referring to the sounding of the *shofar*. It is blown in synagogues to mark the end of the fast at Yom Kippur, and on Rosh Hashanah. By custom the *shofar* is also blown after morning services for the entire month of Elul, which is the month of the Jewish calendar that immediately precedes Rosh Hashanah.

Moses Maimonides (1135–1204 CE) once wrote that the sound of the *shofar* says, "Wake up from your slumber! Examine your deeds, and turn to repentance, remembering your Creator, you sleepers, who forget the truth while caught up in the fads and follies of the time and take a good look at yourselves!"

Psalm 29

Psalm 29 is probably one of the most ancient poems of the Psalter and is considered by many biblical scholars to be an Israelite version of an even older Canaanite poem of praise to Baal. It was added to the Friday-night Sabbath liturgy (along with Psalms 95–99, which it follows) by the Kabbalists of Safed in the sixteenth century. These mystics associated the psalm with the Sabbath because the Hebrew word *kol* (voice, or thunder) is used seven times in nine verses, which they interpreted as referring to the seven days of creation. These psalms were chosen as an introduction to the hymn *L'khah Dodi*, written by Rabbi Solomon ben Moses Halevi Alkabetz (c. 1505–1584) because they emphasized the sovereignty of God, which they interpreted in Kabbalistic terms.

❧ The Firm Ground for Hope: A Ritual for Planting Humans and Trees

HEATHER MURRAY ELKINS, WITH ASSISTANCE FROM DAVID WOOD

THE SPRINGTIME

This is a story about a community's ritual life as it relates to organic and architectural structures. This particular narrative of tree blessings begins the year before the Ground of Hope conference is held at Drew University in Madison, New Jersey. After planting a new tree as a sign of commitment to what is to come, the seminary community gathers in the shade of the four giant oaks that will be destroyed in order to build a new wing with an elevator. We are sacrificing these ancestors for the sake of those who have been excluded by our inaccessible physical structures.

We take our trees seriously. The university has been planted in a place called "The Forest" since the late 1700s. Faculty often sign "In the Forest" in their book dedications. A giant acorn decorates the space above the altar in our chapel, higher than the cross. We unashamedly sing to our trees, mark them for death and new life with acorn banners. We read Wendell Berry's poems from *A Timbered Choir*. Our alums remember acorns thrown on deans' heads, shade offered for study, courtship, and prayer. We ask the trees to bless us as we say goodbye. More than a little tree-hugging goes on after the ritual is over.

THE FALL

The Ground for Hope conference is coming to a close. We gather outside in front of the glass doors of the new atrium, a ground where oaks once thrived. The sun is bright, the air brisk. There are three slender dogwoods waiting to be planted by a holy human community. These are traditionally linked with the narrative of Christ's Passion. They are also native to the Garden State.

Containers holding water have been placed beside each tree-well, as well as a clear-glass pitcher, the crown of thorns chalice, a large clay bowl with embedded branches of thorns, and a tripod of wooden poles tied with a rainbow string. The tripod is a reminder of an old Korean practice of providing heat. It was used by prisoners, and the very poor. Three sticks are tied together, a scrap piece of metal holds the fire, suspending it between the sticks. Four ritual objects have been prepared: a thorn of honey locust, an acorn from the Vanderbilt University campus, a handful of sand from Down Under, and a leaf from the Boddhi Tree.

THE BLESSING BEGINS

The community is invited to remove their shoes, to stand bare-soled on holy ground. Many remove their shoes and socks as the Woody Guthrie song "Holy Ground" is sung. I take a handful of earth and offer it to the community to remind us of the Genesis story of humans made from earth, and holy breath. I do this to remind us that this is the Garden State and that earth is our beginning and our end. We are marked with a dusty sign of mortality. Earth to earth. Dust to dust. I blow on the dirt, place some of it on my cheek, and invite others to do the same.

I will be the celebrant, David Wood the gift-giver of ritual objects, and Luke Higgins the musician, but the community and the trees will be primary performers of this ritual by singing, telling stories, and getting grounded.

> *A reading adapted from Isaiah 61:1–4*
> The spirit of the Living GOD is upon us,
> because the Holy One has anointed us;
> GOD has sent us to bring good news to the oppressed,
> to proclaim liberty to the captives,
> and release to the prisoners;
> to proclaim the year of GOD's favor,
> and the day of judgment of our Creator;
> to comfort all who mourn in Zion—
> to give them a garland instead of ashes,
> the oil of gladness instead of mourning,

the mantle of praise instead of a faint spirit.
They will be called oaks of righteousness,
the planting of the Holy One, to display GOD's glory.
They shall build up the ancient ruins,
they shall raise up the former devastations;
they shall repair the ruined cities,
the devastations of many generations.

The Stories of the Trees

We begin with the stories of the trees. David Wood recites the story of the origin of the word "tree," which is also the story of his own name. He begins with the entry for "true" from *The American Heritage Dictionary* (4th ed., 2000). The words "true" and "tree" are joined at the root, etymologically speaking. In Old English, the words looked and sounded much more alike than they do now: "tree" was *tr̄ow* and "true" was *trōᵉwe*. The first of these comes from the Germanic noun, **trewam*; the second, from the adjective, **treuwaz*. Both of these Germanic words ultimately go back to an Indo-European root, **deru–* or **dreu–*, appearing in derivatives referring to wood and, by extension, firmness. Truth may be thought of as something firm; so, too, can certain bonds between people, like "trust," another derivative of the same root. A slightly different form of the root, **dru–*, appears in the word "druid," a type of ancient Celtic priest. He reminds us that his own name is etymologically **dru-wid-*, or "strong seer." We tell him that those of us at Drew often refer to ourselves as "drew-ids."

The story of the dogwood follows and the head of Drew's Department of Buildings and Grounds is asked why they were chosen. "They're native to New Jersey, they're hardy, and they bloom," is the beginning of the answer. I remind the community of the legend of the dogwood, once one of the greatest trees in creation. It was believed to be the tree once used to make crosses for the crucifixion of Jesus, as well as others who were seen as threats to the empire. In the myth, the dogwood is so grieved and shamed by the violence that it begs its Maker to remove its "treeness" and let it be a simple bush. It asks to be transformed so it can never be used as a weapon again. The dogwood's prayer is answered, and as a witness, its blossoms form the shape of a cross.

The story of the African Tree-planting Movement in Kenya is told, and a selection from Wangari Maathai,"Trees for Democracy," from the *New York Times*, December 10, 2004, is read:

As a member of the National Council of Women of Kenya in the early 1970s, I listened as women related what they wanted but did not have enough of: energy, clean drinking water, and nutritious food. My response was to begin planting trees with them to help heal the land and break the cycle of poverty. Trees stop soil erosion, leading to water conservation and increased rainfall. Trees provide fuel, material for building and fencing, fruits, fodder, shade and beauty.

As the reading ends, Dean Beach spontaneously describes what she has recently seen with her own eyes, the transformation of Kenya through the tree-planting work of women, led by Wangari Maathai.

David Wood holds up the thorn that he has carried from Tennessee. By synergistic coincidence, the same kind of thorn is embedded in the pottery bowl from Drew that holds the water for the new trees. It is a giant, three-pronged spike in the shape of a cross from a honey locust tree at Yellow Bird Sculpture Park in Tennessee. For David, the thorn symbolizes an intersection between paganism and Christianity. The honey locust grows these thorns as a form of protection, yet trees everywhere are being sacrificed. It appears to him to be a natural symbol of an arboreal Passion.

David's next gift is an acorn from Vanderbilt University, where he is a faculty member. It is a living insignia for that university as well as for Drew. The acorn carved in our chapel is mirrored in this gift from a southern alma mater. David buries the acorn in a small plastic container (complete with a small commemorative document). We hope it may germinate when the plastic decays in 60–80 years. Just as the dogwoods are coming to the end of their life-span, an oak may grow to replace those sacrificed to make way for the new building

The last gift is a leaf and a handful of sand. The leaf comes from a cutting taken from the actual Boddhi Tree, thought to be that under which the Buddha was born. A sister of this leaf was buried in a chronopod and tree-replacement ceremony on Kuusiluoto Island, near Helsinki,

Finland, in October 2003.[1] David buries this leaf with the sand brought from the site of a time-capsule buried on the coast of Australia, near Sydney, in November 2000. The sand also represents our farthest-traveled participant: from Australia, outside Melbourne.

The Planting

The community is now invited to plant the dogwood trees. They use the shovels, their hands, and even their bare feet. There is laughter, but also some signs of grief since the earthy link between planting trees and the burial of a loved one is very close and clear.

I invite anyone in the community who has lost someone they love in the last year to step forward and pick up one of the containers of water. I want those who have recently shed tears to have the opportunity to pour the first living water for these new trees, these holy roots of our common human life. Three people come forward and pick up the water containers and begin pouring. Others wait to take their place. Some brush away tears with fingers stained by earth, leaving behind a holy human sign on their faces. "From dust we come, to dust we shall return." There is silence as the last of the water is poured into the soil, and then a chorus of birds begins. We join with the winged singers using the words of an ancient song of the psalmist as our benediction: "And all the trees of the field will clap their hands, while you go out with joy . . ."

❧ Musings from White Rock Lake: Poems

KAREN BAKER-FLETCHER

HA SHEM

Hurricane
Why
Hurricane
Sigh
A touch on
Our shoulders
In early
July
At first
So dry
Then so humid
Weather awry
The birds all timid
In shade of trees
Then diving like turtles
Without grace or ease
Dipping low
For dark and cool
Seeking oasis
In tepid pool
Past breeding
And nesting

Just wrestling
With heat
Past season
To Season
Black and
Orange
Webbed feet
Sense
Adam's empire's
The reason
For warnings
And warmings
In morning's
First "Hi"
Such sorrow
Little regret
For ignoring
Heaven's eye
Before emperors
In ties
Butt naked with lies
Muted earth's cries
In ascent to hell's throne
Deep in the bog of
Imperialist dung
Plaguing
Peasant
Pietas'
Hearts
All flung and wrung
By the tragic folly
of empire's feast
With blasphemous prayers
Mocking the least
Hailing techno-whizzers of Oz
Flaccid grey-skinned priests
Spinning numbers into gods

While laboring serfs grieve
The debris on God's altars
While creation groans
With constant plea
From bayou to bay
And from bay to the sea
And the winds prophesy
What a meager repentance
Such blasphemous offerings
And what mocking remembrance
Of secret waters
Rippling with mirth
Spirit hovering
Over depths
Meeting earth
Dancing in love
With wind
From above
Panting
Then Pulsing
'Til the first cry
Of birth
Alpha
And
Omega
Thus speaks
Ha-Shem.

WITH YOU

Walking on water
I feel
I am
Walking on water
With you.

Wailing on water
I feel

I am in fresh air
Swept with dew.

Walking on water
I feel
I am away from
Lake shore's edge.

Walking on water
I feel
I am
Fallen tree,
Supporting bridge.

WHO AM I?

I am not Sam
I am
Or green eggs and ham.

Who am I?
I am not black
Or white
Or Creek or Asian
Or Ghanaian or Senegalese
Or Latina or Latina
Or Latina American or Mexican
Or Brazilian or Puerto Rican.

Who am I?
I am water
And I am earth
I am mud and green
Brown and grey
Vermillion and clay
With mist for breathin'

And laughter blue
And tears that rain
And dance
With wind in rushes.

I am earth
And breeze
And heat
And ice
And
Womb.

I am water
And dust
And I am God's
beloved
And I am me.

STARLIGHT

Starlight,
Starbright,
Starlight in between.

Starlight
Starbright,
Starlight in the green.

Dancing
Prancing
Blinking
Twinkling
Not so high above.

Having fun
In soughs of wind

Rippling lake
And noonday sun
Kiss ever and again.

SIGNS

Ducks bask
On reeds afloat
Sailors none
Are now aboat
Turtles nap on shaded log
One upon a dam, cool rock
It only we,
Us foolish
Folk
Who trespass
Signs
Of sky and loch.

NOTES

PREFACE | LAUREL KEARNS AND CATHERINE KELLER

1. The term "tipping point" was originally coined in the early 1960s by Morton Grodzins to name "white flight," the dramatic point when white families would move out en masse of a gradually integrating neighborhood. (Morton Grodzins, *The Metropolitan Area As a Racial Problem* [Pittsburg: University of Pittsburg Press, 1958]). It has been lately popularized and merged with the "butterfly effect" of chaos theory.
2. Ross Gelbspan, *The Boiling Point* (New York: Basic, 2004).
3. This echoes the title of Jay McDaniel's book *With Roots and Wings: Christianity in an Age of Ecology and Dialogue* (Maryknoll, N.Y.: Orbis, 1995).
4. http://www.wakeuplaughing.com
5. As impressively collected, for example, in the Harvard World Religion and Ecology volumes, coordinated by Mary Evelyn Tucker and John Grim of the Forum on Religion and Ecology. More information is available at http://www.religionandecology.org, which features a full listing of the titles published in the Harvard Series.
6. Further information about David Wood's art project is available at http://www.vanderbilt.edu/chronopod/. See also Heather Elkins and David Wood, "The Firm Ground for Hope: A Ritual for Planting Humans and Trees," in the present volume.

INTRODUCTION | CATHERINE KELLER AND LAUREL KEARNS

1. There is good news on many campuses as the concept of sustainable or green campuses and education has spread. Many colleges and universities have built LEED-certified green buildings, worked to reduce their CO_2

emissions, and electricity consumption, switched to producing or purchasing some of their power from alternative sources, and examined their production of waste and sources of food.

2. It is worth noting that as academic leaders in the field, such as Rosemary Radford Ruether, Sallie McFague, John B. Cobb Jr., Larry Rasmussen, Mary Evelyn Tucker, and John Grim have retired or left to continue their work elsewhere, they have not been replaced by scholars whose focus is as ecological. Many of the organizations established in the 1990s or before to work on faith-based environmental activism are struggling to support their staff and find enough funding.

3. There is a wide array of religious organizations working on environmental issues. See the Web sites of the National Religious Partnership for the Environment (www.nrpe.org) and its constituent members. See also the Forum on Religion and Ecology (www.religionandecology.org), the Web of Creation (www.webofcreation.org) and the Interfaith Climate Change Network (www.protectingcreation.org) for links to many other organizations. See also Laurel Kearns, "Cooking the Truth: Faith, Science, the Market, and Global Warming," in the present volume, for a more detailed discussion of religious environmental organizations.

4. *The Earth Charter: Values and Principles for a Sustainable Future*. For copies of the brochure or information on the initiative, consult Earth Charter Web site, www.earthcharter.org.

5. Bruno Latour, *Politics of Nature: How to Bring the Sciences into Democracy*, trans. Catherine Porter (Cambridge, Mass.: Harvard University Press, 2004).

6. See David Wood, "Specters of Derrida: On the Way to Econstruction," in the present volume; see also his discussion of Derrida on Marx, "The Eleventh Plague: Environmental Destruction," in *The Step Back: Ethics and Politics after Deconstruction* (Albany: State University of New York Press, 2005).

7. Michael Shellenberger and Ted Nordhaus, "The Death of Environmentalism? Global Warming Politics in a Post-environmental World," *Social Policy* 35, no. 3 (2005): 19–31.

8. John B. Cobb Jr. and Herman Daly, *For the Common Good: Redirecting the Economy toward Community, the Environment and a Sustainable Future* (Boston: Beacon, 1994).

9. This, for example, is the charge levied against William Cronon and colleagues in *UnCommon Ground: Rethinking the Human Place in Nature* (New York: Norton, 1996), a work itself stemming from a conference on the dilemma of naïve realism in popular ecological discourse vis-à-vis the antirealism of strong constructivism. In addition to Cronon's volume, the authors

in the volume *Reinventing Nature: Responses to Postmodern Deconstruction*, published contemporaneously, explore the resonances of constructivism and nature (Michael E. Soule and Gary Lease, eds. *Reinventing Nature: Responses to Postmodern Deconstruction* [Washington, D.C.: Island Press, 1995]).

10. Bruno Latour, *Pandora's Hope: Essays on the Reality of Science Studies* (Cambridge, Mass.: Harvard University Press, 1999), 21; the book is dedicated not coincidentally to Donna Haraway and her cyborgs, among others.

11. *"There is nothing outside of the text* [there is no outside-text; il n'ya pas de hors-texte]." The translator is at pains to prevent the inevitable misreading. Jacques Derrida, *Of Grammatology*, trans. Gyatri Chakravorty Spivak (Baltimore: Johns Hopkins University Press, 1976[1974]), 158.

12. Derrida has been at pains to counter the misunderstanding of his "pas de hors texte": "I never cease to be surprised by critics who see my work as a declaration that there is nothing beyond language" (as quoted in John D. Caputo, *The Prayers and Tears of Jacques Derrida: Religion Without Religion* [Indianapolis: Indiana University Press, 1997], 16–17).

13. Derrida, *Of Grammatology*, 9.

14. Indeed, it has queer affinities with the science studies approach of Donna Haraway. The latter's constructivism, rooted in biology, may remain coy with regard to the ecological crisis, but stands with environmentalism against the corporate control of life and the presumptions of the science establishment it funds. Donna Haraway, *ModestWitnessFeminism@Second _Millennium.FemaleMan©_Meets_OncoMouse™: Feminism and TechnoScience* (New York and London: Routledge, 1997).

15. It, of course, depends on how material life is conceived, as much liberation theology and others concerned about the poor focus on certain material conditions à la Marx, as purely anthropocentric.

16. Latour, *Politics of Nature*.

17. Leonardo Boff, *Cry of the Earth, Cry of the Poor* (Maryknoll, N.Y.: Orbis, 1997); Ivone Gebara, *Longing for Running Water: Ecofeminism and Liberation*. trans. David Molineaux (Minneapolis, Minn.: Augsburg Fortress, 1999); Karen Baker-Fletcher, *Sisters of Dust, Sisters of Spirit: Womanist Wordings on God and Creation* (Minneapolis, Minn.: Augsburg Fortress, 1998).

18. Larry Rasmussen, *Earth Ethics, Earth Community* (Maryknoll, N.Y.: Orbis, 1998); James Cone, "Whose Earth Is It Anyway?" in *The Emergence of a Black Theology of Liberation, 1968–1998* (Boston: Beacon, 2000), 136–45; Sallie McFague, *Super, Natural Christians: How We Should Love Nature* (Minneapolis, Minn.: Augsberg Fortress, 1997).

19. John B. Cobb Jr and Herman E. Daly, *For the Common Good: Redirecting the Economy Toward Community, the Environment, and a Sustainable Future* (Boston: Beacon, 1994); John B. Cobb Jr., *The Earthist Challenge to Economism: A Theological Critique of the World Bank* (New York: Palgrave Macmillan, 1998); John B. Cobb Jr., *Sustaining the Common Good: A Christian Perspective on the Global Economy* (Cleveland: Pilgrim, 1995).

20. In this collection, Keller, McDaniel, Spencer, Baker-Fletcher, Lee, McDaniel, Higgins, Gorman, and Muraca are among those who in their own work develop the process theological heritage.

21. Latour, *Politics of Nature*, 220.

22. Thomas Berry, *Dream of the Earth* (San Francisco: Sierra Club, 1988).

23. Sharon Betcher's ecopneumatology, as espoused in the present volume, notes the meaning of "you're grounded" when spoken by a parent to a child.

24. Jürgen Moltmann, *The Spirit of Life: A Universal Affirmation* (Minneapolis: Augsburg Fortress, 1992), 40–43.

25. The Forum on Religion and Ecology's Web site is available at http://www.religionandecology.org.

ECOTHEOLOGY AND WORLD RELIGIONS | JAY MCDANIEL

I would like to thank Margi Ault-Duell, Laurel Kearns, and Dhawn Martin for their editorial help with this piece.

1. Alfred North Whitehead, *Process and Reality, Corrected Edition*, ed. David Ray Griffin and Donald Sherburne (New York: Free Press, 1978), and idem, *Adventures of Ideas* (New York: Macmillan, 1933).

2. Jay B. McDaniel, and Donna Bowman, *Handbook of Process Theology* (St. Louis: Chalice, 2006). See also John Cobb and David Griffin, *Process Theology: An Introductory Exposition* (Louisville: Westminster-John Knox, 1977).

3. See also Jay B. McDaniel, *With Roots and Wings: Christianity in an Age of Ecology and Dialogue* (Maryknoll, N.Y.: Orbis, 1995).

4. See the State University of New York Press series Constructive Postmodern Thought for some examples of work influenced by process thought.

5. More information about the Earth Charter is available at http://www.earthcharter.org.

6. John B. Cobb Jr. and Herman E. Daly, *For the Common Good: Redirecting the Economy Toward Community, the Environment, and a Sustainable Future* (Boston: Beacon, 1994).

7. Jay B. McDaniel, *Living From the Center: Spirituality in an Age of Consumerism* (St. Louis: Chalice, 2000).

8. See also Jay B. McDaniel, *Gandhi's Hope: Learning from World Religions as a Path to Peace* (Maryknoll, N.Y.: Orbis, 2005).

9. Hans Kung, *Christianity and the World Religions: Paths to Dialogue With Islam, Hinduism, and Buddhism* (Garden City, N.Y.: Doubleday, 1986)

10. See also Jay B. McDaniel, *Of God and Pelicans: A Theology of Reverence for Life* (Louisville: Westminster-John Knox, 1989).

TALKING THE WALK: A PRACTICE-BASED ENVIRONMENTAL ETHIC AS GROUNDS FOR HOPE | ANNA L. PETERSON

1. Willett Kempton, James S. Boster, and Jennifer A. Hartley, *Environmental Values in American Culture* (Cambridge, Mass.: MIT Press, 1995).

2. See S. Lester, "Recycling—A Modern Success Story," *Everyone's Backyard* (Winter 1996): 9; Frederick Allen and Gregg Sekscienski, "Greening at the Grassroots: What Polls say about Americans' Environmental Commitment," *EPA Journal* 18, no. 4 (September–October 1992): 52–53; Nationwide poll conducted by Roper Starch Worldwide in March 1995, reported in *Friends of the Earth* (March–April 1996): 4; Frederick Allen and Roy Popkin, "Environmental Polls: What They Tell Us," *EPA Journal* 14, no. 6 (July–August 1988): 10; and John Gillroy and Robert Shapiro, "The Polls: Environmental Protection," *Public Opinion Quarterly* 50 (1986): 270–79.

3. More information is available at http://www.dukenews.duke.edu/2005/09/nicholaspoll.html.

4. See Lynn White, "Historical Roots of Our Ecological Crisis," *Science* 155: 1203–7; J. Baird Callicott, *Earth's Insights: A Survey of Ecological Ethics from the Mediterranean Basin to the Australian Outback* (Berkeley: University of California Press, 1994).

5. Matthias Finger, "From Knowledge to Action? Exploring the Relationships Between Environmental Experiences, Learning, and Behavior," *Journal of Social Issues* 50, no. 3 (Fall 1994): 141–60.

6. This reflects, further, the well-established division in our culture between intellectual and physical labor, which suggests that a transformation of the task of environmental ethics might also entail the transformation of environmental ethicists into organic intellectuals of a Gramscian sort, involved in practical as well as scholarly pursuits.

7. Karl Marx, *The German Ideology: Part I*, in *The Marx-Engels Reader*, 2nd ed., ed. Robert Tucker (New York: Norton, 1978), 149.

8. Michael Taussig, *Defacement: Public Secrecy and the Labor of the Negative* (Stanford: Stanford University Press, 1998), 226.

9. Anthony Weston, "Non-Anthropocentrism in a Thoroughly Anthropocentrized World," *Trumpeter* 8, no. 3 (1991): 2 (also available at http://trumpeter.athabascau.ca/contents/v8.3/weston.html).

10. David Abram, *The Spell of the Sensuous: Perception and Language in a More-Than-Human World* (New York: Pantheon, 1996), 264.

11. "Sustainability and Politics." Interview with Wes Jackson by Robert Jensen, *Counterpunch*, July 10, 2003, available at http://www.counterpunch.org/jensen07102003.html.

12. William R. Jordan III, *The Sunflower Forest: Ecological Restoration and the New Communion with Nature* (Berkeley and Los Angeles: University of California Press, 2003), 5.

13. Raymond Williams, *Resources of Hope: Culture, Democracy, Socialism*, ed. Robin Gable, with an introduction by Robin Blackburn (London: Verso, 1989), 242.

14. Jim Cheney and Anthony Weston, "Environmental Ethics as Environmental Etiquette," *Environmental Ethics* 21, no. 2 (Summer 1999): 115–17.

15. Ibid., 118.

16. Gustavo Gutiérrez, *A Theology of Liberation* (Maryknoll, N.Y.: Orbis, 1973), 15, 11.

17. Juan Luis Segundo, *The Liberation of Theology* (Maryknoll, N.Y.: Orbis, 1976), 8.

18. Cheney and Weston, "Environmental Ethics as Environmental Etiquette," 119.

19. Ibid., 125.

20. Jackson, "Sustainability and Politics."

21. Weston, "Non-Anthropocentrism," 1.

22. Karl Marx, "Theses on Feuerbach," in Robert Tucker, ed., *The Marx-Engels Reader*, 145.

23. Ibid., 144.

24. Karl Marx, "Contribution to the Critique of Hegel's *Philosophy of Right*: Introduction," in Tucker, *The Marx-Engels Reader*, 54.

25. Ibid.

26. Within environmental philosophy itself, we may also build on naturalistic models, such as the work of Holmes Rolston, which see an objective value in nature.

TALKING DIRTY: GROUND IS NOT FOUNDATION | CATHERINE KELLER

1. Compare Bruno Latour's discussion of "the proliferation of hybrids"—chemical-biological-physical-political-economic entities in "networks" that

are "simultaneously real, like nature, narrated, like discourse, and collective, like society" (*We Have Never Been Modern*, trans. Catherine Porter [Cambridge, Mass.: Harvard University Press, 1993], 6). The recollection of the chaosmic beginnings of our collective world serves, then, as one grounding narrative among many, for the assembly of humans, nonhumans, and the growing set of hybrids between us.

2. Judith Butler, *Gender Trouble: Feminism and the Subversion of Identity* (New York: Routledge, 1999), and *Bodies that Matter: On the Discursive Limits of "Sex"* (New York: Routledge, 1993).

3. William Cronon, ed., *UnCommon Ground: Rethinking the Human Place in Nature* (New York: Norton, 1996).

4. Bruno Latour, *Politics of Nature: How to Bring the Sciences into Democracy* (Cambridge, Mass.: Harvard University Press, 2004), 220.

5. Ibid., 19.

6. Ibid., 18.

7. For the specifically theological effect of postcolonial theory, with its alternative to the increasingly dysfunctional rhetoric of identity politics, see Catherine Keller, Michael Nausner, and Mayra Rivera, eds., *Postcolonial Theologies: Divinity and Empire* (Collegeville, Minn.: Liturgical Press, 2004).

8. Jacques Derrida, *Of Grammatology*, trans. Gayatri Chakravorty Spivak (Baltimore: Johns Hopkins University Press, 1976 [1974]), 23.

9. Even those who bother to police ecoactivist texts for their (indubitable) essentialisms, romanticisms, and positivisms mean to be, ah, constructive, as is the case with the aforementioned anthology, *UnCommon Ground*; see also Noel Sturgeon, *Ecofeminist Natures: Race, Gender, Feminist Theory and Political Action* (New York: Routledge, 1997); and Andrew Ross, *Strange Weather: Culture, Science, and Technology in the Age of Limits* (New York: Haymarket, 1991).

10. I here mean Bruno Latour and Donna Haraway, and, with (even) less focus on ecology, Michel Serres and Isabelle Stengers. Promising evidence of a new wave that might comprise veritably ecofeminist poststructuralist theological thought comes in the form of Anne Elvey's *An Ecological Feminist Reading of the Gospel of Luke: A Gestational Paradigm* (Lewiston, N.Y.: Edwin Mellen, 2005), Marion Grau, *Of Divine Economy: Refinancing Redemption* (New York: T. & T. Clark, 2004), as well as forthcoming titles from Anne Daniell (Fordham University Press) and Sharon Betcher (Augsburg Fortress).

11. As David Wood is demonstrating in our midst, anticipated by the ecophenomenology proposed in his *The Step Back: Ethics and Politics After Deconstruction* (Albany, N.Y.: State University of New York Press, 2005). Also see

intriguing hints in Gayatri Chakravorty Spivak, *Critique of Postcolonial Reason: Toward a History of the Vanishing Present* (Cambridge, Mass.: Harvard University Press, 1999), which has been theologically harvested by Mayra Rivera in *A Touch of Transcendence: a Latina Postcolonial Contribution to Philosophical Theology* (Louisville: Westminster-John Knox, forthcoming).

12. I don't want to spend this paper tracking instances of this effect; for evidence, one might just glance at the list of chapters in Mark C. Taylor's major anthology, *Critical Terms for Religious Studies* (Chicago: University of Chicago Press, 1998). None of the twenty-two articles make reference to nonhumans. One might also see Nicholas Royle, ed., *Deconstructions: A User's Guide* (New York: Palgrave Macmillan, 2000).

13. Derrida, *Of Grammatology,* 23; emphasis added.

14. Into this homogenized origin-ground Derrida then collapses an entire sequence of texts (Rabbi Eliezer, Rousseau, Jaspers) referring to a vague "Book of Nature" as directly or indirectly God's writing (16ff.). Yet one can read some elements of that tradition as discerning in the matter of creation the codes of a cosmic wisdom that destabilizes the disembodied ontotheology of the classical tradition. Cf. Elliot Wolfson, *Language, Eros, Being: Kabbalistic Hermeneutics and Poetic Imagination* (New York: Fordham University Press, 2005) on "Book of Splendor" and "Book of Nature."

15. David Wood, "Specters of Derrida: On the Way to Econstruction" in the present volume.

16. Luce Irigaray, *The Forgetting of Air in Martin Heidegger,* trans. Mary Beth Mader (Austin: University of Texas Press, 1999), 74.

17. Ibid., 85ff.

18. One can hardly dispute the historical function of the metaphysical language of ground to homogenize or ontologize beings in Being; one might, however, argue that the Eckhartian *Grund* [OG, *grunt*], the Schellingian "unruly ground," and even from a certain angle Tillich's "ground of being" release great energies of differentiation—even in the neo-Platonic force field, the mystical ground bottoms into the *Ungrund* in which it *grounds* creaturely life.

19. Gilles Deleuze and Felix Guattari, *A Thousand Plateaus: Capitalism and Schizophrenia,* trans. Brian Masumi (Minneapolis: University of Minneapolis Press, 1987), 11, 54, 56. See also Luke Higgins, "Toward a Deleuze-Guattarian Micropneumatology of Spirit-Dust" in the present volume.

20. William Bryant Logan, *Dirt: The Ecstatic Skin of the Earth* (New York: Riverhead, 1995), 7. But to set the record straight: my geologist niece Jennifer

Zinn assures me that we "do know a thing or two about dirt. In fact we know alot. There are many different types of soils. Even paleosols (old buried soil—unconsolidated)" (Personal e-mail communication, November 20, 2006).

21. Logan, *Dirt*, 8.

22. See biologist and theorist Stuart Kauffman, *At Home in the Universe: The Search for Laws of Self-Organization and Complexity* (New York: Oxford University Press, 1995); and *Investigations* (New York and Oxford: Oxford University Press, 2000).

23. Logan, *Dirt*, 8.

24. My *Face of the Deep* (New York and London: Routledge, 2004) is one long meditation on the chaos of beginnings, including a chapter on the earth *tohu va bohu* as anticipation of chaos theory. Yet the primary imagery is *tehomic*/aqueous rather than terrestrial/grounding.

25. Logan, *Dirt*, 9.

26. "Earthquake Conversations," *Scientific American* 15, no. 2 (2005): 89.

27. Logan, *Dirt*, 11.

28. As Logan comments: "No two molecules of humus may be alike. Though no one has difficulty recognizing a humus molecule, it is quite likely unique because it works upon fractal principles. For humus, similarity is rampant, but identity nonexistent" (Logan, *Dirt*, 16).

29. In the mild and succinct wording of *Scientific American*, citing the IPCC report, "Climate Change 2001," the world will warm between 1.4 and 5.8 degrees Celsius by 2100. "The mild end of that range—a warming rate of 1.4 degrees per 100 years—is still 14 times faster than the one degree per 1,000 years that historically has been the average rate of natural change on a global scale. Should the higher end of the range occur, then the rate of climatic change could be nearly 60 times faster than natural average conditions, which could lead to change that many would consider dangerous. Change at this rate would almost certainly force many species to attempt to move their ranges, just as they did from the ice age/interglacial transition between 10,000 and 15,000 years ago. Not only would species have to respond to climatic change at rates 14 to 60 times faster, but few would have undisturbed, open migration routes as they did at the end of the ice age and the onset of the interglacial era. The negative effects of this significant warming—on health, agriculture, coastal geography and heritage sites, to name a few—could be severe" (*Scientific American* 15, no. 2 [2005]: 13).

30. See Anne Daniell, "Divining New Orleans: Sophialogical Evocations for the Redemption of Place," in the present volume.

31. José Saramago, *The Gospel According to Jesus Christ*, trans. Giovanni Pontiero (New York: Harcourt Brace, 1991). Of course there is an intriguing, Job-like collusion between this God and his devil.

32. Mark I. Wallace, *Finding God in the Singing River: Christianity, Spirit, Nature* (Minneapolis: Augsburg Fortress, 2005), 36.

33. Jürgen Moltmann, *The Spirit of Life: A Universal Affirmation* (Minneapolis: Augsburg Fortress, 1992).

34. *Gaia and God: An Ecofeminist Theology of Earth Healing* (San Francisco: HarperSanFrancisco, 1994).

35. Anne Primavesi, *Gaia's Gift: Earth, Ourselves and God after Copernicus* (New York and London: Routledge, 2003), 134. In her essay in the present volume, "The Pre-Original Gift—and Our Response to It," her pre-original gift displaces the familiar faith by which "our return of thanks to God for Gaia's gift is conceived of as earning an eternal reward in heaven: conditional on our 'good' behavior *on* earth—but not *towards* earth."

36. "Jesus is God as dust. God as dust is Immanuel, God who is with us in our joy, our suffering, our bodiliness, our spiritual growth and struggles." Karen Baker-Fletcher refers to Rosemary Radforth Ruether's observation that the Lord's Prayer contains "remnants of the Jewish Jubilee tradition, connecting heaven with the renewal of earth and its peoples" (Baker-Fletcher, *Sisters of Dust, Sisters of Spirit: Womanist Wordings on God and Creation* [Minneapolis: Augsburg Fortress, 1998], 17–18).

37. *New York Times*, February 9, 2006. The first line, following a quotation of Colossians 1:16, reads: "As evangelical Christians, we believe we're called to be stewards of God's creation, and after considerable study, reflection, and prayer we are now convinced it's time for our country to help solve the problem of global warming." Signers include Rick Warren; writers include Jim Ball of the Evangelical Environmental Network (www.christiansandclimate.org).

38. See Jay McDaniel's essay in the present volume, "Ecotheology and World Religions," as well as his *Ghandi's Hope: Learning from World Religions as a Path to Peace* (Maryknoll, N.Y.: Orbis, 2005), and prior works all practice this radically ecological ecumenism as the basis for world peace between humans.

39. Indeed, to possess some of us; see Keller, *Face of the Deep*.

40. "Now that the emergence of nature no longer comes into play to paralyze the progressive composition of the common world, we have to become capable of convoking the collective that will be charged from now on, as its name indicates, with 'collecting' the multiplicity of associations of humans and nonhumans" (Latour, *Politics of Nature*, 55).

41. The quotation is drawn from the preamble to *The Earth Charter: Values and Principles for a Sustainable Future* (www.earthcharter.org). All quotations from the Earth Charter are to be found at this site.

42. Moltmann, *Spirit of Life*, 10.

43. Gilles Deleuze, *The Fold: Leibniz and the Baroque*, trans. Tom Conley (Minneapolis: University of Minnesota Press, 1993), 80.

ECOFEMINIST PHILOSOPHY, THEOLOGY, AND ETHICS: A COMPARATIVE VIEW | ROSEMARY RADFORD RUETHER

1. See Carol J. Adams, *Ecofeminism and the Sacred* (New York: Continuum, 1994), xi.

2. See Val Plumwood, *Feminism and the Mastery of Nature* (New York: Routledge, 1993), 104–40.

3. On the essentialist debate about ecofeminism, see, for example, Mary Mellor, *Feminism and Ecology* (New York: New York University Press, 1997), 44–70. See also Stephanie Lahar, "Ecofeminist Theory and Grassroots Politics," in *Ecological Feminist Philosophies*, ed. Karen J. Warren (Bloomington: Indiana University Press, 1996), 11–12.

4. Mellor speaks of this view as "affinity ecofeminism" (*Feminism and Ecology*, 56–58, 75–77). The complexity of this "affinity" can be seen in an essay such as that by Charlene Spretnak, "Earthbody and Personal Body as Sacred," in Adams, *Ecofeminism and the Sacred*, 261–80.

5. See discussion of the relativity of culture-nature hierarchies and its inapplicability to tribal and peasant peoples in Heather Eaton and Lois Ann Lorentzen, *Ecofeminism and Globalization: Exploring Culture, Context and Religion* (Lanham, Md.: Rowman and Littlefield, 2003), especially chapters 3 and 4, 41–71.

6. For an effort to imagine an ecofeminist society, see my *Gaia and God: An Ecofeminist Theology of Earthhealing* (San Francisco: HarperSanFrancisco, 1992), 258–68.

7. *Staying Alive* (London: Zed Books, 1989) was published originally in Delhi, India, by Kali for women. *The Violence of the Green Revolution: Third World Agriculture, Ecology & Politics* (London: Zed Books, 2006) originally appeared with The Other India Press. *Monocultures of the Mind: Perspectives on Biodiversity & Biotechnology* (London, 1993) was published in the West by Zed Books, and *Biopiracy: The Plunder of Nature & Knowledge* (1997) and *Stolen Harvest* (2001) by South End Press, Boston. *Ecofeminism* (1993) appeared with several world presses, including Zed.

8. See *Staying Alive*, 1, 15.

9. See Susan Harding, *The Science Question in Feminism* (Ithaca: Cornell University Press, 1986), and Evelyn F. Keller, *Reflections on Gender and Science* (New Haven: Yale University Press, 1985).

10. In this context she cites Brian Easlea, *Science and Sexual Oppression: Patriarchy's Confrontation with Women and Nature* (London: Weidenfeld and Nicholson, 1981).

11. See her chapter on "Women in the Food Chain" in *Staying Alive*, 96–178, and *The Violence of the Green Revolution*.

12. *Staying Alive*, 38–54.

13. See the critique of these themes in Hinduism in chapter 2 of Rosemary Ruether, *Integrating Ecofeminism, Globalization and World Religions* (Lanham, Md.: Rowman and Littlefield, 2004).

14. This critique was made by Aruna Gnanadason in a conversation at a World Council of Churches consultation on ecofeminism in Geneva in July 2003.

15. *Intuiciones Ecofeministas* (Madrid: Editorial Trotto, 2000).

16. Ibid., 22–24.

17. Ivone Gebara, "A Cry for Life from Latin America," in *Spirituality of the Third World: A Cry for Life*, ed. K. C. Abraham and Bernadette Mbuy-beya (Maryknoll, N.Y.: Orbis, 1994), 109–18.

18. "Ecofeminism and Panentheism," interview by Mary Judy Ress, *Creation Spirituality* (November–December, 1993): 9–11.

19. Ivone Gebara, *Longing for Running Water: Ecofeminism and Liberation* (Minneapolis: Augsburg Fortress, 1999), 25–65. See also my *Women Healing Earth, Third World Women on Ecology, Feminism and Religion* (Maryknoll, N.Y.: Orbis, 1996), 13–23.

20. Gebara, *Longing for Running Water*, 82–92.

21. See her reflections on "The Trinity and the Problem of Evil" in my *Women Healing Earth*, 19–22.

22. See particularly her chapter on "Women's Experience of Salvation" in *Out of the Depths: Women's Experience of Evil and Salvation* (Minneapolis: Augsburg Fortress, 2002), 109–44.

23. Carolyn Merchant, *The Death of Nature: Women, Ecology and the Scientific Revolution* (San Francisco: HarperSanFrancisco, 1990).

24. Ibid., 1.

25. For Bacon's sexual imagery, see Merchant, *Death of Nature*, 168–72.

26. Ibid., 290.

27. Carolyn Merchant, *Reinventing Eden: The Fate of Nature in Western Culture* (New York and London: Routledge, 2003).

28. Ibid., 167–70.
29. Ibid., 26–36.
30. David Korten, speaking at the Call to Action Conference, Milwaukee, November 7–8, 2003.
31. Merchant, *Reinventing Eden*, 205–20.
32. Thomas Berry and Brian Swimme, *The Universe Story* (San Francisco: HarperSanFrancisco, 1992), 240–61.

COOKING THE TRUTH: FAITH, SCIENCE, THE MARKET, AND GLOBAL WARMING | LAUREL KEARNS

Special thanks are due to the Louisville Institute for the sabbatical grant that supported the initial research for this project and that of Rebecca Gould, my coauthor on other works related to this project, who along with Matt Immergut, Rick Bohannon, and Matt Westbrook, deserve thanks for their related research and review of this article.

1. The United States Council of Catholic Bishops, the National Council of Churches Eco-Justice Working Group, the Evangelical Environmental Network, and the Coalition on the Environment and Jewish Life.
2. In this text, I will use the terms "global warming" and "global climate change" equally, recognizing that many of those who prefer the term "climate change" wish to imply that such change is due to the natural climatic cycles of the planet, and not caused by human activity. The term "global warming" is preferred by those who see a definite warming trend happening both in ground temperatures and ocean temperatures, which has long-reaching effects on an extensive range of global climate conditions, including making it cooler in some places due to increased rainfall and cloudiness.
3. "If God Is With Me All the Time, Does that Include the Auto Mall?" available at http://www.gbgm-umc.org/NCNYEnvironmentalJustice/transportation.htm. All URLs given here and below were valid as of 2006.
4. In addition to the Oregon Institute on Science and Medicine (OISM), the Union of Concerned Scientists lists the following organizations skeptical of global climate change: George Marshall Institute, which worked with OISM on the petition project; the Greening Earth Society, founded by the Western Fuels Association; the Center for the Study of Carbon Dioxide and Global Change, founded by two brothers who were at one time on the payroll for Western Fuels. Both of the following organizations contend that global warming will bring about positive change in the world: the Science and Environmental Policy Project (SEPP), which holds that the science on global warming is shaky, and that such warming may not be a bad thing;

and the Global Climate Coalition, a huge corporate-sponsored endeavor founded in 1989. British Petroleum (BP) was the first major corporation to pull out of the coalition, followed by General Motors and Texaco and a host of others. See "Prominent Skeptics Organizations," available at http://www.ucsusa.org/global_warming/science/skeptic-organizations.html.

5. There is much debate among religious activists over the usage of the construct "environment" versus "ecology." The first term is seen to embody a sense that nature and the environment are other than human, whereas the latter is seen to incorporate all entities, the human and the more-than-human, in a web of relationships.

6. See Laurel Kearns, "Saving the Creation: Christian Environmentalism in the United States," *Sociology of Religion* 57 (1996): 55–70, and "Noah's Ark Goes to Washington: A Profile of Evangelical Environmentalism," *Social Compass* 44, no. 3 (September 1997): 349–66.

7. Some of this research and writing is being conducted with Rebecca Gould.

8. Fortunately, this ignorance is being eroded as of the spring–summer of 2006, with increasing bad news from the science community about the fast pace of global warming's effects, such as ice melt, and with the attention garnered by former Vice-President Al Gore's movie *An Inconvenient Truth*.

9. I also have chosen to focus on the United States because the majority of my research on religious environmentalism over the past two decades has been focused on events in the United States.

10. This is because CO_2 is emitted in the burning of fossil fuels, which is used to produce electricity or to run automobiles and other internal combustion engines.

11. See Kearns, "Saving the Creation," for a chart elaborating some of the differences across the spectrum of Christian environmental activism. These are ideal types.

12. The Christian Environmental Council, representative of over forty evangelical Christian organizations, similarly presented then President Clinton and Vice-President Gore with a resolution calling for the United States to take a leadership role. See James Ball, *Planting a Tree This Afternoon: Global Warming, Public Theology and Public Policy*, Crossroads Monograph Series on Faith and Public Policy, no. 1 (18) (1998), 69. Efforts by religious groups to influence the global deliberations on the Kyoto Protocol continue. At the 2005 United Nations Climate Change Conference, a "Spiritual Declaration on Climate Change" was issued. See the Montreal Climate Change Conference report at http://www.wcc-coe.org/wcc/what/jpc/climatechange-cop11-report.html#3.

13. That number shifts as some campaigns become inactive. For instance, in 2006 the Web site (http://www.protectingcreation.org) lists only eleven official ICC state campaigns, but also lists related activities in fifteen states.

14. Extensive media-relations training was available for state coordinators at the NCC-EJWG 2001 conference in Washington, D.C., held in conjunction with COEJL, as part of the "Let There Be Light" campaign. Over two hundred attendees were encouraged to lobby on Capitol Hill; compact fluorescent light bulbs were given to every Senate and House member visited by conferees. At a rally on the lawn of the Capitol concerning President Bush's proposed energy bill, several high-ranking religious leaders spoke, as did Senators Joseph Leiberman (D-CT) and Olympia Snow (R-ME) demonstrating the bipartisan nature of the task.

15. Senator James Inhofe (R-OK), then chairman of the Committee on Environment and Public Works, stated in a speech on July 28, 2003, that "much of the debate over global warming is predicated on fear, rather than science" and "after studying the issue over the last several years, I believe that the balance of the evidence demonstrates that natural variability is the overwhelming factor influencing climate." He concluded that "I have offered compelling evidence that catastrophic global warming is a hoax. That conclusion is supported by the painstaking work of the nation's top climate scientists" (text of the speech is available at http://inhofe.senate.gov/press app/record.cfm?id=206907.) (Fred Barnes. in his *Rebel-in-Chief: Inside the Bold and Controversial Presidency of George W. Bush* (New York: Crown Forum, 2003) comments that "though he didn't say so publicly, Bush is a dissenter on the theory of global warming. . . . He avidly read Michael Crichton's 2004 novel *State of Fear,* whose villain falsifies scientific studies to justify draconian steps to curb global warming. . . . Early in 2005, political adviser Karl Rove arranged for Crichton to meet with Bush at the White House. They talked for an hour and were in near-total agreement. The visit was not made public for fear of outraging environmentalists all the more" (22–23). Over 120 scientists have claimed that the Bush administration "cooked the truth" by pressuring them to alter reports or to delete references to "climate change" and "global warming" or from a range of documents, such as press releases and communications with Congress. Peter N. Spotts, "Has the White House Interfered on Global Warming Reports?" *Christian Science Monitor,* January 31, 2007.

16. Literature handed out at the "Let There Be Light" conference.

17. Christine McCarthy McMorris, "What Would Jesus Drive?" *Religion in the News* 6, no. 1 (Spring 2003), available at http://www.trincoll.edu/depts/csrpl/RINVol6No1/What%20Would%20Jesus.htm.

18. This campaign was in large part initiated by Drew PhD Jim Ball. See Alexander Lane, "Evangelist: What Would Jesus Drive?" [Newark, N.J.] *Star Ledger*, February 12, 2006, and Laurie Goodstein, "Living Day to Day by a Gospel of Green," *New York Times*, March 8, 2007 for more details about Jim Ball.

19. A recent campaign aimed at Detroit took place in September 2005. Some hope that it played a role in Ford announcing the production of hybrid vehicles for a significant proportion of their models.

20. Available at http://www.christiansandclimate.org.

21. Warren is the author of *The Purpose Driven Life*, which has sold nearly 14 million copies. Rick Warren, *The Purpose Driven Life: What On Earth Am I Here For?* (Grand Rapids, Mich.: Zondervan, 2002).

22. Laurie Goodstein, "Evangelical Leaders Join Global Warming Initiative," *New York Times*, February 8, 2006. For a listing of all of the signatories, see http://www.christiansandclimate.org.

23. The Reverend Rosemary Bray McNatt, a graduate of Drew University Theological School. The Reverend Gerald Durley of Providence Missionary Baptist Church in Atlanta has more recently taken a lead in getting African-Americans to be involved in responding to climate change. After seeing *The Great Warming*, Durley handed out CFC lightbulbs to his 1000 + congregation, routinely speaks to them about climage change, and helped organize a 2007 National African Earth Day Summit in Atlanta featuring Senator Barack Obama and Congressman John Lewis. He has also challenged other black preachers in Atlanta to preach on Earth Day.

24. Rabbi Arthur Waskow likes to refer to it as "global scorching." See Rebecca Gould and Laurel Kearns, "An Overview of the Eco-Justice Movement," *Earth Letter* (Fall 2005). See the Web site of the Eco-Justice Working Group for a list of denominational statements, available at http://www.nccecojustice.org/anthohome.htm.

25. In June of 2006, the Presbyterian Church (USA) voted to discontinue funding on environmental issues as a separate program. This came as quite a surprise to many. A separate group, Presbyterians for Restoring Creation, is growing and has over 400 members, but it is outside of the denomination's structures.

26. GreenFaith is not part of the Interfaith CCN because New Jersey's Democratic senators are very proenvironmental, and because the state pledged to meet the Kyoto Protocol emissions standards. As part of the pledge, businesses, educational institutions, and religious groups voluntarily sign covenants of sustainability to meet the goals set forth at Kyoto. To aid in

this, the state has an ambitious program to provide funding for alternative energy, including reimbursing contractors for 70 percent of the cost of solar energy installation. I was fortunate to be an early beneficiary of this program. The funds are available to private homeowners, businesses, and religious, health, and educational institutions. Through this program, GreenFaith has facilitated 25 solar installations on religious buildings.

27. Green Mountain Energy has since pulled out of New Jersey. See Fletcher Harper's chapter elsewhere in the present volume for more information on GreenFaith, of which I am a board member.

28. This agreement came early in the history of the NRPE. Currently, each constituent group is working on the issue, with the NCC-EJWG and COEJL cooperating on the ICCN.

29. There is a Jewish-Christian coalition on endangered species called the Noah Alliance (information available at http://www.noahalliance.org). They have been strategic in countering the efforts of Congress to weaken the Endangered Species Act. Others have formed on foresty issues.

30. This is quite clear on the front page of ICCN's Web site, available at http://www.protectingcreation.org.

31. McKibben was also one of the founders of the "What Would Jesus Drive?" campaign to reduce automobile fuel consumption. See the discussion of "What Would Jesus Drive?" campaigns outside of auto dealerships in Lynn, Massachusetts, in "As Support Widens, Environmental Groups Get Religion," *Boston Globe,* July 8, 2001.

32. "Greening the Church," Tipple-Vosberg Lecture Series, Drew University Theological School, October 1999.

33. Bill McKibben, "The Comforting Whirlwind: God and the Environmental Crisis," in "The Cry of Creation: A Call for Climate Justice" (Seattle: Earth Ministry Publications), 3–5.

34. "Religious Leaders' Statement on Energy Conservation," available at http://www.protectingcreation.org/documents/EnergyStatement.html.

35. The first formulation of an "Eleventh Commandment" is credited to Walter Lowdermilk, who gave a speech on Radio Jerusalem in June 1939 entitled "The Eleventh Commandment." The text of his commandment read: "Thou shalt inherit the holy earth as a faithful steward, conserving its resources and productivity from generation to generation. Thou shalt safeguard thy fields from soil erosion, thy living waters from drying up, thy forests from desolation, and protect the hills from overgrazing by thy herds, that thy descendents may have abundance forever. If any shall fail in this stewardship of the land, thy fruitful fields shall become sterile stony ground

and wasting gullies, and thy descendants shall decrease and live in poverty or perish from off the face of the earth" (as quoted in Roderick Nash, *The Rights of Nature: A History of Environmental Ethics* [Madison: University of Wisconsin Press, 1989], 97–98). An Eastern Orthodox group called the 11th Commandment Fellowship was formed in the 1980s. They later joined others in founding the North American Coalition on Christianity and Ecology (NACCE). See Kearns, "Saving the Creation," and Laurel Kearns and Matthew Immergut, "Fred Kreuger," in the *Encyclopedia of Religion and Nature* (Bristol, U.K., and New York: Thoemmes-Continuum, 2005), 972–73.

36. RaeAnn Slaybaugh, "Conserving Energy on Your Campus: Doing What You Can to Save the Earth's—and Your Church's—Resources Is an Issue of Ethics, Stewardship," available at http://www.churchbusiness.com/.

37. This is a very abbreviated form of an argument based on Weber's thesis in the *Protestant Ethic*. Interestingly, Weber commented that the ethos he tried to understand might drive society until the "last ton of fossilized coal is burnt." *The Protestant Ethic and the Spirit of Capitalism*. Trans. Talcott Parsons. New York: Charles Scribner's Sons, 1958) 181.

38. Harvey Cox, "Mammon and the Culture of the Market," in *Meaning and Modernity: Religion, Polity and the Self*, ed. Richard Madsen, William M. Sullivan, Ann Swidler, and Steven M. Tipton (Berkeley: University of California press, 2002), 124–35.

39. See, for instance, Tony Campolo, *How to Rescue the Earth Without Worshipping Nature: A Christian's Call to Save Creation* (Nashville: Thomas Nelson, 1992). Campolo has remained active in religious initiatives on climate change issues. Similar issues about paganism and environmentalism are present in Judaism.

40. Christopher P. Toumey, *God's Own Scientists: Creationists in a Secular World* (New Brunswick, N.J.: Rutgers University Press, 1994).

41. Andy Crouch, "Environmental Wager: Why Evangelicals Are—but Shouldn't Be—Cool Toward Global Warming," *Christianity Today* 49, no. 8 (August 2005): 66, and James Schlesinger, "The Theology of Global Warming," *Wall Street Journal*, August 8, 2005.

42. Put in historical perspective, this is not necessarily anything new, for the acceptance of scientific "truths" by religious adherents has been an ongoing battle.

43. I am referring here in part to Clifford Geertz's poetic way of discussing the difference between religion in more traditional cultures, where one *is held* by their religious views, and in more modern, secularizing societies, where

one *holds* religious beliefs, and can choose not to hold them. See Clifford Geertz, *Islam Observed: Religious Development in Morocco and Indonesia* (Chicago: University of Chicago Press, 1971).

44. "Petition Project" available at http://www.oism.org/pproject.

45. Arthur Robinson, "Learn to Think Scientifically: Each Person Needs to Distinguish Propaganda from Truth," in The Robinson Self-Teaching Homeschool Curriculum., available at http://www.robinsoncurriculum .com/view/rc/s31p997.

46. The Center for Media and Democracy states on its Web site that "when questioned in 1998, OISM's Arthur Robinson admitted that only 2,100 signers of the Oregon Petition had identified themselves as physicists, geophysicists, climatologists, or meteorologists, 'and of those the greatest number are physicists.' This grouping of fields concealed the fact that only a few dozen, at most, of the signatories were drawn from the core disciplines of climate science—such as meteorology, oceanography, and glaciology—and almost none were climate specialists" ("Oregon School of Science and Medicine," available at http://www.sourcewatch.org/index.php?title = Oregon_Institute_of_Science_and_Medicine#Funding). The Union of Concerned Scientists also investigated the petition. Their expressed skepticism about the relevant scientific credentials was returned, as seen in this quote from Dean's World Online magazine: "The fact of the matter is that there are 'scientific groups,' like the Union of Concerned Scientists, that issue broad statements about subjects that few, or even none, of their members are qualified to speak on" (text available at http://www.deanes-may.com/archives/006235.html).

47. A similar move is seen in ornithologist Jerome Jackson's challenge to the report that the ivory-billed woodpecker, thought to be extinct, has been found in the Southeast United States. He termed the report "faith-based ornithology" (James Gorman, "Ivory Bill Report is Called 'Faith-Based Ornithology,'" *New York Times*, January 24, 2006.

48. Reference to statement of the National Center for Policy Analysis that "Global Warming Has Become A European Religion," available at www.n-cpa.org/hotlines/global/pd040201e.html.

49. Janet Guyon, "A Big-Oil Man Gets Religion on Global Warming," *Time* (March 6, 2000).

50. Lindzen's remarks were published in many newspapers. See the text at the Center for Defense of Free Enterprise, available at http://www.cdfe.org/global_warming_religion.htm. His speech was promoted by the George Marshall Institute.

51. Dorothy Smith, *Writing the Social: Critique, Theory and Investigations* (Toronto: University of Toronto Press, 1999), 175–77.

52. Steve Fuller, "The Re-enchantment of Science: A Fit End to the Science Wars?" in *After the Science Wars*, ed. Keith Ashman (New York: Routledge, 2001), 189. For a longer discussion, see Fuller's *Science* (Minneapolis: University of Minnesota Press, 1997).

53. H. Sterling Burnett, "Global Warming: Religion or Science?" available at National Conservative Weekly's Human Events Online http://www .humaneventsonline.com/article.php?id = 8326.

54. Bruno Latour, *Politics of Nature: How to Bring the Sciences into Democracy* (Cambridge, Mass.: Harvard University Press, 2004).

55. David Demeritt, "The Construction of Global Warming and the Politics of Science," *Annals of the Association of American Geographers* 91, no. 2 (2001): 307–17.

56. Sandra Harding, *The Science Question in Feminism* (Ithaca: Cornell University Press, 1986). See also her *Is Science Multicultural: Postcolonialisms, Feminisms and Epistemologies* (Indianapolis: Indiana University Press, 1998).

57. Schlesinger seems to have missed the irony in his example of Galileo and Copernicas, whose theories were rejected by the Church much as contemporary climate science is rejected by a portion of the Church.

58. Paul Rauber, "Eco-thug: Helen Chenoweth—anti-environment congress-woman from Idaho," *Sierra* 81 (May–June 1996): 28.

59. Rose French, "Baptists Warn that Environmental Politics Could Divide Evangelicals," Associated Press, July 15, 2006, available at http://www .dallasnews.com/sharedcontent/dws/dn/religion/stories/071506dnrelgreen baps.46a6f93.html.

60. One of the founders of the "wise-use" movement, Ron Arnold, commented that "we're out to destroy environmentalism," as quoted in David Helvarg, *The War Against the Greens: The "Wise-Use" Movement, the New Right, and Anti-Environmental Violence* (San Francisco: Sierra Club, 1994).

61. Phillip DeVous, "Unholy Alliance: Radical Environmentalists and the Churches," *Acton Commentary*, April 10, 2002.

62. ICES board members E. Calvin Beisner and Robert Royal have published influential books, which warn of "the use and abuse of religion in environmental debates" and are critical of evangelical involvement, such as that of Cal DeWitt and the EEN. E. Calvin Beisner, *Where Garden Meets Wilderness: Evangelical Entry into the Environmental Debate* (Grand Rapids, Mich.: Acton Institute/William B. Eerdmans, 1997), and Robert Royal, *The Virgin and the Dynamo: The Use and Abuse of Religion in Environmental Debates* (Grand Rapids, Mich.: Ethics and Public Policy Institute/William B. Eerdmans, 1999).

63. "Cornwall Declaration on Environmental Stewardship," in Michael Barkey, ed., *Environmental Stewardship in the Judeo-Christian Tradition: Jewish, Catholic and Protestant Wisdom on the Environment* (Grand Rapids, Mich.: Acton Institute, 2000), xi–xv.

64. "About the ICES," available at http://www.stewards.net/About.htm. There is a new organization, Interfaith Stewardship Alliance (ISA), launched in 2006 and backed by the Acton Institute for the Study of Religion and Liberty, which also utilizes the Cornwall Declaration and recruits members for the "Cornwall Network." There is quite an overlap of board members between the two groups, and Calvin Beisner is the national director.

65. Alan Cooperman, "Evangelicals Will Not Take Stand on Global Warming," *Washington Post*, February 2, 2006.

66. Beisner is a board member of the Committee for a Constructive Tomorrow (CFACT), whose Web site "boldly proclaims that the West values competition, progress, and its wildlife, but even more importantly, its people" (available at http://www.cfact.org/site/default.asp). Both CFACT and Acton produce hundreds of radio spots on environmental issues. Many of the major groups, such as Center for the Defense of Free Enterprise and Acton, fund conferences and have active publishing programs.

67. "Evangelical Declaration of Creation Care," in *The Care of Creation: Focusing Concern and Action*, ed. Robert J. Berry (Downers Grove, Ill.: InterVarsity Press, 2000), 17–22.

68. All quotes are from the web-posted versions of these two documents, available at http://www.stewards.net/CornwallDeclaration.htm and http://www.creationcare.org/resources/declaration.php. For an introduction to the "wise-use" movement, see Laurel Kearns, "Wise-Use Movement," in *Encyclopedia of Religion and Nature*, 1755–58.

69. William Cronon, "The Trouble with Wilderness," in William Cronon, ed., *UnCommon Ground: Rethinking the Human Place in Nature* (New York: Norton, 1996), 69–90. The article was used by many on the Right to discredit any attempt to set aside wilderness areas, arguing that wilderness is a human construct.

70. John Ross Edward Bliese, *The Greening of Conservative America* (Boulder, Colo.: Westview, 2001).

71. Since the funders of the evangelical efforts were the same funders of other NRPE efforts, part of the conservative response was to attack the funders as liberal and to point out what issues they funded with which evangelicals would disagree.

72. Robert Wuthnow, *The Restructuring of American Religion* (Princeton: Princeton University Press, 1988).

73. Interview with Rob Gorman, Houma, Louisiana, July 2001.

74. Christian Smith, ed., *Disruptive Religion: The Force of Faith in Social Movement Activism* (Oxford: Routledge, 1996).

75. E. Calvin Beisner, "Christian Economics: A System Whose Time has Come?", in *Christian Perspectives on Economics*, ed. Robert N. Mateer (Lynchburg, Va.: CEBA, 1989).

76. Under the directorship of former New Jersey governor Christine Todd Whitman, the EPA did urge the United States to respond to global warming. This recommendation was quickly withdrawn, and Whitman eventually resigned over her inability to make any headway with the Bush Administration. Under her governorship, New Jersey had already begun implementing many programs to counter emissions and curb energy usage.

77. Exxon donated $50,000 to Acton in 2005. www.exxonmobile.com/corporate/files/corporate/giving05_policy.pdf.

78. James Proctor, "In ——— We Trust: Science, Religion and Authority," in *Science, Religion and the Human Experience*, ed. James Proctor (New York: Oxford University Press, 2005).

79. Bruno Latour, "Thou Shalt Not Freeze Frame," in Proctor, *Science, Religion and the Human Experience*, 27–48, quotation on 45.

80. Harvey Cox, "Mammon and the Culture of the Market," in Madsen et al., *Meaning and Modernity: Religion, Polity and the Self*, 124–35.

81. James Proctor, "Introduction: Rethinking Science and Religion," in Proctor, *Science, Religion and the Human Experience*, 3–23.

82. There is no real choice, if one part of the choice is the future livability of the planet, as ex Vice-President Al Gore not so jokingly questions in *An Inconvenient Truth*. He draws upon an illustration presented at the 1992 UN conference in Rio de Janeiro that shows a set of scales with money/gold on one scale and the planet earth on the other, as if one could choose.

83. Anne Swidler, "Culture in Action: Symbols and Strategies," *American Sociological Review* 51 (April 1986): 273–86.

84. Anthony Wallace, "Revitalization movements," *American Anthropology* 58 (1956): 264, 281.

85. Max Weber, *The Protestant Ethic*, 181.

ECOSPIRITUALITY AND THE BLURRED BOUNDARIES OF HUMANS, ANIMALS, AND MACHINES | GLEN A. MAZIS

1. Aldo Leopold, *A Sand County Almanac, and Sketches Here and There* (New York: Oxford University Press, 1987), 204.

2. Theodore Roszak, *The Voice of the Earth* (New York: Simon and Schuster, 1992), 181, 323.

3. Gary Kowalski, *The Soul of Animals* (Walpole, N.H.: Stillpoint, 1991); see especially chapters 5, 6, and 8.

4. Glen Mazis, "The Famous Artificial Heart Experiment," *Many Mountains Moving: A Literary Journal of Diverse Contemporary Voices* 4, no. 3: 97.

5. Maurice Merleau-Ponty, *Phenomenology of Perception*, trans. Colin Smith (New York: Routledge and Kegan Paul, 1962), 67.

6. Ibid., 68

7. Donna Haraway, *The Companion Species Manifesto: Dogs, People and Significant Otherness* (Chicago: Prickly Paradigm Press, 2003), 30.

8. Eugen Herrigal, *Zen and the Art of Archery* (New York: Vintage, 1989), 63.

9. Catherine Keller, *Face of the Deep: A Theology of Becoming* (New York: Routledge, 2003), 236.

10. Donna Haraway, *Modest_Witness@Second_Millennium.FemaleMale©_Meets _OncoMouse™* (New York: Routledge, 1977), 270.

11. George Page, *Inside the Animal Mind* (New York: Broadway, 1999), 60–61.

12. Donald R. Griffin, *Animal Minds: Beyond Cognition to Consciousness* (Chicago: University of Chicago Press, 2001), 60.

13. Martin Heidegger, *Being and Time*, trans. John Macquarrie and Edward Robinson (New York: Harper and Row, 1962), 136.

14. Ibid., 21–22; and, Griffin, *Animal Minds*, 29–31.

15. Martin Heidegger, *The Fundamental Concepts of Metaphysics: World, Finitude, Solitude*, trans. William McNeill and Nicholas Walker (Bloomington: Indiana University Press, 1995), 248.

16. Ibid., 269.

17. Ibid., 249.

18. Ibid., 259.

19. Ibid., 273.

20. Ibid., 264.

21. Ibid., 250–51.

22. Ibid., 267.

23. Kowalski, *Soul of Animals*, 16.

24. Heidegger, *Fundamental Concepts of Metaphyics*, 242.

25. Merleau-Ponty, *Phenomenology of Perception*, 68.

26. Ibid., 214.

27. Ibid., 140–41.

28. Maurice Merleau-Ponty, *The Visible and the Invisible*, trans. Alphonso Lingis (Evanston: Northwestern University Press, 1968), 248.

29. Ibid., 174.

30. Ibid., 142.

31. See my *Earthbodies: Rediscovering Our Planetary Senses* (Albany: State University of New York Press, 2002), in which I argue that humans, by virtue of their embodiment, have always been incorporating aspects of the built environment around them into their "body schema," such that a club or a car, a hoe or a computer, becomes the way our bodies make their way through the surround.

32. Martin Heidegger, *Discourse on Thinking*, trans. John Anderson and E. Hans Freund (New York: Harper and Row, 1966), 55.

33. Bruno Latour, *Politics of Nature: How to Bring the Sciences into Democracy*, trans. Catherine Porter (Cambridge, Mass.: Harvard University Press, 2004), 223.

34. Anne Foerst, *God in the Machine* (New York, Dutton, 2004), 4.

35. Ibid., 96–97.

36. Ibid., 99.

37. Ibid., 146.

38. Ibid.

39. Ibid., 7.

40. Andy Clark, *Natural-Born Cyborgs: Minds, Technologies, and the Future of Human Intelligence* (New York: Oxford University Press, 2003), 114.

41. Ibid., 114.

42. Ibid., 33.

43. Ibid., 78.

44. Ibid., 115–19. The owl monkeys are part of research being done at Duke University Medical Center in North Carolina.

45. Ibid., 95–96.

46. Michael Chorost, *Rebuilt: How Becoming Part Computer Made Me More Human* (Boston: Houghton Mifflin, 2005), 6–7.

47. Ibid., 79.

48. Ibid., 88.

49. Ibid., 78.

50. Ibid., 90.

51. Ibid., 155–56.

52. Foerst, *God in the Machine*, 35.

53. Ibid.

54. Ibid., 188–89.

55. Louis Dupré, *Religious Mystery and Rational Reflection* (Grand Rapids, Mich.: Eerdmans, 1998), 143.

GETTING OVER "NATURE": MODERN BIFURCATIONS,
POSTMODERN POSSIBILITIES | BARBARA MURACA

The present essay originated as an internal dialogue among several competing voices (and different languages), each of them drawn to the very idea of thinking of "nature" as an ambiguous background for ecology discourses. Both my theoretical engagement with different philosophical traditions and my practical experience in Local Agenda 21 processes, where the recourse to the term "nature" was both unavoidable and fruitful, were active participants in this internal dialogue. The number of perspectives was further expanded by moving from Italy, where I studied Gianni Vattimo's postmodern approach, to Germany, where I work in a tension between the Habermasian context and my interest in process thought. I am grateful for the occasion to reconstruct a path among these voices, permeable boundaries for theoretical journeys.

1. Bruno Latour, *Politics of Nature: How to Bring the Sciences into Democracy*, trans. Catherine Porter (Cambridge, Mass.: Harvard University Press, 2004), 19.
2. Ibid., 3.
3. Gianni Vattimo, *La fine della Modernità* (Milan: Garzanti, 1985), 172 ff.
4. Latour, *Politics of Nature*, 225.
5. "If we are interested in the relationship of the economy to the environment . . . the pre-analytic vision of conventional economics is actively misleading. It has already assumed away the environment and made it external to the analysis. Indeed, economists often refer explicitly to the natural environment as an externality. Having assumed away the environment to begin with, many economists dismiss the possibility that there might be environmental limits to economic growth." (Herman Daly and David C. Korten, "A case of job protection most economists have overlooked," *PCD Forum* (1994): Article 5 [http://www.pcdf.org/1994/05daly&d.htm]).
6. As long as our cycle of production was fairly small compared to the carrying capacity of our ecosystem, the exclusion of nature from any matter of concern was a mere question of specific relevance or interest. As Herman Daly points out, we are moving from an "empty world" to a "full world," that is, "from a world where inputs to and outputs from the economy are unconstrained, to a world in which they are increasingly constrained by the depletion and pollution of a finite environment" (Daly, *Beyond Growth* [Boston: Beacon Press, 1996], 8). Therefore the "external" field is getting threatened in its very existence. Our production system has been pushing nature more and more out of the boundaries of relevance into a thin marginal space around our man-made world, where regeneration seems to be increasingly difficult.
7. Robert M. Solow, "The Economics of Resources or the Resources of Economics," *American Economic Review* 64, no. 2: 11.
8. The approach of Deep Ecology, for example, can respond to these needs. As Latour critically points out, however, Naess is "a good representative of

this philosophy of ecology that does feel the metaphysical limits of the division between nature and humanity, but that strives to 'go beyond' the 'limits of Western philosophy' instead of delving into the political origins of this division" (Latour, *Politics of Nature*, 257).

9. "To take one example in a thousand, we are all familiar with the ravages of social Darwinism, which borrowed its metaphors from politics, projected them onto nature itself, and then reimported them into politics in order to add the seal of an irrefragable natural order to the domination of the wealthy" (ibid., 33).

10. Vandana Shiva, "Science, Nature, and Gender," in *Women, Knowledge, and Reality*, ed. Ann Garry and Marilyn Pearsall (New York: Routledge, 1996), 275. Within ecofeminism, the criticism of Western conceptions of nature intended as passive, dead matter plays a major role. As Rosemary Radford Ruether, referring to Carolyn Merchant, explains, "nature is alive, holistic, and interconnected. Nature has its own self-organizing patterns of life. Humans need to connect with nature, not as dead objects, but rather as active subjects with which they must learn to partner" (Rosemary Radford Ruether, "Ecofeminist Philosophy, Theology, and Ethics: A Comparative View," in the present volume).

11. Carol Bigwood, *Earth Muse: Feminism, Nature, and Art* (Philadelphia: Temple University Press, 1993), 12.

12. Catherine Keller, "Postmodern 'Nature,' Feminism, and Community," in *Theology for Earth Community*, ed. Dieter T. Hessel (Maryknoll, N.Y.: Orbis, 1996), 95.

13. "Bringing-forth [*Hervorbringen*] brings hither out of concealment forth into unconcealment" (Martin Heidegger, *The Question Concerning Technology and Other Essays* [New York: Harper Colophon, 1977], 11).

14. Ibid., 34.

15. To say that things are instruments or means does not automatically imply their reduction to mere means. It is explicitly art that opens the complexity hidden behind mere means. The very idea of an unquestionable presence-at-hand is intimately connected with the instrumental function of things in the world. While technology refers to means as loosed from any reference (*Bezug*) to the background of their emerging, thus reducing them to their being at disposal (*vorhanden*), art opens a place for ulterior, irreducible linkages. See Martin Heidegger, "The Origin of the Work of Art," in Heidegger, *Off the Beaten Track* (New York: Cambridge University Press, 2002).

16. David E. Tabachnick, "*Techné*, Technology, and Tragedy," in *Techné: Research in Philosophy and Technology* 7, no. 3 (Spring 2004): 101. While referring

to Tabachnick's description of Heidegger, I do not endorse his main thesis about the basic similarity between *techné* in a classic sense and modern technology. Tabachnick claims, in reference to Thucydides and Sophocles, that *techné* already implied domination over nature. Even if this is true, it does not necessarily lead to a reduction to presence-at-hand, which takes place within modernity. As Whitehead similarly shows, the isolation of things as mere objects and instruments is based on a specific concept of matter as merely passive and extended. This concept is a product of modern science tout court. Things-as-instruments may very well be found in ancient Greece as well; it is the *mere* presence-at-hand that requires modern metaphysics.

17. Talking of fore- and background could lead to misunderstandings. The *"Grund"* as ground recalls the *"Ab-grund,"* which is at the same time the abyss and the nonground as something never given, yet linked as the horizon of what emerges.

18. *Heraus-fordern* can be literally translated as "forcing, challenging something to come out."

19. The term *Bestand* is currently used in German to translate "stock" in a strict economic sense, as the definition of what capital is: "a stock that yields a flow," that is, a fixed reserve that can generate different kinds of utilities and services.

20. Heidegger, *Question Concerning Technology and Other Essays*, 21.

21. Ibid., 19.

22. I refer here to a more general attitude that traversed the newly emerged natural sciences in the early modern time. Bacon himself did not endorse torture. He did make use of terms of vexation with reference to nature and to the need to compel it to reveal its secrets through experiments. It was more likely Leibniz, however, who explicitly used the rack analogy to praise Bacon's experimental sciences; see Mathews De Madariaga, *Francis Bacon: The History of a Character Assassination* (New Haven and London: Yale University Press, 1996).

23. Latour, *Politics of Nature*, 64.

24. Heidegger, *Question Concerning Technology and Other Essays*, 18.

25. It is not, of course, a matter of going back to the ancient approach to nature to recover the intimate connection between *physis* and *techné*. Rather, the very path through technology as enframing (*Gestell*) can lead beyond itself if accepted as destining and provenance. The real danger implied by reducing everything to mere "stock" or "standing-reserve" (*Bestand*) is at the same time the place for salvation (ibid., 25).

26. Latour, *Politics of Nature*, 19.

27. "Since a babe was born in a manger, it may be doubted whether so great a thing has happened with so little stir" (Alfred North Whitehead, *Science and the Modern World* [New York: Free Press, 1967 (1925)], 2).

28. Alfred North Whitehead, *Function of Reason* (Boston: Beacon Press, 1971 [1929]), 43.

29. Whitehead, *Science and the Modern World*, 17.

30. Ibid., 3.

31. Whitehead acknowledges the incredible potential connected with the rise of the modern sciences. In spite of their narrow and extremely abstract consideration of nature, they opened a path for further development after the fixed and inflexible patterns of scholasticism. Eventually the sciences achieved a comparable rigidity (and dominating power), which turned them into the "obscurantism" of our time; see Whitehead, *Function of Reason*, 44.

32. Whitehead, *Science and the Modern World*, 17, 53. For Whitehead this is a narrow perspective, resulting from abstraction: "For us the red glow of the sunset should be as much part of nature as the molecules and electric waves by which men of science would explain the phenomenon" (Alfred North Whitehead, *Concept of Nature* [Cambridge, Mass.: Prometheus Books, 2004 (1920)], 29).

33. In this reduction of nature to bare matter, Vandana Shiva, referring to Harding and other feminists, sees the achievement of the patriarchy's project: "During the last few years feminist scholarship has begun to recognize that the dominant science system emerged as a liberating force not for humanity as a whole, but as a masculine and patriarchal project which necessarily entailed the subjugation of both nature and women" ("Science, Nature, and Gender," 268).

34. Whitehead, *Concept of Nature*, 20. While "nature" as original matter plays the same role as Aristotle's concept of "substance," being as well the substrate for attributions, it is quite different from his concept of *hylé* (matter): "nature" as matter in the modern sense is less a potentiality for actualization than a bare entity already existing *actualiter* and determined by a series of particular qualities, like extension, impermeability, and geometric structure. See Michael Hampe, *Die Wahrnehmung der Organismen* (Göttingen: Vandenhoeck and Ruprecht 1990), 23.

35. Whitehead, *Concept of Nature*, 26. The plural is essential at this point. Whitehead explicitly does not speak of *a* theory of bifurcation. There is no *one* specific transformation of modern thought as a given destiny we have to

deal with. The development of the modern sciences has traversed many different layers and did not completely eliminate different approaches. Bacon, the father of modern science, is a wonderful example of this complexity: while he established the dramatic use of induction, he was still mainly interested in qualitative classification of phenomena and did not yet follow the Galilean mathematical approach to nature (Whitehead, *Science and the Modern World*, 45).

36. Ibid., 54.

37. Hampe, *Die Wahrnehmung der Organismen*, 25.

38. Whitehead, *Science and the Modern World*, 54.

39. Ibid., 55.

40. Hampe, *Die Wahrnehmung der Organismen*, 39.

41. Ibid., 168.

42. In this sense, Locke and Hume offered a fecund contribution. The justification of the bifurcation itself was now carried out by philosophy and could be dismissed by natural sciences, which then became free to focus on their "given" object, without having to explain its origin.

43. Seyla Benhabib, *Situating the Self: Gender, Community and Postmodernism in Contemporary Ethics* (New York: Routledge, 1992), 206.

44. Jan Flax, *Thinking Fragments: Psychoanalysis, Feminism, and Postmodernism in the Contemporary West* (Berkeley: University of California Press, 1990), 32.

45. As Vattimo maintains: "Even Lyotard, in order to explain that they (*the meta-narratives*) no longer have value, needs to tell a story. He falls back again upon a *narrational legitimation*" (Gianni Vattimo, *Tecnica ed esistenza* [Milano: Bruno Mondadori, 2002], 65).

46. Interestingly enough, plot and weave are one and the same word in Italian.

47. Whitehead, *Science and the Modern World*, 65, 91.

48. Alfred North Whitehead, *Symbolism: Its Meaning and Effect* (New York: Fordham University Press, 1985 [1927]), 39.

49. Alfred North Whitehead, *Modes of Thought* (New York: Free Press, 1968 [1938]), 98.

50. Latour, *Politics of Nature*, 43. Latour offers an intriguing analysis of the interdependency between multiculturalism and the assumption of nature as the common background, to which different cultures are supposed to relate. However, "non-Western cultures *have never been interested* in nature; they have never adopted it as a category; they have never found a use for it" (ibid.).

51. According to the new-liberal economic paradigm, "outer nature is considered as valueless objectivity, bare matter, inextinguishable or scarce source,

therefore as an object for knowledge, as a substrate for technological manipulation and production aims. Hence, . . . [the paradigm] is based on a mechanistic and atomistic conception of nature as it has been developed and supported throughout modernity by Descartes, Bacon, Kant and Marx" (Konrad Ott and Ralf Döring, *Theorie und Praxis starker Nachhaltigkeit* [Marburg: Metropolis, 2004], 106).

52. So-called core-shamanism is only one example of this temptation: "By not imitating any specific cultural tradition, but rather by training in underlying cross-cultural principles, core-shamanism is especially suited for utilization by Westerners who desire a relatively culture-free system that they can adopt and integrate into their contemporary lives" (Michael Harner, "Science, Spirits, and Core Shamanism," *Shamanism* 12, no. 1 [Spring–Summer 1999], available at http://www.shamanism.org/articles/1027871950.html).

53. I use the term "obligational" here in the sense in which Whitehead refers to the ongoing relations that connect all events and things in the universe. As an internal connection, it bears a certain degree of "obligation," of acting upon future possibilities. "Universality of truth arises from the universality of relativity, whereby every particular actual thing lays upon the universe the obligation of conforming to it" (Whitehead, *Symbolism*, 39). Yet this obligation takes the form of persuasion, lure, proposal, rather than being an inescapable destiny of derivation.

54. Gianni Vattimo, *Nichilismo ed emancipazione* (Milano: Garzanti, 2003), 39.

55. Gianni Vattimo, "Ontologia dell'attualità," in *Filosofia 87* (Roma-Bari: Laterza, 1988), 202.

56. Gianni Vattimo, *La fine della Modernità* (Milan: Garzanti, 1985), 180.

57. As Vattimo himself acknowledges, the term *Verwindung* was used by Heidegger only in a marginal way. By focusing on this concept for the development of his own "ontology of actuality," however, he renders it extremely fecund for further philosophical discussion. See ibid., 172.

58. *Sich ver-wählen.*

59. *Sich ver-sprechen.*

60. Vattimo, *La fine della Modernità*, 181.

61. Ibid., 180.

62. Whitehead would call it a fallacy of misplaced concreteness more than a mere error. Modern sciences are not sheer errors, but they eventually got trapped by the fallacy of their own abstractions. The fallacy consists precisely in substituting for the complexity of the concrete in its nonreducible internal relations the simple abstraction of isolated categories of explanation, which are then taken as the whole story. There is nothing wrong with abstraction as such as long as it remains inscribed within its limited and

narrow perspective and does not become a substitute for concreteness: "Science has always suffered from the vice of overstatement. In this way conclusions true within strict limitations have been generalized dogmatically into a fallacious universality" (Whitehead, *Function of Reason*, 27).

63. John Dewey presents Alexander Technique in his introduction to one of Alexander's books as follows: "After studying over a period of years Mr. Alexander's method in actual operation, I would stake myself upon the fact that he has applied to our ideas and beliefs about ourselves and about our acts exactly the same method of experimentation and of production of new sensory observations, as tests and means of developing thought, that have been the source of all progress in the physical sciences; and if, in any other plan, any such use has been made of the sensory appreciation of our attitudes and acts, if in it there has been developed a technique for creating new sensory observations of ourselves, and if complete reliance has been placed upon these findings, I have never heard of it. In some plans there has been a direct appeal to 'consciousness' (which merely registers bad conditions); in some, this consciousness has been neglected entirely and dependence placed instead upon bodily exercises, rectifications of posture, et cetera. But Mr. Alexander has found a method for detecting precisely the correlations between these two members, physical-mental, of the same whole, and for creating a new sensory consciousness of new attitudes and habits. It is a discovery which makes whole all scientific discoveries, and renders them available, not for our undoing, but for human use in promoting our constructive growth and happiness" (F. Matthias Alexander, *Constructive Conscious Control of the Individual* [London: Victor Gollancz, 1987], available at http://www.alexandercenter.com/jd/johndeweyccci.html).

64. Vattimo indeed intends Heidegger's *An-denken*—remembering the destiny of modern metaphysics—precisely in this sense, by connecting it to the idea of recovering (*Verwindung*). See Vattimo, *La fine della Modernità*, 181.

65. Latour, *Politics of Nature*, 109–11.

66. See Alicia H. Puleo, "Feminismo y Ecologismo," available at http://www.singenerodedudas.com/Archivos/000666.php.

67. "Thanks to the moralists, we can keep porous the fragile membrane that separates the collective from what it must be able to absorb in the future if it wants to produce a common world, a well-formed universe, a cosmos" (Latour, *Politics of Nature*, 160).

68. Latour asserts that contemporary ecology makes any concept of nature impossible. It certainly makes impossible a fixed, static, once-and-for-all established concept. Nevertheless, in a shifted way, we can still speak about nature. See Latour, *Politics of Nature*, 27.

69. Whitehead, *Concept of Nature*, 32.

70. "Again, the topic of every science is an abstraction from the full concrete happenings of natures" (Whitehead, *Modes of Thought*, 143).

TOWARD AN ETHICS OF BIODIVERSITY: SCIENCE AND THEOLOGY IN ENVIRONMENTALIST DIALOGUE | KEVIN J. O'BRIEN

1. Assuming that all theologians are driven at least in part by a moral impulse and all Christian ethicists are to some extent theologians, I intend the term "moral theology" to encompass the work of both Christian ethicists and theologians with both Catholic and Protestant backgrounds. I use "environmentalist" to refer to explicit advocacy on behalf of the nonhuman world as the context of human life. I am aware that this overemphasizes the distinction between human beings and our environments, and I am sympathetic to proposals for a more "ecological" ethics that incorporates both human beings and our natural environments. In this essay, however, I reserve "ecological" for reference to the scientific study of the processes and interconnections of life in its contexts, and accept the more common "environmentalist" to identify the ethical stances I am discussing.

2. Readers of this volume will be well aware that this definition creates as many questions as it answers, chief among them being what is meant by "nature." As I hope is demonstrated in this essay, I do not naïvely believe that there is a simple and straightforward definition of "nature" or of "biodiversity" as a characteristic of it. I recognize that "nature" is a deeply constructed concept contingent upon social and cultural context, but also believe theologically that it references a real creation that exists outside of human experience and informs our social constructions. For one take on the thorniness of such definitions, see Kate Soper, *What Is Nature? Culture, Politics, and the Non-Human* (Cambridge, Mass.: Oxford, 1995).

3. For example, in an address to the American Institute of Biological Sciences, conservation biologist Peter Raven argues for such a definition and dismisses any attempt to distinguish biodiversity from other aspects of ecological and environmentalist research: "There will be no adequate preservation of biodiversity without the attainment of ecological stability—sustainability—throughout the world. Such stability, in turn, will be attained only when there is a stable human population and when the needs of individual human beings everywhere are a matter of concern for us all" (Peter Raven, "The Politics of Preserving Biodiversity," *BioScience* 40, no. 10 [1990]: 770). Biodiversity is here presented as a thread connected to all

other environmental issues, and attention to it is a way to increase broader attention to the complexity and degradation of the natural world as a whole.

4. Throughout this essay I refer to "humanity" and "human beings" as a singular category. This is partly an affirmation of our unity as a species and partly a simplification for sake of space. It is important, however, to note that I do not intend to imply that the diversity within our species is unimportant or ethically irrelevant. This would be a gross oversimplification, particularly problematic insofar as it ignores the fact that the dominance of "human beings" over the nonhuman world is really the dominance of a relatively few people over many other human beings as well as vast portions of the natural world. In the larger research project of which this is one part, the importance of making scalar distinctions in the ways we understand all species, including our own, is much more central, but such exploration is beyond the scope of this essay.

5. Gen. 1:31. All biblical citations in this paper are from the New Revised Standard Version. Perhaps an even more fruitful source for a theological ethics of biodiversity would be the story of Noah and the preservationist commitment it implies. Still another resource arises in the last chapters of Job, in which God celebrates the intimate knowledge of animal behavior and the particular characteristics of two mysterious and majestic megafauna: Leviathan and Behemoth. Carol Newsom argues that this speech makes fundamentally ethical points, based on a "moral sense of nature" inherent in it. See Carol A. Newsom, "The Moral Sense of Nature: Ethics in Light of God's Speech to Job," *Princeton Seminary Bulletin* 15, no. 1 (1994).

6. Thomas Aquinas, *Summa Theologiae* (New York: Blackfriars, 1964), 47.1.ans.

7. James M. Gustafson, *A Sense of the Divine: The Natural Environment from a Theocentric Perspective* (Cleveland: Pilgrim, 1994), 99.

8. See Gustafson's introduction in H. Richard Niebuhr, *The Responsible Self: An Essay in Christian Moral Philosophy* (New York: Harper and Row, 1963), 14. It is clear, given the context of this quote, that Gustafson sees his approach to ethics as one that he has in large part inherited from Niebuhr rather than devised entirely himself.

9. Gustafson writes that any theological or ethical claim to "authority independent of particular premises of descriptions, explanations, and interpretations of events which [theologians] address" is, by definition, "not 'naturalistic'" (James M. Gustafson, *Intersections: Science, Theology, and Ethics* [Cleveland: Pilgrim, 1996], 1). It is important to note here that naturalism as I understand it does not necessarily *preclude* these other approaches to

theology. It emphasizes and prioritizes naturalistic explanations, but does not claim that naturalistic explanations are the only valid ones, or that all truths must be first explained naturalistically. It simply notes that there are naturalistic explanations for all phenomena, and argues that an ethics that remains fully accountable to the sphere of shared human experience of the world is worth doing.

10. James M. Gustafson, *An Examined Faith: The Grace of Self-Doubt* (Minneapolis, Minn.: Augsburg Fortress, 2004), 71.

11. Gustafson, *A Sense of the Divine*, 123.

12. In a recent book, Lisa Sideris offers a compelling critique of environmentalist ethicists and theologians who assume "ecology" as the basis of their ethical claims while ignoring the more disturbing implications of evolution. Whitney Bauman offers another example of such caution in his critique of Thomas Berry and Brian Swimme's *Universe Story* as an overly holistic and simplistic interpretation of cosmological and ecological perspectives. See Lisa H. Sideris, *Environmental Ethics, Ecological Theology, and Natural Selection* (New York: Columbia University Press, 2003), and Whitney Bauman, "A Social Ecofeminist Evaluation of Holism in the *Universe Story*," paper presented at the Fifth Trans-disciplinary Theological Colloquium (Drew University, New Jersey, 2005).

13. Edward O. Wilson and Frances M. Peter, *BioDiversity* (Washington, D.C.: National Academy Press, 1988).

14. For a critical perspective on this approach to ecology and Frederic Clements's centrality in it, see especially Daniel B. Botkin, *Discordant Harmonies: A New Ecology for the Twenty-First Century* (New York: Oxford University Press, 1990).

15. Dennis E. Jelinski, "There Is No Mother Nature—There Is No Balance of Nature: Culture, Ecology and Conservation," *Human Ecology* 33, no. 2 (2005): 271–88.

16. Donald Worster, *Nature's Economy: A History of Ecological Ideas*, 2nd ed. (New York: Cambridge University Press, 1994), chap. 17.

17. David Hawksworth and Mary Kalin-Arroyo, "Magnitude and Distribution of Biodiversity," in Vernon Heywood and Robert Watson, eds., *Global Biodiversity Assessment* (New York: Cambridge University Press, 1995), 107–92.

18. These three levels—genetic, species, and ecosystem—are widely but not universally accepted as the most important at which to measure biodiversity. As far as I know, the first published use of these three was in a book published by the Wilderness Society, but it claims to reflect an already existing tendency among ecologists. See Elliott Norse et al., *Conserving Biological Diversity in Our National Forests* (Washington, D.C.: The Wilderness

Society, 1986), 2. More contemporary examples of the popularity of these levels include Kevin J. Gaston and John I. Spicer, *Biodiversity: An Introduction*, 2nd ed. (Malden: Blackwell, 2004), 3–8, and Ian R. Swingland, "Definition of Biodiversity," in Simon Levin, ed., *Encyclopedia of Biodiversity* (San Diego: Academic Press, 2001).

19. Tsunemi Kubodera and Kyoichi Mori, "First-Ever Observations of a Live Giant Squid in the Wild," *Proceedings of the Royal Society* 272, no. 1575 (2005).

20. Bernard Wood and Paul Constantino, "Human Origins: Life at the Top of the Tree," in Joel Cracraft and Michael J. Donoghue, eds., *Assembling the Tree of Life* (New York: Oxford University Press, 2004), 526.

21. See especially Gaston and Spicer, *Biodiversity: An Introduction*, chap. 2.

22. Walter V. Reid and Kenton Miller, *Keeping Options Alive: The Scientific Basis for Conserving Biodiversity* (Washington, D.C.: World Resources Institute, 1989). Their literature review reveals that human activity is thought to increase extinction rates "between 100 and 1000 times."

23. David Tilman, "The Ecological Consequences of Changes in Biodiversity: A Search for General Principles," *Ecology* 80, no. 5 (1999): 1471.

24. Peter M. Vitousek and others, "Human Domination of Earth's Ecosystems," *Science* 277 (1997): 494.

25. Ibid., 495, 496, 498.

26. Sallie McFague, *The Body of God: An Ecological Theology* (Minneapolis: Augsburg Fortress, 1993), 185.

27. Dieter Hessel, "Introduction: Eco-Justice Theology After Nature's Revolt," in Dieter T. Hessel, ed., *After Nature's Revolt: Eco-Justice and Theology* (Minneapolis: Augsburg Fortress, 1992), 12.

28. Arguments to the contrary would suggest that there is an "anthropic principle" inherent in creation, a series of events that, over evolutionary time, made the emergence and thriving of the human species possible. This is certainly one possible interpretation of naturalistic data, but I am not convinced that the planet is clearly *more* suited to human beings than it is to any of the millions of other species now living. Roger Shinn makes this point well, asking why we would not instead assert a "viral principle" to explain the fact that "nature supports viruses that combat human life and health, forces that by all signs have a far greater staying power than we" (Roger Shinn, "The Mystery of the Self and the Enigma of Nature," in *Christianity in the 21st Century*, ed. Deborah A. Brown [New York: Crossroad, 2000], 98).

29. H. Paul Santmire, "The Genesis Creation Narratives Revisited: Themes for a Global Age," *Interpretation* 45, no. 4 (1991): 374–75.

30. See for instance Calvin B. DeWitt et al., *Caring for Creation: Responsible Stewardship of God's Handiwork* (Grand Rapids, Mich.: Baker, 1998), Drew Christiansen and Walter Grazer, *And God Saw That It Was Good: Catholic Theology and the Environment* (Washington, D.C.: United States Catholic Conference, 1996), and Sean McDonagh, *To Care for the Earth* (Santa Fe: Bear and Company, 1986).

31. Jay B. McDaniel, "The Garden of Eden, the Fall, and Life in Christ," in Mary Evelyn Tucker and John Grim, eds., *Worldviews and Ecology: Religion, Philosophy, and the Environment* (Maryknoll, N.Y.: Orbis, 1994), 74–75 (emphasis in original).

32. Reed F. Noss and Allen Cooperrider, *Saving Nature's Legacy: Protecting and Restoring Biodiversity* (Washington, D.C.: Island, 1994), 28–29.

33. I hope that I have articulated my naturalistic approach sufficiently enough that it is clearly distinct from creationism and other arguments that theology and science should or must have exactly matching worldviews. In my opinion, the most fundamental mistake creationists make is not the claim that God created the world or did so in a certain time frame, but rather the assertion that these resolutely *theological* claims should be taken seriously in *scientific* conversations.

34. For a book-length study of scientific advocacy for biodiversity preservation, see David Takacs, *The Idea of Biodiversity: Philosophies of Paradise* (Baltimore: Johns Hopkins University Press, 1996).

35. Edward O. Wilson, *The Diversity of Life* (Cambridge, Mass.: Harvard University Press, 1992), 286, 289.

36. See for instance Gretchen C. Daily, *Nature's Services: Societal Dependence on Natural Ecosystems* (Washington, D.C.: Island Press, 1997), and Gordon H. Orians et al., eds., *The Preservation and Valuation of Biological Resources* (Seattle: University of Washington Press, 1990).

37. David Tilman and Joel E. Cohen, "Biosphere II and Biodiversity—the Lessons So Far," *Science* 274 (1996).

38. Kent H. Redford and Brian D. Richter, "Conservation of Biodiversity in a World of Use," *Conservation Biology* 13, no. 6 (1999): 1254.

39. James R. Miller, "Biodiversity Conservation and the Extinction of Experience," *TRENDS in Ecology and Evolution* 20, no. 8 (2005): 430.

40. Paul Ehrlich, "Bioethics: Are Our Priorities Right?" *BioScience* 53, no. 12 (2003): 1211.

41. Anna Peterson, *Being Human: Ethics, Environment, and Our Place in the World* (Berkeley: University of California Press, 2001), 233.

42. Larry Rasmussen, "Eco-Justice: Church and Community Together," in Dieter T. Hessel and Larry L. Rasmussen, eds., *Earth Habitat: Eco-Injustice and the Church's Response* (Minneapolis: Augsburg Fortress, 2001), 8.

INDIGENOUS KNOWING AND RESPONSIBLE LIFE IN THE WORLD | JOHN GRIM

1. See Julian Berger, *The Gaia Atlas of First Peoples* (London: Gaia Books, 1990), and R. H. Barnes, Andrew Gray, and Benedict Kingsbury, eds. *Indigenous Peoples of Asia* (Ann Arbor, Mich.: Association for Asian Studies Monographs, 1995), monograph no. 48.

2. See also Sadruddin Aga Khan and Hassan bin Talal, *Indigenous Peoples, A Global Quest for Justice: A Report for the Independent Commission on International Humanitarian Affairs* (London: Zed, 1987).

3. I am using "negritude" here in the sense developed by the Martinique poet Aimé Césaire in writings such as *Discourse on Colonialism* (New York: New York University Press, 2000 [1955]).

4. Debroah Bird Rose et al., eds., *Country of the Heart: An Indigenous Australian Homeland* (Canberra: Aboriginal Studies Press for the Australian Institute of Aboriginal and Torres Strait Islander Studies, 2002), 15.

5. Richard Peet and Michael Watts, *Liberation Ecologies: Environment, Development, Social Movements* (London and New York: Routledge, 1996), 7.

6. See, for example, Stan Stevens, ed., *Conservation Through Cultural Survival: Indigenous Peoples and Protected Areas* (Washington, D.C.: Island, 1997); and Mac Chapin, "A Challenge to Conservationists," in *World Watch* 17, no. 6 (November–December 2004): 17–31.

7. See John Grim, ed., *Indigenous Traditions and Ecology: The Interbeing of Cosmology and Community* (Cambridge, Mass.: Harvard Divinity School Center for the Study of World Religions, 2001); Marie Battiste, ed., *Reclaiming Indigenous Voice and Vision* (Vancouver: University of British Columbia Press, 2000); and Waziyatawin A. Wilson, ed., "The Recovery of Indigenous Knowledge," special issue of *American Indian Quarterly* 28, nos. 3–4 (Summer–Fall, 2004): 359–633.

8. Roy Ellen and Helen Harris, "Introduction," in *Indigenous Environmental Knowledge and Its Transformations: Critical Anthropological Perspectives*, Roy Ellen, Peter Parkes, Alan Bicker, eds. (Amsterdam: Harwood Academic Publishers, 2000), 6–20.

9. See, for example, the arguments made by Darrell A. Posey, "Intellectual Property Rights and the Sacred Balance: Some Spiritual Consequences from

the Commercialization of Traditional Resources," and Thomas Greaves, "Contextualizing the Environmental Struggle," in Grim, *Indigenous Traditions and Ecology*, 3–23, 25–46.

10. Arun Agrawal, "Dismantling the divide between indigenous and scientific knowledge," *Development and Change* 26 (1995): 415.

11. See, for example, Gerald R. Alfred, *Heeding the Voices of Our Ancestors: Kahnawake Mohawks Politics and the Rise of Nationalism* (Oxford: Oxford University Press, 1995); M. Annette Jaimes, ed., *The State of Native America: Genocide, Colonization and Resistance* (Boston: South End, 1992); Franke Wilmer, *The Indigenous Voice in World Politics* (Newbury Park, Mass.: Sage, 1993); and Linda T. Smith, *Decolonizing Methodologies: Research and Indigenous Peoples* (London: Zed, 1999); Radha Jhappan, "Global Community? Supranational Strategies of Canada's Aboriginal Peoples," *Journal of Indigenous Studies* 3, no. 1 (1992): 59–97; and Douglas E. Sanders, *The Formation of the World Council for Indigenous Peoples* (Copenhagen: International Working Group for Indigenous Affairs, 1977), document 29.

12. Pamela J. Stewart and Andrew Strathern, "Indigenous knowledge confronts development among the Duna of Papua New Guinea," in Alan Bicker, Paul Sillitoe, and Johan Pottier, eds., *Development and Local Knowledge: New Approaches to Issues in Natural Resources Management, Conservation, and Agriculture* (London and New York: Routledge, 2004): 51–63.

13. Gregory Cajete, *Native Science: Natural Laws of Interdependence* (Santa Fe: Clear Light, 2000), 2.

14. Beverly R. Ortiz, "Wild Gardens: How Native Americans Shaped Local Landscapes," *BayNature* (January–March, 2006): 23–24.

15. Ellen and Harris, "Introduction," 23.

16. Gisday Wa and Delgam Uukw, *The Spirit of the Land: The Opening Statement of the Gitksan and Wets'uwetén Hereditary Chiefs in the Supreme Court of British Columbia* (Gabrola, 1987). Quoted from David Suzuki and Peter Knudtson, eds., *Wisdom of the Elders: Sacred Native Stories of Nature* (New York and Toronto: Bantam, 1992), 158.

17. Henry S. Sharp, *Loon: Memory, Meaning, and Reality in a Northern Dene Community* (Lincoln and London: University of Nebraska Press, 2001), 63.

18. Robin Ridington, *Little Bit Know Something: Stories in a Language of Anthropology* (Iowa City: University of Iowa Press, 1990), 127–28.

19. Marlene B. Castellano, "Updating Aboriginal Traditions of Knowledge," in George Dei, Bud Hall, and Dorothy Rosenberg, eds., *Indigenous Knowledge in Global Contexts* (Toronto: University of Toronto Press, 2000), 21–36.

20. See Sarah Whitecalf, H. Christopher Wolfart, and Freda Anenakew, *kinêhiyâwiwininaw nêhiyawêwin: The Cree Language Is Our Identity* (Winnipeg: University of Manitoba Press, 1993).

21. James Henderson, "Ayukpachi: Empowering Aboriginal Thought," in Battiste, *Reclaiming Indigenous Voice and Vision*, 263.

22. For further reading see Vandana Shiva, *Protect or Plunder?: Understanding Intellectual Property Rights* (New York: Zed, 2002) or *Biopiracy: The Plunder of Nature and Knowledge* (Cambridge, Mass.: South End Press, 1997).

23. Wade Davis, *Shadows in the Sun: Travels to Landscapes of "Spirit and Desire"* (New York: Broadway, 1998).

24. Jeremy Narby, *The Cosmic Serpent: DNA and the Origins of Knowledge* (New York: Jeremy P. Tarcher/Putnam, 1998).

25. Javier G. Silva, "Religion, Ritual, and Agriculture among the Present-Day Nahua of Mesoamerica," in John Grim, *Indigenous Traditions and Ecology*, 304.

26. *Codice Florentino: Book VI y X da la Colección Palatina da la Biblioteca Medicea Laurenziana*, 3 vols. (Mexico: Secretaria de Gobernación, Archivo general de la Nación, 1979).

27. Marie Battiste and James Henderson, *Protecting Indigenous Knowledge and Heritage* (Saskatoon: Purich, 2000), 35, cited in Deborah McGregor, "Coming Full Circle: Indigenous Knowledge, Environment, and Our Future," in "The Recovery of Indigenous Knowledge," special issue of *American Indian Quarterly* 28, nos. 3–4 (Summer–Fall 2004): 390.

28. Linda Smith, "Kaupapa Maori Research," in Battiste, *Reclaiming Indigenous Voice and Vision*, 234–35.

THE PREORIGINAL GIFT—AND OUR RESPONSE TO IT | ANNE PRIMAVESI

1. Emmanuel Levinas, *Autrement Qu'Être ou Au-Delà de L'Essence* (Paris: Kluwer Academic, 1978), 24.

2. Ibid.

3. Marcel Mauss, *The Gift* (London: Routledge, 1990), 2–9.

4. Stephen H. Webb, *The Gifting God: a Trinitarian Ethics of Excess* (New York and Oxford: Oxford University Press, 1996), 4.

5. David Abram, *The Spell of the Sensuous: Perception and Language in a More-than-Human World* (New York: Pantheon, 1996), ix.

6. Colin Tudge, *The Secret Life of Trees* (London: Penguin, 2005), 76ff.

7. John Robert McNeill, *Something New Under the Sun: An Environmental History of the Twentieth-Century World* (New York and London: Norton, 2000), 26–31.

8. Anne Primavesi, *Gaia's Gift: Earth, Ourselves, and God after Copernicus* (London and New York: Routledge, 2003), 124.

9. Ibid., 117–22.

10. Arindam Chakrabarti, "Debts and Dwellings: The Vedas and Levinas on the Ethical Metaphysics of Hospitality," in *Das Antlitz des "Anderen,"* ed. S. Fritsch-Oppermann (Loccum: Evangelische Akademie), 127–52, quotation on 145.

11. Ernst Cassirer, *Language and Myth* (New York: Dover, 1946), 10–15; and, Susanne K. Langer, *Philosophy in a New Key: A Study in the Symbolism of Reason, Rite, and Art* (Cambridge, Mass.: Harvard University Press, 1951), 140ff.

12. Ernst Cassirer, *The Philosophy of Symbolic Forms*, vol. 4 (New Haven and London: Yale University Press, 1996), 7–16.

13. Eva L. Jablonka and Marion J. Lamb, *Evolution in Four Dimensions: Genetic, Epigenetic, Behavioral, and Symbolic Variation in the History of Life* (Cambridge, Mass.: Massachusetts Institute of Technology Press, 2005), 193–98; and, Langer, *Philosophy in a New Key*, 26ff.

14. Anne Primavesi, *Sacred Gaia* (London and New York: Routledge, 2000), 7ff.

15. Primavesi, *Gaia's Gift*, 125.

16. Primavesi, *Sacred Gaia*, 160–61, and idem, *Gaia's Gift*, 86, 125–35.

17. Langer, *Philosophy in a New Key*, 131–46; and Jablonka and Marion, *Evolution in Four Dimensions*, 203–6.

18. Robert W. Funk, *The Five Gospels* (New York: Macmillan, 1993), 167.

19. Partha Dasgupta, *Human Well-Being and the Natural Environment* (Oxford: Oxford University Press, 2001), x.

20. Webb, *The Gifting God*, 6.

21. Ibid., 7.

22. Calvin O. Schrag, *God as Otherwise than Being: Towards a Semantics of the Gift* (Evanston, Ill.: Northwestern University Press, 2002), 107.

23. Jacques Derrida, *The Gift of Death* (Chicago: University of Chicago Press, 1995), 29.

24. Primavesi, *Sacred Gaia*, 176ff.

25. Primavesi, *Gaia's Gift*, 112ff.

26. Schrag, *God as Otherwise than Being*, 108.

27. Primavesi, *Gaia's Gift*, 23.

28. Schrag, *God as Otherwise than Being*, 109ff.

29. Derrida, *The Gift of Death*, 109–12.

30. Anne Primavesi, *Making God Laugh: Human Arrogance and Ecological Humility* (Santa Rosa, Calif.: Polebridge, 2004), 17ff.

31. Webb, *The Gifting God*, 71–75; and Primavesi, *Sacred Gaia*, 175ff.

32. John D. Caputo, *The Prayers and Tears of Jacques Derrida* (Bloomington: Indiana University Press, 1997), 176ff.

33. Simone Weil, *Gravity and Grace* (London: Routledge, 1952), 119; emphasis added.

34. Langer, *Philosophy in a New Key*, 143.

35. Derrida, *The Gift of Death*, 29.

36. Primavesi, *Sacred Gaia*, 29–32, 177.

PROMETHEUS REDEEMED? FROM AUTOCONSTRUCTION TO ECOPOETICS | KATE RIGBY

1. See, for example, Catherine Keller, *Face of the Deep: A Theology of Becoming* (London and New York: Routledge, 2003), and Mark I. Wallace, *Finding God in the Singing River: Christianity, Spirit, Nature* (Minneapolis, Minn.: Augsburg Fortress, 2005). See also Anne Elvey, *An Ecological Feminist Reading of the Gospel of Luke: A Gestational Paradigm* (Lewiston, Maine: Edwin Mellen, 2005), and Mark Manolopoulos, *If Creation Is a Gift: Towards an Eco/theo/logical Aporetics* (PhD diss., Monash University, 2003). Many thanks to Catherine Keller and David Wood for their invaluable comments on an earlier incarnation of this essay as a contribution to the 2005 Ecosophia Symposium at Drew University, New Jersey.

2. J. W. von Goethe, "Prometheus," in *Goethe: Selected Poems*, ed. Christopher Middleton and trans. Michael Hamburger (Boston: Suhrkamp/Insel, 1983), 28–31. Further citations from this work will appear in the text.

3. "Who helped me / Against the Titan's arrogance? / Who rescued me from death, / From slavery? / Did not my holy and glowing heart, / Unaided, accomplish all?" ("Prometheus," ll. 27–32).

4. Immanuel Kant, "Answer to the Question: What is enlightenment?" (1784), in *Practical Philosophy*, ed. and trans. Mary J. Gregor (Cambridge: Cambridge University Press, 1996), 17.

5. See, for example, Pheng Cheah's brilliant critique of Judith Butler in "Mattering," *Diacritics* 26, no. 1 (1996): 108–39. With reference to Butler's *Bodies that Matter: On the Discursive Limits of Sex* (London: Routledge, 1993), Cheah observes that "matter is invested with dynamism and said to be open to contestation only because the matter concerned is the product of socio-historical forms of power, that is, *of the human realm*" (113). The result is to replace natural determinism with cultural determinism, as "the constraints of structure or construction on transformative rational agency seem to replicate the limitations or weightiness of nature" (108).

6. James Watson, who, with Francis Crick, famously discovered the structure of DNA, and subsequently inaugurated the human genome project, on "DNA," screened on Australian Broadcasting Corporation national television, August 11, 2005. Watson is dedicated to the use of genetic engineering for eugenic purposes.

7. Val Plumwood, "Nature as Agency and the Prospects for a Progressive Naturalism," *Capitalism, Nature, Socialism* 12, no. 4 (December 2001): 3.

8. Freya Mathews, *For Love of Matter: A Contemporary Panpsychism* (Albany: State University of New York Press, 2003).

9. Freya Mathews, *Reinhabiting Reality: Towards a Recovery of Culture* (Albany: State University of New York Press, 2005), 12.

10. Mathews, *For Love of Matter*, 170–71.

11. Jane Bennett, "The Force of Things: Steps Toward an Ecology of Matter," *Political Theory* 32, no. 3 (June 2004): 351.

12. Ibid., 375, 365.

13. Wallace, *Finding God*, 109.

14. Katherine Hayles, "Searching for Common Ground," in Michael E. Soulé and Gary Lease, ed., *Reinventing Nature: Responses to Postmodern Deconstruction* (Washington D.C.: Island Press, 1995), 53.

15. Elizabeth Grosz, *Time Travels: Feminism, Nature, Power* (London: Allen and Unwin, 2005), 4, and idem, *The Nick of Time: Politics, Evolution and the Untimely* (London: Allen and Unwin, 2004).

16. Grosz, *Time Travels*, 48.

17. Ilya Prigogine and Isabelle Stengers, *Order out of Chaos: Man's New Dialogue with Nature* (London: Fontana, 1990), 50.

18. Keller, *Face of the Deep*, 195.

19. Much of this new work is referenced in Kate Rigby, "The Rebirth of Nature," in *Topographies of the Sacred: The Poetics of Place in European Romanticism* (Charlottesville and London: University of Virginia Press, 2004), 17–52.

20. Keller, *Face of the Deep*, 222. I would like to share this generous reading, but I cannot help wondering whether Keller passes too quickly over the connotations of the Hebrew *rada*, which is often translated as "dominion" in Genesis 1:26, while eliding the even more dominological implications of "subdue" (*kabas*) in 1:28. According to Theodore Hiebert, the former was used primarily for rule over Israel's enemies, while the latter describes "the actual act of subjugation, or forcing another into a subordinate position." He nonetheless adds (in a model example of ecological contextualization) that the "particular harshness of the term for the human-earth relationship

in Genesis 1 may be best understood in the context of the particular harshness of subsistence agriculture in the Mediterranean highlands that provided the livelihood of the priests' constituency" (Hiebert, "The Human Vocation: Origins and Transformations in Christian Traditions," in *Christianity and Ecology: Seeking the Well-Being of Earth and Humans*, ed. Dieter T. Hessel and Rosemary Radford Ruether (Cambridge, Mass.: Harvard University Center for the Study of World Religions, 2000), 137.

21. In the interests of a more literal rendering, I have adapted Vernon Watkins's translation in Middleton's edition of Goethe's *Selected Poems*, 83–85. Further citations from this work will appear in the text.

22. Anne Primavesi, *Gaia's Gift* (London: Routledge, 2003).

23. I take the concept of "double death" from Australian anthropologist Deborah Bird Rose in *Reports from a Wild Country: Ethics for Decolonisation* (Sydney: University of New South Wales Press, 2004), 175–76.

24. Tony Kelly, "Meditating on Death," *St. Mark's Review* 155 (Spring 1993): 33–34. See also Kelly's wonderful work of ecotheology, *An Expanding Theology: Christian Faith in a World of Connections* (Sydney: E. J. Dwyer, 1993), which deserves to be better known.

25. Vergil, *Aeneid*, 1:462, quoted in Kelly, "Meditating on Death," 31.

26. My information on this work comes from Jason Starr's fascinating documentary film, "What the Universe Tells Me: Unravelling the Mysteries of Mahler's Third Symphony" (2003).

27. J. W. von Goethe, *Faust: Part Two*, trans. David Luke (Oxford: Oxford University Press, 1994).

28. Erazim Kohák, *The Embers and the Stars: A Philosophical Enquiry into the Moral Sense of Nature* (Chicago: Chicago University Press, 1984), 5.

29. This felicitous phrase comes from Rod Giblett, *Postmodern Wetlands: Culture, History, Ecology* (Edinburgh: Edinburgh University Press, 1996), 105.

30. Luke translates *das ewig Weibliche* as "eternal Womanhood."

31. For an extended, and somewhat more critical, reading of *Faust* in relation to Goethean science, see Kate Rigby, "Freeing the Phenomena: Goethean Science and the Blindness of Faust," *ISLE* 7, no. 2 (2000): 24–42. See also Gernot Böhme's ecophilosophical reading, *Goethes Faust als philosophischer Text* (Kusterdingen: Die Graue Edition, 2005).

32. J. W. von Goethe, *Scientific Studies*, ed. and trans. Douglas Miller (Princeton: Princeton University Press, 1988), 171.

33. My emphasis on 'falling short' counters Heidegger's poetics of adequation and is informed by the work of Jean-Louis Chrétien, especially *The Call and*

the Response, trans. Anne A. Davenport (New York: Fordham University Press, 2004). See also Kate Rigby, "Earth, World, Text: The (Im)possibility of Ecopoiesis," *New Literary History* 35, no. 3 (Summer 2004): 427–42.

TOWARD A DELEUZE-GUATTARIAN MICROPNEUMATOLOGY OF SPIRIT-DUST | LUKE HIGGINS

1. Sallie McFague, *The Body of God: An Ecological Theology* (Minneapolis: Augsburg Fortress, 1993).

2. Ibid., 20.

3. Bruno Latour, *Politics of Nature: How to Bring the Sciences Into Democracy*, trans. Catherine Porter (Cambridge, Mass.: Harvard University Press, 2004).

4. Latour outlines a new bicameral structure for politics that would divide itself along different lines than those of "fact" and "value." The division of labor is, rather, split between a process of accounting for the beings of the world (with what entities are we dealing?) and a process of composing them into a collective world (how might we all live together?). Each of these phases would involve *both* scientists and politicians. See Latour, *Politics of Nature*, 91–127.

5. Gilles Deleuze and Felix Guattari, *A Thousand Plateaus: Capitalism and Schizophrenia*, trans. Brian Masumi (Minneapolis: University of Minnesota Press, 1987). Employing thinkers such as Deleuze and Guattari for the constructive formulation of an ecotheology may appear suspect given these thinkers' open hostility to religious thought systems. I believe their critique of transcendence, however, does not ring the death knell for faith-religion but, rather, can lead to a new way of envisioning the divine. Instead of identifying the divine with the transcendent "plane of organization," as they do, I will instead look to their notion of the "plane of immanence" for insights into the divine nature. This proposal will be developed particularly in the second section of this essay.

6. Latour coins the word "pluriverse": "Since the word "*uni*-verse" has the same deficiency as the word "nature" (for unification has come about without due process), the expression "pluriverse" is used to designate propositions that are candidates for common existence before the process of unification in the common world" (Latour, *Politics of Nature*, 246).

7. Deleuze and Guattari, *A Thousand Plateaus*, 8.

8. Ibid., 40.

9. Gilles Deleuze and Felix Guattari, *What is Philosophy?*, trans. Hugh Tomlinson and Graham Burchell (New York: Columbia University Press, 1994), 118.

10. Deleuze and Guattari, *A Thousand Plateaus*, 40.

11. Deleuze-Guattari's critique of mimesis may correlate to some degree with Alfred North Whitehead's critique of substance-accident metaphysics in his doctrine of internal relations. Shared by both (sets of) thinkers is a commitment to the idea that what we are on the most basic level is what we become through our relationships. See Alfred N. Whitehead, *Process and Reality, Corrected Edition*, ed. David Ray Griffin and Donald Sherburne (New York: Free Press, 1978).

12. "The becoming-animal always involves a pack, a band, a population, a peopling, in short, a multiplicity" (Deleuze and Guattari, *A Thousand Plateaus*, 239).

13. Ibid., 274–75.

14. Ibid., 305.

15. Deleuze and Guattari, *A Thousand Plateaus*, 251.

16. Ibid., 253.

17. See note 5, above.

18. Karen Baker-Fletcher, *My Sister, My Brother: Womanist and Xodus God-talk* (Maryknoll, N.Y.: Orbis, 1997), 86.

19. Mark Wallace, *Fragments of the Spirit: Nature, Violence, and the Renewal of Creation* (Harrisburg, Pa.: Trinity Press International, 2002), 153.

SPECTERS OF DERRIDA: ON THE WAY TO ECONSTRUCTION | DAVID WOOD

1. I hint at this in my "Thinking with Cats," in *Animal Philosophy*, ed. Peter Atterton and Matthew Calarco (London and New York: Continuum, 2004).

2. See my "Comment ne pas manger: Deconstruction and humanism," in *Animal Others*, ed. Peter Steeves (Albany: State University of New York Press, 1999), and Jacques Derrida, "The Animal that Therefore I Am (More to Follow)," trans. David Wills, *Critical Inquiry* 28 (Winter 2002): 369–418.

3. It seems, too, that many Christian fundamentalists are allergic to this term. It is not clear whether their difficulty comes from its historical connection with liberal politics that they reject, or from their resistance to the need to refocus relations of obligation and dependency away from God and toward the natural world. This scene is increasingly complicated by the emergence within evangelical groups of a commitment to "creation care." The question of "humanism" is a vexed one. There are those who seem to have proposed that as the earth was better off, and would be better off again, with 10 percent of the current human population. This view has been attributed to poet Gary Snyder, who certainly did argue for "A healthy and

spare population of all races, much less in number than today" (Snyder, "Four Changes" in *Four Changes* [Chicago: Robert Shapiro, 1969 (1970)]), and who also wrote: "The whole population issue is fraught with contradictions, but the fact stands that by standards of planetary biological welfare there are already too many human beings" (Snyder, *A Place in Space* [New York: Counterpoint, 1996]). But it takes a particularly attenuated mind to conclude from this that Snyder is advocating the forced decimation of the human species. What is certainly worthy of discussion is whether the world is a better place, other things being equal, if there are more (happy) people in it. At issue is whether or when it makes sense to add up happinesses.

4. Jacques Derrida, *Specters of Marx: The State of the Debt, the Work of Mourning, and the New International*, trans. Peggy Kamuf (New York and London: Routledge, 1994), 85.

5. These "plagues" are: (1) Unemployment (in a new sense, social inactivity); (2) exclusion of the stateless; (3) economic war; (4) contradictions of the free market; (5) burden of national debt; (6) arms industry; (7) nuclear proliferation; (8) interethnic wars; (9) mafia and drug cartels (phantom states); and (10) limits of concept and practice of international law.

6. The seven plagues are: (1) Land: ugly and painful sores broke out on the people who had the mark of the beast and worshiped his image (Rev. 16:2); (2) Sea: it turned into blood like that of a dead man, and every living thing in the sea died (Rev. 16:3); (3) Rivers and Springs: they became blood (Rev. 16:4); (4) Sun: the sun was given power to scorch people with fire. They were seared by the intense heat and they cursed the name of God, who had control over these plagues, but they refused to repent and glorify him (Rev. 16:8); (5) Throne of the beast: his kingdom was plunged into darkness. Men gnawed their tongues in agony and cursed the God of heaven because of their pains and their sores, but they refused to repent of what they had done (Rev. 16:10); (6) Great river Euphrates: its water was dried up to prepare the way for the kings from the East (Rev. 16:12); (7) Air: The great city split into three parts, and the cities of the nations collapsed. God remembered Babylon the Great and gave her the cup filled with the wine of the fury of his wrath. Every island fled away and the mountains could not be found. From the sky huge hailstones of about a hundred pounds each fell upon men. And they cursed God on account of the plague of hail, because the plague was so terrible (Rev. 16:19).

7. For the record, Derrida himself quickly accepted this suggestion when I first made it in my paper "Globalization and Freedom," presented at the "Returns of Marx" conference, Paris, March 2003.

8. "Philosophy in a Time of Terror," with Giovanni Borradori, in *Philosophy in a Time of Terror: Dialogues with Jürgen Habermas and Jacques Derrida* (Chicago: University of Chicago Press, 2003). He points out that the hijackers used U.S. planes, U.S. flight-training facilities, and belonged to Bin Laden's organization, which was originally trained and financed by the United States. Moreover, these attacks were a response to perceived U.S. economic and military activity.

9. Such as "Structure, Sign and Play in the Human Sciences," in Jacques Derrida, *Writing and Difference* trans. Alan Bass (Chicago: University of Chicago Press, 1978).

10. Martin Heidegger, *Being and Time*, trans. Joan Stambaugh (Albany: State University of New York Press, 1996).

11. We invaded Iraq for its oil (2003), and we are entering the age of oil-shortage. Yet an accelerating need for energy suggests that we will need to find cheap substitutes for oil, which may make the whole quest for oil-dominance redundant.

12. Quoted in Cyril Barrett, *Wittgenstein on Ethics and Religious Belief* (Oxford: Basil Blackwell, 1991).

13. See Jacques Derrida, "Faith and Knowledge," in Derrida, *Acts of Religion* (London and New York: Routledge, 2002), and Jacques Derrida, *The Gift of Death* (Chicago: University of Chicago Press, 1995).

14. There are now activists pursuing this agenda, such as the Reverend Jim Ball, for example, and his Evangelical Environmental Network.

15. The extreme version of this view is captured by the idea that Jesus will return to save us when we have chopped down the last tree.

16. This is itself an extension of Nietzsche's sense that we have not got rid of God if we still believe in grammar, that is, in the idea of some logical or formal schema underlying the world. Our claim about what would be a misuse of God is not meant to rule out other grounds for belief. It is parallel to Sartre's argument against Husserl (in *The Transcendence of the Ego* [New York: Noonday, 1957]) that not only do we not need a transcendental ego to unify the field of consciousness, it would get in the way of the actual practical ways consciousness is unified.

17. Michel Foucault, "Nietzsche, Genealogy, History," in *Language, Counter-Memory, Practice* (Ithaca: Cornell University Press, 1977).

18. Global Biodiversity Strategy, WRI, IUCN, United Nations Environmental Program, 1992. A daily average loss of 137 species has been estimated elsewhere.

19. For an excellent analysis of the conceptual and political incoherence of the ecofascism "smear," see David Orton's *Green Web Bulletin* #68, "Ecofascism: What Is It? A Left Biocentric Analysis."

20. "Operation Iraqi Freedom" makes this clear.

21. Even Genesis offers another vision: "Then the LORD God took the man and put him in the garden of Eden to tend ["dress," KJV] and keep it" (2:15). This greatly modifies the force of "have dominion" and "subdue it" from Genesis 1:26 and 1:28! Tend (Hebrew, *'abad*) means "to work or serve," and thus in reference to the ground or a garden, it can be defined as "to till or cultivate." It possesses the nuance seen in the KJV's "dress": implying adornment, embellishment, and improvement. "Keep" (Hebrew, *shamar*) means "to exercise great care over." In the context of Genesis 2:15, it expresses God's wish that humankind, in the person of Adam, "take care of," "guard," or "watch over" the garden. A caretaker maintains and protects his charge so that he can return it to its owner in as good or better condition than when he received it. This note draws substantially from Richard T. Ritenbaugh, "The Bible and the Environment: Prophecy Watch," *Forerunner* (February 1999).

22. Gen. 1:28

23. Catherine Keller, *Face of the Deep* (London: Routledge, 2003), 137.

24. In Heidegger, *Being and Time*.

25. See "Language," in *Poetry, Language, Thought* (New York: Harper and Row, 1971), 210.

26. Jacques Derrida, *Of Grammatology* (Baltimore: Johns Hopkins University Press, 1976), 158.

27. Thomas Berry, *The Great Work* (New York: Bell Tower, 1999), 100.

28. The worry about deploying the value of freedom in this way obviously extends to both military conquest and subordination of other peoples, and to slavery and patriarchy, each of which have been justified in terms of subordinating a slave mentality to one that can exercise leadership.

29. "Darwin may have had this third option in mind when he wrote his friend Joseph Hooker: 'It is really laughable to see what different ideas are prominent in various naturalists' minds, when they speak of "species"; in some, resemblance is everything and descent of little weight—in some, resemblance seems to go for nothing, and Creation the reigning idea—in some, sterility an unfailing test, with others it is not worth a farthing. It all comes, I believe, from trying to define the indefinable'" (C. Darwin to J. Hooker, December 24, 1856, quoted in Marc Ereshefsky, s.v., "Species," *Stanford Encyclopedia of Philosophy*, available online at http://plato.stanford.edu/entries/species/, as of March 17, 2006.

30. I can imagine someone arguing that now that we have conquered genetic sequencing, we do not need actual other species any more. If we ever need them, we can recreate them. Three objections: if species disappear before we discover them, as mostly happens, then they will take their code with them into oblivion; the technology of gene mapping is *not* the same as life-creation—how would we know what life-forms to recreate?; and, finally, all this presupposes sufficient continuing health of the planet to sustain these labs.

31. Is not human freedom an *essential* difference? Who is not tempted to say this? And yet we do need to ask: essential for what? For legitimating colonial, patriarchal, speciesist violence when we claim to be delivering freedom to our fellow humans? Can we really separate our ontological intuitions from their political efficacy? And do we really know what we mean by "freedom"?

32. In Derrida, *Writing and Difference*.

33. Steven M. Wise, *Rattling the Cage: Toward Legal Rights for Animals* (New York: Perseus, 2000), argues the case for rights for animals, and legal personhood for chimps and bonobos.

34. Jacques Derrida, *The Other Heading*, trans. Pascale-Anne Brault and Michael B. Naas (Bloomington: Indiana University Press, 1992).

35. In this essay, I mean by "living beings" both plants and animals, as well as all other creatures from single-celled organisms on upward. I admit to mammalocentric tendencies, but I am fighting to overcome them. And I am assuming, too, that there are other proper objects of environmental concern, such as species-survival, biodiversity, habitats, ecological health, and the like.

36. We would also want to welcome the destabilizing of our assured sense of the human that we cannot avoid confronting in the writing of Nietzsche, for whom "man is a rope, tied between beast and overman— a rope over an abyss. . . . What is great in man is that he is a bridge and not an end" (Friedrich Nietzsche, *Thus Spake Zarathustra* [Harmondsworth: Penguin, 1961], prologue 4). But see also Donna Haraway's speculations in "A Cyborg Manifesto: Science, Technology, and Socialist-Feminism in the Late Twentieth Century," in *Simians, Cyborgs and Women: The Reinvention of Nature* (New York and London: Routledge, 1991), 149–81.

37. Jacques Derrida, "Force of Law: The Mystical Foundation of Authority," in *Deconstruction and the Possibility of Justice*, ed. Drucilla Cornell et. al. (London and New York: Routledge, 1992), 27.

38. Richard Sylvan and David Bennett, *The Greening of Ethics* (Tucson: University of Arizona Press, 1995), 169.

39. For a brilliant and sustained critique of such uses of discount rates, and the limits of classical economic theory in general, see Herman E. Daly and John B. Cobb, Jr., *For the Common Good: Redirecting the Economy toward Community, the Environment, and a Sustainable Future* (Boston: Beacon Press, 1994).

40. Derrida, *Of Grammatology*, 158–59.

41. He has had other critics, of course, who attacked him from the other direction. Jean-Luc Ferry's *The New Ecological Order* is a diatribe against what he claims are the fascistic implications of the antihumanism of the Generation of '68, in particular Deleuze and Guatarri, and Serres, but also Derrida and Foucault. (See a response by Verena Conley in *Ecopolitics* (London: Routledge, 1996). There seems to be a confusion with some of the extreme pronouncements of Deep Ecology.

42. By a "nonreductive" materialism, I mean one that takes seriously the complex ways in which things and processes are organized, ways that cannot be reduced to vertical structures of domination. And which allow for new possibilities of eventuation. It becomes clear that certain *spiritual* traditions, as well as process theology, have long functioned as shelters against the tendency to facile reductionism. The effort in this essay to draw together deconstruction and environmentalism at the same time opens a dialogue with the best theological efforts to address our ecological situation.

43. See Bruno Latour, "Is there a Non-Modern Style?" *Domus* (January 2004).

44. Jacques Derrida, *Limited Inc.* (Evanston, Ill.: Northwestern University Press, 1988).

45. By a logic that strangely reverses this structure, there is a certain convergence between what Derrida is saying about Rousseau, and what Rousseau himself already says, as if Derrida's "supplement" to Rousseau ends up showing that Rousseau knew it all along, and did not need Derrida to bring it out.

46. Bill McKibben, *The End of Nature* (New York: Random House, 1989).

47. Donald Worster, *The Wealth of Nature: Environmental History and the Ecological Imagination* (New York: Oxford University Press, 1993), 3.

48. The concept of a "democracy-to-come" is advanced, for example, in Derrida's *The Politics of Friendship* (London: Verso, 1997).

49. "Politics and Friendship: A Discussion with Jacques Derrida," Centre for Modern French Thought, University of Sussex, United Kingdom, December 1, 1997.

50. Derrida, *Specters of Marx*, 85.

51. "Should Trees Have Standing?"[1972], in Chistopher Stone, *Should Trees Have Standing? and Other Essays* (Dobbs Ferry, N.Y.: Oceana, 1996).

52. Currently, for example, there seems to be a genuine conflict between two bodies representing the interests of trees (and offering certification of environmentally sustainable practices), the Sustainable Forestry Initiative (SFI) and the Forest Stewardship Council (FSC). In fact, it soon becomes clear that the former actually represents the interests of paper and logging companies, while the latter attempts to represent the interests of the whole ecological space in which trees are grown.

53. Reference here to a "parliament of the living" is a response to Bruno Latour's invocation of a "parliament of things." I share some of his suspicion about the distinction between nature and culture, but I do think that living beings are the source of all *interest*, and thus of what needs to be represented in a broader parliament. See Bruno Latour, *Politics of Nature: How to Bring the Sciences into Democracy*, trans. Catherine Porter (Cambridge, Mass.: Harvard University Press, 2004).

54. Derrida, *Writing and Difference*, 117.

55. See, for example, the interview with Tom Regan, author of *The Case for Animal Rights* (1983). "As animal advocates, we have a reason to get up in the morning. A reason to rest at night. And that is to be a voice for the voiceless" (*Satya* [August 2004]), which can be accessed at www.satyamag.com/aug04/regan.html (accessed March 13, 2007).

SACRED-LAND THEOLOGY: GREEN SPIRIT, DECONSTRUCTION, AND THE QUESTION OF IDOLATRY IN CONTEMPORARY EARTHEN CHRISTIANITY | MARK I. WALLACE

I am grateful to Catherine Keller and Laurel Kearns of Drew University, Ellen Ross of Swarthmore College, Roger Latham of Continental Conservation, and the participants and audience at EcoSophia: The Fifth Transdisciplinary Theological Colloquium, Drew University, October 1–3, 2005, for their many insightful comments and criticisms.

1. "Pseudo-Titus," in Bart D. Ehrman, *Lost Christianities: Books that Did Not Make It into the New Testament* (Oxford: Oxford University Press, 2003), 239.

2. The figuration of the Spirit as the "green face" of God is a metaphor that offers new vision into the reality of Godself. Good metaphors (and I hope this metaphor is a candidate for such an honorific) are not merely ornaments of speech but eye-opening juxtapositions of two terms—in this case, "green face" and "God"—that generate novel redescriptions of reality. Enabling, productive metaphors bring together two dissimilar terms or ideas to redescribe the world in a manner that is fresh and alive; they illuminate

dimensions of existence unavailable to human understanding apart from this type of tension-ridden imagery. The argument for the reality-illuminating referential function of metaphorical discourse is advanced in Paul Ricoeur's *The Rule of Metaphor: Multi-disciplinary Studies of the Creation of Meaning in Language*, trans. Robert Czerny with Kathleen McLaughlin and John Costello (Toronto: University of Toronto Press, 1977).

3. As I perform a retrieval of the Spirit's *earthen* identity in this article, I also hope to recover the Spirit's *female* identity. As God's indwelling, corporeal presence within the created order, the Spirit is variously identified with feminine and maternal characteristics in the biblical witness. In the Bible the Spirit is envisioned as God's helping, nurturing, inspiring, and birthing presence in creation. The mother Spirit Bird in the opening creation song of Genesis, like a giant hen sitting on her cosmic nest egg, hovers over the earth and brings all things into life and fruition. Catherine Keller lyrically evokes the image of the brooding divine Mother Bird as a powerful statement of God embodying Godself in living systems. In becoming matter, she writes, the Spirit, "far from effecting a spiritual disembodiment, a flight from the earth, suggests in its very birdiness a dynamism of embodiment: lines of flight *within* the world. Moreover, the etymological connotation of brooding has always emitted the mythical associations of the mother bird laying the world-egg. Despite its precarious biblical legitimacy, the egg has tucked itself into the long history of interpretation—as, for instance, in [the medieval theologian] Hildegard's image of the universe as a cosmic egg" (Keller, *Face of the Deep: A Theology of Becoming* [London: Routledge, 2003], 233.) As Keller puts it, the "very birdiness" of the Spirit's progenerative activity plunges us into the earthy depths of the Spirit's "flight *within* the world"—her this-worldly love and passion for the integrity of the "world-egg" she has produced. In turn, this same hovering Spirit Bird, as a dove that alights on Jesus as he comes up through the waters of his baptism, appears in all four of the Gospels' baptismal accounts to signal God's approval of Jesus' public work. In this article I will take the liberty of referring to the Spirit as "she" in order to recapture something of the biblical understanding of God as feminine Spirit within the created order.

4. All things are enfleshments of God's being and reality in this model of the natural world as God's body. See Sallie McFague, *The Body of God: An Ecological Theology* (Minneapolis: Augsburg Fortress, 1993), and Grace M. Jantzen, *God's World, God's Body* (Louisville: Westminster-John Knox, 1984).

5. See an expansion of this thesis in my *Finding God in the Singing River: Christianity, Spirit, Nature* (Minneapolis: Augsburg Fortress, 2005), from which some of the material in this article has been adapted.

6. Plato, *Timaeus* 42–49, 89–92.

7. Matt. 19:2.

8. Peter Brown, *The Body and Society: Men, Women, and Sexual Renunciation in Early Christianity* (New York: Columbia University Press, 1988), 160–89.

9. See Sallie McFague, *Models of God: Theology for an Ecological, Nuclear Age* (Philadelphia: Fortress, 1987), 169–72.

10. McFague, *The Body of God*, 141–50.

11. Ibid., 143.

12. The hope for a recovery of Christian love and passion for flesh and the body is to go back to the future, to retrieve the Bible's fecund earth symbols for God as the beginning of a new ecological Christianity. Deep strains within Christian spirituality are marked by indifference (or even hostility) to "this world" in favor of "the world to come." But not all Christian thinkers have suffered from this debilitating dualism. In the thirteenth century, St. Francis of Assisi celebrated the four cardinal elements, along with human beings and animal beings, as members of the same cosmic family parented by a caring creator God. St. Francis's poetry is suffused with biophilic earth imagery. "Be praised my lord for Brother Wind and for the air and cloudy days / Be praised my lord for Sister Water because she shows great use and humbleness in herself and preciousness and depth / Be praised my lord for Brother Fire through whom you light all nights upon the earth / Be praised my lord because our sister Mother Earth sustains and rules us and raises food to feed us" (St. Francis of Assisi, "Be Praised My Lord with All Your Creatures," in *Earth Prayers from Around the World*, ed. Elizabeth Roberts and Elias Amidon [San Francisco: HarperSanFrancisco, 1991], 226–27).

13. I have drawn my knowledge about the Crum Creek watershed from Roger Latham, "The Crum Woods in Peril: Toward Reversing the Decline of an Irreplaceable Resource for Learning"; "Crum Creek Watershed: A Protection Guide," Chester-Ridley-Crum Watersheds Association pamphlet; and, "Crum Creek 1995," report by the Advanced Research Biology Students of Conestoga High School, Pennsylvania, under the direction of Norman E. Marriner.

14. The Thoreau quotation is from Hazelden Meditations, *Wisdom to Know* (Center City, Minn.: Hazelden Foundation, 2005), 29.

15. To say that God is *especially* present in healthy bioregions is not to say that God is absent from landscapes scarred by environmental abuse. While God as Spirit is everywhere, it is also the case that there are places where God is hidden, in retreat, or "eclipsed," as Martin Buber puts it, because of

human conduct. In *Eclipse of God* (New York: Harper and Row, 1952), Buber writes that it is possible for human beings to drive away God's presence through their actions. Buber's sensibility here informs my model of the relation of God and the earth.

16. See John D. Caputo and Michael J. Scanlon, eds., *God, the Gift, and Postmodernism* (Bloomington: Indiana University Press, 1999); Steven Connor, *Postmodernist Culture: An Introduction to Theories of the Contemporary* (Cambridge: Basil Blackwell, 1989); David Harvey, *The Condition of Postmodernity: An Inquiry into the Origins of Cultural Change* (Cambridge: Basil Blackwell, 1990); Allan Megill, *Prophets of Extremity: Nietzsche, Heidegger, Foucault, Derrida* (Berkeley: University of California Press, 1985); and Merold Westphal, "Blind Spots: Christianity and Postmodern Philosophy," *Christian Century* (June 14, 2003): 32–35.

17. On the question of religion, postmodernism, and the environment, see Max Oelschlaeger, ed., *Postmodern Environmental Ethics* (Albany: State University of New York Press, 1995); David Ray Griffin, *God and Religion in the Postmodern World: Essays in Postmodern Theology* (Albany: State University of New York Press, 1989); Bruno Latour, *Politics of Nature: How to Bring the Sciences into Democracy*, trans. Catherine Porter (Cambridge, Mass.: Harvard University Press, 2004); and Michael E. Zimmerman, *Contesting Earth's Future: Radical Ecology and Postmodernity* (Berkeley: University of California Press, 1994).

18. See the proceedings of this seminar in *UnCommon Ground: Rethinking the Human Place in Nature*, ed. William Cronon (New York: Norton, 1996).

19. William Cronon, "Introduction: In Search of Nature," in *UnCommon Ground*, 25–26.

20. George Sessions, "Reinventing Nature? The End of Wilderness? A Response to William Cronon's *UnCommon Ground*," *Wild Earth* 6 (Winter 1996–97): 46. See the collection of articles in this number of *Wild Earth* for a biocentric reply to Cronon and other postmodern environmental writers.

21. Dave Foreman, "Around the Campfire," *Wild Earth* 6 (Winter 1996–97): 4.

22. Cronon, "Foreword to the Paperback Edition," in *UnCommon Ground*, 21.

23. John B. Cobb Jr., *Is It Too Late? A Theology of Ecology* (Berkeley: Bruce, 1972).

24. John B. Cobb Jr. and Herman E. Daly, *For the Common Good: Redirecting the Economy toward Community, the Environment, and a Sustainable Future* (Boston: Beacon, 1989).

25. Among his other writings that bring together process philosophy, especially the work of Alfred North Whitehead, and Christian theology, see Cobb's *A Christian Natural Theology Based on the Thought of Alfred North Whitehead* (Philadelphia: Westminster, 1965).

26. John B. Cobb Jr., "Protestant Theology and Deep Ecology," in *Deep Ecology and World Religions: New Essays on Sacred Ground*, ed. David Landhis Barnhill and Roger S. Gottlieb (Albany: State University of New York Press, 2001), 223. Other environmental theologians make a similar point. James A. Nash says that while "only the Creator is worthy of worship, all God's creatures are worthy of moral consideration" (*Loving Nature: Ecological Integrity and Christian Responsibility* [Nashville: Abingdon, 1991], 96).

27. Cobb, "Protestant Theology and Deep Ecology," 224.

28. Cobb and Daly, *For the Common Good*, 384.

29. Most religious environmental authors agree with Cobb that deep green Christianity goes too far in erasing the line of distinction they aver separates humankind from otherkind. For such thinkers it is inconceivable, in terms of both value and ethics, to imagine a world in which human beings are not both fundamentally different from and in some basic sense superior to other life-forms. In an otherwise insightful plea for Christian ecotheology, Steven Bouma-Prediger argues that "insofar as [deep ecology] proponents claim that all organisms have equal value and worth, it is unclear how to adjudicate competing interests or goods. . . . How can one consistently put into practice such a position? . . . [In] acting we presuppose a [human-centered] scale or hierarchy of values. Better to be honest about what that axiological scale is than to pretend that all organisms are of equal value" (Bouma-Prediger, *For the Beauty of the Earth: A Christian Vision for Creation Care* [Grand Rapids, Mich.: Baker Academic, 2001], 132). My suggestion is otherwise: that we act honestly, indeed, by weaning ourselves away from this traditional humanist value scale. We are all equal—all living beings and other entities in the biosphere—and we all depend upon one another for meeting our vital needs in the food-web. Decisions about resource-allocations should focus on how best to preserve the integrity of this web through sustainable predation patterns without appealing to a value hierarchy with human needs at the top of the hierarchy. Our needs do not trump the needs of other communities of beings. Or, to put it another way, all of our needs come first because we all depend upon each other for our daily survival. When we put our desires first, what we forget is that what we truly need is the preservation of a series of interdependent, healthy green belts across the planet for our, and our biological neighbors', present and future sustenance. Environmental ethical decisions should not be made with primary reference to human needs, but in consideration of the health of entire ecosystems and their residential populations of plants and animals. Biocentric rather than anthropocentric criteria, ironically and wonderfully, ensure

the good life for all of us. A robust and healthy food-web is the primary value that should guide resource-allocation decisions; this value, not Cobb's and Bouma-Prediger's benign humanism, is the core value upon which human health, and the health of all other beings, is best secured.

30. Aldo Leopold, *A Sand County Almanac* (New York: Ballantine, 1970), 252–53.

31. Ibid., 262.

32. See, for example, Cobb and Daly, *For the Common Good*, 382–406.

GROUNDING THE SPIRIT: AN ECOFEMINIST PNEUMATOLOGY | SHARON BETCHER

1. Nelle Morton, "A Word We Cannot Yet Speak," in *The Journey Is Home* (Boston: Beacon, 1985), 87.

2. Linda Holler, "Thinking with the Weight of the Earth: Feminist Contributions to an Epistemology of Concreteness," *Hypatia* 5 (1990): 1–23.

3. Marie Isaacs, *The Concept of Spirit: A Study in Hellenistic Judaism and its Bearing on the New Testament* (London: Heythrop College, 1976).

4. Sallie McFague, *The Body of God: An Ecological Theology* (Minneapolis: Augsburg Fortress, 1993), 14.

5. Mary Daly, *Pure Lust: Elemental Feminist Philosophy* (Boston: Beacon, 1984), 48.

6. Michael Lodahl, *Shekhinah/Spirit: Divine Presence in Jewish and Christian Religion* (New York: Paulist, 1992), 50, 67–69.

7. McFague, *The Body of God*, 9.

8. Catherine Keller, "Postmodern 'Nature,' Feminism and Community," in *Theology for Earth Community: A Field Guide*, ed. Dieter T. Hessel (Maryknoll, N.Y.: Orbis, 1996), 101.

9. Kathleen Sands, having lifted the lid on propositional absolutes, employs the notion of "the tragic" to name the experience of the competing goods and goals of life-communities, the plurality of values and truths. See *Escape from Paradise: Evil and Tragedy in Feminist Theology* (Minneapolis: Augsburg Fortress, 1994).

10. H. Paul Santmire, *The Travail of Nature: The Ambiguous Ecological Promise of Christian Theology* (Philadelphia: Fortress, 1985), 9.

11. Augustine, *Confessions of St. Augustine*, trans. F. J. Sheed (New York: Sheed and Ward, 1943), 326. Santmire—even while suggesting that Augustine's thought presents us with "the flowering of the ecological promise of classical theology"—reads the psychic ambiguity that has marked Christians' relation with nature as being pivotally situated within and through Augustine's thought (73, 75).

12. Tertullian, "On Baptism" and "On the Veiling of Virgins," in *The Ante-Nicene Fathers: The Writings of the Fathers Down to a.d. 325*, ed. Alexander

Roberts and James Donaldson (Grand Rapids, Mich.: Eerdmans, 1982), vol. 4.

13. Martin Luther, *Luther's Works: Lectures on Romans*, vol. 25, ed. Hilton C. Oswald (St. Louis: Concordia, 1972), 362. Luther's emphasis on human justification by faith—with nature as but a theatrical stage-set thereof—prepared the way for the anthropocentrism of the modern era. See Santmire, *The Travail of Nature*, 121–33.

14. Luther, *Works*, 361.

15. Ibid., 363.

16. Keller, "Postmodern 'Nature,'" 150.

17. Luce Irigaray, *Sexes and Genealogies*, trans. Gillian C. Gill (New York: Columbia University Press, 1993), 109.

18. Nicolai Berdyaev, *The Meaning of History* (New York: Charles Scribner's Sons, 1936), 117, 116.

19. Cited in Melanie May, *A Body Knows: A Theopoetics of Death and Resurrection* (New York: Continuum, 1995), 73.

20. Jürgen Moltmann, *Spirit of Life: A Universal Affirmation* (Minneapolis: Augsburg Fortress, 1992), 1, 8–10. See also Steven Bouma-Prediger, *The Greening of Theology: The Ecological Models of Rosemary Radford Ruether, Joseph Sittler, and Jürgen Moltmann* (Atlanta: Scholar's Press, 1995), 106–9.

21. Elizabeth Ann Johnson, *Women, Earth, and Creator Spirit* (New York: Paulist, 1993), 2.

22. Wendell Berry, *The Unsettling of America: Culture and Agriculture* (New York: Avon, 1977), 108–9.

23. Julia Kristeva, *Powers of Horror: An Essay on Abjection* (New York: Columbia University Press, 1982), 57.

24. Jay McDaniel, *Of God and Pelicans: A Theology of Reverence for Life* (Louisville: Westminster-John Knox, 1989), 22; emphasis added.

25. Berry, *The Unsettling of America*, 22.

26. Luce Irigaray, "The 'Mechanics' of Fluids" (chap. 6), in *This Sex Which Is Not One*, trans. Catherine Porter (Ithaca: Cornell University Press, 1985), 106.

27. Jürgen Moltmann, *God in Creation: A New Theology of Creation and the Spirit of God* (Minneapolis: Augsburg Fortress, 1993), 71, 69, 68.

28. Irigaray, *Sexes and Genealogies*, 80, 81.

29. Luce Irigaray, *I Love to You: Sketch of a Possible Felicity in History* (New York and London: Routledge, 1996).

30. Elisabeth A. Grosz, *Volatile Bodies: Toward a Corporeal Feminism* (Bloomington: Indiana University Press, 1994), 192.

31. Kristeva, *Powers of Horror*, 113–14.

32. Ibid., 116.

33. Wolfhart Pannenberg, *Toward a Theology of Nature: Essays on Science and Faith*, ed. Ted Peters (Louisville: Westminster-John Knox, 1993), 137.

34. Michael Welker, *God the Spirit*, trans. John F. Hoffmeyer (Minneapolis: Augsburg Fortress, 1994), 164, 331.

35. Irigaray, *Sexes and Genealogies*, 136.

36. Aldo Leopold, *A Sand County Almanac* (New York: Oxford University Press, 1968), employs the concept of "the land pyramid" as a way to enable our thinking of the earth as an energy circuit and, in that vein, a system of interlocking food chains. "Plants absorb energy from the sun. This energy flows through a circuit called the biota, which may be represented by a pyramid consisting of layers. The bottom layer is the soil. A plant layer rests on the soil, an insect layer on the plants, a bird and rodent layer on the insects, and so on up through various animal groups, to the apex layer, which consists of the larger carnivores. . . . Each successive layer depends on those below it for food and often for other services, and each in turn furnishes food and services to those above" (214–15).

37. Dorothee Soelle, "Between Matter and Spirit: Why and in What Sense Must Theology be Materialist?" in *God of the Lowly: Socio-Historical Interpretations of the Bible*, ed. Willy Schottroff and Wolfgang Stegemann (Maryknoll, N.Y.: Orbis, 1984), 98.

38. Wendell Berry, "Christianity and the Survival of Creation," in *Cross Currents* 43, no. 2 (Summer 1993): 157.

39. Berry, *The Unsettling of America*, 107.

40. Elaine Scarry, *The Body in Pain: The Making and Unmaking of the World* (New York: Oxford University Press, 1985), 219.

41. Barbara Newman, *Sister of Wisdom: St. Hildegard's Theology of the Feminine* (Berkeley: University of California Press, 1987).

42. Hildegard of Bingen, *Hildegard of Bingen's Book of Divine Works*, ed. Matthew Fox (Santa Fe: Bear and Company, 1987), 199.

43. Hildegard, *Book of Divine Works*, 212.

44. Ibid., 145.

45. William Bryant Logan, *Dirt: The Ecstatic Skin of the Earth* (New York: Riverhead Books, G. P. Putnam's Sons, 1995), 7.

46. Hildegard, *Book of Divine Works*, 368.

47. Gaston Bachelard, *The Poetics of Space*, trans. Maria Jolas (New York: Orion, 1964), 104.

48. Stuart Kauffman, *At Home in the Universe: The Search for Laws of Self-Organization and Complexity* (New York: Oxford University Press, 1995).

49. Joseph Sittler, "Evangelism and the Care of the Earth," in *Preaching in the Witnessing Community*, ed. Herman G. Stuempfle Jr. (Philadelphia: Fortress, 1973), 63.

50. Welker, *God the Spirit*, xi.

HEARING THE OUTCRY OF MUTE THINGS: TOWARD A JEWISH CREATION THEOLOGY | LAWRENCE TROSTER

1. Arthur Green, *Seek My Face, Speak My Name: A Contemporary Jewish Theology* (Northvale, N.J.: Jason Aronson, 1992), 53.

2. Ibid.

3. Numerous essays from a Kabbalistic perspective are presented in Ari Elon, Naomi Mara Hyman, and Arthur Waskow, eds., *Trees, Earth, and Torah: A Tu B'Shvat Anthology* (Philadelphia: Jewish Publication Society, 1999).

4. Arthur Green, "A Kabbalah for the Environmental Age," in Hava Tirosh-Samuelson, ed., *Judaism and Ecology: Created World and Revealed Word* (Cambridge, Mass.: Harvard University Press, 2002), 11.

5. Hava Tirosh-Samuelson, Ibid., lii–lv, 389–404. The problem of using Kabbalah as a source for environmentalism can be seen in Lawrence Fine's description of the tension in sixteenth-century Kabbalist Isaac Luria's view of the natural world. "On the one hand, his cosmogonic teachings exhibit an anticosmic dualism in which the material world is deprecated in favor of a divine one from which all being derives. . . . Nevertheless, these views did not translate in an utter devaluation of the natural world. On the contrary, the natural world for Luria was a means by which to encounter the divine" (Fine, *Physician of the Soul, Healer of the Cosmos: Isaac Luria and His Kabbalistic Fellowship* [Stanford: Stanford University Press, 2003], 356).

6. Green, "A Kabbalah for Environmental Age," 3.

7. After Jonas died in 1993, the Hastings Center published a special issue of the *Hastings Center Report* 25, no. 7, entitled "The Legacy of Hans Jonas." It included articles on Jonas's philosophy, bioethics, environmental ethics, and political influence.

8. Jonas's writing had an important impact on the Green Party in Germany. See Hans Jonas, *Mortality and Morality: A Search for the Good after Auschwitz*, ed. Lawrence Vogel (Evanston, Ill.: Northwestern University Press, 1996).

9. On the *Sitra Achra*, see Gershom Scholem, *On the Mystical Shape of the Godhead: Basic Concepts in the Kabbalah* (New York: Schocken, 1991), 56–87. On the universe expressing both creativity and creative destruction, see Richard L. Rubenstein, *After Auschwitz: History, Theology, and Contemporary Judaism*, 2nd ed. (Baltimore: Johns Hopkins University Press, 1992), 172–73.

The term "creative destruction" was first utilized in economic theory by Joseph Schumpeter (1883–1950), who adapted the term from the work of fellow economist Werner Sombart (1863–1941), but also from the philosophies of Friedrich Nietzsche and Karl Marx. On the idea that God is both the ground of law and the ground of novelty, see Ian Barbour, *Religion and Science: Historical and Contemporary Issues* (San Francisco: HarperCollins, 1997), 284–87.

10. Neil Gillman, *Sacred Fragments: Recovering Theology for the Modern Jews* (Philadelphia: Jewish Publication Society, 1990), xxvi.

11. Neil Gillman, *The Death of Death: Resurrection and Immortality in Jewish Thought* (Woodstock, Vt.: Jewish Lights Publishing, 1997), 126. Gillman here is speaking about how the concept of resurrection became canonized by being embedded in the *Amidah*, one of the oldest prayers of the core of the liturgy, which religious Jews recite daily.

12. See for example, Reuven Hammer, *Or Hadash: A Commentary on Siddur Sim Shalom for Shabbat and Festivals* (New York: The Rabbinical Assembly and the United Synagogue of Conservative Judaism, 2003), xxi–xxiv.

13. See my "Created in the Image of God: Humanity and Divinity in an Age of Environmentalism," in Martin Yaffe, ed., *Judaism and Environmental Ethics* (Lanham, Md.: Lexington, 2001), 73–79, and "Caretaker or Citizen: Hans Jonas, Aldo Leopold, and the Development of Jewish Environmental Ethics," in Hava Tirosh-Samuelson and Christian Wiese, eds., *Judaism and the Phenomenon of Life: The Legacy of Hans Jonas* (Boston: Brill Academic Publishers).

14. Jonas received his advanced Jewish education at the Hochschule für der Wissenschaft des Judentums (University for the Science of Judaism) in Berlin. At the same time, he studied with Heidegger and Rudolf Bultmann at the University of Marburg, receiving his doctorate under Heidegger in 1930. Jonas was one of a group of Jewish disciples of Heidegger that included Hannah Arendt, Karl Lowith, Herbert Marcuse, and Emmanuel Levinas. According to Richard Wolin, of the group Jonas was the most connected to his Judaism. *Heidegger's Children* (Princeton: Princeton University Press, 2001), 104.

15. In 1933, the German Association for the Blind expelled its Jewish members. Even though Jonas was himself not blind, he was morally outraged by the decision and left Germany—vowing he would only return in the uniform of a conquering army—traveling first to England and then, by 1935, to Palestine.

16. This is best summed up in an article entitled "Contemporary Problems in Ethics from a Jewish Perspective" that Jonas wrote for the journal of the

Central Conference of American Rabbis and is reprinted in Yaffe, *Jewish Environmental Ethics*, 250–63.

17. Ibid., 250.

18. Ibid., 252.

19. Ibid., 253.

20. Ibid., 254.

21. Ibid., 255.

22. Hans Jonas, *The Imperative of Responsibility: In Search of an Ethics for the Technological Age* (Chicago: University of Chicago Press, 1984).

23. Jonas, *The Imperative of Responsibility*, 6.

24. Ibid., 27: "[T]he perception of the *malum* is infinitely easier to us than the perception of the *bonum*; it is more direct, more compelling, less given to differences of opinion or taste, and most of all, obtruding itself without our looking for it."

25. Lawrence Vogel, "Introduction," in Jonas, *Mortality and Morality*, 16.

26. Jonas, *The Imperative of Responsibility*, 36–37.

27. Ibid., x.

28. Jonas, *Mortality and Morality*, 66–67.

29. Ibid., 70–74, 88–92; see also Hans Jonas, *The Phenomenon of Life: Toward a Philosophical Biology*, 2nd ed. (Chicago: University of Chicago Press, 1982), 186.

30. Jonas, *The Imperative of Responsibility*, 8, 136–42.

31. Lawrence Vogel, "Does Environmental Ethics Need a Metaphysical Grounding?" *Hastings Center Report* 25, no. 7: 37.

32. Vogel, "Does Environmental Ethics . . . ," 38.

33. Vogel, "Introduction," in Jonas, *Mortality and Morality*, 19.

34. Ibid., 36.

35. Now printed in Jonas, *Mortality and Morality*, 131–43. This version is a translation of a lecture given in Germany in 1984. It was a revised version of his "The Concept of God After Auschwitz," in Albert H. Friedlander, ed., *Out of the Whirlwind: A Reader of Holocaust Literature* (New York: Union of American Hebrew Congregations), 465–76. This incorporated material from an earlier essay, "Immortality and the Modern Temper," published originally in 1962 but also included in Jonas's *Phenomenon of Life* as well as *Mortality and Morality*, 113–30.

36. Jonas, *Mortality and Morality*, 132.

37. Ibid., 133.

38. Ibid.

39. Ibid., 134.

40. Ibid., 142.

41. Ibid.

42. William Kaufman, *The Evolving God in Jewish Process Theology* (Lewiston, Maine: Edwin Mellen, 1997), 151.

43. Jonas, *Mortality and Morality*, 137.

44. Ibid., 138.

45. Ibid.

46. Ibid., 138–39. See also Charles Hartshorne, *Omnipotence and Other Theological Mistakes* (Albany: State University of New York Press, 1984).

47. Jonas, *Mortality and Morality*, 139.

48. Ibid., 154.

49. Ibid., 156.

50. Elliot Dorff, "Medieval and Modern Theories of Revelation," in David Lieber, Chaim Potok, Harold Kushner, Jules Harlow, Elliot Dorff, and Susan Grossman, eds., *Etz Hayim: Torah and Commentary* (New York: Rabbinical Assembly and United Synagogue of Conservative Judaism, 2001), 1401–5; Norbert M. Samuelson, *Revelation and the God of Israel* (Cambridge: Cambridge University Press, 2002).

51. *Mishnah Torah, Hilkhot Yesodei Ha-Torah* 2:1:

> It is a positive commandment to love this great and awe-inspiring God and also to fear God, as the verse states, "You shall love the Lord your God" (Deut. 6:5), and "You shall fear the Lord your God" (Deut. 6:13). But how does one learn to love and to fear God? When a person contemplates God's works and God's great and marvelous creatures (by thorough study), he sees from them wisdom that is without estimate or end, and is immediately filled with love and praise and longs ardently to know the Holy Name even as David said, "My soul thirsts for God" (Ps. 42:2). And when a person thinks about these mighty matters he draws back and trembles and realizes that he is a minute creature, lowly and dark, capable only of a little knowledge in the presence of perfect knowledge, as David said, "When I consider Your heavens . . . what is man, that You are mindful of him? Mortal man that you have taken note of him?" (Ps. 8:3–4).

52. Christian theologians and scientists in the sixteenth through eighteenth centuries often used a theological construct called the "Two Books of God," the Book of Nature and the Book of Scripture, to describe the relationship between the truth of Scripture and the truth of science. The origin of the

metaphor may be from a commentary of Augustine to Psalm 45:7 but the definitive origin is from late medieval pulpit rhetoric. See Ernst Robert Curtius, *European Literature and the Latin Middle Ages* (Princeton: Princeton University Press, 1990 [1953]), 315–26. For the use of the metaphor by Christian scientists, see John Hedley Brooke, *Science and Religion: Some Historical Perspectives* (Cambridge: Cambridge University Press, 1999 [1991]), 22, and James R. Moore, "Geologists and Interpreters of Genesis in the Nineteenth Century," in David C. Lindberg and Ronald L. Numbers, eds., *God and Nature: Historical Essays on the Encounter between Christianity and Science* (Berkeley: University of California Press, 1986), 322–50. For a modern Christian theologian's attempt to revive the metaphor see Roland Mushat Frye, "Two Books of God," in *Theology Today* 39, no. 2 (July 1982): 260–66. Byron L. Sherwin, "Judaism, Technology and the "New Science," *Proceedings of the Rabbinical Assembly*, Volume LXIII, 2001, pp. 78–89, claims that the two-book metaphor is found in the work of the medieval Jewish philosopher Levi ben Gershon (known also as Gersonides, 1288–1344) but his evidence is not convincing. Professor Menachem Fisch, in a private communication, asserts that there is no Jewish theologian who ever used this construct either in its concrete metaphor or in its idea of there being two sources of revelation. He believes, and I concur, that Maimonides came closest to expressing the idea that the natural world could be a source of revelation.

53. Jonas, *Mortality and Morality*, 165–97.
54. This is similar to the Participatory Anthropic Theory (PAP) of physicist John Wheeler. See Paul Davies, *The Mind of God: The Scientific Basis for a Rational World* (New York: Simon and Schuster, 1993), 223–25.
55. Jonas, *Mortality and Morality*, 167.
56. See Paul Davis, *The Fifth Miracle: The Search for Origin and the Meaning of Life* (New York: Simon and Schuster, 1999).
57. Here Jonas parts company with Whitehead. See *Mortality and Morality*, 211n4. Jonas finds Whitehead to be "overbold" and "not covered by any datum of our experience, which allows us to discover or suspect traces of subjectivity only in high-level formations of organisms."
58. Ibid., 172.
59. Ibid. Unlike Jonas, John Haught, in *God After Darwin: A Theology of Evolution* (Boulder, Colo.: Westview, 2000), 168–84, believes in a kind of logos that exists in the universe even before the emergence of life. He accepts and follows Teilhard de Chardin and Whitehead, whom Jonas explicitly rejects. See Haught's Boyle lecture, "Darwin, Design and the Promise of Nature," available at http://www.stmarylebow.co.uk/docs/BoyleBooklet.pdf.

60. Hans Jonas, "Tool, Image and Grave: On What Is beyond the Animal in Man," in idem, *Mortality and Morality*, 75–86.

61. Jonas, *Mortality and Morality*, 175.

62. Ibid., 192.

63. I first dealt with this issue in a different way in my "Asymmetry, Negative Entropy and the Problem of Evil," *Judaism* 34, no. 4 (Fall 1985), 453–461. For a biblical theology that takes into account the forces of chaos, see Jon D. Levenson, *Creation and the Persistence of Evil: The Jewish Drama of Divine Omnipotence* (San Francisco: Harper and Row, 1988).

64. Haught, *God After Darwin*, 184.

65. Jonas, *Mortality and Morality*, 188–89.

66. Norbert M. Samuelson, *Judaism and the Doctrine of Creation* (Cambridge: Cambridge University Press, 1994), 238–40.

67. Haught, *God After Darwin*, 43. I will leave the last for another time but there should be a discussion among postmodern theologians about their vision of redemption. In a review of modern Jewish concepts of eschatology, Neil Gillman has shown how postmodern theologians tend to view the traditional idea of resurrection as a symbol, which points toward a future hope beyond time and space. See Gillman, *The Death of Death*, 215–41.

68. Midrash Rabbah to Ecclesiastes 7:13 (no. 1): "Consider God's doing! Who can straighten what has been twisted?"

69. Hans Jonas, "The Outcry of Mute Things," in idem, *Mortality and Morality*, 203. This is from a speech that Jonas gave on January 30, 1993, in Italy on the occasion of his being given the Premio Nonino Prize. He passed away six days later upon his return to the United States.

CREATIO EX NIHILO, TERRA NULLIUS, AND THE ERASURE OF PRESENCE | WHITNEY A. BAUMAN

1. Don Cupitt, *Creation out of Nothing* (London: SCM, 1990), 96.

2. Catherine Keller, *Face of the Deep: A Theology of Becoming* (New York and London: Routledge, 2003).

3. Charles Hartshorne, *Omnipotence and Other Theological Mistakes* (Albany: State University of New York Press, 1984), 71.

4. This not to mention the critique of anthropocentrism that Keller derives from process thought, especially in her discussion of the book of Job in *Face of the Deep*, 124–140 (chap. 7).

5. This, I think, is most noticeable in her understanding of a plurisingular reality and subsequent theology: "Such a theology, neither monistic nor

dualistic, prepares a pluralism not of many separate ones but of plurisingularities, of interdependent individuations, constantly coming, flowing, *through* one another" (Keller, *Face of the Deep*, 179).

6. Keller follows Edward Said's distinction between "origins" (an absolute-acontextual term) and "beginnings" (an always relative and contextual term) in her analysis of *creatio ex nihilo*. See *Face of the Deep*, 5, 10, 159.

7. The term is Val Plumwood's: Val Plumwood, *Environmental Culture: The Ecological Crisis of Reason* (New York and London: Routledge, 2002), 104.

8. Keller, *Face of the Deep*, 6.

9. Ibid., 148.

10. For a concise introduction to the concept of *terra nullius* and its legal use, see Stuart Banner, "Why *Terra Nullius*? Anthropology and Property Law in Early Australia," in *Law and History Review* 23, no. 1 (Spring 2005): 95–132. See also Henry Reynolds, *The Law of the Land* (New York: Viking, 1987).

11. Val Plumwood, *Environmental Culture*, 104.

12. Ibid., 101.

13. Max Horkheimer and Theodor Adorno, *Dialectic of Enlightenment*, trans. by Edmund Jephcott (Stanford: Stanford University Press, 2002), 4.

14. Plumwood, *Environmental Culture*, 19, defines the term "monological epistemology" particularly well.

15. Catherine Keller, *God and Power: Counter-Apocalyptic Journeys* (Minneapolis: Augsburg Fortress, 2005), 50: "Omnipotence . . . creates *ex nihilo*: it has always already defeated the chaos."

16. Sandra Harding discusses the "backgrounding" logic of some of these concepts in *Is Science Multicultural: Postcolonialisms, Feminisms, and Epistemologies* (Bloomington: Indiana University Press, 1998), 27–35. Jane Flax argues against the danger of these "innocent" assumptions (à la Foucault) in "The End of Innocence," in *Feminists Theorize the Political*, Judith Butler and Joan W. Scott, eds. (New York and London: Routledge, 1992), 445–63. Specifically concerning the "Enlightenment self-understanding," she writes that "all difference, disorder will be brought within the beneficent sovereignty of the One" (453–54).

17. For example: Gerhard May, *Creatio ex Nihilo: The Doctrine of "Creation out of Nothing" in Early Christian Thought*, trans. by A. S. Worrall (Edinburgh: T. and T. Clark, 1994); Keller, *Face of the Deep*; and Sjoerd L. Bonting, *Creation and Double Chaos: Science and Theology in Discussion* (Minneapolis: Augsburg Fortress, 2005).

18. Carolyn Merchant, *Reinventing Eden: The Fate of Nature in Western Culture* (New York and London: Routledge, 2003), 79–82.

19. John Locke, *Two Treatises of Government: A Critical Edition*, ed. Peter Laslett (Cambridge: Cambridge University Press, 1966). Reference to this argument of dominion being extended to all can be found in *Treatise* 1, §24–30.

20. Jeremy Waldron, *God, Locke, and Equality: Christian Foundations of John Locke's Political Thought* (Cambridge: Cambridge University Press, 2002), 22.

21. Locke, *Treatise* 1, §29: "God says unto *Adam* and *Eve*, Have Dominion . . . these words were not spoken till *Adam* had his Wife, must not she thereby be Lady, as well as he Lord of the World?" As Waldron notes, however, Locke makes a distinction between *political* and *private* (or marital) life and only supports the political equality of women. There are many passages that suggest the "natural" subordination of women in marriage-private life (Waldron, *God, Locke, and Equality*, 35–43).

22. Locke, *Treatise* 1, §30.

23. Waldron, *God, Locke, and Equality*, 64.

24. Ibid., 206. See also Laslett's comments in the note to *Treatise* 2, §24. The literature around Locke and slavery is complicated. I only argue here that the complications might be a result of his understanding of individual human property, which mimics the omnipotence (united will and act) of the Creator God *ex nihilo*.

25. Locke, *Treatise* 2, §34.

26. Ibid., §37.

27. John Locke, *An Essay Concerning Human Understanding*, ed. by Gary Fuller, Robert Stecker, and John Wright (New York and London: Routledge, 2000), book 4, §4.

28. Ibid., book 4, §10.

29. Locke, *Treatise* 2, §26–27.

30. Merchant, *Reinventing Eden*, 80.

31. Locke, *Essay*, book 4, §10.

32. Merchant, *Reinventing Eden*, 82.

33. Horkheimer and Adorno, *Dialectic*, 6.

34. Plumwood, *Environmental Culture*, 214.

35. Stuart Banner, "Why *Terra Nullius*?", 33.

36. Emmerich de Vattel, *The Law of Nations: Or, Principles of the Law of Nature, Applied to the Conduct and Affairs of Nations and Sovereigns*, ed. and trans. Joseph Chitty (Philadelphia: T. and J. W. Johnson Law, 1853).

37. Vattel, *Law of Nations*, §§1.

38. Ibid., §§78.

39. For a good description of this process, and how enclosures created a labor pool for the impending Industrial Revolution, see Carolyn Merchant,

"Farm, Fen, and Forest: European Ecology in Transition," in idem, *The Death of Nature: Women, Ecology and the Scientific Revolution* (San Francisco: HarperSanFrancisco, 1980), 42–68.

40. Vattel, *Law of Nations*, §§81.

41. Ibid., §§205 (emphasis mine).

42. Ibid., §§209.

43. Plumwood, *Environmental Culture*, 104.

44. Jared Diamond, *Guns, Germs and Steel* (New York: Norton, 1999), 78. See also David E. Stannard, *The Conquest of the New World: American Holocaust* (Oxford: Oxford University Press, 1992), 57–96.

45. Catherine Keller, *God and Power: Counter-Apocalyptic Journeys*, 29.

46. George W. Bush, "White House Press Conference," March 29, 2001, text available at http://www.whitehouse.gov/news/releases/2001/03/20010329 .html (emphasis mine).

47. I thank Catherine Keller for pointing this out to me in her review of an earlier draft of this chapter.

48. Zygmunt Bauman, *Globalization: The Human Consequences* (New York: Columbia University Press, 1998), 19.

49. Ibid., 19. Keller speaks of destruction-denial of space by colonial attitudes, as well: see her chapter "Everywhere and Nowhere; Postcolonial Positions," in *God and Power*, 97–112; and her chapter "De-Colon-izing Spaces," in *Apocalypse Now and Then: A Feminist Guide to the End of the World* (Boston: Beacon, 1996), 140–80.

50. Merchant, *Reinventing Eden*, 86. Merchant writes of the stewardship ethic: "Just as God was the caretaker, steward, and wise manager of the natural world, so humans had responsibility to imitate that mandate" (ibid.).

51. For instance, Merchant's "partnership ethic": "A partnership ethic holds that the greatest good for the human and nonhuman communities is in their mutual living interdependence" (ibid., 223).

52. Keller, *Face of the Deep*, 49: According to Keller, *ex nihilo* sets up such a dualistic framework, even in attempting to avoid it: "According to the logic of *ex nihilo*, one is either good or evil, corporeal or incorporeal, eternal or temporal, almighty or powerless, propertied or inferior" (ibid.). See also the chapter on Barth where she notes that the dualism between matter and spirit that *ex nihilo* seeks to avoid becomes an even stronger dualism between Creator and nothingness (ibid., 84–102).

53. Sharon Betcher, "Monstrosities, Miracles, and Mission: Religion and the Politics of Disablement," in Catherine Keller, Michael Nausner, and Mayra Rivera, eds., *Postcolonial Theologies: Divinity and Empire* (St. Louis: Chalice, 2004), 97.

54. Keller, *Face of the Deep*, 10.

55. See Keller's "Talking Dirty: Ground Is Not Foundation," elsewhere in the present volume.

56. Gayatri Chakravorty Spivak, *Death of a Discipline* (New York, NY: Columbia University Press), 71.

57. Keller on Spivak in Keller, *God and Power*, 130.

58. See Primavesi, "The Preoriginal Gift—and Our Response to It," elsewhere in the present volume.

59. Susan Armstrong, "An Outline of a Theology of Difference," in Donald A. Crosby and Charley D. Hardwick, eds., *Religious Experience and Ecological Responsibility* (New York: Peter Lang, 1996), 91.

60. Keller, *Face of the Deep*, 80.

61. This term is Donna Haraway's. See Donna Haraway, *The Companion Species Manifesto: Dogs, People, and Significant Otherness* (Chicago: PricklyParadigm Press, 2003), 2: "In layers of history, layers of biology, layers of natureculture, complexity is the name of our game" (ibid.)

62. Plumwood, *Environmental Culture*, 188–95. Val Plumwood uses the term "negotiation" as a metaphor for dialogical communication and action with the rest of the natural world.

SURROGATE SUFFERING: PARADIGMS OF SIN, SALVATION, AND SACRIFICE WITHIN THE VIVISECTION MOVEMENT | ANTONIA GORMAN

1. Cobbe was founder and president of the Society for the Protection of Animals Liable to Vivisection, later known as the Victoria Street Society. According to Richard D. French, the society "became the most important and politically influential of the antivivisection societies." The society's success was due almost entirely to Cobbe's indefatigable efforts to recruit socially prominent and well-known personalities to her cause. See Richard D. French, *Antivivisection and Medical Science in Victorian Society* (Princeton: Princeton University Press, 1975), 88.

2. Ibid., 369.

3. Edward Maitland as quoted in French, *Antivivisection*, 391.

4. For examples of liberative and oppressive interpretations of atonement Christology, see, respectively, Leonardo Boff, *Ecology and Liberation: A New Paradigm* (Maryknoll, N.Y.: Orbis, 1996), and Rita Nakashima Brock and Rebecca Ann Parker, *Proverbs of Ashes: Violence, Redemptive Suffering, and the Search for What Saves Us* (Boston: Beacon, 2001).

5. See, for example, Peter Singer, *Animal Liberation* (New York: Avon, 1990); Tom Regan, *The Case for Animal Rights* (Berkeley and Los Angeles: University of California Press, 1983); and Josephine Donovan and Carol J. Adams, eds., *Beyond Animal Rights: A Feminist Caring Ethic for the Treatment of Animals* (New York: Continuum, 1996).

6. William E. Phipps, *Darwin's Religious Odyssey* (Harrisburg, Pa.: Trinity Press International, 2002).

7. Harriet Ritvo, *The Animal Estate: the English and Other Creatures in the Victorian Age* (Cambridge, Mass.: Harvard University Press, 1987), 39.

8. Phipps, *Darwin*, 108.

9. Ibid., 68.

10. Max Horkheimer and Theodor W. Adorno, *Dialectic of Enlightenment: Philosophical Fragments*, ed. W. Allen Ashby (Amherst, Mass.: Prometheus, 1997), 1.

11. See Warren Ashby, ed., *A Comprehensive History of Western Ethics: What Do We Believe?* (Amherst, Mass.: Prometheus, 1997), 337.

12. Ibid.

13. Ibid., 364.

14. James Turner, *Reckoning with the Beast: Animals, Pain, and Humanity in the Victorian Mind* (Baltimore: Johns Hopkins University Press, 1980), 4.

15. Kate Rigby, *Topographies of the Sacred: The Poetics of Place in European Romanticism* (Charlottesville and London: University of Virginia Press, 2004), 11.

16. Ibid., 4, 17.

17. William Wordsworth, "The Ruined Cottage" as quoted in Rigby, *Topographies*, 18.

18. Friedrich Schleiermacher, "Talks on Religion to its Cultured Despisers" (1799), as quoted in Rigby, *Topographies*, 51.

19. Rigby, *Topographies*, 50.

20. Ibid., 24–25.

21. Ibid., 5, 30–31, 36–37.

22. Ibid., 49.

23. According to Phipps, the term "panentheism" (from the Greek, "all [is] in God") was first coined by the nineteenth-century German philosopher Karl Krause. This term can be distinguished from "pantheism," which contends that God is totally immanent in the world, and "deism," which believes that God entirely transcends the world. See Phipps, *Darwin*, 188.

24. John B. Cobb, Jr., *Grace and Responsibility: A Wesleyan Theology for Today* (Nashville: Abingdon, 1995), 36–37.

25. Ibid., 64.

26. Turner, *Reckoning*, 5–6.

27. Cobb, *Grace and Responsibility*, 53.

28. Lewis G. Regenstein, *Replenish the Earth: A History of Organized Religion's Treatment of Animals and Nature—Including the Bible's Message of Conservation and Kindness Toward Animals* (New York: Crossroad, 1991), 91.

29. In 1641, Massachusetts passed a law against cruelty to domestic animals, but Martin's Act became the first national legislation against cruelty. See ibid., 92.

30. See http://www.oldbaileyonline.org/history/crime/policing.html.

31. Turner, *Reckoning*, 43.

32. Ibid., 55.

33. Ritvo, *Animal Estate*, 132.

34. As quoted in Turner, *Reckoning*, 55.

35. Ibid., 58.

36. French, *Antivivisection*, 18, 23–24, 38–39.

37. Webster's defines "iatrogenic" as "induced inadvertently by a physician or his treatment." Here the term is used to indicate medical harm directly or indirectly caused to humans by the practice of vivisection.

38. Jean Swingle Greek, DVM, and C. Ray Greek, MD, *What Will We Do If We Don't Experiment on Animals? Medical Research for the Twenty-first Century* (Victoria, Canada: Trafford, 2004).

39. Ibid., 34.

40. Lord Lister as quoted in Coral Lansbury, *The Old Brown Dog: Women, Workers, and Vivisection in Edwardian England* (Madison: University of Wisconsin Press, 1985), 91.

41. Lansbury, *Old Brown Dog*, 59.

42. Ibid., 141.

43. Ibid., 52–58.

44. See, for instance, Immanuel Jokobovits, "The Medical Treatment of Animals in Jewish Law," *The Journal of Jewish Studies* 7 (1956): 207–14; and G. E. Paget, "The Ethics of Vivisection," *Theology: A Monthly Review* 78 (January 1975): 355–61.

45. Canterbury Animal Respect Network for a Green Environment, "Death by Vivisection, Part I," available at http://www.carn-age.org.uk/vivisl.html.

46. Turner, *Reckoning*, 103, 106.

47. Ritvo, *Animal Estate*, 160.

48. French, *Antivivisection*, 392.

49. Ibid., 266.

50. Lansbury, *Old Brown Dog*, 25.

51. Turner, *Reckoning*, 24–33, 123.

52. French, *Antivivisection*, 215–19, 283, 408–12.

53. Lansbury, *Old Brown Dog*, 84.

54. Ibid., 82.

55. Claude Bernard, *An Introduction to the Study of Experimental Medicine*, trans. Henry Copley Green, AM (New York: Dover, 1957 [1865]), 102.

56. Ibid.

57. Jeremy Bentham, "Introduction to the Principles of Morals and Legislation" (1789), as quoted in Turner, *Reckoning*, 13.

58. Bernard, *Introduction*, 99.

59. Ibid., 15, 18.

60. Ibid., 169.

61. See, for instance, Singer, *Animal Liberation*, and Regan, *Animal Rights*.

62. See, for example, Carol J. Adams and Josephine Donovan, ed., *Animals and Women: Feminist Theoretical Explorations* (Durham, N.C.: Duke University Press, 1995), 79–80.

63. Brock and Parker, *Ashes*, 8, 31.

64. Lansbury, *Old Brown Dog*, 172.

65. Carol J. Adams, "Woman-Battering and Harm to Animals," in Adams and Donovan, *Animals and Women*, 55–85.

66. Brock and Parker, *Ashes*, 25.

67. Marion Grau, *Of Divine Economy: Refinancing Redemption* (New York: T. & T. Clark, 2004), 164–65.

68. Carolyn Merchant, Carol J. Adams, Sallie McFague, John B. Cobb Jr., Jay McDaniel, among others, have already done an excellent and invaluable job in this regard and I direct the reader to their works for more in-depth insights into the intrinsic value of the nonhuman world.

69. Wolfhart Pannenberg, "Eternity, Time and the Trinitarian God" (Princeton: Center of Theological Inquiry, 2002), reprinted and available at http://www.ctinquiry.org/publications/reflections_volume_3/pannenberg.htm.

70. Augustine, *Of Man's Perfection in Righteousness* (chap. 20), trans. Peter Holmes, reprinted and available at http://www.logoslibrary.org/augustine/perfecton/20.html.

71. Alfred North Whitehead, *Science and the Modern World: Lowell Lectures, 1925* (New York: Free Press, 1967), 49.

72. Ibid., 69.

73. Ibid., 49–50.

74. Ibid., 51.

75. Ibid., 91.

76. See, for instance, Genesis 2:7; Psalms 104:29; and Acts 17:28.

77. Donna J. Haraway, *Modest witness@SecondMillenium. FemaleMan Meets OncoMouse™: Feminism and Technoscience* (New York and London: Routledge, 1997), 37.

78. Bill McKibben, *The Comforting Whirlwind: God, Job, and the Scale of Creation* (Grand Rapids, Mich.: Eerdmans, 1994).

79. Catherine Keller, *Face of the Deep: A Theology of Becoming* (New York and London: Routledge, 2003), 124.

80. Ibid., 140.

THE HOPE OF THE EARTH: A PROCESS ECOESCHATOLOGY FOR SOUTH KOREA | SEUNG GAP LEE

A slightly amended version of this chapter, entitled "A Process Eschatological Eco-ethics for the Korean Churches and Christians," appeared in the *Korean Journal of Christian Studies* 48 (October 2006): 177–99.

1. Situated on South Korea's western coast about 250km southwest of Seoul, "Saemangeum" is the world's most important staging site for the fast-declining Spoon-billed Sandpiper. It has been reported that the Saemangeum tidal flat area has fully realized the possibilities of biodiversity, especially given its open ecological interconnections to the ecosystems of Korean, Chinese, and Japanese coasts, rivers, and islands. The fight against the complete reclamation of Saemangeum's 40,100 hectares of tidal-flats and shallows—the largest such project of its kind ever undertaken in the world—is one of the most important and urgent conservation issues in northern Asia. The completion of a 33-km-long seawall across the free-flowing estuaries of the Mangyeung and Dongjin Rivers has doomed hundreds of thousands of shorebirds by removing a key feeding area along the East Asian-Australasian Flyway. An appeal was made to the new president of South Korea, in 2003, and agreed to by the participants of the "NGO and Local Community World Conference on Wetlands" held in Valencia, Spain, November 15–16, 2002. Yet in December 2005 the Saemangeum project was approved in its original plan by the highest low court of the government. Details are available at http://www.konetic.or.kr/DBService/news_url.asp?category=news&key=13577 5 &collection=envnews.

2. Wonkee Park, "Wounded Earth and Ways of Environmental Healing," *Christian Thought* 459 (March 1997): 11.

3. The conservative churches in South Korea, which are fundamentalist in nature, tend to reject and ignore the responsibility of South Korean Christianity in the ecological crisis of the country. Rather, they want to argue that Buddhism in South Korea, with its temples and activities in the mountains, is more responsible for the ecological destruction. Also, these churches are more concerned with protecting their basic doctrines from recent theological developments concerning the ecological crisis, for example ecofeminism. See ibid., 13.

4. David Cho, *The Five-Fold Gospel and the Three-Fold Blessing* (Seoul: Word of God Publishers, 1998). This church is a part of the Korean Assemblies of God. Rev. Yongkee Cho leads his flock with a theology of the "Five-Fold Gospel and the Three-Fold Blessing," as the title of the book implies.

5. Lynn White Jr., "The Historical Roots of Our Ecological Crisis," *Science* 155 (March 10, 1967): 1203–7.

6. Keun Hwan Kang, "Roots and Presence of the Types of Faith in the Korean Churches," *Christian Thought* 546 (June 2004): 24.

7. Regarding South Korean Christianity's attitudes to society, Kyoung Jae Kim comments that Korean Protestant churches have been divided into two bodies and have developed in the mutual tension of that relationship. According to Kim, "One is the conservative part regarding themselves as 'evangelists.' Remaining outside the WCC, they are interested in individual salvation and church prosperity. The 'Pure Gospel Central Church' led by Rev. Cho Yong Ki may be categorized by that conservative religious society. The other body is a progressive Protestant society which supports KNCC's ecumenical theology and its doctrine. Trying to follow the social role of the Gospel, they have been interested in human rights, eco-environmental movements, unification activity, and laborer-peasant rights movements" (Kyoung Jae Kim, "A Theological Appraisal of Korean Pentecostal Pneumatology and Its Movement: A Comparison of the Five-Fold Gospel with the Baar statement of WCC" (2005), available at http://soombat.org/wwwb/CrazyWWWBoard/cgi?db = article&mode = read&num = 47& page = 1&ftype = 6&fval = &backdepth = 1 (website in Korean).

8. Clarence L. Bence, "Processive Eschatology: A Wesleyan Alternative," available at http://wesley.nnu.edu/theojrnal/11–15/14–04.htm.

9. Ibid.

10. John B. Cobb Jr., *Is It Too Late? A Theology of Ecology* (Beverly Hills: Bruce, 1972), 118, 126. Through his writing, Cobb has played a significant role in articulating the ecological implications of process theology. One of his principal arguments has been the inherent interconnections of God, humanity, and the natural world.

11. According to the tradition, Taoism originated with Lao Tzu, born circa 604 B.C., who was a solitary recluse and remained practically unknown. His ideas are espoused in the *Tao Te Ching*. See *Tao Te Ching*, trans. and commentary by Herrymon Maurer (London: Wildwood House Limited, 1986).

12. The word *Tao* literally means "path" or "way." But more specifically, it is the "way of the cosmos." It has a twofold aspect, namely *Mu* (nonbeing) and *Yu* (being). Hence, *Tao* is the Mystery, both transcendent and immanent.

13. Jung Young Lee, *The Theology of Change: A Christian Concept of God in an Eastern Perspective* (Maryknoll, N.Y.: Orbis, 1979), 44.

14. Sang-Won Doh, "A Cosmological Return in Theology," in *East Wind: Taoist and Cosmological Implications of Christian Theology*, ed. Charles Courtney (Lanham, Md.: University Press of America, 1997), 11–12.

15. "The Chipko movement was [composed of] a group of villagers in the Uttarakhand region of India who opposed commercial logging. The movement is best known for its tactic of hugging trees to prevent them [from] being cut down. This gave rise to the term 'tree hugger' for environmentalists. Also, it was notable in that the movement was led by women who were influenced by Mahatma Gandhi. The name of the movement comes from the Hindi word for 'embrace', as the villagers hugged the trees, and prevented the contractors from felling them" (Wikipedia [the free Internet encyclopedia]).

16. Isabel Carter Heyward, *The Redemption of God: A Theology of Mutual Relation* (Lanham, Md.: University Press of America, 1982).

17. Catherine Keller, "Why Apocalypse, Now?", unpublished manuscript.

18. Bokin Kim, "Sot'aesan's *Eun* as a Theoretical Foundation for Environmental Ethics," available at http://www.geocities.com/Athens/Forum/7602/sg5.html. After his awakening in 1916, Sot'aesan, the founder of Won Buddhism, a reformed sect of Buddhism in Korea, undertook a critical analysis of the Buddhist tradition. Sot'aesan intended to restore the original teaching of Siddhartha Gotama, known as "the Buddha," and create a reformed Buddhism based on the true spirit of Buddhism. The Buddha's teaching, *pratityasamupada*, in which all beings originate as dependent beings, was reformulated in Sot'aesan's teaching of *Eun,* in which all beings are mutually indebted, and thus beneficent to each other.

19. Gregory James Moses, "Process Relational Ecological Theology: Problems and Prospects" (July 2000), available at http://members.optusnet.com.au/~gjmoses/ecothlfr.htm.

20. "Matthew 25 has had a profound effect upon the development of Western humanitarianism, in that (despite the official doctrine of impossibility) people took seriously the idea that doing something to their fellow human beings meant doing something to God" (David R. Griffin, "A Process Theology of Creation," *Mid-Stream* 13 nos. 1–2 [Fall–Winter 1973–74]: 70).

21. Ibid., 70.

22. Sallie McFague, *Models of God: Theology for an Ecological, Nuclear Age* (London: SCM, 1987), 78.

23. Sallie McFague, *The Body of God* (Minneapolis, Minn.: Augsburg Fortress, 1993), 197.

24. Ibid., 197–98.

25. *Minjung* theology (literally "theology of the mass of the people") is a Korean version of liberation theology whose main thrust is toward the liberation of persons from social injustice, economic exploitation, political oppression, and racial discrimination, and that teaches that Jesus Christ is the liberator of these oppressed people. The major papers from a conference on *Minjung* theology, October 22, 1979, were edited by Yong-Bock Kim, director of the Christian Institute for the Study of Justice and Development, in Seoul, South Korea, and published as *Minjung Theology: People as the Subjects of History* (Maryknoll, N.Y.: Orbis Books, 1983).

26. Sang Sung Lee, *The Korean Church as People's Movement*, PhD diss., Drew University, 1998, 182.

27. Marjorie H. Suchocki, *God-Christ-Church: A Practical Guide to Process Theology* (New York: Crossroad, 1989), 198.

28. Ibid.

29. Rosemary Radford Ruether, *Sexism and God-talk: Toward a Feminist Theology*, (Boston: Beacon, 1983), 254–56.

30. John B. Cobb Jr., "Jürgen Moltmann's Ecological Theology in Process Perspective," *Asbury Theological Journal* 55, no. 1 (Spring 2000): 122.

31. John B. Cobb Jr., "Prophetic and Apocalyptic Hope," *Creative Transformation* 9, no. 1 (Fall 1999): 20.

32. Catherine Keller, "Talk about the Weather: The Greening of Eschatology," in *Ecofeminism and the Sacred*, ed. Carol J. Adams (New York: Continuum, 1993), 46.

33. Catherine Keller, "Eschatology, Ecology, and a Green Ecumenacy," in *Reconstructing Christian Theology*, ed. Rebecca S. Chopp and Mark Lewis Taylor (Minneapolis: Augsburg Fortress, 1994), 327–28.

34. Ibid.

35. Keller, "Talk about the Weather," 47.

36. Ibid.

37. Ted Peters, *God—the World's Future: Systematic Theology for a Postmodern Era* (Minneapolis: Augsburg Fortress, 1992), 369.

38. Keller, "Talk about the Weather," 46.

39. Ibid., 48.

40. Keller, "Eschatology, Ecology, and a Green Ecumenacy," 330.

41. Catherine Keller, *Apocalypse Now and Then: A Feminist Guide to the End of the World* (Boston: Beacon, 1996), 274.

42. Keller, "Eschatology, Ecology, and a Green Ecumenacy," 330.

43. Ibid., 342.

44. Keller, "Talk about the Weather," 48.

RESTORING EARTH, RESTORED TO EARTH: TOWARD AN ETHIC FOR REINHABITING PLACE | DANIEL T. SPENCER

1. Larry Rasmussen, *Earth Community, Earth Ethics* (Maryknoll, N.Y.: Orbis, 1998).

2. Daniel T. Spencer, *Gay and Gaia: Ethics, Ecology, and the Erotic* (Cleveland: Pilgrim, 1996).

3. Eleanor Haney, in her excellent ecofeminist work *The Great Commandment*, argues for a similar concept she terms "eco-social location" so as to signal the need to consider both the ecological and social dimensions of each of our locations. I have chosen to continue to use the term "ecological location" as an ongoing reminder that human social dynamics are always contained within—and are a subset of—the broader ecological realities in which we live. See Eleanor Haney, *The Great Commandment: A Theology of Resistance and Transformation* (Cleveland: Pilgrim, 1998).

4. "lgbt": Lesbian-Gay-Bisexual-Transgendered.

5. These perspectives are privileged not because the oppressed are somehow inherently or morally superior, but because reality viewed from their locations is more likely to generate critical knowledge. As Donna Haraway argues, "The standpoints of the subjugated are not 'innocent' positions. On the contrary, they are preferred because in principle they are least likely to allow denial of the critical and interpretative core of all knowledge. . . . 'Subjugated' standpoints are preferred because they seem to promise more adequate, sustained, objective, transforming accounts of the world" (Donna Haraway, "Situated Knowledges: The Science Question in Feminism and the Privilege of Partial Perspective," in *Simians, Cyborgs, and Women: The Reinvention of Nature* [New York and London: Routledge, 1991], 191).

6. Haraway suggests two helpful ways to address this. First, "articulation" rather than "representation." That is, rather than seeking an "objective" representative for the voices of nonhuman nature, those human actors with intimate knowledge of, and involvement with, these nonhuman agents articulate accountable, locatable, and partial perspectives from within this relational matrix. Second, "a praxis of affinity" rather than "identity." Since all of our locations and subjectivities are both shifting and multidimensional, so will be the ways we see and interpret the world. See Spencer, *Gay and Gaia*, 95–99.

7. Donna Haraway, *Primate Visions: Gender, Race, and Nature in the World of Modern Science* (New York and London: Routledge, 1989), 8.

8. Larry Rasmussen, *Earth Community, Earth Ethics*, 5

9. Ibid., 15–16.

10. "The SER Primer on Ecological Restoration," a publication of the Science and Policy Working Group of the Society for Ecological Restoration, 1st ed., available at http://www.ser.org/content/ecological_restoration_primer.asp.

11. Eric Higgs. *Nature by Design: People, Natural Process, and Ecological Restoration* (Cambridge, Mass.: MIT Press, 2003), 109.

12. William R. Jordan III, *The Sunflower Forest: Ecological Restoration and the New Communion with Nature* (Berkeley: University of California Press, 2003), 3. For a thoughtful examination of the tensions between preservationists and restorationists in environmentalism, see G. Stanley Kane, "Restoration or Preservation? Reflections on a Clash of Environmental Philosophies," in *Beyond Preservation: Restoring and Inventing Landscapes*, ed. A. Dwight Baldwin Jr., Judith De Luce, and Carl Pletsch (Minneapolis: University of Minnesota Press, 1994), 69–84.

13. Robert Eliot, "Faking Nature," *Inquiry* 25 (1982): 81–93.

14. Eric Katz, "The Big Lie: Human Restoration of Nature," *Research in Philosophy and Technology* 12 (1992): 231–43.

15. Andrew Light and Eric Higgs, "The Politics of Ecological Restoration," *Environmental Ethics* 18 (1996): 227–47.

16. Adam Rissien, unpublished Master's Thesis, Department of Environmental Studies, University of Montana, 2006.

17. See also Eric Higgs, "What is 'Good' Ecological Restoration?" *Conservation Biology* 11, no. 2 (1997): 338–48.

18. Given global climate change, it may no longer be feasible to restore some ecosystems to an earlier set of largely predisturbance conditions. See Laurel Kearns, "Cooking the Truth: Faith, Science, the Market, and Global Warming," elsewhere in the present volume.

19. Albert Borgmann, *Technology and the Character of Contemporary Life* (Chicago: University of Chicago Press, 1984).

20. Albert Borgmann, *Holding on to Reality: The Nature of Information at the Turn of the Millennium* (Chicago: University of Chicago Press, 1999).

21. Higgs, *Nature by Design*, 190.

22. See Larry Rasmussen, *Moral Fragments and Moral Community: A Proposal for Church in Society* (Minneapolis, Minn.: Augsburg Fortress, 1993), for Rasmussen's discussion of Borgmann's analysis of focal practices for moral community.

23. Higgs, *Nature by Design*, 12–13.

24. "Peak oil" refers to the theory of geophysicist M. King Hubbert that worldwide oil production will peak sometime between 2000 and 2010, after which it will decline, leading to energy shortages and a rapid escalation in the costs of production and consumption of fossil fuels.

25. See, for example, James H. Kunstler, *The Long Emergency: Surviving the Converging Catastrophes of the Twenty-First Century* (New York: Grove-Atlantic, 2005).

26. William Jordan, *The Sunflower Forest*, 21.

27. Ibid., 4–5.

28. Ibid., 197.

29. Joseph Sittler, "A Theology for Earth," *Christian Scholar* 37 (September 1954): 369–74. An edited version of Sittler's essay is also reproduced in *Worldviews, Religion, and the Environment: A Global Anthology*, ed. Richard Foltz (Belmont, Calif.: Thomson Wadsworth, 2003), 16–19.

30. Sittler, in Foltz, *Worldviews*, 19.

31. For example, see Rosemary Radford Ruether, *Gaia and God: An Ecofeminist Theology of Earth Healing* (San Francisco: HarperCollins, 1992); Anne Primavesi, *From Apocalypse to Genesis: Ecology, Feminism and Christianity* (Minneapolis, Minn.: Augsburg Fortress, 1991), *Sacred Gaia* (New York and London: Routledge, 2000), and *Gaia's Gift* (New York and London: Routledge, 2003); and Catherine Keller, *Apocalypse Now and Then: A Feminist Guide to the End of the World* (Boston: Beacon, 1997).

32. For an excellent analysis of the multidimensional nature of these reworkings of religion, see Mary Evelyn Tucker, *Worldly Wonder: Religions Enter Their Ecological Phase* (La Salle, Ill: Open Court, 2003).

33. Ibid., 10–11.

34. For an example of the cosmological context of restoration, see Brian Swimme and Thomas Berry, *The Universe Story: From the Primordial Flaring Forth to the Ecozoic Era—A Celebration of the Unfolding of the Cosmos* (San Francisco: HarperCollins, 1992).

35. Catherine Keller, "Talk about the Weather: The Greening of Eschatology," in *Ecofeminism and the Sacred*, ed. Carol J. Adams (New York: Continuum, 1993), 30–49.

36. Ibid., 36; emphasis in original.

37. See Sallie MacFague, "Imaging a Theology of God's Nature: The World as God's Body," in *Liberation Theology: An Introductory Reader*, ed. Curt Cadorette, Marie Giblin, Marilyn Legge, and Mary H. Snyder (Maryknoll, N.Y.: Orbis, 1992), 285.

38. See the work of Michael L. Humphreys for the development of a partnership ethic as a critique of, and replacement for, a more traditional Christian stewardship ethic. Michael L. Humphreys, "Partnership Ethics: Developing a New Paradigm for Christian Eco-social Environmental Ethics—A Study of the Columbia River Pastoral Letter Project," unpublished PhD diss., Drew University, 2006.

39. Tucker, *Worldly Wonder*, 50, 52.

CARIBOU AND CARBON COLONIALISM: TOWARD A THEOLOGY OF ARCTIC PLACE | MARION GRAU

In Memoriam: Jonathon Solomon (1932–2006), Gwich'in Elder and Chief. Solomon is credited with making the exploitation of the Arctic National Wildlife Refuge a national and international issue and for insisting that human-rights be part of the debate. At a time when nobody understood the significance of the threat to the refuge, he single-handedly defined it and organized the Gwich'in villages around it. Thanks to his vision and initiative, the Arctic National Wildlife Refuge remains protected and watched over.

1. Ken Madsen, *Under The Arctic Sun: Gwich'in, Caribou, and the Arctic National Wildlife Refuge* (Englewood, Colo.: Westcliffe, 2002), 174.

2. See Steven C. Dinero, "'The Lord Will Provide': The History and Role of Episcopalian Christianity in Nets'aii Gwich'in Social Development—Arctic Village, Alaska," *Indigenous Nations Studies Journal* 4, no. 1 (Spring 2003): 7.

3. The formulation is Catherine Keller's, in reference to theological interpretations of political actions; see Catherine Keller, *God and Power: Counter-Apocalyptic Journeys* (Minneapolis: Augsburg Fortress, 2005), 55ff.

4. Mark I. Wallace, *Finding God in the Singing River: Christianity, Spirit, Nature* (Minneapolis: Augsburg Fortress, 2005), 57.

5. Gilles Deleuze and Felix Guattari, *A Thousand Plateaus: Capitalism and Schizophrenia*, trans. Brian Massumi (Minneapolis: University of Minnesota Press, 1987), 12.

6. If we fail to look at complex problems and to eliminate the false separations between ecology, sustainability, indigenous sovereignty, and colonial expansions of faith, culture, and economy, we will fall short in crafting hopeful alternatives. Below, I make use of the work of economists, ecologists, energy specialists, historians, geographers, and journalists, all of whom have viewpoints that bear directly on the issues under discussion.

7. There are those who argue that this "war" is simply a skirmish to distract so-called liberals from paying attention to what U.S. neoconservatives and oil industry lobbyists really want: tax breaks, energy subsidies, and wide-scale deregulation. Should that, in fact, be the case, we have good cause to broaden the scope of our vision. David M. Standlea mentions Ralph Nader as one of those who have argued that "ANWR was a smokescreen and a brilliant strategy" to achieve other long-held goals of Big Oil. See David M. Standlea, *Oil, Globalization, and the War for the Arctic Refuge* (Albany: State University of New York Press, 2006), 68, 101.

8. Anne M. Daniell, "Incarnating Theology in an Estuary-Carnival Place: New Orleans in the Pontchartrain Basin" (Ph.D. diss., Drew University, 2005), 12.

9. This is also suggested in Standlea, *Oil, Globalization*, 11.

10. Evon Peter is a former chief of the Neetsaii Gwich'in from Arctic Village, and is currently chairperson of Native Movement (nativemovement.org). The full text of his comments is available at http://www.oilonice.org/toolkit/html/community.html.

11. Dinero, "'The Lord Will Provide,'" 3.

12. Personal communication with Bishop of the Episcopal Diocese of Alaska, the Right Rev. Mark MacDonald.

13. Dinero, "'The Lord Will Provide,'" 4.

14. See, for example, Walter R. Borneman, *Alaska: Saga of a Bold Land* (New York: HarperCollins, 2003), 18; Jared Diamond, *Guns, Germs, and Steel: The Fates of Human Societies* (New York: Norton, 1997), 44–45; and Spencer Wells, *The Journey of Man: A Genetic Odyssey* (New York: Random House, 2002), 139–44.

15. Karl W. Luckert, *The Navajo Hunter Tradition* (Tucson: University of Arizona Press, 1975), 9.

16. Borneman, *Alaska*, 534.

17. Marla Cone, "Dozens of Words for Snow, None For Pollution," *Ode* 3, no. 9 (November 2005): 47.

18. See also Whitney Bauman, "*Creatio ex Nihilo, Terra Nullius,* and the Erasure of Presence," elsewhere in the present volume.

19. See an account of the history of ANCSA in John Strohmeyer, *Extreme Conditions: Big Oil and the Transformation of Alaska* (Anchorage, Alaska: Cascade, 1997), and a survey of native communities after ANCSA in Thomas R. Berger, *Village Journey: The Report of the Alaska Native Review Commission* (New York: Hill and Wang, 1985).

20. Jamie Wilson, "Senate blocks attempt to allow oil drilling in Alaskan wildlife reserve," *Guardian*, December 22, 2005.

21. Steven C. Dinero, "The Real Cost of Drilling," *Christian Science Monitor*, October 8, 1999.

22. Sonia Shah, *Crude: The Story of Oil* (New York: Seven Stories, 2004), 59.

23. The aftermath of the Exxon Valdez disaster stands as a call to realize the long-term diminishment of an ecosystem after an oil-spill. Oil companies promise to do the development "right," but, as Riki Ott has shown, corporate data is routinely doctored and distorted: the gains overestimated, the risks underestimated. In addition, ExxonMobil has repeatedly broken the promises that were made in helping to restore the area. See Riki Ott, *Sound Truth and Corporate Myths: The Legacy of the Exxon Valdez Spill* (Cordova, Alaska: Dragonfly Sisters, 2005).

24. See Paul Roberts, *The End of Oil: On the Edge of a Perilous New World* (Boston: Houghton Mifflin, 2004). The main point of debate among experts at this point is no longer *if* oil-production is going to peak but *when*, and what the eventual consequences could be. Multiple opinions are represented in World Watch Institute, "Peak Oil Forum," in *World Watch: Visions for a Sustainable World* 19, no. 1.

25. See these books about the oil industry: Shah, *Crude*; Roberts, *The End of Oil*; and Joseph J. Romm, *The Hype About Hydrogen: Fact and Fiction in the Race to Save the Climate* (Washington, D.C.: Island, 2004), among many others.

26. Andrew Taylor, "Senate Panel drops spending cuts, seeks to revive Arctic drilling plan," *Boston Globe*, March 8, 2006.

27. See the activist Web site Exxpose Exxon at http://www.exxposeexxon.com.

28. See Ott, *Sound Truth and Corporate Myths*.

29. After encountering missionaries or traders, many Gwich'in became sedentary, switching from their previous semi-nomadic habits. This increased their vulnerability to changes in the behavior of the caribou herd. See Gwich'in Steering Committee, the Episcopal Church, and Richard J. Wilson, "A Moral Choice For the United States: The Human Rights Implications for the Gwich'in of Drilling in the Arctic National Wildlife Refuge (ANWR)" (2005), available at http://www.gwichinsteeringcommittee.org/GSChumanrightsreport.pdf, 17.

30. Alaska Native Oil and Gas Working Group, "Oil and the Alaska Native Claims Settlement Act" (2002), available at http://www.treatycouncil.org/Alaska%20Native%20Working.pdf.

31. See Jared Diamond, *Collapse: How Societies Chose to Fail or Succeed* (New York: Viking, 2005).

32. Vine Deloria describes "reservation life" as a place where Native Americans are "not allowed to be Indians" but "cannot become whites," an often irresolvable tension that has resulted in "alcoholism and suicide." While the Gwich'in do not live on a reservation, many of the same phenomena are known among them. See Vine Deloria, *God is Red: A Native View of Religion* (Golden, Colo.: Fulcrum, 1994), 242.

33. Ms. Kassi's reference is to the murder of Ken Saro-Wiwa, a Nigerian activist protesting Shell Oil's development plans in that country. See Madsen, *Under The Arctic Sun*, 174.

34. The quoted phrase is Charlene Spretnak's, as quoted in Wallace, *Finding God in the Singing River*, 60.

35. Deloria, *God is Red*, 266.

36. Rick Bass, *Caribou Rising: Defending the Porcupine Herd, Gwich'in Culture, and the Arctic National Wildlife Refuge* (San Francisco: Sierra Club, 2004), 60.

37. Dinero, "'The Lord Will Provide,'" 7, and personal communication with Bishop of the Episcopal Diocese of Alaska, the Right Rev. Mark MacDonald.

38. Bass, *Caribou Rising*, 72.

39. Dinero, "'The Lord Will Provide,'" 12–13.

40. Much of the content of this document is based on the UN Charter of Human Rights.

41. On this issue, see also Aaron Sachs, *Eco-Justice: Linking Human Rights and the Environment* (Washington, D.C.: Worldwatch Institute, 1995).

42. Wallace, *Finding God in the Singing River*, 145.

43. Bass, *Caribou Rising*, 60.

44. Barry Lopez, *Arctic Dreams* (New York: Vintage, 1986), 169.

45. Susan Joy Hassol, *Impacts of a Warming Arctic*, Arctic Climate Impact Assessment (Cambridge: Cambridge University Press, 2004), 72, and also available at http://www.acia.uaf.edu/pages/overview.html.

46. Evon Peter, "The People and the Caribou Are One," *Voices from the Earth* 6, no. 1 (Spring 2005). The text is also available at http://www.sric.org/voices/2005/v6n1/caribou.html.

47. Bass, *Caribou Rising*.

48. Paraphrased from handwritten notes taken in the course entitled "Re/Locating Theologies," and given in the Fall 2005 at the Graduate Theological Union, Berkeley, California.

49. As here: http://www.juneauempire.com/anwr/lifeblood.shtml.

50. Berger, *Village Journey*, 95.

51. Vizenor, as quoted in Sharon D. Welch, *After Empire: The Art and Ethos of Enduring Peace* (Minneapolis: Augsburg Fortress, 2004), 65.

52. Donna J. Haraway, *Simians, Cyborgs, and Women: The Reinvention of Nature* (New York and London: Routledge, 1991), 154.

53. Clara Sue Kidwell, Homer Noley, George E. Tinker, and Jace Weaver, eds., *A Native American Theology* (Maryknoll, N.Y.: Orbis, 2001), 120.

54. Lewis Hyde, *Trickster Makes This World: Mischief, Myth, and Art* (New York: Farrar, Straus and Giroux, 1998), 10.

55. Borneman, *Alaska*, 464, 467f.

56. Winona LaDuke, *Recovering the Sacred: The Power of Naming and Claiming* (Cambridge, Mass.: South End, 2005), 249–53.

57. Beth Gorham, "U.S. Senate blocks attempt to allow oil drilling in Alaska wildlife refuge," *Canadian Press*, December 22, 2005.

58. Daniell, "Incarnating Theology," 8.

DIVINING NEW ORLEANS: INVOKING WISDOM FOR THE REDEMPTION OF PLACE | ANNE DANIELL

This chapter contains portions, revised, of an article that recently appeared in *Environmental Ethics*—"Toward a Materialist Environmental Ethic." *Environmental Ethics* 28. no. 4 (Winter 2006): 375–393.

1. A traditional epithet for New Orleans, of which the exact origin is unknown. It invokes the idea that New Orleans is an environment where one can let go of one's cares.

2. A flourish on Luke 13:34/Matthew 13:37, Jesus' lament over Jerusalem, in the persona of Wisdom.

3. "Louisiana Governor: 'We will rebuild,'" September 15, 2005, available online at http://www.cnn.com/2005.

4. Amy Wold, "'Toxic soup' concerns—all hype?: Experts debate true risks in N.O.," in the *Advocate* (Baton Rouge), February 13, 2006. The precise spots in which soil samples were taken, and the manner of interpreting the results, played roles in how these different groups assessed the situation. For their part, the EPA and the DEQ emphasized that while water and soil samples revealed contaminants, the levels of toxicity did not differ significantly from pre-Katrina samples taken in the same general areas. Environmental and neighborhood association groups, on the other hand, emphasized that the type and degree of contamination was different during and immediately following the flooding.

5. When Hurricane Rita made landfall near the Louisiana-Texas border, about 250 miles west of New Orleans, its storm bands caused further weakening of the levees and renewed inundation in New Orleans' Lower Ninth Ward.

6. The title Mayor Ray Nagin gave to his initial repopulation and rebuilding plan.

7. Mark Fischetti, "Drowning New Orleans," *Scientific American*, October 2001, 78.

8. Center for Cooperative Research: Hurricane Katrina, available at http://www.cooperativeresearch.org/timeline.jsp?timeline = hurricane_katrina & startpos = 0#hurricane_katrina_3109.

9. See Darwin Spearing, *Roadside Geology of Louisiana* (Missoula: Mountain, 1995).

10. For some time now, Louisiana's federal senators and representatives (former and current), especially John Breaux, Mary Landrieu, and David Vitter, have pushed for national funding to restore Louisiana's coastal wetlands. While a few minor projects have gotten underway, overall coastal restoration has been massively underfunded.

11. "President Discusses Hurricane Relief in Address to the Nation." Text available at http://www.whitehouse.gov/news/releases/2005/09/2005 0915-8.html.

12. Ibid.

13. Now, a year after the city's evacuation, these demographics have changed again and likely will continue to do so. More African American residents have returned. There has also been growth in the Hispanic population, as many Hispanic immigrants have moved in to do the grueling work of gutting homes, clearing mounds of trash, and rebuilding homes. Mayor Nagin's infamous Martin Luther King Jr. Day "chocolate city" remark was perhaps not so bigoted as some want to believe. Though it may have been uttered with political calculation (Nagin was positioning himself for re-election, and needed the African-American vote), it could also be heard as a statement of hope about bringing back the city's majority, who had not yet found a way to return.

14. Native American influence on African-based traditions in the circum-Caribbean is not well documented, but instead is widely presumed. As such, it is part of the mythology of New Orleans' West African–based cultural performance groups, such as the Mardi Gras Indians.

15. Danny Duncan Collum, "America's Holy City," *Sojourners* 34, no. 10: 13. While I agree that cultural and religious vibrancy often are found in economically impoverished communities, I at the same time feel uncomfortable automatically linking economic poverty with richness of spirit. Still, it

is true that much that has given shape to a spirit of place in New Orleans derives from the musical and cultural performance traditions of historically poor and marginalized communities.

16. Andrei Codrescu, *New Orleans, Mon Amour: Twenty Years of Writing From the City* (New York: Algonquin, 2006), 71.

17. For more on the concepts of "place," "sacred place," "spirit of place," and a New Orleans "sense of place," see Anne Daniell, *Incarnation Theology in an Estuary-Carnival Place: New Orleans in the Pontchartrain Basin*, PhD Diss., Drew University, October 2005. Wendell Berry's essays have influenced my own understanding of "spirit of place" (see especially "The Work of Local Culture"in *What Are People For?* [San Francisco: North Point, 1990]). Contemporary theologians who have touched upon spirit of place or theology of place include Sallie McFague (*The Body of God* [Minneapolis: Augsburg Fortress, 1993]), Jürgen Moltmann (*God in Creation* [Minneapolis: Augsburg Fortress, 1985]), and Catherine Keller (*Apocalypse Now and Then* [Boston: Beacon, 1996]). The few theologians whose major works specifically address "theology of place" include Geoffrey R. Lilburne (*A Sense of Place: A Christian Theology of the Land* [Nashville: Abingdon, 1989]) and John Inge (*A Christian Theology of Place* [Burlington: Ashgate, 2003]).

18. A number of religious and theological traditions have embodied this dual aspect of wisdom in one religious figure, such as Mahayana Buddhism's Kuan-Yin, the bodhisattva of wisdom and compassion, and Hildegard of Bingen's *Caritas*, or divine love, often interchangeable with *Sapientia* (Latin for wisdom).

19. For more on Wisdom as a theological figure see Denis Edwards's *Jesus, the Wisdom of God: An Ecological Theology* (Maryknoll, N.Y.: Orbis, 1995) and Elizabeth Johnson's *She Who Is: The Mystery of God in Feminist Theological Discourse* (New York: Crossroad, 1992). Roland Murphy's *The Tree of Life: An Exploration of Biblical Wisdom Literature* (Grand Rapids, Mich.: Eerdmans, 1996) discusses the various roles wisdom plays in the Hebrew Bible.

20. In Hebrew, *Hokmah*; in Greek, *Sophia*; in Latin, *Sapientia*.

CONSTRUCTING NATURE AT A CHAPEL IN THE WOODS | RICHARD R. BOHANNON II

1. See Anna Peterson's discussion of "chastened constructionism" for a more detailed elaboration; Anna L. Peterson, *Being Human: Ethics, Environment, and Our Place in the World* (Berkeley: University of California Press, 2001).

2. This sacralization need not be in an explicit view of a church being a "sacred" place—an idea no doubt foreign to many in the evangelical world

that is most relevant to this essay. Rather, the expenditure of a large sum of money and resources for a building to be used for explicitly religious purposes, even when it is privately funded, is perhaps itself enough to lend a building a special, religious significance. That is, I would argue that simply the erection of a building for religious purposes already lends the building a normative weight within the community building it; money that could have been used for causes central to the community's mission (for example, feeding the poor or saving souls) was instead invested in a sanctuary.

3. E. Fay Jones, *Outside the Pale: The Architecture of Fay Jones* (Fayetteville: University of Arkansas Press, 1999), 42.

4. E. Fay Jones, a student of Frank Lloyd Wright, was awarded the American Institute of Architects (AIA) Gold Medal in 1990, arguably the highest honor an architect can receive from colleagues. Thorncrown Chapel itself won a prestigious 1981 AIA Honor Award, and was also nominated by an AIA panel of architects in 1991 as *the* most outstanding American architectural work of the 1980s, ahead of such seminal works as Maya Lin's Vietnam Memorial in Washington, D.C., and Richard Meier's High Museum of Art in Atlanta. Most recently, Thorncrown Chapel has received the AIA 2006 Twenty-Five Award, given each year to a building that has "stood the test of time." While clearly part of the United States's architectural establishment, Jones (a graduate of and professor at the University of Arkansas) was largely an outsider among the elite centers of power in the architectural field, being from a relatively unremarkable architecture school, owning a small firm well outside of architecture's geographic centers (such as Chicago or New York), and having built no large projects.

Fay Jones designed several religious buildings reminiscent in form to Thorncrown, including the glass-walled Cooper Memorial Chapel in nearby Bella Vista, Arkansas (1988), and the entirely wall-less Pine Eagle, a small, interfaith chapel at a camp in southern Mississippi (1991), among others. See Robert Adam Ivy Jr., *Fay Jones* (Washington, D.C.: American Institute of Architects, 2001). After his retirement, his firm, now Maurice Jennings and David McKee, Architects, has gone on to design several additional churches and chapels, each with the requisite gabled roof, intricate truss work, and large, clear-glass windows.

5. Thorncrown Chapel, "About Us," available at http://www.thorncrown.com/aboutus.htm.

6. A fear of worshipping creation is, of course, commonplace within discussions of environmentalism among evangelicals. See Laurel Kearns, "Saving the Creation: Christian Environmentalism in the United States," *Sociology of Religion* 57 (1996): 55–70.

7. See, for example, Bruno Latour, *Pandora's Hope: Essays on the Reality of Science Studies* (Cambridge, Mass.: Harvard University Press, 1999), 174–215.

8. Jonathan Murdoch, "Ecologising Sociology: Actor-Network Theory, Co-construction and the Problem of Human Exemptionalism," *Sociology* 35, no. 1 (2001): 120.

9. The role of benches in human interaction becomes especially apparent when you compare the seating arrangements of different Christian traditions. The arrangement of benches facing each other, such as in a Quaker meeting, for instance, when compared to rows of benches all facing forward, such as at Thorncrown, both reflects and fosters their different types of social organization. At Thorncrown the congregation is much like an audience, facing a religious leader and witnessing religious activities at the front of the chapel (whether it be a wedding or a sermon), whereas in a silent Quaker meeting the congregants face one another, with no clear leader. See Richard Giles, *Re-Pitching the Tent: Re-Ordering the Church Building for Worship and Mission*, 3rd ed. (Collegeville, Minn.: Liturgical Press, 2004).

10. Bruno Latour, *Reassembling the Social: An Introduction to Actor-Network-Theory* (New York: Oxford University Press: 2005), 9, 69.

11. Harry W. Paul, "The Pasteurization of France" [Review], *American Journal of Sociology* 96, no. 1 (1990): 233.

12. Latour, *Reassembling*, 71.

13. Ibid., 68.

14. Pierre Bourdieu, *Outline of a Theory of Practice*, trans. Richard Nice (Cambridge: Cambridge University Press: 1977), 91.

15. For Bourdieu, a social field is "a network, or a configuration, of objective relations between positions. These positions are objectively defined . . . by their present and potential situation . . . in the structure of the distribution of species of power (or capital) whose possession commands access to the specific profits that are at stake in the field, as well as by their objective relation to other positions" (Pierre Bourdieu and Loïc J. D. Wacquant, *An Invitation to Reflexive Sociology* [Chicago: University of Chicago Press, 1992], 97). For the difference between the methodologies of Bourdieu and Latour, see, for example, Pierre Bourdieu, *Science of Science and Reflexivity*, trans. Richard Nice (Chicago: University of Chicago Press, 2004), 26–30.

16. One's "habitus" is a nonconscious symbolic map of reality, of what is possible and what is desirable, which in part functions similar to what Peter Berger calls a "social theodicy," an explanation of how and why perceived reality is what it is. See Peter Berger, *The Sacred Canopy: Elements of a Sociological Theory of Religion* (New York: Anchor, 1990), 54–59.

17. Bourdieu and Wacquant, *Invitation*, 100.

18. See, for example, Carolyn Merchant, *The Death of Nature: Women, Ecology, and the Scientific Revolution* (San Francisco: HarperSanFrancisco, 1990); Ariel Salleh; *Ecofeminism as Politics: Nature, Marx, and the Postmodern* (London: Zed, 1997); and Robert D. Bullard, *Dumping in Dixie: Race, Class, and Environmental Quality*, 3rd ed. (Boulder: Westview, 2000).

19. Donna Haraway, *Primate Visions: Gender, Race, and Nature in the Modern World* (New York and London: Routledge, 1989).

20. Using the term "nonhumans" instead of "nature" does not solve the ethical language problems of defining humans in dichotomy to the rest of the world, of course. As Bruno Latour argues, however, it might prove useful specifically for research. He remarks, "There is nowhere something which is nonhuman. It is a concept and it is a practical concept to do research, and it does vastly better research than the object/subject dichotomy" ("A Strong Distinction Between Humans and Nonhumans Is No Longer Required for Research Purposes: A Debate between Bruno Latour and Steve Fuller," ed. Collin Baron, *History of the Human Sciences* 16, no. 2 [2003]: 80).

21. "Wedding Prices and Information," Thorncrown Chapel Web site, available at http://www.thorncrown.com/rules.htm. Interestingly, the Cooper Memorial Chapel, built by E. Fay Jones in neighboring Bella Vista, Arkansas, with a remarkably similar design and setting, hosts weddings for people of any (or no) religious tradition. Plans are also being developed to convert a house designed by Jones, also in northwest Arkansas, into a synagogue. See Adam Wallworth, "Proposal would turn E. Fay Jones house into Jewish synagogue," *Northwest Arkansas Times*, January 21, 2006.

22. Ivy, *Fay Jones*, 32.

23. Trees have had an ambiguous, and often negative, history within Christianity—from saints chopping down sacred groves to cathedrals emulating the forest in their ceiling's stonework. See Nicole A. Roskos, "Felling Sacred Groves: Appropriation of a Christian Tradition for Antienvironmentalism," elsewhere in the present volume.

24. Similarly, Phil Macnaghten and John Urry have shown how idealized notions of a very humanized "countryside" developed in England in direct relationship to the industrial growth of towns and cities. See Phil Macnaghten and John Urry, *Contested Natures* (London: Sage, 1998), 174.

25. Jones writes: "There are a lot of these transitional areas where you're trying to string out these inside-out relationships in a horizontal way. In these areas, there can be no typical openings; there must be an extension of ceiling materials from inside to outside, and stone floor materials going from inside to outside without interruption" (Jones, *Outside the Pale*, 70).

26. Erik Swyngedouw, "The City As a Hybrid: On Nature, Society and Cyborg Urbanization," *Capitalism, Nature, Socialism* 7, no. 2 (1996): 65–80.

27. There is a strong tradition in modern architecture, of course, of explicitly referring to buildings as machines, such as Le Corbusier advocating for "House-Machines" and "House-Tools." See Le Corbusier, *Towards a New Architecture*, trans. John Rodker (Mineola, N.Y.: Dover, 1986), 7, 263.

28. Donna Haraway, *Simians, Cyborgs, and Women: The Reinvention of Nature* (New York and London: Routledge, 1991), 151–53.

29. For Latour, all science (at least as practiced in the contemporary West) assumes a basic antidemocratic stance. He draws this out by distinguishing between "Science" (with a capital "S") and "the sciences," using Plato's allegory of the cave as a metaphor, in an effort both to critique how Western Science has developed while also salvaging scientific practice and insights from a purely relativist deconstruction. Science (as opposed to the sciences) functions much like the enlightened subjects who arise out of Plato's cave, ascertain the true nature of reality, and then return to the cave so that they might dispense their knowledge upon those bound by the chains of ignorance. Science, then, creates a sort of cult of experts who alone are able to understand nature-reality, and whose facts and statements cannot be disputed. See Bruno Latour, *Politics of Nature: How To Bring the Sciences into Democracy*, trans. Catherine Porter (Cambridge, Mass.: Harvard University Press, 2004), 10–14.

30. Ibid., 53.

31. Ibid., 59.

32. Fay Jones is following the tradition of his mentor, Frank Lloyd Wright, with this notion of organic architecture, or designing buildings that are (in Wright's words) "integral and consistent with the laws of Nature; the love of human-nature square with human life" (Frank Lloyd Wright, *An Organic Architecture: The Architecture of Democracy* [Cambridge, Mass.: MIT Press, 1970], 8).

33. "Thorncrown Chapel Weddings," Thorncrown Chapel Web site, available at http://www.thorncrown.com/weddings.htm.

34. Latour, *Reassembling*, 68.

35. He writes, "My principal objection to wilderness is that it may teach us to be dismissive or even contemptuous of . . . humble places and experiences" (William Cronon, "The Trouble with Wilderness; or, Getting Back to the Wrong Nature," in *UnCommon Ground: Rethinking the Human Place in Nature*, ed. William Cronon [New York: Norton, 1996], 86–87).

36. Jones, *Outside the Pale*, 42.

FELLING SACRED GROVES: APPROPRIATION OF A CHRISTIAN TRADITION FOR ANTIENVIRONMENTALISM | NICOLE A. ROSKOS

1. Judith M. Hadley, *The Cult of Asherah in Ancient Israel and Judah* (Cambridge: Cambridge University Press, 2000), 78. See also Larry Rasmussen, *Earth Community, Earth Ethics* (Maryknoll, N.Y.: Orbis, 1996).

2. Ibid., 82.

3. The New Revised Standard Version translates Asherim as "sacred poles" : "But this is how you must deal with them: break down their altars, smash their pillars, hew down their sacred poles."(Deut. 7:5).

4. W. F. Tamblyn, "British Druidism and the Roman War Policy," *American Historical Review* 15, no. 1 (October 1909): 25.

5. Ibid.

6. Robert Pogue Harrison, *Forests: The Shadow of Civilization* (Chicago: University of Chicago Press, 1992), 55.

7. Ibid., 57.

8. Charles. J. Glacken, *Traces on the Rhodian Shore* (Berkeley: University of California Press, 1967), 310.

9. Robert Markus, *Gregory the Great and His World* (Cambridge: Cambridge University Press, 1997), 184

10. The World Tree was one of many oak pillars representing a connection between the heavens and earth for the Saxons. Irminsal corresponds with the Old Norse World Tree, Yggdrasil, or "Odin's steed," the ash tree upon which took place Odin's self-sacrifice. S.V. "Irminsal," Wickipedia, the free online encyclopedia, available at http://en.wikipedia.org/wiki/Irminsul.

11. Thomas F. X. Noble and Ephraim Emerton, eds., *The Letters of Saint Boniface* (New York: Columbia University Press, 2001), xiv, and David Chidester, *Christianity: A Global History* (San Francisco: HarperSanFrancisco, 2000).

12. Robert Atwell and Christopher L. Webber, eds., *Celebrating the Saints: Devotional Readings for Saints' Days* (London: Morehouse, 2001).

13. The authors to be discussed below are all aligned against modern movements of forest preservation and put forth their agenda through evocations, and celebration, of the tree-felling saints.

14. There are many evangelicals who would not agree with Beisner. See, for example, the Evangelical Environmental Network (EEN), whose Web site is available at http://www.een.org.

15. E. Calvin Beisner, *Where Garden Meets Wilderness: Evangelical Entry into the Environmental Debate* (Grand Rapids, Mich.: Acton Institute for the Study of Religion and Liberty, 1997), xix.

16. Ibid., xv.

17. Ibid., 53–57.

18. An excellent overview of this movement is provided in Laurel Kearns, "Wise-Use Movement," in *Encyclopedia of Religion and Nature* (Bristol, U.K.: Thoemmes-Continuum, 2005), 1755–58.

19. Acton defines itself in very different terms: "With its commitment to pursue a society that is free and virtuous, the Acton Institute for the Study of Religion and Liberty is a leading voice in the national environmental and social policy debate. The Acton Institute is uniquely positioned to comment on the sound economic and moral foundations necessary to sustain humane environmental and social policies." The Acton Institute is a nonprofit, ecumenical think tank located in Grand Rapids, Michigan. The Institute works internationally to "promote a free and virtuous society characterized by individual liberty and sustained by religious principles." More of this is available at at http://www.acton.org/press/releases.php?release = 76.

20. Further information on ICES and the Cornwall Declaration is available at http://www.stewards.net/About.htm. ICES says that "religious, economic, and scientific traditions . . . are now under (the) assault" of "liberal environmental advocacy." See also Kearns in this volume.

21. The full text of the Cornwall Declaration is available from the Acton Institute Web site, http://www.acton.org/policy/environment/cornwall.html.

22. Ibid.

23. Richard T. Wright, "Tearing Down the Green: Environmental Backlash in the Evangelical Sub-Culture," available at http://www.asa3.org/asa/pscf6–95wright.html.

24. World Watch Institute, "State of the World, 2003," chap. 2 ("Watching Birds Disappear"), available at http://www.worldwatch.org/node/3613.

25. Robert Whelan, *The Cross and the Rainforest: A Critique of Radical Green Spirituality* (Grand Rapids, Mich.: Acton Institute for the Study of Religion and Liberty, 1997), 39.

26. Ibid., 38.

27. Ibid.

28. Whelan cites the United Nations Food and Agriculture program in general support of his argument, but offers no statistics.

29. R. S. Whaley, "An Introduction to Forest Issues," Religious Campaign for Forest Conservation, available at http://www.creationethics.org/index.cfm?fuseaction = webpage&page_id = 8.

30. Janet N. Abromavitz, "Taking a Stand: Cultivating a New Relationship with the World's Forests," Worldwatch [Institute], paper 140, available at http://www.worldwatch.org/pubs/ paper/140.

31. The Acton Institute for the Study of Religion and Liberty, available at http://www.acton.org/tfavs/ and http://www.acton.org/programs/students/novak.

32. Joan Andersen, "The Christmas Tree," available at http://www.users.erols.com/bcccsbs/christmas/tree.htm.

33. George N. Appell, "The Unexpected Consequences of Clear-Cutting in Borneo and Maine," *The Forest Ecology Network* 4, no. 1 (Winter 2000).

34. Whelan, *The Cross and the Rainforest*, 40.

35. Ibid.

36. Sebastien Rouillard, "Parthenie, our Historie de la tres auguste eglises de Chartres," in Robert Banner, ed., *Chartres Cathedral* (London: Thames and Hudson, 1969).

37. Harrison, *Forests*, 178.

38. Ibid.

39. Harvey Cox, "Mammon and the Culture of the Market," in Richard Madsen, William M. Sullivan, Ann Swidler, and Steven M. Tipton, eds., *Meaning and Modernity: Religion, Polity and the Self* (Berkeley: University of California Press, 2002), 124–35. See also idem, "The Market as God," originally published in the *Atlantic*, and available at http://www.theatlantic.com/issues/99mar/marketgod.htm; and David Loy, "Religion and the Market," available at http://www.religiousconsultation.org/loy.htm.

40. Beldan Lane, "Open the Kingdom for a Cottonwood Tree," *Christian Century* 29 (October 1997).

41. Ibid.

42. More information about both organizations is available at the Opening the Book of Nature Web site, http://www.bookofnature.org.

43. Matthew Immergut, PhD Diss., Drew University, 2006.

44. Ibid.

45. More information is available at the Au Sable Institute's Web site, http://www.ausable.org/au.newsletter.sp2004.sheldon.cfm.

ETHICS AND ECOLOGY: A PRIMARY CHALLENGE OF THE DIALOGUE OF CIVILIZATIONS | MARY EVELYN TUCKER

1. Ten volumes resulted from these conferences as well as a Web site administered through the university, available at http://www.environment.harvard.edu/religion.

2. Mihaly Csikszentmihalyi, *The Evolving Self* (New York: HarperCollins, 1993), 18.

3. Ibid., 19.

4. Ibid., 25.

5. The full text of the document is available at http://www.earthcharter.org/files/charter/charter.pdf.

6. Agenda 21 Earth Summit: United Nations program of Action from Rio. New York: United Nations, 1992.

7. http://www.earthcharter.org/files/charter/charter.pdf.

8. Ibid.

THE FIRM GROUND FOR HOPE: A RITUAL FOR PLANTING HUMANS AND TREES | HEATHER MURRAY ELKINS, WITH ASSISTANCE FROM DAVID WOOD

1. More information is available at http://www.vanderbilt.edu/chronopod.

MUSINGS FROM WHITE ROCK LAKE: POEMS | KAREN BAKER-FLETCHER

The poem "Ha Shem" is reprinted by permission of Karen Baker Fletcher, *Dancing with God: The Trinity from a Womanist Perspective* (St. Louis: Chalice Press, 2006), 52–54.

CONTRIBUTORS

Karen Baker-Fletcher is Associate Professor of Systematic Theology at Southern Methodist University. She is the author of several books, among them: *Dancing with God: The Trinity from a Womanist Perspective* (Chalice, 2006), and *Sisters of Dust, Sisters of Spirit: Womanist Wordings on God and Creation* (Augsburg Fortress, 1998).

Whitney A. Bauman is a PhD candidate in theology at the Graduate Theological Union in Berkeley, California. He is currently finishing his dissertation, "From *Creatio ex Nihilo* to *Terra Nullius*: The Colonial Mind and the Colonization of Creation," which focuses on developing a postcolonial understanding of the Christian doctrine of creation. Whitney is also on the Steering Committee of the Theological Roundtable on Ecological Ethics and Spirituality (TREES) at the Graduate Theological Union.

Sharon Betcher is Associate Professor of Theology at the Vancouver School of Theology. A constructive theologian with an emphasis on pneumatological dimensions, she has published articles in the areas of postcolonial theologies, disabilities and religion, and progressive or regenerative Christianities, as well as ecotheology. She is the author of *Spirit and the Politics of Disenchantment* (Fortress Press, 2007).

Richard R. Bohannon II is a PhD student in Religion and Society at Drew University, where his research focuses on Christian environmentalism and the built environment in the United States. Under the guidance of

GreenFaith in New Jersey, he is currently directing Building in Good Faith, an effort to develop a comprehensive "green" guide to religious architecture.

Anne Daniell is an independent scholar living and writing in New Orleans. She received her PhD from Drew University in 2005 with an emphasis in Ecological Theology. Her dissertation explored a "spirituality of place," looking specifically at New Orleans and the Pontchartrain Basin. Her publications include *Process and Difference: Between Cosmological and Post-structuralist Postmodernisms* (SUNY Press, 2002), a volume coedited with Catherine Keller.

Heather Murray Elkins is Professor of Worship, Preaching, and the Arts at the Theological School and Graduate Division of Religion at Drew University. She began her teaching career as an instructor in the first bilingual independent school on the Navaho Reservation and she has served as a local church pastor, a truck-stop chaplain, a university chaplain, and an academic dean. Her courses at Drew include feminist studies in liturgy and preaching as well as Appalachian Studies of health, land, and the arts. Her most recent books are *The Holy Stuff of Life* (Pilgrim Press, 2006) and *Wising Up: Ritual Resources for Women of Faith in Their Journey of Aging* (Pilgrim Press, 2006).

Antonia Gorman is a PhD candidate in Theological and Religious Studies at Drew University. She currently is writing her dissertation, a constructive theology of salvation that brings together feminist, process, and ecological theologies with insights from the animal-protection movement. She is a Will Herberg Scholar and a winner of the Mulder Prize for academic excellence. Her publications include an essay in the book *The Way of Compassion*, as well as multiple contributions to the periodicals *Satya* and the *Religious Observer*.

Marion Grau, a native of Germany, is Associate Professor of Theology at the Church Divinity School of the Pacific, a member school of the Graduate Theological Union in Berkeley, California, where she has taught since 2001. Her essays have appeared in *Strike Terror No More* (Chalice, 2002) and *Postcolonial Theologies: Divinity And Empire* (Chalice, 2004). She is the author of *Of Divine Economy: Refinancing Redemption* (T. and T. Clark/Continuum, 2004) and coeditor with Rosemary Radford Ruether of *Interpreting the Postmodern: Responses to Radical Orthodoxy* (T. and T. Clark/Continuum, 2006).

John Grim is a Senior Lecturer and Senior Scholar at Yale University. With Mary Evelyn Tucker, he is Coordinator of the Forum on Religion and Ecology and series editor of World Religions and Ecology, from Harvard Divinity School's Center for the Study of World Religions. He has been a Professor of Religion at Bucknell University and at Sarah Lawrence College, where he taught courses in Native American and indigenous religions, world religions, and religion and ecology. He is the author *The Shaman: Patterns of Religious Healing Among the Ojibway Indians* (University of Oklahoma Press, 1983) and, with Mary Evelyn Tucker, coeditor of *Worldviews and Ecology* (Orbis, 1994). He was editor of "Religion and Ecology: Can the Climate Change?" a special issue of *Daedalus* (2001). He is currently President of the American Teilhard Association.

Fletcher Harper, an Episcopal priest, is Executive Director of GreenFaith, an interfaith environmental coalition based in New Jersey. He served as a parish priest for ten years prior to joining GreenFaith, where he preaches and speaks regularly to houses of worship around the state about the moral, religious basis for protecting the environment.

Luke Higgins received his Masters of Divinity from Pacific School of Religion in Berkeley, California, and is currently working on his PhD at Drew University. His interests lie at the borders of process theology, ecotheology, and poststructuralism. He also works as a youth minister at a U.C.C. church in Cedar Grove, New Jersey.

Laurel Kearns is Associate Professor of Sociology of Religion and Environmental Studies in the Theological School and Graduate Division of Religion at Drew University. She has published articles on Christian environmental activism, environmental justice and racism, "greening" the sociology of religion, as well as "The Context for Eco-theology," in *The Blackwell Companion to Modern Theology* (Basil Blackwell, 2004). She served as assistant editor of Christianity articles as well as contributed several articles to the *Encyclopedia of Religion and Nature* (Continuum, 2005). Her passion for ecology stems from her growing up on Sanibel Island, Florida.

Catherine Keller is Professor of Constructive Theology in the Theological School and Graduate Division of Religion at Drew University. Her publications span a wide theopoetic-theopolitical spectrum: she is the coeditor (with Anne Daniell) of *Process and Difference: Between Cosmological and Poststructuralist Postmodernism* (State University of New York Press, 2002) and, with Mayra Rivera and Michael Nausner, of *Postcolonial Theologies: Divinity, Hybridity and Empire* (Chalice, 2004), as well as the author of five other books, including: *From a Broken Web: Separation, Sexism and Self* (Beacon, 1986); *Apocalypse Now and Then: A Feminist Guide to the End of the World* (Beacon, 1996); *Face of the Deep: A Theology of Becoming* (Routledge, 2003); and, most recently, *God and Power: Counter-Apocalyptic Journeys* (Augsburg Fortress, 2005). *On the Mystery*, a work for beginners in theology, is forthcoming.

Seung Gap Lee received his doctorate in Process Theology and Ecotheology from Drew University. His dissertation is entitled "The Hope of the Earth: A Process Eschatological Ecoethics." Teaching "Modern Society and Christianity," "Theology and Ethics," and "Recent Theologies" at numerous schools and seminaries in South Korea, he attempts to apply his reflections and interpretations of process thought to the theology and practice of the South Korean churches and Christians who are being challenged to reform themselves in order to more meaningfully respond to their changing worlds.

Glen A. Mazis is Professor of Philosophy and Humanities at Penn State, Harrisburg, and coordinator of their Interdisciplinary Humanities Masters Program. His books include: *Emotion and Embodiment: Fragile Ontology* (Lang, 1993); *The Trickster, Magician and Grieving Man: Returning Men to Earth* (Inner Traditions, 1994); and *Earthbodies: Rediscovering Our Planetary Senses* (State University of New York, 2002). His current book project is *Humans/Animals/Machines: Blurred Boundaries—Dangers and Promise.*

Jay McDaniel is Professor of Religion and Director of the Steel Center for the Study of Religion and Philosophy at Hendrix College, Conway,

Arkansas. His books include: *With Roots and Wings: Christianity in an Age of Ecology and Dialogue* (Orbis, 1995); *Living from the Center: Spirituality in an Age of Consumerism* (Chalice, 2000); and *Gandhi's Hope: Learning from Other Religions as a Path to Peace* (Orbis, 2005). He is also coeditor of *A Handbook of Process Theology* (Chalice, 2006). Born in San Antonio, Texas, he traces his own interest in ecological theology to the influence in his life of the Guadalupe River, where he spent many summers, and also to the companionship of dogs. He is married with two children, and an active member of the United Methodist Church. His hobbies are playing the guitar and watching basketball games.

Barbara Muraca is a doctoral candidate in Environmental Ethics at the University of Greifswald, Germany. She holds a master's degree in theoretical philosophy from the University of Turin, Italy. Her doctoral research is being supported by the German foundation Hans-Boeckler-Stiftung. Her academic interests include process philosophy and theology and also hermeneutics, postmodern thought, feminist theory, and ecology. She has published several articles about sustainability, ecology, and process thought, both in English and in German, and is coeditor of the bilingual book *Nachhaltigkeit ist machbar—La sostenibilita è possibile* (VAS Verlag, 2005), about Local Agenda 21 processes and environmental communication.

Jane Ellen Nickell, an ordained United Methodist minister, is Chaplain at Allegheny College in Meadville, Pennsylvania, and is pursuing a PhD degree in sociology of religion at Drew University. She has taught and written articles on ecology and religion, including ways that congregations can bring nature into worship.

Kevin J. O'Brien is Assistant Professor of Religion at Pacific Lutheran University in Tacoma, Washington. He received his doctorate from the Ethics and Society Program of Emory University's Graduate Division of Religion. His dissertation, "An Ethics of Biodiversity: Moral Theology, Ecology, and Environmentalism," develops and reflects on the intersections between science and ethics and between social and environmentalist ethics.

Anna L. Peterson is Professor of Religion at the University of Florida. She publishes and teaches on social and environmental ethics and Latin American religion. Her publications include *Martyrdom and the Politics of Religion: Progressive Catholicism in El Salvador's Civil War* (SUNY Press, 1997); *Being Human: Ethics, Environment, and Our Place in the World* (University of California, 2001); and *Seeds of the Kingdom: Utopian Communities in the Americas* (Oxford University Press, 2005).

Anne Primavesi, formerly Research Fellow in Environmental Theology at the University of Bristol and Birckbeck College, University of London, currently holds Research Fellowships at Westar Institute, Santa Rosa, California, and is a Founding Fellow of the Lokahi Foundation, London. Her publications include *Sacred Gaia: Holistic Theology and Earth System Science* (Routledge, 2000); *Gaia's Gift: Earth, Ourselves, and God After Copernicus* (Routledge, 2003); seminal studies of the theological implications of James Lovelock's scientific theory; and definitive entries on Gaia, ecofeminism, and ecological theology in a variety of reference works, including *The Blackwell Companion to the Bible and Culture* (Blackwell, 2006).

Rosemary Radford Ruether received her MA and PhD from the Claremont Graduate University in 1960 and 1965 in Roman History and Classics and Patristics, respectively. She was the Georgia Harkness Professor at Garrett Theological Seminary from 1976–2000. She presently is completing a position as the Carpenter Professor of Feminist Theology at the Graduate Theological Union in Berkeley, California, and is a visiting scholar at the Claremont Graduate University. She is author or editor of forty-four books. The most recent are: *Goddesses and the Divine Feminine* (University of California Press, 2005); *Integrating Ecofeminism, Globalization, and World Religions* (Rowman and Littlefield, 2005); and a three-volume *Encyclopedia of Women and Religion in North America*, coedited with Rosemary Keller (Indiana University Press, 2006).

Kate Rigby (FAHA) is Associate Professor in Comparative Literature and Cultural Studies at Monash University in Australia. She is a coeditor of *PAN (Place, Activism, Nature)*, and has published widely in the areas of ecocriticism, ecophilosophy, and ecology and religion. Her most recent book is *Topographies of the Sacred: The Poetics of Place in European Romanticism* (University of Virginia Press, 2004).

Nicole A. Roskos received her doctorate in Ecological and Feminist Theology from Drew University. Her dissertation is entitled "Falling Nature: An Ecofeminist Theology of Fall Narratives." She has been an environmental activist for over a decade, a professor of religious studies, and is currently painting figures in landscapes in the Catskill Mountains of New York.

Daniel T. Spencer is Assistant Professor of Environmental Studies at the University of Montana, where he has taught since 2002. From 1993–2002 he was Professor of Religion and Ethics at Drake University in Des Moines, Iowa. His publications include *Gay and Gaia: Ethics, Ecology and the Erotic* (Pilgrim, 1996), as well as contributions to the *Encyclopedia of Religion and Nature*. He is currently working on a collection of essays that connect ecological location and ethics.

Lawrence Troster is the Jewish Chaplain and Associate of the Institute of Advanced Theology at Bard College in Annandale-on-Hudson, New York. He is also the Director of the Fellowship Program for GreenFaith, an interfaith environmental coalition in New Jersey, and the Rabbinic Fellow of the Coalition on the Environment and Jewish Life. He has served as the rabbi of several congregations in New Jersey and Canada. He received his bachelor's degree from the University of Toronto and his master's degree and rabbinic ordination from the Jewish Theological Seminary of America in New York City. Rabbi Troster is also the cochair of the Interfaith Partnership for the Environment of the United Nations Environment Program. He has published numerous articles on theology, environmentalism, liturgy, bioethics, and the relationship of science and religion.

Mary Evelyn Tucker is a cofounder and codirector of the Forum on Religion and Ecology. With her husband, John Grim, she organized the Harvard conference series on world religions and ecology and together they were series editors for the ten volumes that resulted from the conferences. She is the author of *Worldly Wonder: Religions Enter Their Ecological*

Phase (Open Court, 2003), *Moral and Spiritual Cultivation in Japanese Neo-Confucianism* (State University of New York Press, 1989), and *The Philosophy of Qi* (Columbia University Press, 2007). She has coedited several volumes on ecology and worldviews, Buddhism, Confucianism, and Hinduism. She is currently a Research Associate at the Harvard-Yenching Institute and the Reischauer Institute at Harvard University and senior lecturer and senior researcher at Yale University.

Mark I. Wallace is Associate Professor in the Department of Religion at Swarthmore College, Pennsylvania. Among other works, he has authored: *Finding God in the Singing River: Christianity, Spirit, Nature* (Augsburg Fortress, 2005), and *Fragments of the Spirit: Nature, Violence, and the Renewal of Creation* (Continuum, 1996; Trinity, 2002). He is a member of the Constructive Theology Workgroup and recently received an Andrew W. Mellon New Directions Fellowship for a research sabbatical in Costa Rica.

David Wood is Professor of Philosophy at Vanderbilt University, where he teaches twentieth-century continental philosophy and environmental philosophy. He has authored or edited some sixteen books, including *The Deconstruction of Time* (Northwestern University Press, 2000); *Thinking After Heidegger* (Polity, 2002); *The Step Back* (State University of New York Press, 2005) and *Time After Time* (Indiana University Press, 2007). He co-directs Vanderbilt's "small green think tank" on ecology and spirituality, and he is an earth-artist.